Delay Ordinary and Partial Differential Equations

This book is devoted to linear and nonlinear ordinary and partial differential equations with constant and variable delay. It considers qualitative features of delay differential equations and formulates typical problem statements. Exact, approximate analytical and numerical methods for solving such equations are described, including the method of steps, methods of integral transformations, method of regular expansion in a small parameter, method of matched asymptotic expansions, iteration-type methods, Adomian decomposition method, collocation method, Galerkin-type projection methods, Euler and Runge-Kutta methods, shooting method, method of lines, finite-difference methods for PDEs, methods of generalized and functional separation of variables, method of functional constraints, method of generating equations, and more.

The presentation of the theoretical material is accompanied by examples of the practical application of methods to obtain the desired solutions. Exact solutions are constructed for many nonlinear delay reaction-diffusion and wave-type PDEs that depend on one or more arbitrary functions. A review is given of the most common mathematical models with delay used in population theory, biology, medicine, economics, and other applications.

Delay Ordinary and Partial Differential Equations contains much new material previously unpublished in monographs. It is intended for a broad audience of scientists, university professors, and graduate and postgraduate students specializing in applied and computational mathematics, mathematical physics, mechanics, control theory, biology, medicine, chemical technology, ecology, economics, and other disciplines.

Individual sections of the book and examples are suitable for lecture courses on applied mathematics, mathematical physics, and differential equations for delivering special courses and for practical training.

Advances in Applied Mathematics
Series Editor: Daniel Zwillinger

Introduction to Quantum Control and Dynamics
Domenico D'Alessandro

Handbook of Radar Signal Analysis
Bassem R. Mahafza, Scott C. Winton, Atef Z. Elsherbeni

Separation of Variables and Exact Solutions to Nonlinear PDEs
Andrei D. Polyanin, Alexei I. Zhurov

Boundary Value Problems on Time Scales, Volume I
Svetlin Georgiev, Khaled Zennir

Boundary Value Problems on Time Scales, Volume II
Svetlin Georgiev, Khaled Zennir

Observability and Mathematics
Fluid Mechanics, Solutions of Navier-Stokes Equations, and Modeling
Boris Khots

Handbook of Differential Equations, Fourth Edition
Daniel Zwillinger, Vladimir Dobrushkin

Experimental Statistics and Data Analysis for Mechanical and Aerospace Engineers
James Middleton

Advanced Engineering Mathematics with MATLAB®, Fifth Edition
Dean G. Duffy

Handbook of Fractional Calculus for Engineering and Science
Harendra Singh, H. M. Srivastava, Juan J. Nieto

Advanced Engineering Mathematics
A Second Course with MATLAB®
Dean G. Duffy

Quantum Computation
Helmut Bez and Tony Croft

Computational Mathematics
An Introduction to Numerical Analysis and Scientific Computing with Python
Dimitrios Mitsotakis

Delay Ordinary and Partial Differential Equations
Andrei D. Polyanin, Vsevolod G. Sorokin, Alexei I. Zhurov

https://www.routledge.com/Advances-in-Applied-Mathematics/book-series/CRCADVA
PPMTH?pd=published,forthcoming&pg=1&pp=12&so=pub&view=list

Delay Ordinary and Partial Differential Equations

Andrei D. Polyanin
Vsevolod G. Sorokin
Alexei I. Zhurov

CRC Press is an imprint of the
Taylor & Francis Group, an **informa** business

A CHAPMAN & HALL BOOK

First edition published 2024
by CRC Press
6000 Broken Sound Parkway NW, Suite 300, Boca Raton, FL 33487-2742

and by CRC Press
4 Park Square, Milton Park, Abingdon, Oxon, OX14 4RN

CRC Press is an imprint of Taylor & Francis Group, LLC

© 2024 Taylor & Francis Group, LLC

Reasonable efforts have been made to publish reliable data and information, but the author and publisher cannot assume responsibility for the validity of all materials or the consequences of their use. The authors and publishers have attempted to trace the copyright holders of all material reproduced in this publication and apologize to copyright holders if permission to publish in this form has not been obtained. If any copyright material has not been acknowledged please write and let us know so we may rectify in any future reprint.

Except as permitted under U.S. Copyright Law, no part of this book may be reprinted, reproduced, transmitted, or utilized in any form by any electronic, mechanical, or other means, now known or hereafter invented, including photocopying, microfilming, and recording, or in any information storage or retrieval system, without written permission from the publishers.

For permission to photocopy or use material electronically from this work, access www.copyright.com or contact the Copyright Clearance Center, Inc. (CCC), 222 Rosewood Drive, Danvers, MA 01923, 978-750-8400. For works that are not available on CCC please contact mpkbookspermissions@tandf.co.uk

Trademark notice: Product or corporate names may be trademarks or registered trademarks and are used only for identification and explanation without intent to infringe.

Library of Congress Cataloging-in-Publication Data

Names: Polyanin, A. D. (Andrei Dmitrievich), author. | Sorokin, Vsevolod
  G., author. | Zhurov, Alexei I., author.
Title: Delay ordinary and partial differential equations / Andrei D.
  Polyanin, Vsevolod G. Sorokin, Alexei I. Zhurov.
Description: First edition. | Boca Raton : CRC Press, 2024. | Series:
  Advances in applied mathematics | Author's name contains typo on title
  page of galley. Should read Alexei I. Zhurov. | Includes bibliographical
  references and index.
Identifiers: LCCN 2023007961 | ISBN 9780367486914 (hardback) | ISBN
  9781032549866 (paperback) | ISBN 9781003042310 (ebook)
Subjects: LCSH: Delay differential equations. | Differential equations,
  Partial.
Classification: LCC QA371 .P564 2024 | DDC 515/.35--dc23/eng20230715
LC record available at https://lccn.loc.gov/2023007961

ISBN: 978-0-367-48691-4 (hbk)
ISBN: 978-1-032-54986-6 (pbk)
ISBN: 978-1-003-04231-0 (ebk)

DOI: 10.1201/9781003042310

Typeset in Nimbus Roman
by KnowledgeWorks Global Ltd.

Contents

Preface   xi

Notations and Remarks   xv

Authors   xvii

1. Delay Ordinary Differential Equations   1
- 1.1. First-Order Equations. Cauchy Problem. Method of Steps. Exact Solutions   1
  - 1.1.1. Preliminary Remarks   1
  - 1.1.2. First-Order ODEs with Constant Delay. Cauchy Problem. Qualitative Features   1
  - 1.1.3. Exact Solutions to a First-Order Linear ODE with Constant Delay. The Lambert W Function and Its Properties   4
  - 1.1.4. First-Order Nonlinear ODEs with Constant Delay That Admit Linearization or Exact Solutions   12
  - 1.1.5. Method of Steps. Solution of the Cauchy Problem for a First-Order ODE with Constant Delay   13
  - 1.1.6. Equations with Variable Delay. ODEs with Proportional Delay   17
  - 1.1.7. Existence and Uniqueness of Solutions. Suppression of Singularities in Solving Blow-Up Problems   23
- 1.2. Second- and Higher-Order Delay ODEs. Systems of Delay ODEs   28
  - 1.2.1. Basic Concepts. The Cauchy Problem   28
  - 1.2.2. Second-Order Linear Equations. The Cauchy Problem. Exact Solutions   29
  - 1.2.3. Higher-Order Linear Delay ODEs   34
  - 1.2.4. Linear Systems of First- and Second-Order ODEs with Delay. The Cauchy Problem. Exact Solutions   38
- 1.3. Stability (Instability) of Solutions to Delay ODEs   42
  - 1.3.1. Basic Concepts. General Remarks on Stability of Solutions to Linear Delay ODEs   42
  - 1.3.2. Stability of Solutions to Linear ODEs with a Single Constant Delay   43
  - 1.3.3. Stability of Solutions to Linear ODEs with Several Constant Delays   50
  - 1.3.4. Stability Analysis of Solutions to Nonlinear Delay ODEs by the First Approximation   52
- 1.4. Exact and Approximate Analytical Solution Methods for Delay ODEs   55
  - 1.4.1. Using Integral Transforms for Solving Linear Problems   55
  - 1.4.2. Representation of Solutions as Power Series in the Independent Variable   64

	1.4.3.	Method of Regular Expansion in a Small Parameter	69
	1.4.4.	Method of Matched Asymptotic Expansions. Singular Perturbation Problems with a Boundary Layer	71
	1.4.5.	Method of Successive Approximations and Other Iterative Methods	74
	1.4.6.	Galerkin-Type Projection Methods. Collocation Method	80

2. Linear Partial Differential Equations with Delay 85

2.1. Properties and Specific Features of Linear Equations and Problems with Constant Delay ... 85
    2.1.1. Properties of Solutions to Linear Delay Equations ... 85
    2.1.2. General Properties and Qualitative Features of Delay Problems . 91
2.2. Linear Initial-Boundary Value Problems with Constant Delay ... 91
    2.2.1. First Initial-Boundary Value Problem for One-Dimensional Parabolic Equations with Constant Delay ... 91
    2.2.2. Other Problems for a One-Dimensional Parabolic Equation with Constant Delay ... 96
    2.2.3. Problems for Linear Parabolic Equations with Several Variables and Constant Delay ... 102
    2.2.4. Problems for Linear Hyperbolic Equations with Constant Delay 107
    2.2.5. Stability and Instability Conditions for Solutions to Linear Initial-Boundary Value Problems ... 109
2.3. Hyperbolic and Differential-Difference Heat Equations ... 113
    2.3.1. Derivation of the Hyperbolic and Differential-Difference Heat Equations ... 113
    2.3.2. Stokes Problem and Initial-Boundary Value Problems for the Differential-Difference Heat Equation ... 115
2.4. Linear Initial-Boundary Value Problems with Proportional Delay ... 119
    2.4.1. Preliminary Remarks ... 119
    2.4.2. First Initial-Boundary Value Problem for a Parabolic Equation with Proportional Delay ... 120
    2.4.3. Other Initial-Boundary Value Problems for a Parabolic Equation with Proportional Delay ... 122
    2.4.4. Initial-Boundary Value Problem for a Linear Hyperbolic Equation with Proportional Delay ... 124

3. Analytical Methods and Exact Solutions to Delay PDEs. Part I 127

3.1. Remarks and Definitions. Traveling Wave Solutions ... 127
    3.1.1. Preliminary Remarks. Terminology. Classes of Equations Concerned ... 127
    3.1.2. States of Equilibrium. Traveling Wave Solutions. Exact Solutions in Closed Form ... 130
    3.1.3. Traveling Wave Front Solutions to Nonlinear Reaction-Diffusion Type Equations ... 134
3.2. Multiplicative and Additive Separable Solutions ... 139
    3.2.1. Preliminary Remarks. Terminology. Examples ... 139

      3.2.2. Delay Reaction-Diffusion Equations Admitting Separable Solutions 141
      3.2.3. Delay Klein–Gordon Type Equations Admitting Separable Solutions 148
      3.2.4. Some Generalizations 149
3.3. Generalized and Functional Separable Solutions 154
      3.3.1. Generalized Separable Solutions 154
      3.3.2. Functional Separable Solutions 159
      3.3.3. Using Linear Transformations to Construct Generalized and Functional Separable Solutions 162
3.4. Method of Functional Constraints 166
      3.4.1. General Description of the Method of Functional Constraints . . 166
      3.4.2. Exact Solutions to Quasilinear Delay Reaction-Diffusion Equations 167
      3.4.3. Exact Solutions to More Complicated Nonlinear Delay Reaction-Diffusion Equations 177
      3.4.4. Exact Solutions to Nonlinear Delay Klein–Gordon Type Wave Equations 194

4. Analytical Methods and Exact Solutions to Delay PDEs. Part II   **201**
4.1. Methods for Constructing Exact Solutions to Nonlinear Delay PDEs Using Solutions to Simpler Non-Delay PDEs 201
      4.1.1. The First Method for Constructing Exact Solutions to Delay PDEs. General Description and Simple Examples 201
      4.1.2. Using the First Method for Constructing Exact Solutions to Nonlinear Delay PDEs 203
      4.1.3. The Second Method for Constructing Exact Solutions to Delay PDEs. General Description and Simple Examples 207
      4.1.4. Employing the Second Method to Construct Exact Solutions to Nonlinear Delay PDEs 209
4.2. Systems of Nonlinear Delay PDEs. Generating Equations Method . . . 212
      4.2.1. General Description of the Method and Application Examples . 212
      4.2.2. Quasilinear Systems of Delay Reaction-Diffusion Equations and Their Exact Solutions 215
      4.2.3. Nonlinear Systems of Delay Reaction-Diffusion Equations and Their Exact Solutions. 218
      4.2.4. Some Generalizations 222
4.3. Reductions and Exact Solutions of Lotka–Volterra Type Systems and More Complex Systems of PDEs with Several Delays 225
      4.3.1. Reaction-Diffusion Systems with Several Delays. The Lotka–Volterra System 225
      4.3.2. Reductions and Exact Solutions of Systems of PDEs with Different Diffusion Coefficients ($a_1 \neq a_2$) 226
      4.3.3. Reductions and Exact Solutions of Systems of PDEs with Equal Diffusion Coefficients ($a_1 = a_2$) 239

4.3.4. Systems of Delay PDEs Homogeneous in the Unknown Functions 247
4.4. Nonlinear PDEs with Proportional Arguments. Principle of Analogy of Solutions ... 250
 4.4.1. Principle of Analogy of Solutions ... 250
 4.4.2. Exact Solutions to Quasilinear Diffusion Equations with Proportional Delay ... 253
 4.4.3. Exact Solutions to More Complicated Nonlinear Diffusion Equations with Proportional Delay ... 257
 4.4.4. Exact Solutions to Nonlinear Wave-Type Equations with Proportional Delay ... 263
4.5. Unstable Solutions and Hadamard Ill-Posedness of Some Delay Problems 270
 4.5.1. Solution Instability for One Class of Nonlinear PDEs with Constant Delay ... 270
 4.5.2. Hadamard Ill-Posedness of Some Delay Problems ... 271

5. Numerical Methods for Solving Delay Differential Equations   273

5.1. Numerical Integration of Delay ODEs ... 273
 5.1.1. Main Concepts and Definitions ... 273
 5.1.2. Qualitative Features of the Numerical Integration of Delay ODEs 275
 5.1.3. Modified Method of Steps ... 278
 5.1.4. Numerical Methods for ODEs with Constant Delay ... 279
 5.1.5. Numerical Methods for ODEs with Proportional Delay. Cauchy Problem ... 283
 5.1.6. Shooting Method (Boundary Value Problems) ... 287
 5.1.7. Integration of Stiff Systems of Delay ODEs Using the Mathematica Software ... 290
 5.1.8. Test Problems for Delay ODEs. Comparison of Numerical and Exact Solutions ... 293
5.2. Numerical Integration of Delay PDEs ... 296
 5.2.1. Preliminary Remarks. Method of Time-Domain Decomposition 296
 5.2.2. Method of Lines—Reduction of a Delay PDE to a System of Delay ODEs ... 297
 5.2.3. Finite Difference Methods ... 302
5.3. Construction, Selection, and Usage of Test Problems for Delay PDEs .. 309
 5.3.1. Preliminary Remarks ... 309
 5.3.2. Main Principles for Selecting Test Problems ... 309
 5.3.3. Constructing Test Problems ... 310
 5.3.4. Comparison of Numerical and Exact Solutions to Nonlinear Delay Reaction-Diffusion Equations ... 316
 5.3.5. Comparison of Numerical and Exact Solutions to Nonlinear Delay Klein–Gordon Type Wave Equations ... 322

6. Models and Delay Differential Equations Used in Applications — 327
6.1. Models Described by Nonlinear Delay ODEs 327
    6.1.1. Hutchinson's Equation—a Delay Logistic Equation 327
    6.1.2. Nicholson's Equation 330
    6.1.3. Mackey–Glass Hematopoiesis Model 333
    6.1.4. Other Nonlinear Models with Delay 336
6.2. Models of Economics and Finance Described by ODEs 340
    6.2.1. The Simplest Model of Macrodynamics of Business Cycles . . 340
    6.2.2. Model of Interaction of Three Economical Parameters 341
    6.2.3. Delay Model Describing Tax Collection in a Closed Economy . 342
6.3. Models and Delay PDEs in Population Theory 343
    6.3.1. Preliminary Remarks 343
    6.3.2. Diffusive Logistic Equation with Delay 344
    6.3.3. Delay Diffusion Equation Taking into Account Nutrient Limitation 345
    6.3.4. Lotka–Volterra Type Diffusive Logistic Model with Several Delays 346
    6.3.5. Nicholson's Reaction-Diffusion Model with Delay 347
    6.3.6. Model That Takes into Account the Effect of Plant Defenses on a Herbivore Population 349
6.4. Models and Delay PDEs Describing the Spread of Epidemics and Development of Diseases 350
    6.4.1. Classical SIR Model of Epidemic Spread 350
    6.4.2. Two-Component Epidemic SI Model 353
    6.4.3. Epidemic Model of the New Coronavirus Infection 354
    6.4.4. Hepatitis B Model 355
    6.4.5. Model of Interaction between Immunity and Tumor Cells . . . 357
6.5. Other Models Described by Nonlinear Delay PDEs 358
    6.5.1. Belousov–Zhabotinsky Oscillating Reaction Model 358
    6.5.2. Mackey–Glass Model of Hematopoiesis 359
    6.5.3. Model of Heat Treatment of Metal Strips 360
    6.5.4. Food Chain Model 361
    6.5.5. Models of Artificial Neural Networks 362

References   **365**

Index   **399**

Preface

Linear and nonlinear differential equations with delay (ordinary and partial) or, simply, delay differential equations* are often used for mathematical modeling of phenomena and processes in various areas of theoretical physics, mechanics, control theory, biology, biophysics, biochemistry, medicine, ecology, economics, and technical applications.

Let us list a few factors that lead one to introduce delay into mathematical models described by differential equations. For example, in biology and biomechanics, delays are due to the limited speed of transmission of nerve and muscle reactions in living tissues. In medicine, when one deals with the spread of infectious diseases, the delay time is determined by the incubation period (the time interval between initial contact with an infectious agent and appearance of the first signs or symptoms of the disease). In population dynamics, delays arise because individuals participate in reproduction only after reaching a certain age. In control theory, delays are usually associated with the finite speed of signal propagation and the limited speed of technological processes.

The presence of a delay in mathematical models and differential equations is a complicating factor, which, as a rule, leads to a narrowing of the stability region of the solutions obtained. Studying and solving ordinary differential equations (ODEs) with delay is comparable in complexity to studying and solving partial differential equations (PDEs) without delay.

The book details qualitative features of delay differential equations and presents typical statements of initial value and initial-boundary value problems for them. Exact, approximate analytical, and numerical methods for solving such equations are described. In addition to differential equations with constant delay, equations with proportional delay (of the pantograph type) are studied, as well as more complex equations with a general variable delay or several delays. The presentation of the theoretical material comes with examples of the practical application of the methods to obtain desired solutions.

The book reviews the most common mathematical models with delay used in population theory, biology, medicine, and other applications.

Analytical solutions to Cauchy-type linear problems for first- and second-order ODEs and systems of ODEs with constant or proportional delays are presented. Some classes of nonlinear first-order delay ODEs that admit linearization or exact solutions are considered. The issues of stability and instability of solutions to ODEs with delay are discussed.

The most common analytical and numerical methods for solving initial and boundary value problems for ODEs with constant or variable delay are described.

---

*In the literature, there is also a longer alternative name: differential equations with delayed argument.

These include the method of steps, methods of integral transforms, method of regular expansion in a small parameter, method of matched asymptotic expansions, iterative methods, Adomian decomposition method, homotopy analysis method, collocation method, Galerkin-type projection methods, Euler and Runge–Kutta methods, shooting method, methods based on the use of the Mathematica package, and more.

We use the method of separation of variables to obtain Fourier series solutions in space variables of linear initial-boundary value problems for parabolic and hyperbolic PDEs with constant or proportional delay and different boundary conditions. Numerical methods for solving initial-boundary value problems for linear and nonlinear delay PDEs are also presented. The most attention is paid to the method of lines, which relies on reducing a delay PDE to a system of delay ODEs. Finite-difference methods based on an implicit scheme, a weighting scheme, a scheme of increased order of accuracy, and more are considered. The time domain decomposition method, which generalizes the method of steps used to solve delay ODEs, is also discussed. We formulate the basic principles for constructing and selecting test problems for assessing the adequacy and estimating the accuracy of numerical methods and approximate analytical methods for solving delay PDEs.

The general solutions to nonlinear delay PDEs cannot be obtained even in the simplest cases. Therefore, when studying such equations, one usually has to search and analyze their particular solutions, usually called *exact solutions*.

The book pays much attention to the description and practical application of methods for constructing exact solutions to nonlinear equations of mathematical physics with delay. These are the methods of generalized and functional separation of variables, the method of functional constraints, the method of generating equations, the principle of the analogy of solutions, and others. Notably, the vast majority of analytical methods that successfully allow one to find exact solutions of nonlinear partial differential equations without delay are either inapplicable to constructing exact solutions of nonlinear PDEs with constant or variable delay or have a minimal area of applicability. Equations of mathematical physics with two independent variables and a delay have the following essential qualitative features: (i) PDEs with constant delay do not admit self-similar solutions, unlike PDEs without delay, many of which do, and (ii) PDEs with proportional delay in either independent variable do not have traveling wave solutions, unlike simpler PDEs without delay, which often have.

The book considers many nonlinear reaction-diffusion and wave-type equations with delay dependent on one or several arbitrary functions or involving several free parameters. Such equations are the most difficult to analyze, and their exact solutions can be used to test numerical and approximate analytical methods for solving related initial-boundary value problems and estimate the errors of the methods.

The book contains much new material that has not previously been published in monographs.

The authors tried to avoid using special terminology whenever possible to maximize the circle of potential readers with different mathematical backgrounds. Therefore, some results are described in a schematic and simplified manner, which suffices for practical applications. Many sections can be read independently, making it easier to work with the material. A detailed table of contents allows the reader to find the desired information quickly.

The authors thank A. V. Aksenov for the discussions and valuable remarks.

The authors hope that the book will be helpful for a wide range of scientists, university professors, and graduate and postgraduate students specializing in applied mathematics, mathematical physics, computational mathematics, mechanics, control theory, biology, biophysics, biochemistry, medicine, chemical technology, and ecology. In addition, individual sections of the book, methods and examples can be used in teaching applied mathematics, mathematical physics, and functional differential equations to deliver special courses and perform practical exercises.

Authors

Notations and Remarks

Latin Characters

a, a_1, a_2 diffusion coefficients (dimensional or dimensionless) in reaction-diffusion type equations;

$C_1, C_2, \ldots$ arbitrary constants;

$\cos_d(t, \tau)$ delayed cosine function, see formula (1.2.2.3);

$\cos_s(t, p)$ stretched cosine function, $\cos_s(t, p) = \sum_{n=0}^{\infty} (-1)^n p^{n(2n-1)} \dfrac{t^{2n}}{(2n)!}$;

$\exp_d(t, \tau)$ delayed exponential function, $\exp_d(t, \tau) \equiv \sum_{k=0}^{[t/\tau]+1} \dfrac{[t - (k-1)\tau]^k}{k!}$;

$\exp_s(t, p)$ stretched exponential function, $\exp_s(t, p) \equiv \sum_{n=0}^{\infty} p^{\frac{n(n-1)}{2}} \dfrac{t^n}{n!}$;

$\operatorname{Im} A$ imaginary part of the complex number A;

p, q scaling parameters of arguments ($0 < p < 1$ and $0 < q < 1$ for time-proportional delay differential equations);

$\operatorname{Re} A$ real part of the complex number A;

$\sin_d(t, \tau)$ delayed sine function, see formula (1.2.2.4);

$\sin_s(t, p)$ stretched sine function, $\sin_s(t, p) \equiv \sum_{n=0}^{\infty} (-1)^n p^{n(2n+1)} \dfrac{t^{2n+1}}{(2n+1)!}$;

t time (independent variable);

u unknown function (dependent variable) at the current time t; for equations with two independent variables, $u = u(x, t)$;

$W = W(z)$ Lambert W function, defined implicitly by $z = We^W$ ($z = x + iy$ is a complex variable);

$W_p = W_p(x)$ principal branch of the Lambert W function ($x \geq -1/e$, $W_p \geq -1$);

$W_n = W_n(x)$ second branch of the Lambert W function ($-1/e \leq x < 0$, $W_n \leq -1$);

w unknown function at a preceding time, $w = u(t - \tau)$ (for ODEs with constant delay) or $w = u(pt)$ (for ODEs with proportional delay, $0 < p < 1$); for PDEs in two independent variables, $w = u(x, t - \tau)$, $w = u(x, pt)$, or $w = u(px, qt)$;

w_k unknown functions at preceding times, $w_k = u(t - \tau_k)$ (for ODEs with constant delays, $k = 1, \ldots, m$); for PDEs in two independent variables, $w_k = u(x, t - \tau_k)$;

x, y space variable (Cartesian coordinates) or the real and imaginary parts of the complex number $z = x + iy$;

$x_1, \ldots, x_n$ Cartesian coordinates in an n-dimensional space;

$\mathbf{x}$ n-dimensional vector, $\mathbf{x} = (x_1, \ldots, x_n)$.

Greek Characters

Δ Laplace operator:

$\Delta = \frac{\partial^2}{\partial x^2} + \frac{\partial^2}{\partial y^2}$ in the two-dimensional space,

$\Delta = \sum\limits_{k=1}^{n} \frac{\partial^2}{\partial x_k^2}$ in the n-dimensional space;

τ delay time ($\tau > 0$), which can be constant or time dependent, $\tau = \tau(t)$;

$\tau_1, \ldots, \tau_m$ delay times.

Short Notations for Derivatives and Operators

Partial derivatives of a function $u = u(x, t)$:

$$u_x = \frac{\partial u}{\partial x}, \quad u_t = \frac{\partial u}{\partial t}, \quad u_{xx} = \frac{\partial^2 u}{\partial x^2}, \quad u_{xt} = \frac{\partial^2 u}{\partial x \partial t}, \quad u_{tt} = \frac{\partial^2 u}{\partial t^2}, \quad \ldots, \quad u_x^{(n)} = \frac{\partial^n u}{\partial x^n}.$$

Ordinary derivatives of a function $f = f(t)$:

$$f'_t = \frac{df}{dt}, \quad f''_{tt} = \frac{d^2 f}{dt^2}, \quad f'''_{ttt} = \frac{d^3 f}{dt^3}, \quad f''''_{tttt} = \frac{d^4 f}{dt^4}, \quad f_t^{(n)} = \frac{d^n f}{dt^n} \quad \text{for} \quad n > 4.$$

Sometimes alternative notations are also used for the first two derivatives:

$$f'(t) = f'_t \quad f''(t) = f''_{tt}.$$

Diffusion term of a partial differential equation in the n-dimensional case:

$$\text{div}\,[f(u)\nabla u] = \sum_{k=1}^{n} \frac{\partial}{\partial x_k}\left[f(u)\frac{\partial u}{\partial x_k}\right], \quad \text{where } f(u) \text{ is the diffusion coefficient.}$$

Remarks

1. The book often uses the abbreviations ODE (or ODEs) and PDE (or PDEs), which stand for an 'ordinary differential equation' (or 'ordinary differential equations') and a 'partial differential equation' (or 'partial differential equations').

2. Any arbitrary functions (usually denoted by f and g) included in the considered ODEs and PDEs with delay are considered to be continuous.

3. If a formula or a solution involves derivatives of some functions, it is assumed that these derivatives exist.

4. If a formula or a solution involves indefinite or definite integrals, it is assumed that these integrals exist.

5. In formulas and solutions that involve expressions like $\frac{f(t)}{a-2}$, it is often omitted but implied that $a \neq 2$.

6. The symbols ▶ and ◀ mark the beginning and the end of an example given in the text.

Authors

Andrei D. Polyanin, D.Sc., Ph.D., is a well-known scientist of broad interests and is active in various areas of mathematics, theory of heat and mass transfer, hydrodynamics, and chemical engineering sciences. He is one of the most prominent authors in the field of reference literature on mathematics. Professor Polyanin graduated with honors from the Department of Mechanics and Mathematics at the Lomonosov Moscow State University in 1974. Since 1975, Professor Polyanin has been working at the Ishlinsky Institute for Problems in Mechanics of the Russian (former USSR) Academy of Sciences, where he defended his Ph.D. in 1981 and D.Sc. degree in 1986. He is an author of more than 30 books and over 270 articles and holds three patents. His books include A. D. Polyanin and V. F. Zaitsev, *Handbook of Exact Solutions for Ordinary Differential Equations*, CRC Press, 1995 (2nd edition in 2003); A. D. Polyanin and A. V. Manzhirov, *Handbook of Integral Equations*, CRC Press, 1998 (2nd edition in 2008); A. D. Polyanin, *Handbook of Linear Partial Differential Equations for Engineers and Scientists*, Chapman & Hall/CRC Press, 2002 (2nd edition, co-authored with V. E. Nazaikinskii, in 2016); A. D. Polyanin, V. F. Zaitsev, and A. Moussiaux, *Handbook of First Order Partial Differential Equations*, Taylor & Francis, 2002; A. D. Polyanin and V. F. Zaitsev, *Handbook of Nonlinear Partial Differential Equations*, Chapman & Hall/CRC Press, 2004 (2nd edition in 2012); A. D. Polyanin and A. V. Manzhirov, *Handbook of Mathematics for Engineers and Scientists*, Chapman & Hall/CRC Press, 2007, A. D. Polyanin and V. F. Zaitsev, *Handbook of Ordinary Differential Equations: Exact Solutions, Methods, and Problems*, CRC Press, 2018, A. D. Polyanin and A. I. Zhurov, *Separation of Variables and Exact Solutions to Nonlinear PDEs*, CRC Press, 2022.

Vsevolod G. Sorokin, Ph.D., is an actively working scientist in the field of mathematical physics, partial differential equations with delay, and numerical methods for delay differential equations. He graduated from the Department of Applied Mathematics at the Bauman Moscow State Technical University in 2014 and received there his Ph.D. in mathematical modelling and numerical methods in 2018. Since 2018, Doctor Sorokin has been working at the Ishlinsky Institute for Problems in Mechanics of the Russian Academy of Sciences. Doctor Sorokin has published over 25 research articles and the book *Delay Differential Equations: Properties, Methods, Solutions and Models* by A. D. Polyanin, V. G. Sorokin, and A. I. Zhurov, 2022 (in Russian).

Alexei I. Zhurov, Ph.D., is a noted scientist in the fields of mathematical physics, partial differential equations, and nonlinear mechanics. Dr. Zhurov graduated with honors from the Faculty of Airphysics and Space Research at the Moscow Institute of Physics and Technology in 1990. He is a member of staff of the Ishlinsky Institute for Problems in Mechanics of the Russian Academy of Sciences, where he received his Ph.D. in theoretical and fluid mechanics in 1995 and is a senior research scientist there since 1999. Since 2001, he has joined Cardiff University as a research scientist in the area of biomechanics and morphometrics. Doctor Zhurov has published over 120 research articles and five books, including *Solution Methods for Nonlinear Equations of Mathematical Physics and Mechanics* by A. D. Polyanin, V. F. Zaitsev, and A. I. Zhurov, Fizmatlit, 2005 (in Russian) and A. D. Polyanin and A. I. Zhurov, *Separation of Variables and Exact Solutions to Nonlinear PDEs*, CRC Press, 2022.

1. Delay Ordinary Differential Equations

1.1. First-Order Equations. Cauchy Problem. Method of Steps. Exact Solutions

1.1.1. Preliminary Remarks

The simplest spatially homogeneous processes with aftereffect are described by delay ordinary differential equations (delay ODEs). The analysis and solution of such equations are commeasurable in complexity with those of partial differential equations without delay. Currently, the theory of delay ODEs and other functional differential and differential-difference equations, is reasonably well elaborated (e.g., see [24, 37, 42, 71, 138, 144, 146, 182, 205, 275–277, 283, 284, 347, 348, 389, 482]). Based on the cited literature and other sources, the present chapter briefly outlines the most important theoretical findings on ODEs with constant or variable delay. These include qualitative features of such equations, exact and approximate solutions to linear and nonlinear delay ODEs, statements of and analytical solution methods for main problems, and theorems on the existence, uniqueness and stability of solutions.

Remark 1.1. Section 5.1 deals with numerical methods for integrating delay ODEs, while Section 6.1 treats mathematical models of various processes based on delay ODEs.

1.1.2. First-Order ODEs with Constant Delay. Cauchy Problem. Qualitative Features

Equations with a single constant delay. Cauchy problem. We will look at first-order delay ordinary differential equations of the form

$$u'_t = f(t, u, w), \quad w = u(t - \tau), \quad t > t_0, \tag{1.1.2.1}$$

where $u = u(t)$ is the unknown function, t is time, f is a given continuous function, $\tau > 0$ is a constant delay, and t_0 is some constant that will be called the *initial time*. If f is implicitly independent of t, then equation (1.1.2.1) is called *autonomous*.

Delay ordinary differential equations (1.1.2.1) and related more complicated delay equations arise in numerous applications and disciplines, including control theory [226, 268, 477, 552, 585], neurodynamics [193, 265, 317], laser physics [18, 290, 535], radio physics [59, 131], nuclear physics [110, 186, 197], mathematical ecology and biology [64, 94, 180, 182, 209, 259, 264, 283, 287, 460, 604], medicine [4, 191, 331, 358, 469, 533], and economics [93, 96, 246, 326, 465, 605].

The *Cauchy problem* for the delay ODE (1.1.2.1) is stated as follows: find a solution of this equation that satisfies the initial condition

$$u = \varphi(t) \quad \text{for} \quad t_0 - \tau \le t \le t_0, \tag{1.1.2.2}$$

where $\varphi(t)$ is a given continuous function. The limiting value $\tau = 0$ in (1.1.2.1)–(1.1.2.2) corresponds to the Cauchy problem for an ODE without delay (e.g., see [255, 350, 423] for the properties and solution methods for such problems).

Qualitative features. Solution smoothing. Below we note some specific features that distinguish Cauchy problems for delay ODEs from those for ODEs without delay.

First, the initial condition is set on a closed interval $E_{t_0} = \{t_0 - \tau \le t \le t_0\}$ rather than at a single point $t = t_0$, as for equations without delay. Most commonly, one uses $t_0 = 0$ as the initial times. Sometimes, $t_0 = \tau$ can also be used. One seeks a solution continuous at t_0, which implies that $u(t_0 + 0) = \varphi(t_0)$.

Secondly, even though the functions φ and f have continuous derivatives up to an indefinitely high order, the solution $u(t)$ of the initial problem (1.1.2.1)–(1.1.2.2) will generally have jump discontinuities at $t = t_0 + (k-1)\tau$, where $k = 1, 2, \ldots$, in the kth-order derivatives. However, the lower-order derivatives will be continuous at these points. This phenomenon is known as 'solution smoothing' or, sometimes, 'propagation of discontinuities in derivatives'.

Let us consider problem (1.1.2.1)–(1.1.2.2). On the interval $t_0 \le t \le t_0 + \tau$, we have

$$u'_t = f(t, u(t), \varphi(t - \tau)).$$

On the preceding interval, $t_0 - \tau \le t \le t_0$, the first derivative is calculated using the initial condition (1.1.2.2):

$$u'_t = \varphi'_t(t).$$

Then, to the right of t_0, we get

$$u'_t(t_0 + 0) = f(t_0, \varphi(t_0), \varphi(t_0 - \tau)),$$

and to the left of t_0,

$$u'_t(t_0 - 0) = \varphi'_t(t_0).$$

Clearly, the continuity of u'_t at t_0 can only be ensured through a special selection of the initial function φ by requiring that

$$\varphi'(t_0) = f(t_0, \varphi(t_0), \varphi(t_0 - \tau)).$$

Therefore, the derivative $u'_t(t)$ is generally discontinuous at t_0.

The first derivative of the solution is continuous at $t_0 + \tau$. Indeed, it follows from (1.1.2.1) that $u'_t(t) = f(t, u(t), w(t))$, with the right-hand side being a continuous function of t at $t_0 + \tau$, since $w(t) = u(t - \tau)$ is continuous at this point. However, the second derivative

$$u''_{tt} = f_t + f_u u'_t + f_w w'_t$$

is discontinuous at $t_0 + \tau$, since $w'_t = u'_t(t - \tau)$ is discontinuous at this point. This follows from the fact that $w'(t_0+\tau) = u'_t(t_0)$, with the discontinuity of u'_t at t_0 shown above for the generic case. Yet at $t = t_0 + 2\tau$, the derivative $u''_{tt}(t)$ is continuous, since $w'_t(t)$ and $w(t)$ are both continuous at this point.

Continuing the reasoning, we notice that the derivative $u^{(k)}(t)$ is discontinuous at $t_0 + (k-1)\tau$, while the lower-order derivatives are all continuous, provided that f is a sufficiently smooth function.

▶ **Example 1.1.** Consider the following Cauchy problem for a linear delay ODE with a simple initial condition [555]:

$$u'(t) = u(t-1), \qquad t > 0; \qquad (1.1.2.3)$$
$$u(t) = 1, \qquad -1 \leq t \leq 0. \qquad (1.1.2.4)$$

It follows from (1.1.2.4) that $u'(t) = 0$ on $-1 \leq t \leq 0$. At the same time, we see from (1.1.2.3) in view of condition (1.1.2.4) that $u'(t) = 1$ on $0 < t \leq 1$. Hence, $u'(t)$ is discontinuous at $t = 0$.

Now let us look at the point $t = k$, where k is an integer. On differentiating (1.1.2.3) k times, we get

$$u^{(k+1)}(t) = u^{(k)}(t-1).$$

By induction we find that

$$u^{(k+1)}(t) = u'(t-k).$$

It follows that the derivative $u^{(k+1)}$ is discontinuous at $t = k$. ◀

Equations with several constant delays. Equation (1.1.2.1) is the simplest functional differential equation with a single delay. First-order ordinary differential equations with several delays are more complicated and can be written as

$$u'_t = f(t, u, w_1, \ldots, w_m), \quad w_k = u(t - \tau_k), \quad k = 1, \ldots, m, \qquad (1.1.2.5)$$

where $\tau_k > 0$ are given numbers.

The Cauchy problem for equation (1.1.2.5) is stated as follows: find a solution to this equation that satisfies the initial condition

$$u = \varphi(t) \quad \text{at} \quad t_0 - \tau_{\max} \leq t \leq t_0, \qquad (1.1.2.6)$$

where $\varphi(t)$ is a given continuous function and $\tau_{\max} = \max\limits_{1 \leq k \leq m} \tau_k$ is the maximum delay.

Equations of neutral or advanced type. First-order functional differential equations of the form

$$F(t, u, u'_t, w, w'_t) = 0, \quad w = u(t-\tau), \qquad (1.1.2.7)$$

also arise in the literature. These involve two derivatives, u'_t and w'_t, and are called *equations of neutral type* or *neutral differential equations*. More general equations can include several delays, $\tau_1, \ldots, \tau_m$.

Equations of advanced type or *advanced differential equations* are equations (1.1.2.7) of a special form that contain w'_t but do not involve u'_t.

Neutral and advanced differential equations are quite rare in applications and, therefore, are not discussed in the present book. For theoretical treatment of such equations, see, for example, [37, 144, 205, 275–277].

1.1.3. Exact Solutions to a First-Order Linear ODE with Constant Delay. The Lambert W Function and Its Properties

Exponential solutions to a first-order linear ODE with constant delay. The Lambert W function. Let us consider the first-order linear homogeneous ordinary differential equation with constant coefficients and a constant delay

$$u'_t = au + bw, \quad w = u(t - \tau), \tag{1.1.3.1}$$

where a and b are real constants. Just as with ODEs without delay, equation (1.1.3.1) has exponential exact solutions

$$u(t) = \exp(\lambda t). \tag{1.1.3.2}$$

On substituting (1.1.3.2) into (1.1.3.1) and canceling by $e^{\lambda t}$, we arrive at a *characteristic equation* for determining the parameter λ:

$$\lambda - a - be^{-\lambda \tau} = 0. \tag{1.1.3.3}$$

Equation (1.1.3.3) involves three free parameters: a, b, and τ. For $\tau = 0$, which corresponds to an ODE without delay, equation (1.1.3.3) is a linear algebraic equation with the only root $\lambda = a + b$. The presence of delay, $\tau > 0$ (with $b \neq 0$), changes the situation qualitatively, since (1.1.3.3) becomes a transcendental equation with indefinitely many complex conjugate roots, $\lambda_m = \operatorname{Re} \lambda_m \pm i \operatorname{Im} \lambda_m$, $i^2 = -1$, $m = 0, 1, 2, \ldots$, and, possibly, one or two real roots at certain values of the parameters. By the superposition principle, any linear combination of exponentials (1.1.3.2), with $\lambda = \lambda_k$ being different roots of the characteristic equation (1.1.3.3), is a solution of the linear delay ODE (1.1.3.1).

First, let us find conditions under which the characteristic equation (1.1.3.3) has real roots. Solutions of equation (1.1.3.3) can be described using the *Lambert function* $W = W(z)$. For complex $z = x + iy$, it is defined implicitly by the transcendental equation

$$We^W = z. \tag{1.1.3.4}$$

For properties of this function and its applications, see, for example, [114, 139, 524, 568, 585].

The constant λ in the exponential solution (1.1.3.2) of the characteristic equation (1.1.3.3) can be expressed in terms of the Lambert function as

$$\lambda = a + \frac{1}{\tau} W(x), \quad x = b\tau e^{-a\tau}. \tag{1.1.3.5}$$

1.1. First-Order Equations. Cauchy Problem. Method of Steps. Exact Solutions

The Lambert W function on the real axis. For real $z = x$, the function $W(x)$ is single-valued for $x \geq 0$ and two-valued on the interval $(-1/e, 0)$. For $x \geq -1/e$ and $W \geq -1$, the single-valued branch of the Lambert function, which is conventionally called the *principal branch*, will be denoted $W_\mathrm{p}(x)$. It is also known as the *positive branch*. The other branch of $W(x)$, which is characterized by the inequalities $-1/e \leq x < 0$ and $W \leq -1$, will be denoted $W_\mathrm{n}(x)$*.

Figure 1.1 displays the two branches of the Lambert function $W(x)$, W_p and W_n, on the ray $-1/e \leq x < \infty$. The logarithmic function $\ln(1 + x)$ is also shown for comparison.

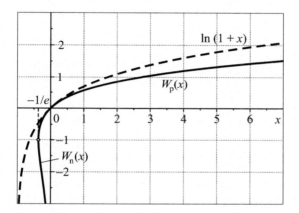

Figure 1.1. The real branches of the Lambert W function, $W_\mathrm{p}(x)$ and $W_\mathrm{n}(x)$.

In parametric form, the real branches $W_\mathrm{p}(x)$ and $W_\mathrm{n}(x)$ are defined as

$$x = se^s, \quad W_\mathrm{p} = s, \quad -1 \leq s < +\infty;$$
$$x = se^s, \quad W_\mathrm{n} = s, \quad -\infty < s \leq -1.$$

The following Taylor series expansion, convergent for $|x| < 1/e$, holds:

$$\begin{aligned}
W_\mathrm{p}(x) &= \sum_{n=1}^{\infty} (-1)^{n-1} \frac{n^{n-1}}{n!} x^n \\
&= x - x^2 + \frac{3}{2}x^3 - \frac{8}{3}x^4 + \frac{125}{4}x^5 - \cdots.
\end{aligned} \qquad (1.1.3.6)$$

---

*The real branches of the Lambert W function, W_p and W_n, are frequently denoted W_0 and W_{-1} (e.g., see [114, 240]). However, the same notation, with a different meaning, is sometimes used to denote complex branches of the Lambert function [114], which may lead to confusion. The present book uses the more convenient notation W_p for the principal branch, introduced in [369] (the subscript 'p' stands for *principal* or *positive*). The second branch is denoted W_n (the subscript 'n' stands for *negative*).

The following asymptotic formulas are true [114, 369]:

$$W_p(x) = \zeta_1 - \ln \zeta_1 + \frac{\ln \zeta_1}{\zeta_1} + \frac{\ln^2 \zeta_1}{2\zeta_1^2} - \frac{\ln \zeta_1}{\zeta_1^2} + O\left(\frac{\ln^3 \zeta_1}{\zeta_1^3}\right) \quad \text{as} \quad x \to +\infty,$$
$$\zeta_1 = \ln x; \tag{1.1.3.7}$$

$$W_n(x) = \zeta_2 - \ln \zeta_2 - \frac{\ln \zeta_2}{\zeta_2} - \frac{\ln^2 \zeta_2}{2\zeta_2^2} - \frac{\ln \zeta_2}{\zeta_2^2} + O\left(\frac{\ln^3 \zeta_2}{\zeta_2^3}\right) \quad \text{as} \quad x \to -0,$$
$$\zeta_2 = \ln(-1/x). \tag{1.1.3.8}$$

Properties of the Lambert W function and its values at some points:

$$W_p(xe^x) = x \ (x \geq -1), \quad \ln W_p(x) = \ln x - W_p(x) \ (x > 0),$$
$$W_p(x \ln x) = \ln x \ (x \geq e^{-1}), \quad W_p(-\ln x/x) = -\ln x \ (0 < x \leq e),$$
$$W_n(xe^x) = x \ (x \leq -1), \quad W_n(x \ln x) = \ln x \ (x \leq e^{-1}),$$
$$W_p(-1/e) = -1, \quad W_p(0) = 0, \quad W_p(e) = 1, \quad W_p(e^{1+e}) = e.$$

The principal branch of the Lambert W function on the ray $0 \leq x < \infty$ is well approximated by the simple explicit formula [557]:

$$W_p(x) = \ln(1+x)\left[1 - \frac{\ln(1 + \ln(1+x))}{2 + \ln(1+x)}\right]. \tag{1.1.3.9}$$

It is accurate to two asymptotic terms as $x \to 0$ and $x \to \infty$ (see formulas (1.1.3.6) and (1.1.3.7)). The relative error of the approximate formula (1.1.3.9) does not exceed 10^{-2} for any positive x.

In the range $-e^{-1} \leq x \leq 1$, the approximate formula [557]:

$$W_p(x) = \frac{ex}{1 + \left[(2ex+2)^{-1/2} + (e-1)^{-1} - 2^{-1/2}\right]^{-1}}$$

also holds true. Its relative error does not exceed 10^{-3} in this range.

For other formulas suitable to approximate different portions of the branches of the Lambert W function, see [33, 528].

With the above properties of the Lambert W function and formula (1.1.3.5), one can easily find conditions under which the characteristic equation (1.1.3.3) has real roots. The results are summarized in Table 1.1.

The Lambert W function in the complex plane. The Lambert function $W(z)$ has infinitely many branches, $W_m = W_m(z)$ $(m = 0, \pm 1, \pm 2, \dots)$, in the complex plane $z = x + iy$ $(i^2 = -1)$.

The following asymptotic formula holds [114]:

$$W_m = \ln z - \ln \ln z + 2\pi i m + (1+i)o(1) \quad \text{as} \quad z \to \infty. \tag{1.1.3.10}$$

Substituting $z = x + iy$ and $W = \xi + i\eta$ into (1.1.3.4) and rearranging using Euler's formula $e^{iy} = \cos y + i \sin y$, we arrive at the system of transcendental equations

$$e^\xi(\xi \cos \eta - \eta \sin \eta) = x,$$
$$e^\xi(\xi \sin \eta + \eta \cos \eta) = y. \tag{1.1.3.11}$$

Table 1.1. The number of real roots of the characteristic of equation (1.1.3.3) at different values of the determining parameters of the delay ODE (1.1.3.1).

Determining conditions	Number of real roots	Range of roots
$-e^{a\tau-1}\tau^{-1} < b < 0$	Two roots, λ_1 and λ_2	$a - \tau^{-1} < \lambda_1 < a$, $\lambda_2 < a - \tau^{-1}$
$b \geq 0$	One root, λ_1	$\lambda_1 > a$ if $b > 0$, $\lambda_1 = a$ if $b = 0$
$b = -e^{a\tau-1}\tau^{-1}$	One root, λ_1 (double)	$\lambda_1 = a - \tau^{-1}$
$b < -e^{a\tau-1}\tau^{-1}$	No roots	

By setting $y = 0$ in (1.1.3.11), we will study the complex values of the Lambert W function on the real axis x. We note that the change of variable η to $-\eta$ preserves the equations (1.1.3.11). This means that the roots of W corresponding to real x are complex conjugate. Therefore, it only suffices to look at the case $\eta \geq 0$.

For $y = 0$, the second equation in (1.1.3.11) has two solutions. One is trivial, $\eta = 0$; it leads to real values of the Lambert W function, which were discussed above. The other solution, which determines complex values of the Lambert W function, can be written as

$$\xi = -\eta \cot \eta. \qquad (1.1.3.12)$$

The right-hand side of (1.1.3.12) tends to infinity as the points $\eta = n\pi$, $n = \pm 1$, $\pm 2, \ldots$, are approached. For $n\pi < \eta < (n+1)\pi$ with $n = 0, 1, 2, \ldots$, relation (1.1.3.12) describes branches of the Lambert W function in the complex plane for $\eta \geq 0$. These curves are depicted in Figure 1.2.

On substituting (1.1.3.12) into the first equation of (1.1.3.11) and rearranging, we obtain the relation

$$-\frac{\eta}{\sin \eta} \exp(-\eta \cot \eta) = x. \qquad (1.1.3.13)$$

It defines implicitly the imaginary part η of the Lambert W function as a function of x. With formulas (1.1.3.12) and (1.1.3.13), we represent the complex-valued branches $W_m = W_m(x)$, $x < 0$, in the parametric form

$$W_0 = \xi_0 + i\eta_0,$$
$$\xi_0 = -s \cot s, \quad \eta_0 = s, \quad x = -\frac{s}{\sin s} \exp(-s \cot s), \quad |s| < \pi;$$
$$W_m = \xi_m + i\eta_m, \quad m = \pm 1, \pm 2, \ldots, \qquad (1.1.3.14)$$
$$\xi_m = -s \cot s, \quad \eta_m = s \operatorname{sign} m, \quad x = -\frac{s}{\sin s} \exp(-s \cot s),$$
$$2|m|\pi < s < (2|m|+1)\pi.$$

The point with coordinates $x = -e^{-1}$ and $\xi_0 = -1$, which corresponds to $s = \eta_0 = 0$, is assigned to belong to W_0.

Figure 1.3 displays a few real and imaginary branches of the Lambert W function as functions of x for $x < 0$ (solid lines). These are obtained using formulas (1.1.3.14) by setting numeric values of the real parameter s in appropriate intervals. It is

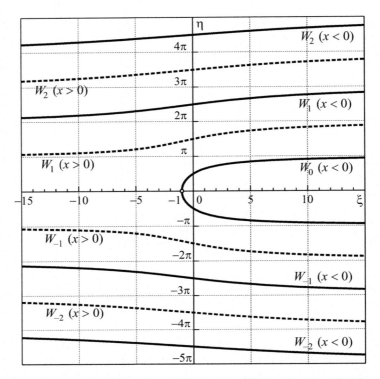

Figure 1.2. Complex-valued branches of the Lambert W function described by formula (1.1.3.12); $\xi = \operatorname{Re} W(x)$, $\eta = \operatorname{Im} W(x)$, $y = 0$. Solid lines indicate branches for $x < 0$, while dashed lines show branches with $x > 0$.

apparent that $\xi_m \to -\infty$ ($m = \pm 1, \pm 2, \ldots$) as $x \to -0$. In addition, as the branch number's modulus $|m|$ increases, the real part of the Lambert W function decreases, while the absolute value of the imaginary part of W increases. The real parts of $W_{\pm m}(x)$ vanish at $x_m = -\frac{\pi}{2} - 2\pi|m|$, where $m = 0, 1, \ldots$ For $-\pi/2 < x < 0$, the real parts of all branches of the Lambert W function are negative.

For $x > 0$, the complex-valued branches $W_m = W_m(x)$ can be represented in parametric form as

$$W_m = \xi_m + i\eta_m, \quad m = \pm 1, \pm 2, \ldots;$$

$$\xi_m = -s \cot s, \quad \eta_m = s \operatorname{sign} m, \quad x = -\frac{s}{\sin s} \exp(-s \cot s); \qquad (1.1.3.15)$$

$$(2|m| + 1)\pi < s < (2|m| + 2)\pi.$$

Furthermore, Figure 1.3 displays a few real and imaginary branches of the Lambert W function as functions of x for $x > 0$ (dashed lines). These are obtained using formulas (1.1.3.15). One can see that $\xi_m \to -\infty$ ($m = \pm 1, \pm 2, \ldots$) as $x \to +0$. Also, as the branch number's modulus $|m|$ increases, the real part of the Lambert W function decreases, while the absolute value of the imaginary part increases. The

1.1. First-Order Equations. Cauchy Problem. Method of Steps. Exact Solutions

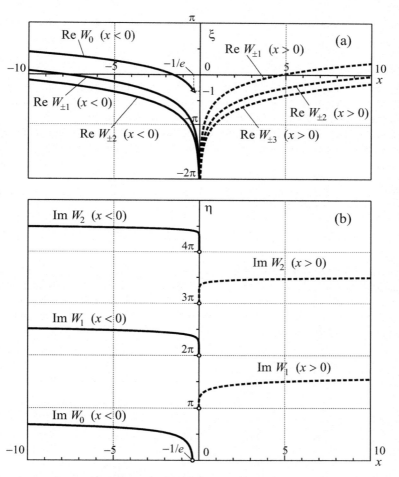

Figure 1.3. Complex-valued branches of the Lambert W function for $y = 0$ as described by formulas (1.1.3.14) and (1.1.3.15): (a) $\xi_m = \operatorname{Re} W_m(x)$, (b) $\eta_m = \operatorname{Im} W_m(x)$. The branches with $x < 0$ are shown in solid lines, while those with $x > 0$ are shown in dashed lines.

real parts of $W_{\pm m}(x)$ vanish at $x_m = \frac{3\pi}{2} + 2\pi(|m| - 1)$. Although the real parts of all branches $W_{\pm m}$ ($m = \pm 1, \pm 2, \ldots$) are negative for $0 < x < 3\pi/2$, there is one real positive root on the principal branch W_p.

By taking the modulus of the real and imaginary parts of the complex representation of the Lambert W function (1.1.3.4), we obtain the following relation for real $z = x$:

$$e^\xi (\xi^2 + \eta^2)^{1/2} = |x|. \tag{1.1.3.16}$$

For a given x, it defines a contour line in the complex plane $W = \xi + i\eta$, where the points of all branches W_k lie. It follows from (1.1.3.16) that the following inequality holds for nonnegative values of the real part of the Lambert W function, $\xi \geq 0$:

$$(\xi^2 + \eta^2)^{1/2} \leq |x| \quad \text{(the equality is attained at } \xi = 0\text{).} \tag{1.1.3.17}$$

This means that all points of the branches W_k located in the half-plane $\xi > 0$ lie inside the circle of radius $|x|$. In other words, for any real x, the real parts of the branches of the Lambert W function are limited by the quantity $|x|$: $\operatorname{Re} W_m \leq |x|$. More precisely, the maximum allowed value of ξ in the positive half-plane $\xi \geq 0$ among all W_k is determined by $\eta = 0$ in (1.1.3.16), which implies that $\xi_{\max} = W_p(|x|)$.

Remark 1.2. *The contour lines defined by the implicit relation (1.1.3.16) can be represented in parametric form as*

$$\xi = s, \quad \eta = \pm\sqrt{x^2 e^{-2s} - s^2}, \quad -\infty < s \leq W_p(|x|). \tag{1.1.3.18}$$

Figure 1.4 depicts a few contour lines, defined by the implicit relation (1.1.3.16) (or parametric formulas (1.1.3.18)), in the complex plane $W = \xi + i\eta$ at $x = \pm 0.5$, ± 1.0, and ± 2.0. Open circles indicate points of intersection between contour lines and respective branches W_m of the Lambert W function for $x > 0$, while solid circles indicate points of intersection for $x < 0$.

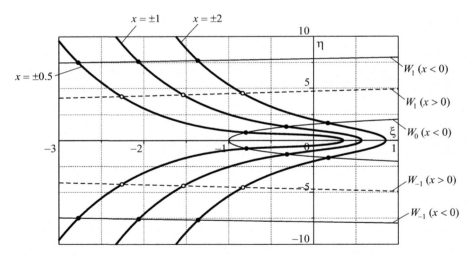

Figure 1.4. Contour lines of the Lambert W function, $|We^W| = |x|$, defined by the implicit relation (1.1.3.16) in the complex plane $W = \xi + i\eta$ at $x = \pm 0.5$, ± 1.0, and ± 2.0. Open circles indicate roots of the Lambert W function for $x > 0$, while solid circles indicate roots for $x < 0$.

Table 1.2 lists a few complex values of the many-valued Lambert W function at four real values of x.

1.1. First-Order Equations. Cauchy Problem. Method of Steps. Exact Solutions

Table 1.2. The values of the Lambert function $W(x)$ at a few real values of x on several first branches $W_m(x)$.

Branches of $W(x)$	$x = -\pi/2$	$x = -1$	$x = 1$	$x = e$
$W_{\pm 1}(x)$	$\pm \frac{\pi}{2} i$	$-0.3181 \pm 1.3372 i$	$-1.5339 \pm 4.3752 i$	$-0.5321 \pm 4.5972 i$
$W_{\pm 2}(x)$	$-1.6043 \pm 7.6472 i$	$-2.0623 \pm 7.5886 i$	$-2.4016 \pm 10.7763 i$	$-1.3940 \pm 10.8680 i$
$W_{\pm 3}(x)$	$-2.1983 \pm 13.9812 i$	$-2.6532 \pm 13.9492 i$	$-2.8536 \pm 17.1135 i$	$-1.8490 \pm 17.1715 i$
$W_{\pm 4}(x)$	$-2.5667 \pm 20.2945 i$	$-3.0202 \pm 20.2725 i$	$-3.1630 \pm 23.4277 i$	$-2.1599 \pm 23.4702 i$
$W_{\pm 5}(x)$	$-2.8349 \pm 26.5974 i$	$-3.2878 \pm 26.5805 i$	$-3.3987 \pm 29.7313 i$	$-2.3966 \pm 29.7648 i$

Approximate complex values of the principal branch $W_0(z)$ of the Lambert W function can be evaluated using the explicit approximation formula [557]:

$$W_0(z) = \frac{2\ln(1+A_1 y) - \ln[1 + A_2 \ln(1 + A_3 y)] + A_4}{1 + [2\ln(1+A_1 y) + 2A_5]^{-1}}, \quad y = \sqrt{2ez+2}, \quad (1.1.3.19)$$

$$A_1 = 0.8842, \quad A_2 = 0.9294, \quad A_3 = 0.5106, \quad A_4 = -1.213, \quad A_5 = 2.344.$$

It provides exact asymptotics near the points $z = 0$ and $z = -e^{-1}$ and at large $|z|$. In the entire complex plane z, the maximum relative error of formula (1.1.3.19) does not exceed 10^{-2} (the principal branches of $\sqrt{z}$ and $\ln z$ must be used for the calculations).

For the different aspects of the numerical computation of the branches of the complex-valued Lambert W function, see, for example, [69, 240].

Some remarks. In general, the coefficient λ in the exponential solution (1.1.3.2) of equation (1.1.3.3) can be expressed via the Lambert W function as (1.1.3.5), where W on the right-hand side is understood as the set of all real and complex branches of the Lambert W function. Each pair of complex conjugate roots $W_{\pm m} = \xi_m \pm i \eta_m$ defines a pair of exponential solutions to the delay ODE (1.1.3.1) of the form

$$\begin{aligned} u_{\pm m}(t) &= e^{(\lambda_{r,m} \pm i \lambda_{i,m})t} = e^{\lambda_{r,m} t}[\cos(\lambda_{i,m} t) \pm i \sin(\lambda_{i,m} t)], \\ \lambda_{r,m} &= a + \tau^{-1} \xi_m(x), \quad \lambda_{i,m} = \tau^{-1} \eta_m(x), \quad x = b\tau e^{-a\tau}, \end{aligned} \quad (1.1.3.20)$$

which are obtained using formulas (1.1.3.2) and (1.1.3.5). Since the delay ODE (1.1.3.1) is linear and homogeneous, the real and imaginary parts of the complex solutions (1.1.3.20),

$$\begin{aligned} u_m^{(1)}(t) &= \operatorname{Re} u_{\pm m}(t) = e^{\lambda_{r,m} t} \cos(\lambda_{i,m} t), \\ u_m^{(2)}(t) &= \operatorname{Im} u_{\pm m}(t) = e^{\lambda_{r,m} t} \sin(\lambda_{i,m} t), \end{aligned} \quad (1.1.3.21)$$

are real solutions of the original equation (1.1.3.1).

The following two simple statements hold true:

1°. If the inequalities $a < 0$ and $0 < b < -a$ hold, all roots of the characteristic equation (1.1.3.3) have a negative real part.

2°. If $b > -a$, the characteristic equation (1.1.3.3) has at least one root with a positive real part.

Remark 1.3. There are more general but more complicated conditions under which the real parts of all roots of the characteristic equation (1.1.3.3) are negative. These conditions are stated below in Subsection 1.3.2 (see the Hayes theorem [37]).

Notably, if

$$a = 0, \quad b = k(-1)^{n+1}, \quad k = \frac{(2n+1)\pi}{2\tau}, \quad n = 0, \pm 1, \pm 2, \ldots, \quad (1.1.3.22)$$

equation (1.1.3.1) has periodic solutions of the form

$$u(t) = \cos(kt + \delta), \quad (1.1.3.23)$$

where δ is an arbitrary constant.

Remark 1.4. The change of variable $u(t) = e^{at}U(t)$ reduces equation (1.1.3.1) to the simpler form

$$U'_t = be^{-a\tau}\bar{U}, \quad \bar{U} = U(t - \tau).$$

Remark 1.5. The first-order linear nonhomogeneous ODE with constant coefficients and a constant delay

$$u'_t = au + bw + c, \quad w = u(t - \tau),$$

can be reduced, with the substitution $u = v - \frac{c}{a+b}$ for $b \neq -a$, to a homogeneous delay ODE of the form (1.1.3.1). If $b = -a$, one should use the change of variable $u = v + kt$ with $k = \frac{c}{1-a\tau}$ to obtain a homogeneous delay ODE.

1.1.4. First-Order Nonlinear ODEs with Constant Delay That Admit Linearization or Exact Solutions

We describe below several simple first-order nonlinear ODEs with constant delay that reduce to linear ODEs with constant delay or admit exact solutions representable in terms of elementary functions. These equations and their solutions can be used to test approximate analytical and numerical methods for solving nonlinear delay ODEs.

Equation 1. The nonlinear ODE with constant delay

$$u'_t = a(t)u + b(t)u^{1/2} + c(t)u^{1/2}w^{1/2}, \quad w = u(t - \tau),$$

can be reduced, with the substitution $u = v^2$ ($v \geq 0$), to a linear ODE with constant delay $v'_t = \frac{1}{2}a(t)v + \frac{1}{2}c(t)\bar{v} + \frac{1}{2}b(t)$, where $\bar{v} = v(t - \tau)$.

Equation 2. The nonlinear ODE with constant delay

$$u'_t = a(t)u + b(t)u^{1-k} + c(t)u^{1-k}w^k, \quad w = u(t - \tau),$$

can be reduced, with the substitution $u = v^{1/k}$, to a linear ODE with constant delay $v'_t = ka(t)kv + kc(t)\bar{v} + kb(t)$, where $\bar{v} = v(t - \tau)$.

Equation 3. The nonlinear ODE with constant delay

$$u'_t = a(t) + b(t)e^{\lambda u} + c(t)e^{\lambda(u-w)}, \quad w = u(t-\tau),$$

can be reduced, with the substitution $v = e^{-\lambda u}$, to a linear ODE with constant delay $v'_t = -\lambda a(t)v - \lambda c(t)\bar{v} - \lambda b(t)$, where $\bar{v} = v(t-\tau)$.

Equation 4. The nonlinear ODE with constant delay

$$u'_t = a(t)u \ln u + b(t)u \ln w + c(t)u, \quad w = u(t-\tau),$$

can be reduced, with the substitution $u = e^v$, to a linear ODE with constant delay $v'_t = a(t)v + b(t)\bar{v} + c(t)$, where $\bar{v} = v(t-\tau)$.

Remark 1.6. Exact solutions to the above nonlinear equations 1 to 4 with constant coefficients a, b, and c can be obtained using the substitutions specified and the results described in Subsection 1.1.3.

Equation 5. The nonlinear ODE with constant delay

$$u'_t = f(u-w), \quad w = u(t-\tau),$$

involves an arbitrary function $f(z)$ and remains the same under the substitution of u with $u + $ const. This equation admits the exact solution $u(t) = bt + C$, where C is an arbitrary constant and b is a root of the transcendental equation $b = f(b\tau)$.

Equation 6. The nonlinear ODE with constant delay

$$u'_t = uf(w/u), \quad w = u(t-\tau),$$

involves an arbitrary function $f(z)$ and remains the same under the substitution of u with const $\cdot$ u. This equation admits the exact solution $u(t) = Ce^{\lambda t}$, where C is an arbitrary constant and λ is a root of the transcendental equation $\lambda = f(e^{-\lambda\tau})$.

1.1.5. Method of Steps. Solution of the Cauchy Problem for a First-Order ODE with Constant Delay

Method of steps for first-order ODEs with constant delay. The Cauchy problem with constant delay (1.1.2.1)–(1.1.2.2) on a finite interval can be solved using the *method of steps*. It suggests that the solution is obtained by successively integrating simpler ODEs without delay on equal segments of length τ: $t_0 + n\tau \leq t \leq t_0 + (n+1)\tau$, $n = 0, 1, 2, \ldots$.

For $n = 0$, we find $w(t) = u(t-\tau) = \varphi(t-\tau)$ on the interval $t_0 \leq t \leq t_0 + \tau$. As a result, we get

$$u'_t = f(t, u, \varphi_0(t-\tau)), \quad t_0 \leq t \leq t_0 + \tau;$$
$$u(t_0) = \varphi_0(t_0).$$

Here, the function φ has been renamed φ_0 for the convenience of subsequent presentation. Assuming that the solution $u = \varphi_1(t)$ of the problem exists on the entire interval $t_0 \leq t \leq t_0 + \tau$, we similarly get

$$u'_t = f(t, u, \varphi_1(t - \tau)), \quad t_0 + \tau \leq t \leq t_0 + 2\tau;$$
$$u(t_0 + \tau) = \varphi_1(t_0 + \tau).$$

In general, the problem for each individual interval is written as

$$u'_t = f(t, u, \varphi_n(t - \tau)), \quad t_0 + n\tau \leq t \leq t_0 + (n+1)\tau, \quad n = 0, 1, 2, \ldots;$$
$$u(t_0 + n\tau) = \varphi_n(t_0 + n\tau).$$

The function $\varphi_n(t)$ is the solution to the Cauchy problem on the preceding interval $t_0 + (n-1)\tau \leq t \leq t_0 + n\tau, n = 1, 2, \ldots$.

Solution of linear problems with constant delay by the method of steps. Below are a few examples illustrating the practical use of the method of steps for solving Cauchy problems described by linear ODEs with constant delay.

▶ **Example 1.2.** Consider the following Cauchy problem for a linear ODE with constant delay subject to a special initial condition:

$$u'_t = bw, \quad w = u(t - \tau), \quad t > 0;$$
$$u = 1 \quad \text{at} \quad -\tau \leq t \leq 0, \qquad (1.1.5.1)$$

where b is a free parameter ($b \neq 0$).

With the method of steps applied to problem (1.1.5.1), the first step gives

$$u'_t = b, \quad 0 < t \leq \tau \quad \text{(equation)};$$
$$u = 1 \quad \text{at} \quad t = 0 \quad \text{(initial condition)}.$$

Integrating yields

$$u = 1 + bt, \quad 0 \leq t \leq \tau.$$

The next step leads to the problem

$$u'_t = b[1 + b(t - \tau)], \quad \tau < t \leq 2\tau \quad \text{(equation)};$$
$$u = 1 + b\tau \quad \text{at} \quad t = \tau \quad \text{(initial condition)}.$$

Its solution is expressed as

$$u = 1 + bt + \tfrac{1}{2}b^2(t - \tau)^2, \quad \tau \leq t \leq 2\tau.$$

By repeating similar computations, one arrives at the following formula (e.g., see [144]):

$$u = 1 + b\frac{t}{1!} + \cdots + b^k\frac{[t - (k-1)\tau]^k}{k!}, \quad (k-1)\tau \leq t \leq k\tau, \qquad (1.1.5.2)$$

where k is any positive integer. ◀

1.1. First-Order Equations. Cauchy Problem. Method of Steps. Exact Solutions

For what follows, it is convenient to introduce the *delayed exponential function*:

$$\exp_d(t, \tau) \equiv \sum_{k=0}^{[t/\tau]+1} \frac{[t - (k-1)\tau]^k}{k!}, \tag{1.1.5.3}$$

where $[A]$ stands for the integer part of the number A, and the subscript d indicates delay. The following properties hold:

$$\exp_d(0, \tau) = 1, \quad \exp_d(t, 0) = e^t, \quad [\exp_d(t, \tau)]'_t = \exp_d(t - \tau, \tau).$$

Figure 1.5 displays the delayed exponential function (1.1.5.3) for $\tau = 0.5, 1.0$, and 2.0. The ordinary exponential function e^t, which corresponds to $\tau = 0$, is shown by a dashed line.

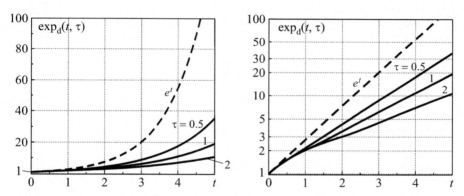

Figure 1.5. Graphs of the delayed exponential function $\exp_d(t, \tau)$ for $\tau = 0.5, 1.0$, and 2.0 in the ordinary Cartesian (left) and logarithmic (right) coordinates. The exponential function e^t is shown by a dashed line; it corresponds to $\tau = 0$.

Solution (1.1.5.2) to problem (1.1.5.1) is expressed in terms of the delayed exponential function (1.1.5.3) as

$$u = \exp_d(bt, b\tau). \tag{1.1.5.4}$$

▶ **Example 1.3.** It is not difficult to verify that the function

$$u(t) = e^{at} \exp_d(\lambda t, \lambda \tau), \quad \lambda = e^{-a\tau} b,$$

is an exact solution to the linear delay ODE with constant coefficients

$$u'_t = au + bw, \quad w = u(t - \tau), \quad t > 0, \tag{1.1.5.5}$$

subjected to the exponential initial condition $u = e^{at}$ at $-\tau \le t \le 0$. ◀

Representations of solutions to linear problems using the delayed exponential function. Solutions to the more general, important problems specified below can be represented using the delayed exponential function (1.1.5.3).

Problem 1. The solution to the Cauchy problem for the linear homogeneous delay ODE (1.1.5.5) subjected to the general initial condition

$$u = \varphi(t) \quad \text{at} \quad -\tau \leq t \leq 0 \tag{1.1.5.6}$$

can be written in a closed form [25]:

$$u(t) = e^{a(t+\tau)} \exp_d(\lambda t, \lambda \tau) \varphi(-\tau)$$
$$+ \int_{-\tau}^{0} e^{a(t-s)} \exp_d(\lambda(t-\tau-s), \lambda\tau)[\varphi'_s(s) - a\varphi(s)]\,ds, \quad \lambda = e^{-a\tau}b. \tag{1.1.5.7}$$

Another representation of the solution to the Cauchy problem for equation (1.1.5.5) with a general initial condition set on the interval $0 \leq t \leq \tau$ can be found in [144].

Problem 2. The solution to the Cauchy problem for the linear nonhomogeneous delay ODE

$$u'_t = au + bw + f(t), \quad w = u(t-\tau), \quad t > 0, \tag{1.1.5.8}$$

subjected to the homogeneous initial condition

$$u = 0 \quad \text{at} \quad -\tau \leq t \leq 0 \tag{1.1.5.9}$$

can be represented using the delayed exponential function as [25]:

$$u(t) = \int_0^t e^{a(t-s)} \exp_d(\lambda(t-s), \lambda\tau) f(s)\,ds, \quad \lambda = e^{-a\tau}b. \tag{1.1.5.10}$$

Remark 1.7. The sum of solutions (1.1.5.7) and (1.1.5.10) is a solution to the linear nonhomogeneous delay ODE (1.1.5.8) with the general initial condition (1.1.5.6).

Solution of nonlinear problems with constant delay by the method of steps. Below we will show how one can construct an exact solution to the Cauchy problem for some classes of nonlinear ODEs with constant delay using the method of steps.

Problem 1. Consider the nonlinear delay ODE

$$u'_t = f(t,w)u + g(t,w), \quad w = u(t-\tau), \tag{1.1.5.11}$$

subjected to the general initial condition (1.1.2.2). In the special case of $f(t,w) = a(t)$ and $g(t,w) = b(t)w + c(t)$, it is a linear ODE with a single delay and with variable coefficients of a general form.

Since equation (1.1.5.11) is linear in u, we obtain in each step the following Cauchy problem for a linear ODE without delay:

$$u'_t = f(t, \varphi_n(t-\tau))u + g(t, \varphi_n(t-\tau)), \quad t_0 + n\tau \leq t \leq t_0 + (n+1)\tau,$$
$$u(t_0 + n\tau) = \varphi_n(t_0 + n\tau), \tag{1.1.5.12}$$

where $n = 0, 1, 2, \ldots$, and $\varphi_n(t)$ is the solution to the Cauchy problem obtained in the preceding step on the interval $t_0 + (n-1)\tau \leq t \leq t_0 + n\tau$; $\varphi_0(t) \equiv \varphi(t)$.

The solution to problem (1.1.5.12) is expressed as (based on the results presented, for example, in [255, 350, 421, 423]):

$$u(t) = e^{F(t)}\left[\varphi_n(t_0 + n\tau) + \int_{t_0+n\tau}^{t} e^{-F(t)}g(\xi, \varphi_n(\xi - \tau))\,d\xi\right],$$
$$F(t) = \int_{t_0+n\tau}^{t} f(\xi, \varphi_n(\xi - \tau))\,d\xi, \quad t_0 + n\tau \leq t \leq t_0 + (n+1)\tau,$$
(1.1.5.13)

where $n = 0, 1, 2, \ldots$

Problem 2. The Cauchy problem for the nonlinear delay ODE

$$u'_t = f(t, w)u + g(t, w)u^k, \quad w = u(t - \tau),$$

and the general initial condition (1.1.2.2) is reduced with the substitution $y = u^{1-k}$ to Problem 1, where the functions f and g should be appropriately renamed.

Problem 3. The Cauchy problem for the nonlinear delay ODE

$$u'_t = f(t, w) + g(t, w)e^{\lambda u}, \quad w = u(t - \tau),$$

and general initial condition (1.1.2.2) is reduced with the substitution $y = e^{-\lambda u}$ to Problem 1, where the functions f and g should be appropriately renamed.

Problem 4. The Cauchy problem for the nonlinear delay ODE

$$u'_t = f(t, w)u + g(t, w)u \ln u, \quad w = u(t - \tau),$$

and general initial condition (1.1.2.2) is reduced with the substitution $u = e^y$ to Problem 1, where the functions f and g should be appropriately renamed.

The method of steps for ODEs with several constant delays. The method of steps is suitable for solving the Cauchy problem for the first-order ODE with several delays (1.1.2.5) and the initial data (1.1.2.6). A solution to this problem is constructed by successively integrating an ODE without delay on the intervals $t_0 + nh \leq t \leq t_0 + (n+1)h$, $n = 0, 1, 2, \ldots$. The integration step is determined by the minimum delay $h = \min\limits_{1 \leq k \leq m} \tau_k$ (e.g., see [37]).

1.1.6. Equations with Variable Delay. ODEs with Proportional Delay

ODEs with variable delay. Pantograph equation. So far, the book has been concerned with ODEs with constant delay. However, many applications deal with more complicated ODEs that involve a variable delay of the form $\tau = \tau(t)$, where $\tau(t)$ is a given positive continuous function that can vanish at one or more isolated points. A variable delay may occur, for example, when the transmission rate of the control signal from one object to another is finite and constant, and these objects are moving away from each other at a constant or variable speed.

Let us first look at ordinary differential equations with a variable delay proportional to the independent variable.

▶ **Example 1.4.** The linear first-order functional differential equation

$$u'_t = au + bw, \quad w = u(pt), \tag{1.1.6.1}$$

with $p > 0$ ($p \neq 1$) is called a *pantograph equation*.

If $0 < p < 1$, equation (1.1.6.1) describes the dynamics of a contact current collector (pantograph) of an electric train [367] and is an important special case of an ODE with variable delay where $\tau(t) = (1-p)t$, since $t - \tau(t) = pt$. The function $u(pt)$ appearing in the pantograph equation (1.1.6.1) differs from $u(t)$ in stretching along the t axis by a factor of $1/p$. ◀

The pantograph equation and related more complex functional-differential equations that involve functions with an extended ($0 < p < 1$) or compressed ($p > 1$) argument arise in different models in biology [34, 128, 141, 206, 207, 591], population dynamics [8], astrophysics [11], mechanics [367], number theory [332], stochastic games [155], graph theory [454], risk and queuing theory [179], and neural network theory [599].

The studies concerned with the analysis and approximate analytical solutions of ODEs with proportional argument include, for example, [26, 166, 216, 232, 233, 260, 310, 367, 375, 451, 590]. Notably, although the majority of the studies deal with the case $0 < p < 1$, the equations in [11, 141, 206, 591] were derived for $p > 1$.

Nonlinear equations of the form

$$u'_t = f(t, u, w), \quad w = u(pt), \quad 0 < p < 1, \tag{1.1.6.2}$$

are also special cases of ODEs with variable delay with $\tau(t) = (1-p)t$. In what follows, differential equations like the one above with a delay proportional to time will be referred to as *equations with proportional delay*.

The Cauchy problem for ODEs with proportional delay. The initial data in the Cauchy problem for equations (1.1.6.1) and (1.1.6.2) with $0 < p < 1$ is specified as follows:

$$u = \varphi(t) \quad \text{at} \quad pt_0 \le t \le t_0. \tag{1.1.6.3}$$

It is apparent that the length of the initial interval, where the initial data (1.1.6.3) is set, depends significantly on the choice of the initial point t_0 and equals $L = (1-p)t_0$. If $t_0 = 0$, then the initial interval degenerates into a single point, $t = 0$. In this case, the initial condition for the ODEs with proportional delay (1.1.6.1) and (1.1.6.2) is set in exactly the same way as for the ODEs without delay at $t = 0$; it is this case that occurs most often in this book and numerous publications.

▶ **Example 1.5.** Consider the Cauchy problem for the simple ODE with proportional delay

$$u'_t = bw, \quad w = u(pt), \quad t > 0; \quad u(0) = c. \tag{1.1.6.4}$$

Its solution can be represented as an infinite series convergent for any t [232, 260]:

$$u(t) = c\exp_s(bt, p), \quad \exp_s(t, p) \equiv \sum_{n=0}^{\infty} p^{\frac{n(n-1)}{2}} \frac{t^n}{n!} \quad (0 < p < 1). \tag{1.1.6.5}$$

The function $\exp_s(t,p)$, whose properties are largely similar for $t \geq 0$ to the ordinary exponential function, will be referred to as the *stretched exponential function*.[*]
The following relations hold:

$$\exp_s(0,p) = 1, \quad \exp_s(t,0) = 1+t, \quad \exp_s(t,1) = e^t,$$
$$[\exp_s(t,p)]'_t = \exp_s(pt,p), \quad [\exp_s(t,p)]^{(n)}_t = p^{\frac{n(n-1)}{2}} \exp_s(p^n t, p), \qquad (1.1.6.6)$$

where $n = 1, 2, \ldots$ In addition, $\exp_s(t,p) > \exp_s(t,q)$ if $p > q$ and $t > 0$.

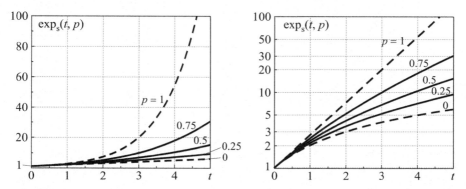

Figure 1.6. Graphs of the stretched exponential function $\exp_s(t,p)$ for $p = 0$, 0.25, 0.50, 0.75, and 1.00 in Cartesian (left) and logarithmic (right) coordinates.

Figure 1.6 displays the stretched exponential function $\exp_s(t,p)$ for three values of the parameter: $p = 0.25, 0.50$, and 0.75. The ordinary exponential function e^t and linear function $1+t$, corresponding to $p=1$ and $p=0$, are shown by dash lines.

The maximum error of the approximate formula for the stretched exponential $\exp_s(t,p)$ obtained by retaining the first five terms of the series (1.1.6.5) (up to $n=4$ inclusive) does not exceed 1% over the range $-1.1 \leq t \leq 2.3$ for $0.2 \leq p \leq 0.8$. ◄

We note below some qualitative features of the solution to the Cauchy problem (1.1.6.4) for $b < 0$. To be specific, we substitute $b = -1$ and $c = 1$ into (1.1.6.5) to obtain

$$u(t) = \exp_s(-t,p) = \sum_{n=0}^{\infty} (-1)^n p^{\frac{n(n-1)}{2}} \frac{t^n}{n!}. \qquad (1.1.6.7)$$

Below are a few properties of the zeros of the function (1.1.6.7) (for details, see [127, 291, 349, 543]).

1°. The function (1.1.6.7) has countably many positive zeros: $0 < t_0 < t_1 < t_2 < \cdots$. The numerical values of the first six roots of the function $\exp_s(-t, 0.5)$ are 1.488, 4.881, 13.560, 34.775, 84.977, and 201.003.

---
[*] In [491, 543], the term *deformed exponential function* was used instead. The term 'stretched exponential function' is more precise, because deformations include both stretching ($0 < p < 1$) and compression ($p > 1$). Furthermore, the term 'deformed exponential function' began to be used earlier in a completely different sense in statistical physics (e.g., see [68, 356]). The cited studies did not use the notation $\exp_s(t,p)$, where subscript s stands for stretching.

2°. For $t > 0$ ($0 < p < 1$), the function (1.1.6.7) describes oscillations with monotonically increasing amplitudes (see Fig. 1.7). For example, for $p = 0.5$, it has the extrema $-0.262, 0.908, -9.139, 223.362$, and $-12{,}313.172$ at $t = 2.976, 9.762, 27.121, 69.551$, and 169.955, respectively.

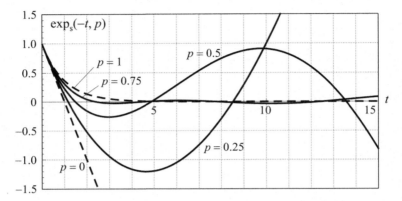

Figure 1.7. Graphs of the stretched exponential function $\exp_s(-t, p)$ at $p = 0, 0.25, 0.50, 0.75,$ and 1.00.

3°. The ratio t_{n+1}/t_n decreases monotonically, with the following limit relation holding: $\lim_{n\to\infty} t_{n+1}/t_n = 1/p$. In particular, for $p = 0.5$, we have $t_2/t_1 = 2.778$, $t_{12}/t_{11} = 2.163$, $t_{52}/t_{51} = 2.038$, $t_{102}/t_{101} = 2.020$, and $t_{202}/t_{201} = 2.010$.

4°. The zeros of the function (1.1.6.7) are described by the following asymptotic formula [543]:
$$t_n = np^{1-n}\bigl[1 + O(n^{-2})\bigr] \quad \text{as} \quad n \to \infty.$$

Nonlinear ODEs with proportional delay admitting exact solutions. Below are a few nonlinear first-order ODEs with proportional delay that have exact solutions representable in terms of elementary functions. These solutions can be used to test approximate analytical and numerical methods for solving nonlinear ODEs with variable delays.

Equation 1. The nonlinear ODE with proportional delay
$$u'_t = au + bw^2, \quad w = u(\tfrac{1}{2}t),$$
admits the exact solution $u(t) = Ce^{(a+bC)t}$ satisfying the initial condition $u(0) = C$, where C is an arbitrary constant.

Equation 2. The nonlinear ODE with proportional delay
$$u'_t = au + bw^{1/p}, \quad w = u(pt),$$
admits the exact solution
$$u(t) = C\exp(\lambda t), \quad \lambda = a + bC^{(1-p)/p},$$
satisfying the initial condition $u(0) = C$, where C is an arbitrary constant.

Equation 3. The nonlinear ODE with proportional delay

$$u'_t = f(w - pu), \quad w = u(pt),$$

involving an arbitrary function $f(z)$, admits the exact solution

$$u(t) = At + C, \quad A = f((1-p)C),$$

which satisfies the initial condition $u(0) = C$, where C is an arbitrary constant.

Equation 4. The nonlinear ODE with proportional delay

$$u'_t = uf(w/u^p), \quad w = u(pt),$$

involving an arbitrary function $f(z)$, admits the exact solution

$$u(t) = Ce^{\lambda t}, \quad \lambda = f(C^{1-p}),$$

which satisfies the initial condition $u(0) = C$, where $C > 0$ is an arbitrary constant.

Equation 5. The nonlinear ODE with proportional delay

$$u'_t = u^k f(w/u), \quad w = u(pt),$$

involving an arbitrary function $f(z)$, admits the exact solution

$$u(t) = at^{\frac{1}{1-k}}, \quad a = \left[(1-k)f\left(p^{\frac{1}{1-k}}\right)\right]^{\frac{1}{1-k}}.$$

For $0 \le k < 1$, this solution satisfies the initial condition $u(0) = 0$.

Equation 6. The nonlinear ODE with proportional delay

$$u'_t = a - bw^2, \quad w = u(\tfrac{1}{2}t),$$

admits the exact solutions

$$u(t) = \sqrt{\frac{2a}{b}} \sin\left(b\sqrt{\frac{a}{2b}}\,t\right) \qquad \text{for } ab > 0,$$

$$u(t) = -\sqrt{-\frac{2a}{b}} \sinh\left(b\sqrt{-\frac{a}{2b}}\,t\right) \qquad \text{for } ab < 0,$$

which satisfy the homogeneous initial condition $u(0) = 0$.

Remark 1.8. *The modified nonlinear equations 1–4 from Subsection 1.1.4 in which the constant delay is replaced with a proportional delay, so that $w = u(pt)$, admit an exact linearization with the same substitutions.*

ODEs with several proportional delays. Let us look at equations of the form

$$u'_t = f(t, u, w_1, \ldots, w_m), \quad w_k = u(p_k t), \quad k = 1, \ldots, m, \qquad (1.1.6.8)$$

for $0 < p_k < 1$ and all k.

The initial data in the Cauchy problem for equation (1.1.6.8) is specified as follows:
$$u = \varphi(t) \quad \text{at} \quad p_{\min} t_0 \leq t \leq t_0, \tag{1.1.6.9}$$
where $p_{\min} = \min\limits_{k=1,\ldots,m} p_k$. If $t_0 = 0$, then the initial interval degenerates into a single point, $t = 0$, in which case the initial condition is set in exactly the same way as for ODEs without delay at $t = 0$.

With the new variables [76, 260]
$$x = \ln t, \quad y(x) = u(t),$$
equation (1.1.6.8) converts to the ODE with m constant delays
$$y'_x = e^x f(e^x, y, y_1, \ldots, y_m), \quad y_k = y(x - \tau_k), \quad \tau_k = \ln \frac{1}{p_k} > 0, \quad k = 1, \ldots, m.$$

Equations with several variable delays. In the literature, there are also more complex functional-differential equations that contain the desired function with one or more delays dependent nonlinearly on time:
$$u'_t = f(t, u, w_1, \ldots, w_m), \quad w_k = u(t - \tau_k(t)), \quad k = 1, \ldots, m, \tag{1.1.6.10}$$
where $\tau_k(t) > 0$ are given functions. Such equations are known as *ODEs with variable delays*.

In the statement of the Cauchy problem for equations with several variable delays (1.1.6.10), the initial condition is written as
$$u = \varphi(t) \quad \text{for} \quad t \in E_{t_0}. \tag{1.1.6.11}$$
The initial interval E_{t_0} consists of the point t_0 and the values of $t - \tau_k(t)$ that are less than t_0 for $t \geq t_0$, so that
$$E_{t_0} = \{t_* \leq t \leq t_0\}, \quad t_* = \min_{1 \leq k \leq m} \min_{t \geq t_0}[t - \tau_k(t)]. \tag{1.1.6.12}$$

▶ **Example 1.6.** For ODEs with several constant delays (1.1.2.5), the use of formula (1.1.6.12) gives the following initial interval:
$$E_{t_0} = \{t_* \leq t \leq t_0\}, \quad t_* = t_0 - \tau_{\max}, \quad \tau_{\max} = \max_{1 \leq k \leq m} \tau_k. \tag{1.1.6.13}$$
Its length, L, is independent of the choice of the initial point t_0 and equals the maximum delay: $L = \max\limits_{1 \leq k \leq m} \tau_k$. In view of (1.1.6.13), the initial condition (1.1.6.11) for equation (1.1.2.5) can be represented as (1.1.2.6). ◀

Remark 1.9. *In the numerical solution of the Cauchy problem for equations with one or more variable delays, the initial interval E_{t_0} is sometimes replaced for simplicity with (any) other interval that certainly contains the initial one; for example, one chooses $(-\infty, t_0]$.*

Remark 1.10. *The studies [135, 136, 144] deal with more complicated ODEs with one or more variable delays, τ_k, that depend not only on the independent variable t but also on the desired function u, so that $\tau_k = \tau_k(t, u)$.*

The method of steps for ODEs with variable delay. The method of steps is also suitable for solving the Cauchy problem for ODEs with a variable delay of the general form [144]:

$$u'_t = f(t, u, w), \quad w = u(t - \tau(t)), \quad \tau(t) > 0.$$

In this case, the initial data is set on the interval (1.1.6.12) with $m = 1$, and the step is chosen equal to $h = \min_{t_0 \leq t \leq T} \tau(t)$, where $[t_0, T]$ is the interval where the solution is sought.

This procedure is easy to generalize to ODEs with several variable delays.

Remark 1.11. *For the pantograph equation (1.1.6.1) and more complex ODE with proportional delay (1.1.6.2), we have $\tau(t) = (1 - p)t$. The method of steps is inapplicable for the Cauchy problem for this equation with the initial point $t_0 = 0$, because $h = 0$.*

1.1.7. Existence and Uniqueness of Solutions. Suppression of Singularities in Solving Blow-Up Problems

Existence and uniqueness of solutions. The method of steps enables one to prove the existence and uniqueness of the solution to the Cauchy problem for constant delay ODEs, since the well-known existence and uniqueness theorems apply to the resulting ODEs without delay (e.g., see [350, 421, 423]). Therefore, a solution, $u = u(t)$, to problem (1.1.2.1)–(1.1.2.2) exists as long as the functions $f = f(t, u, w)$ and $\varphi = \varphi(t)$ are continuous and it is unique if $f(t, u, w)$ has a bounded first-order partial derivative in u (or satisfies the Lipschitz condition in the second argument: $|f(t, u, w) - f(t, z, w)| \leq M|u - z|$, where M is some positive number).

We now consider the more general Cauchy problem for ODE with several delays

$$\begin{aligned} u'_t &= f(t, u, w_1, \ldots, w_m), \quad w_i = u(t - \tau_i(t)), \quad i = 1, \ldots, m, \\ u &= \varphi(t) \quad \text{on interval} \quad E_{t_0}, \end{aligned} \quad (1.1.7.1)$$

where E_{t_0} is the initial interval, whose length is defined by formula (1.1.6.12).

The existence and uniqueness theorem for a solution to this problem is stated as follows (e.g., see [144, 276]).

Theorem. *Suppose that all delays $\tau_i(t)$ in equation (1.1.7.1) are continuous for $t_0 \leq t \leq t_0 + H$ ($H > 0$) and nonnegative. Moreover, suppose the function f is continuous in a neighborhood of the point $(t_0, \varphi(t_0), \varphi(t_0 - \tau_1(t_0)), \ldots, \varphi(t_0 - \tau_m(t_0)))$ and has bounded first-order partial derivatives in all arguments starting from the second (or satisfies the Lipschitz conditions in these arguments) and the initial function $\varphi(t)$ is continuous on E_{t_0}. Then there exists a unique solution, $u = u(t)$, of the Cauchy problem for equation (1.1.7.1) for $t_0 \leq t \leq t_0 + h$, where h is sufficiently small.*

The proof of this theorem is given, for example, in [144]. It is based on applying the contraction mapping theorem.

Remark 1.12. *Equations of the neutral type (1.1.2.7) can also be solved with the method of steps. Unlike delay ODEs, solutions to neutral differential equations cannot be smoothed at the points* $t = t_0 + n\tau$ $(n = 0, 1, 2, \dots)$ *(e.g., see [37, 144]).*

Suppression of singularities in blow-up problems by introducing a delay. For ODEs without delay, there are Cauchy problems whose solutions tend to infinity (have a singularity) as they approach a finite time $t = t_*$. The singular point t_* does not appear in the equation explicitly and is unknown in advance. Such solutions exist on a finite time interval $t_0 \leq t < t_*$ and are known as blow-up solutions [185, 421, 501].

▶ **Example 1.7.** Consider the Cauchy problem for the ODE without delay

$$u'_t = u^2, \quad t > 0; \quad u(0) = 1, \tag{1.1.7.2}$$

which admits the exact solution

$$u = \frac{1}{1-t}. \tag{1.1.7.3}$$

This solution exists on a limited time interval, $0 \leq t < 1$, and has a singularity at $t = t_* = 1$. ◀

Now consider the more general Cauchy problem for the autonomous first-order ODE

$$u'_t = f(u), \quad t > 0; \quad u(0) = a > 0, \tag{1.1.7.4}$$

where $f(u) > 0$ is a continuous function defined for all $u \geq a$.

Sufficient conditions for the existence of a blow-up solution. Suppose that for some $\sigma > 0$, the limiting relation

$$\lim_{u \to +\infty} \frac{f(u)}{u^{1+\sigma}} = s, \quad 0 < s \leq \infty, \tag{1.1.7.5}$$

holds [421]. Then the Cauchy problem (1.1.7.4) has a blow-up solution. If $f(u)$ is differentiable, then (1.1.7.5) can be replaced with the equivalent criterion

$$\lim_{u \to +\infty} \left[u^{-\sigma} f'_u(u) \right] = s_1, \quad 0 < s_1 \leq \infty \quad (\sigma > 0).$$

The studies [10, 405, 406, 501] (see also the literature cited in [405, 406]) describe some numerical methods for solving blow-up problems for nonlinear first-, second-, and higher-order ODEs.

The complication of mathematical models by introducing a delay into the right-hand side of ODEs in blow-up problems can suppress solution singularity (for conditions of singularity existence or absence in solutions to delay ODEs, see, for example, [65, 90, 150]). Below we will give two examples of delay problems that become blow-up problems in the limit case $\tau = 0$ (1.1.7.2) but do not have a solution singularity for $\tau > 0$.

1.1. First-Order Equations. Cauchy Problem. Method of Steps. Exact Solutions

▶ **Example 1.8.** Consider the Cauchy problem for the delay ODE

$$u'_t = w^2, \quad w = u(t-\tau), \quad t > 0; \quad u(t) = 1, \quad -\tau \le t \le 0. \tag{1.1.7.6}$$

The exact solution to problem (1.1.7.6) for $\tau = 1$ on the interval $-1 \le t \le 3$ is

$$u = \begin{cases} 1, & -1 \le t \le 0; \\ 1+t, & 0 < t \le 1; \\ \frac{1}{3}(5+t^3), & 1 < t \le 2; \\ \frac{1}{126}(-158 + 224t + 168t^2 - 70t^3 - 35t^4 + 42t^5 - 14t^6 + 2t^7), & 2 < t \le 3. \end{cases}$$

Figure 1.8a shows the exact solution (1.1.7.3) of problem (1.1.7.2) and numerical solutions of problem (1.1.7.6) at $\tau = 0.1$ and $\tau = 0.5$; the vertical axis is on the logarithmic scale. ◀

▶ **Example 1.9.** Consider another Cauchy problem for the delay ODE

$$u'_t = uw, \quad w = u(t-\tau), \quad t > 0; \quad u(t) = 1, \quad -\tau \le t \le 0. \tag{1.1.7.7}$$

Figure 1.8b displays the exact solution (1.1.7.3) of problem (1.1.7.2) and numerical solutions of problem (1.1.7.7) at $\tau = 0.1$ and $\tau = 0.5$. ◀

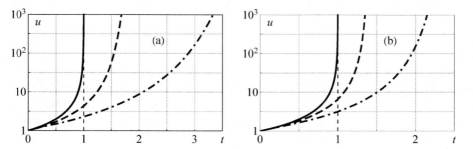

Figure 1.8. Exact solution (1.1.7.3) of the problem without delay (1.1.7.2) (solid line) and numerical solutions of two delay problems: (a) problem (1.1.7.6) and (b) problem (1.1.7.7) at $\tau = 0.1$ (dashed line) and $\tau = 0.5$ (dot-and-dash line).

It is apparent from Figures 1.8a and 1.8b that the introduction of a delay in the above blow-up problems suppresses the solution singularity entirely.

The following statement is true. Suppose the Cauchy problem for the ODE without delay (1.1.7.4) has a blow-up solution. Then the solution to the modified problem with delay

$$u'_t = f(w), \quad w = u(t-\tau), \quad t > 0; \quad u(t) = a, \quad -\tau \le t \le 0,$$

where $\tau > 0$, does not have singularities at finite t.

Suppression of singularities in blow-up problems by introducing a stretching parameter. In blow-up problems for ODEs, introducing a stretching parameter p

in the argument of the unknown function, thus leading to an equation with proportional delay, results in the suppression of solution singularities. Let us discuss this matter in more detail.

Consider the Cauchy problem for the nonlinear equation with proportional delay

$$u'_t = f(w), \quad w = u(pt), \quad t > 0; \quad u(0) = a > 0, \tag{1.1.7.8}$$

where $f(u) > 0$ and $f'_u(u) > 0$ are continuous functions defined for all $u \geq a$, and $0 < p < 1$.

Let problem (1.1.7.8) with $p = 1$ have a blow-up solution with a singular point $t = t_*$. We denote this solution by $v = v(t)$ ($0 \leq t < t_*$).

It can be easily shown that for small t, the solution to problem (1.1.7.8) is representable in the form

$$u(t) = a + f(a)t + \tfrac{1}{2}pf(a)f'_u(a)t^2 + o(t^2). \tag{1.1.7.9}$$

It follows that $u(t) < v(t)$ for small t. Clearly, this inequality will also hold for any range $0 \leq t \leq t^\circ$, where $t^\circ < t_*$.

Consider the sequence of points $t_n = p^{2-n}t^\circ$, where $n = 1, 2, \ldots$ On the first interval, $pt^\circ \leq t \leq t^\circ$, problem (1.1.7.8) does not have singularities. Assume that on the nth interval, $t_n \leq t \leq t_{n+1}$, a solution of problem (1.1.7.8) is known and it does not have singularities. Let us look at the $(n+1)$st interval, $t_{n+1} \leq t \leq t_{n+2}$. On integrating equation (1.1.7.8) from t_{n+1} to t, we get

$$u(t) = u(t_{n+1}) + \int_{t_{n+1}}^{t} f(u(pt)) \, dt. \tag{1.1.7.10}$$

The stretched argument $x = pt$ of the unknown function on the interval $t_{n+1} \leq t \leq t_{n+2}$ ranges over $t_n \leq x \leq t_{n+1}$, where (by assumption) the unknown function does not have singularities. Therefore, the composite function $f(u(pt))$ does not have singularities on the interval $t_{n+1} \leq t \leq t_{n+2}$ either. Neither does the integral on the right-hand side of (1.1.7.10); it can be evaluated as

$$\int_{t_{n+1}}^{t} f(u(pt)) \, dt = \frac{1}{p} \int_{t_n}^{x/p} f(u(x)) \, dx.$$

Since $0 < p < 1$, we get $t_n = p^{2-n}t^\circ \to \infty$ as $n \to \infty$. It follows that the solution to problem (1.1.7.10) does not have singularities on a bounded time interval.

▶ **Example 1.10.** Consider the Cauchy problem for the ODE with proportional delay

$$u'_t = w^2, \quad w = u(pt), \quad t > 0; \quad u(0) = 1, \tag{1.1.7.11}$$

where $0 < p < 1$.

For small t, the solution to problem (1.1.7.11) can be approximated by the polynomial

$$u = 1 + t + pt^2 + p^2(\tfrac{1}{3} + \tfrac{2}{3}p)t^3 + p^4(\tfrac{1}{2} + \tfrac{1}{6}p + \tfrac{1}{3}p^2)t^4, \tag{1.1.7.12}$$

whose error is $O(t^5)$.

Figure 1.9 displays numerical solutions to problem (1.1.7.11) for $p = 0, 0.25, 0.5$, and 0.75 (shown in solid lines) and the approximate solutions computed by formula (1.1.7.12) (dashed lines). The vertical axis is on the logarithmic scale. For $p = 1$, the exact blow-up solution is computed by formula (1.1.7.3). It is apparent that the introduction of stretching suppresses the blow-up singularity entirely in the problem. Notably, for $0 < p \leq 0.5$ and $0 \leq t \leq 2$, the approximate formula (1.1.7.12) provides high accuracy (the relative error at $p = 0.5$ and $t = 2$ is 0.0526).

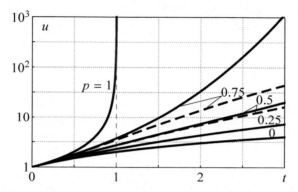

Figure 1.9. Solutions to problem (1.1.7.11) obtained by a numerical integration and using the approximate formula (1.1.7.12) for $p = 0, 0.25, 0.5,$ and 0.75.

◀

▶ **Example 1.11.** Consider the Cauchy problem for another ODE with proportional delay

$$u'_t = uw, \quad w = u(pt), \quad t > 0; \quad u(0) = 1. \tag{1.1.7.13}$$

Figure 1.10 displays numerical solutions to problem (1.1.7.11) for $p = 0, 0.25, 0.5,$ and 0.75. The vertical axis is on the logarithmic scale. For $p = 1$, the exact blow-up solution is computed by formula (1.1.7.3). It is apparent that the introduction of stretching suppresses the blow-up singularity entirely in the problem. ◀

A reasonably general proposition can be stated for more complicated problems described by nonlinear ODEs with proportional delay

$$u'_t = F(u, w), \quad w = u(pt), \quad t > 0; \quad u(0) = a > 0, \tag{1.1.7.14}$$

where $0 < p < 1$.

Proposition. Let $F(u, w)$ be a positive continuous function of two arguments in the domain $D = \{a \leq u < \infty, a \leq w < \infty\}$. Then problem (1.1.7.14) has a blow-up solution if and only if the simpler auxiliary problem for an ODE without a stretched argument

$$u'_t = F(u, a), \quad t > 0; \quad u(0) = a > 0, \tag{1.1.7.15}$$

has a blow-up solution.

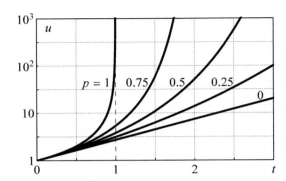

Figure 1.10. Solutions to problem (1.1.7.13) obtained by a numerical integration for $p = 0$, 0.25, 0.5, and 0.75.

Remark 1.13. *This proposition also holds true if the proportional delay equation (1.1.7.14) is replaced by a constant delay equation with $w = u(t - \tau)$ and the initial condition $u(t) = a$ at $-\tau \leq t \leq 0$ ($\tau > 0$).*

The solutions of the problems in Examples 1.10 and 1.11, as well as Examples 1.8 and 1.9, did not have singularities, because the simpler auxiliary problems described by the ODEs $u'_t = a^2$ and $u'_t = au$, respectively, did not have singularity either.

▶ **Example 1.12.** Problem (1.1.7.14) with $F(u, w) = (u/w)^2$ has a blow-up solution, because the simpler auxiliary problem (1.1.7.15) with $F(u, a) = u^2/a^2$ has a blow-up solution. ◀

1.2. Second- and Higher-Order Delay ODEs. Systems of Delay ODEs

1.2.1. Basic Concepts. The Cauchy Problem

In general, an nth-order ordinary differential equation involving k constant delays and solved for the highest derivative can be represented as

$$u_t^{(n)} = F\big(t, u, u'_t, \ldots, u_t^{(n-1)}, w_1, w'_1, \ldots, w_1^{(n_1)}, \ldots, w_k, w'_k, \ldots, w_k^{(n_k)}\big),$$
$$u = u(t), \quad w_i = u(t - \tau_i), \quad \tau_i > 0, \quad i = 1, \ldots, k,$$
(1.2.1.1)

where $n > \max(n_1, \ldots, n_k)$. The symbol $w_i^{(j)}$ denotes the jth derivative of the function $u(z)$ taken at $z = t - \tau_i$. We assume that F is a continuous function of all its arguments.

Suppose that an initial point t_0 is given. Then each τ_i can be associated with an initial set $E_{t_0}^{(i)} = \{t_0 - \tau_i \leq t \leq t_0\}$. The entire initial set is defined as $E_{t_0} = \bigcup_{i=1}^k E_{t_0}^{(i)} = \{t_0 - \tau_{\max} \leq t \leq t_0\}$, where $\tau_{\max} = \max_{1 \leq i \leq k} \tau_i$ is the maximum delay.

The Cauchy problem for the ODE with constant delays (1.2.1.1) is formulated as: find a solution $u = u(t)$ having continuous derivatives up to $u_t^{(n-1)}$ inclusive that satisfies the initial conditions

$$u = \varphi_0(t),$$
$$u_t' = \varphi_1(t),$$
$$\dots$$
$$u_t^{(n-1)} = \varphi_{n-1}(t) \quad \text{at} \quad t_0 - \tau_{\max} \le t \le t_0, \quad (1.2.1.2)$$

where $\varphi_j(t)$ are given continuous functions.

In applications, one usually considers the case where the initial conditions in the Cauchy problem for equation (1.2.1.1) are set in a consistent way with each other using one function $\varphi(t)$. Specifically, one chooses the initial conditions (1.2.1.2) such that

$$\varphi_0(t) = \varphi(t), \quad \varphi_1(t) = \varphi_t'(t), \quad \varphi_j(t) = \varphi_t^{(j)}(t), \quad j = 2, \dots, n-1. \quad (1.2.1.3)$$

In this case, the initial data (1.2.1.2)–(1.2.1.3) are customarily written in the short form

$$u = \varphi(t) \quad \text{at} \quad t_0 - \tau_{\max} \le t \le t_0. \quad (1.2.1.4)$$

For more complicated ODEs with k variable delays in (1.2.1.1), one should set $\tau_i = \tau_i(t)$ ($i = 1, \dots, k$), where $\tau_i = \tau_i(t)$ are given positive continuous functions. The initial set E_{t_0} in the Cauchy problem is then defined in the same way as for first-order ODEs.

If $n = \max(n_1, \dots, n_k)$, then equation (1.2.1.1) is attributed to neutral differential equations. If $n < \max(n_1, \dots, n_k)$, it is an advanced differential equation.

1.2.2. Second-Order Linear Equations. The Cauchy Problem. Exact Solutions

Solution of the Cauchy problem for second-order delay ODEs. Consider the Cauchy problem for a second-order linear nonhomogeneous ODE with a constant delay and consistent initial data of a general form:

$$u''(t) + a^2 u(t - \tau) = f(t), \quad t > 0; \quad (1.2.2.1)$$
$$u = \varphi(t) \quad \text{at} \quad -\tau \le t \le 0. \quad (1.2.2.2)$$

The solution to the Cauchy problem (1.2.2.1)–(1.2.2.2) can be represented using two functions described below [269] (see also [129]).

The *delayed cosine* and *delayed sine* functions are defined as

$$\cos_d(t,\tau) = \begin{cases} 0, & t < -\tau, \\ 1, & -\tau \leq t \leq 0, \\ 1 - \dfrac{t^2}{2!} + \cdots + (-1)^k \dfrac{[t-(k-1)\tau]^{2k}}{(2k)!}, & (k-1)\tau < t \leq k\tau, \end{cases}$$
(1.2.2.3)

$$\sin_d(t,\tau) = \begin{cases} 0, & t < -\tau, \\ t + \tau, & -\tau \leq t \leq 0, \\ t + \tau - \dfrac{t^3}{3!} + \cdots + (-1)^k \dfrac{[t-(k-1)\tau]^{2k+1}}{(2k+1)!}, & (k-1)\tau < t \leq k\tau, \end{cases}$$
(1.2.2.4)

where $k = 1, 2, \ldots$

The delayed cosine and sine functions are particular solutions of the homogeneous equation (1.2.2.1) with $a = 1$ and $f(t) = 0$.

It was shown in [129] that the solution of the Cauchy problem (1.2.2.1)–(1.2.2.2) can be represented as

$$u(t) = \varphi(0)\cos_d(a(t-\tau), a\tau) + a^{-1}\varphi'_t(0)\sin_d(a(t-\tau), a\tau)$$
$$- a\int_{-\tau}^{0} \sin_d(a(t-2\tau-s), a\tau)\varphi(s)\,ds$$
$$+ a^{-1}\int_{0}^{t} \sin_d(a(t-\tau-s), a\tau)f(s)\,ds. \qquad (1.2.2.5)$$

An alternative but less convenient representation of the solution to the Cauchy problem (1.2.2.1)–(1.2.2.2) can be found in [269].

Solution of the Cauchy problem for another second-order delay ODE. Consider the following Cauchy problem for a general second-order linear homogeneous ODE with a constant delay and consistent initial data:

$$u''(t) = -\alpha^2 u(t) + \beta u(t-\tau), \quad t > \tau; \qquad (1.2.2.6)$$
$$u = \varphi(t) \quad \text{at} \quad 0 \leq t \leq \tau. \qquad (1.2.2.7)$$

The study [455] showed that the solution of the Cauchy problem (1.2.2.6)–(1.2.2.7) with $\alpha \neq 0$ in the region $t > \tau$ can be expressed via solutions of two simpler problems as

$$u(t) = \frac{\varphi(\tau) - \gamma\varphi(0)}{1-\gamma} u_1(t) - \frac{\varphi'(\tau) - \gamma\varphi'(0)}{1-\gamma}\left(\frac{\tau}{1-\gamma}u_1(t) - u_2(t)\right)$$
$$+ \frac{\gamma}{1-\gamma}\int_0^\tau \left(\frac{\tau}{1-\gamma}u_1(t) - u_2(t)\right)\varphi''(t)\,dt, \quad \gamma = \frac{\beta}{\alpha^2}, \qquad (1.2.2.8)$$

where $u_1(t)$ and $u_2(t)$ are solutions to problem (1.2.2.6)–(1.2.2.7) for $\varphi(t) \equiv 1$ and $\varphi(t) \equiv t$, respectively.

Below are the auxiliary functions $u_1(t)$ and $u_2(t)$ involved in equation (1.2.2.8) that were obtained by the method of steps in [455].

1°. The solution to problem (1.2.2.6)–(1.2.2.7) with $\varphi(t) \equiv 1$ can be represented on the interval $m\tau \leq t \leq (m+1)\tau$ as

$$u_1(t) = \gamma^m + (1-\gamma) \sum_{k=1}^{m} \gamma^{k-1} \sum_{n=0}^{k-1} A_{k,n} \frac{[\alpha(t-k\tau)]^n}{n!} \cos[\alpha(t-k\tau) - \tfrac{1}{2}\pi n], \tag{1.2.2.9}$$

where $\gamma = \beta/\alpha^2$. The constants $A_{k,n}$ are defined as

$$A_{k,0} = 1, \quad A_{k,n} = \sum_{j=0}^{k-n-1} \frac{n}{n+2j} 2^{-n-2j} C_{n+2j}^j, \quad 1 \leq n < k, \tag{1.2.2.10}$$

with $C_n^j = \frac{n!}{j!(n-j)!}$ being binomial coefficients. Notably, $0 < A_{k,n} \leq 1$.

2°. The solution to problem (1.2.2.6)–(1.2.2.7) with $\varphi(t) \equiv t$ on the interval $m\tau \leq t \leq (m+1)\tau$ can be represented as

$$u_2(t) = \gamma^m(t - m\tau) + \tau \sum_{k=1}^{m} \gamma^{k-1} \sum_{n=0}^{k-1} A_{k,n} \frac{[\alpha(t-k\tau)]^n}{n!} \cos[\alpha(t-k\tau) - \tfrac{1}{2}\pi n]$$

$$+ \frac{1-\gamma}{\alpha} \sum_{k=1}^{m} \gamma^{k-1} \sum_{n=0}^{k-1} B_{k,n} \frac{[\alpha(t-k\tau)]^n}{n!} \sin[\alpha(t-k\tau) - \tfrac{1}{2}\pi n], \tag{1.2.2.11}$$

where $\gamma = \beta/\alpha^2$. The constants $A_{k,n}$ are computed by formulas (1.2.2.10), and the constants $B_{k,n}$ are obtained from

$$B_{k,0} = 2^{1-2k} k C_{2k}^k,$$

$$B_{k,n} = 2^{n+1-2k} \sum_{j=0}^{k-n-1} \frac{n(k-n-j)}{n+2j} C_{n+2j}^j C_{2(k-n-j)}^{k-n-j}, \quad 1 \leq n < k.$$

Solution of the Cauchy problem for a second-order ODE with proportional delay. Consider the Cauchy problem for the linear equation with proportional delay

$$u''_{tt}(t) = au(pt), \quad t > 0; \tag{1.2.2.12}$$

$$u(0) = b, \quad u'_t(0) = c. \tag{1.2.2.13}$$

Following [308], we seek particular solutions of equation (1.2.2.12) in the form

$$u(t) = \exp_s(\beta t, q), \quad \exp_s(t, q) \equiv \sum_{n=0}^{\infty} q^{\frac{n(n-1)}{2}} \frac{t^n}{n!} \quad (0 < q < 1), \tag{1.2.2.14}$$

where $\exp_s(t, q)$ is the stretched exponential function (see Example 1.5), and q and β are parameters to be determined. Applying the last formulas in (1.1.6.6) successively, we find the first two derivatives of the function (1.2.2.14):

$$u'_t = \beta \exp_s(\beta q t, q), \quad u''_{tt} = \beta^2 q \exp_s(\beta q^2 t, q).$$

Substituting the second expression into equation (1.2.2.12) gives

$$\beta^2 q \exp_s(\beta q^2 t, q) = a \exp_s(\beta p t, q).$$

This equation is satisfied if we set

$$\beta^2 q = a, \quad q^2 = p,$$

which leads to two sets of the unknown parameters: $q = \sqrt{p}$ and $\beta = \pm\sqrt{a/\sqrt{p}}$. These values define two linearly independent particular solutions of the ODE with proportional delay (1.2.2.12): $u_{1,2}(t) = \exp_s(\pm a^{1/2} p^{-1/4} t, p^{1/2})$. Therefore, the general solution of the linear homogeneous equation (1.2.2.12) is expressed as [308]:

$$u(t) = C_1 \exp_s(-a^{1/2} p^{-1/4} t, p^{1/2}) + C_2 \exp_s(a^{1/2} p^{-1/4} t, p^{1/2}), \quad (1.2.2.15)$$

where C_1 and C_2 are arbitrary constants.

On substituting (1.2.2.15) into the initial conditions (1.2.2.13), we find the constants C_1 and C_2. This results in an exact solution to the Cauchy problem (1.2.2.12)–(1.2.2.13):

$$\begin{aligned} u(t) &= \tfrac{1}{2}(b - c a^{-1/2} p^{1/4}) \exp_s(-a^{1/2} p^{-1/4} t, p^{1/2}) \\ &+ \tfrac{1}{2}(b + c a^{-1/2} p^{1/4}) \exp_s(a^{1/2} p^{-1/4} t, p^{1/2}). \end{aligned} \quad (1.2.2.16)$$

Formulas (1.2.2.15) and (1.2.2.16) involve the quantity $\sqrt{a}$, which becomes pure imaginary for $a < 0$. We will dwell on this case.

The formal substitution it for t in the formula for the stretched exponential function (1.1.6.5) gives

$$\exp_s(it, p) = \cos_s(t, p) + i \sin_s(t, p), \quad i^2 = -1 \quad (0 < p < 1), \quad (1.2.2.17)$$

where

$$\begin{aligned} \cos_s(t, p) &= \sum_{n=0}^{\infty} (-1)^n p^{n(2n-1)} \frac{t^{2n}}{(2n)!}, \\ \sin_s(t, p) &= \sum_{n=0}^{\infty} (-1)^n p^{n(2n+1)} \frac{t^{2n+1}}{(2n+1)!}. \end{aligned} \quad (1.2.2.18)$$

By analogy with the stretched exponential function, the real functions $\cos_s(t, p)$ and $\sin_s(t, p)$ will be referred to as the *stretched cosine* and *stretched sine*, respectively; $\cos_s(t, p)$ is an even function and $\sin_s(t, p)$ is an odd function. These functions were introduced in [308], where different notations and nomenclature were used. They possess the properties

$$\begin{aligned} \cos_s(0, p) &= 1, \quad \cos_s(t, 1) = \cos t, \\ \sin_s(0, p) &= 0, \quad \sin_s(t, 1) = \sin t, \end{aligned} \quad (1.2.2.19)$$

and can be expressed in terms of the stretched exponential function as

$$\cos_s(t, p) = \frac{\exp_s(it, p) + \exp_s(-it, p)}{2}, \quad \sin_s(t, p) = \frac{\exp_s(it, p) - \exp_s(-it, p)}{2i}. \quad (1.2.2.20)$$

1.2. Second- and Higher-Order Delay ODEs. Systems of Delay ODEs

The following formulas for derivatives hold true:
$$[\cos_s(t,p)]'_t = -\sin_s(pt,p), \qquad [\sin_s(t,p)]'_t = \cos_s(pt,p),$$
$$[\cos_s(t,p)]''_{tt} = -p\cos_s(p^2 t,p), \qquad [\sin_s(t,p)]''_{tt} = -p\sin_s(p^2 t,p). \tag{1.2.2.21}$$

For $p = 1$, these coincide with the formulas for the derivatives of the ordinary trigonometric functions.

It was shown in [308] that for $0 < p \leq 1$, the functions $\cos_s(t,p)$ and $\sin_s(t,p)$ have infinitely many zeros on the real axis. The functions $\cos_s(z,p)$ and $\sin_s(z,p)$ with the complex argument $z = x + iy$ only have zeros on the real axis $x = t$.

Considering the above, the general solution to the linear homogeneous equation (1.2.2.12) for $a < 0$ can be represented as
$$u(t) = C_1 \cos_s(|a|^{1/2} p^{-1/4} t, p^{1/2}) + C_2 \sin_s(|a|^{1/2} p^{-1/4} t, p^{1/2}), \tag{1.2.2.22}$$

where C_1 and C_2 are arbitrary constants. The corresponding solution to the Cauchy problem (1.2.2.12)–(1.2.2.13) is
$$u(t) = b\cos_s(|a|^{1/2} p^{-1/4} t, p^{1/2}) + c|a|^{-1/2} p^{1/4} \sin_s(|a|^{1/2} p^{-1/4} t, p^{1/2}). \tag{1.2.2.23}$$

Likewise, one can obtain exact solutions to the more general second-order linear ODE with proportional delay
$$u''_{tt}(t) + a_1 u'_t(pt) + a_0 u(p^2 t) = 0$$

and the nth-order ODE with proportional delay
$$u_t^{(n)}(t) + a_{n-1} u_t^{(n-1)}(pt) + \cdots + a_1 u'_t(p^{n-1} t) + a_0 u(p^n t) = 0$$

in terms of the stretched exponential function. Particular solutions to these equations should be sought in the form (1.2.2.14) (see [308] for details).

Solution of the Cauchy problem for a second-order ODE with two proportional delays. We now consider the more general Cauchy problem, for a linear ODE with two proportional delays:
$$u''_{tt}(t) = au(t) + bu(pt) + cu(qt), \quad t > 0; \tag{1.2.2.24}$$
$$u(0) = A, \quad u'_t(0) = B, \tag{1.2.2.25}$$

where $0 < p < 1$ and $0 < q < 1$.

A solution to problem (1.2.2.24)–(1.2.2.25) can be sought as a power series in the form of a linear combination of an even and an odd function:
$$u(t) = Au_1(t) + Bu_2(t), \tag{1.2.2.26}$$

where
$$u_1(t) = 1 + \sum_{n=1}^{\infty} \gamma_{2n} t^{2n}, \qquad \gamma_{2n} = \frac{1}{(2n)!} \prod_{k=0}^{n-1} (a + bp^{2k} + cq^{2k}),$$
$$u_2(t) = t + \sum_{n=1}^{\infty} \gamma_{2n+1} t^{2n+1}, \qquad \gamma_{2n+1} = \frac{1}{(2n+1)!} \prod_{k=0}^{n-1} (a + bp^{2k+1} + cq^{2k+1}).$$
$$\tag{1.2.2.27}$$

The functions $u_1(t)$ and $u_2(t)$ satisfy the initial conditions

$$u_1(0) = 1, \quad u_1'(0) = 0; \quad u_2(0) = 0, \quad u_2'(0) = 1.$$

For $a = -1$ and $b = c = 0$, these functions become the cosine and sine, respectively. For $a = 1$ and $b = c = 0$, they become the hyperbolic cosine and hyperbolic sine.

1.2.3. Higher-Order Linear Delay ODEs

General linear ODEs with delays and their properties. In general, an nth-order linear ordinary differential equation with variable coefficients and m variable delays is expressed as

$$u_t^{(n)}(t) + \sum_{i=0}^{n-1} \sum_{j=0}^{m} a_{ij}(t) u_t^{(i)}(t - \tau_j) = f(t), \tag{1.2.3.1}$$

$$\tau_0 = 0, \quad \tau_j = \tau_j(t) > 0, \quad j = 1, \ldots, m,$$

where $a_{ij}(t)$, $\tau_j(t)$, and $f(t)$ are given continuous functions, and $t > t_0$.

If $f(t) \equiv 0$, equation (1.2.3.1) is called *homogeneous* and if $f(t) \not\equiv 0$, it is *non-homogeneous*. Equation (1.2.3.1) can be conveniently written in the short notation

$$L[u] = f(t). \tag{1.2.3.2}$$

The linear differential operator with delay L possesses the properties

$$L[u_1 + u_2] = L[u_1] + L[u_2],$$
$$L[Cu] = CL[u],$$

where C is an arbitrary constant, and $u_1 = u_1(t)$, $u_2 = u_2(t)$, and $u = u(t)$ are arbitrary functions that have continuous derivatives up to order n inclusive.

Linear homogeneous ODEs with delays of the form $L[u] = 0$ possess the following properties [144]:

1°. Any linear homogeneous equation has a trivial solution, $u = 0$.

2°. The linearity and homogeneity of an equation are preserved under a linear and homogeneous transformation $u(t) = h(t)\bar{u}(t)$ of the unknown function, where $h(t)$ is a sufficiently smooth function.

3°. Let $u_1 = u_1(t)$, ..., $u_k = u_k(t)$ be any particular solutions of the linear homogeneous equation $L[u] = 0$. Then the linear combination

$$u = C_1 u_1 + \cdots + C_k u_k,$$

where $C_1, \ldots, C_k$ are arbitrary constants, is also a solution of the equation. This property of linear homogeneous equations is known as the *principle of linear superposition* of solutions.

Let $\{u_k\}$ be an infinite sequence of solutions to the linear homogeneous equation $L[u] = 0$. Then the series $\sum_{k=1}^{\infty} u_k$ is called, regardless of its convergence, a *formal solution* of this equation. If the solutions u_k are classical (n times continuously differentiable functions) and the series $\sum_{k=1}^{\infty} u_k$ and the respective series of the derivatives of u_k are uniformly convergent, then the sum of the series is a classical solution to the homogeneous equation $L[u] = 0$.

Below are the simplest properties of solutions to the linear nonhomogeneous equation (1.2.3.2):

1°. If $\tilde{u}_f(t)$ is a particular solution of the linear nonhomogeneous equation (1.2.3.2) and $\tilde{u}_0(t)$ is a particular solution of the respective linear homogeneous equation, with $f(t) \equiv 0$, then the sum

$$C\tilde{u}_0(t) + \tilde{u}_f(t),$$

where C is an arbitrary constant, is also a solution of the nonhomogeneous equation (1.2.3.2). The following statement also holds: the general solution to a linear nonhomogeneous equation is the sum of the general solution to the respective homogeneous equation and any particular solution to the nonhomogeneous equation.

2°. Let u_1 and u_2 be solutions to nonhomogeneous linear equations with the same left-hand sides but different right-hand sides:

$$L[u_1] = f_1(t), \quad L[u_2] = f_2(t).$$

The function $u = u_1 + u_2$ is a solution to the equation

$$L[u] = f_1(t) + f_2(t).$$

Linear homogeneous ODEs with constant coefficients and constant delays. Consider the following nth-order linear homogeneous ordinary differential equation with constant coefficients and m constant delays:

$$u_t^{(n)}(t) + \sum_{i=0}^{n-1} \sum_{j=0}^{m} a_{ij} u_t^{(i)}(t - \tau_j) = 0, \qquad (1.2.3.3)$$

$$\tau_0 = 0, \quad 0 < \tau_1 < \tau_2 < \cdots < \tau_m,$$

where a_{ij} and τ_j are some real constants, and $t > t_0$.

We look for particular solutions of equation (1.2.3.3) in the exponential form

$$u(t) = \exp(\lambda t), \qquad (1.2.3.4)$$

where λ is the desired constant.

On substituting (1.2.3.4) into (1.2.3.3) and cancelling by $e^{\lambda t}$, we arrive at a *characteristic equation* for determining λ:

$$\Phi(\lambda) = 0, \quad \text{where} \quad \Phi(\lambda) \equiv \lambda^n + \sum_{i=0}^{n-1} \sum_{j=0}^{m} a_{ij} \lambda^i e^{-\lambda \tau_j}. \qquad (1.2.3.5)$$

The function $\Phi(\lambda)$ is called a *characteristic quasi-polynomial*.

Equation (1.2.3.5) is transcendental and it has infinitely many roots. It can have both real roots, λ_k, and complex conjugate roots, $\lambda_k = \alpha_k \pm i\beta_k$, where $i^2 = -1$. To each real or complex root λ_k of the characteristic equation (1.2.3.5), there corresponds one or more solutions to the ordinary differential equation with delays (1.2.3.3). The possible situations are described below:

1°. If a root λ_k is real and has multiplicity 1, that is, $\Phi(\lambda_k) = 0$ and $\Phi'_\lambda(\lambda_k) \neq 0$, then equation (1.2.3.3) has a particular solution (1.2.3.4) at $\lambda = \lambda_k$.

2°. If a root λ_k of the characteristic equation (1.2.3.5) is real and has multiplicity r_k, that is, $\Phi(\lambda_k) = \Phi'_\lambda(\lambda_k) = \cdots = \Phi_\lambda^{(r_k-1)}(\lambda_k) = 0$ and $\Phi_\lambda^{(r_k)}(\lambda_k) \neq 0$, then equation (1.2.3.3) has particular solutions of the form

$$u_k(t) = P_k(t)\exp(\lambda_k t), \quad P_k(t) = \sum_{j=1}^{r_k} A_{kj}t^{j-1}, \qquad (1.2.3.6)$$

where A_{kj} are arbitrary constants.

3°. To a pair of complex conjugate roots $\lambda_k = \alpha_k \pm i\beta_k$ of multiplicity 1 there corresponds a pair of complex solutions $e^{(\alpha_k \pm i\beta_k)t}$ of equation (1.2.3.3) or two real solutions of this equation

$$u_{k1}(t) = \exp(\alpha_k t)\cos(\beta_k t), \quad u_{k2}(t) = \exp(\alpha_k t)\sin(\beta_k t). \qquad (1.2.3.7)$$

To pure imaginary roots $\lambda_k = i\beta_k$ there correspond periodic solutions $u_{k1}(t) = \cos(\beta_k t)$ and $u_{k2}(t) = \sin(\beta_k t)$.

4°. To a pair of complex conjugate roots $\lambda_k = \alpha_k \pm i\beta_k$ of multiplicity r_k there correspond real particular solutions to equation (1.2.3.3) of the form

$$\begin{aligned}u_{k1} &= P_k(t)\exp(\alpha_k t)\cos(\beta_k t), \quad P_k(t) = \sum_{j=1}^{r_k} A_{kj}t^{j-1}, \\ u_{k2} &= Q_k(t)\exp(\alpha_k t)\sin(\beta_k t), \quad Q_k(t) = \sum_{j=1}^{r_k} B_{kj}t^{j-1},\end{aligned} \qquad (1.2.3.8)$$

where A_{kj} and B_{kj} are arbitrary constants.

By virtue of the principle of linear superposition, more complicated particular solutions of equation (1.2.3.3) can be constructed using linear combinations of the particular solutions described in Items 1°–4° that correspond to different roots of the characteristic equation (1.2.3.5).

▶ **Example 1.13.** Let us find conditions under which the nth-order ODE with constant delay

$$u_t^{(n)} = au + bw, \quad w = u(t-\tau), \qquad (1.2.3.9)$$

has periodic solutions.

On substituting $u = e^{i\beta_k}$ into (1.2.3.9) and cancelling by $e^{i\beta_k}$, we get

$$(i\beta_k)^n = a + be^{-i\beta_k \tau}. \qquad (1.2.3.10)$$

1.2. Second- and Higher-Order Delay ODEs. Systems of Delay ODEs

Below we consider the cases of even and odd n separately.

1°. For equations of an even order, with $n = 2m$ ($m = 1, 2, \ldots$), we separate the real and imaginary parts of (1.2.3.10) to obtain

$$(-1)^m \beta_k^{2m} = a + b\cos(\beta_k \tau), \quad \sin(\beta_k \tau) = 0. \tag{1.2.3.11}$$

It follows that

$$(-1)^m (\pi k/\tau)^{2m} = a + (-1)^k b, \quad k = 1, 2, \ldots. \tag{1.2.3.12}$$

The parameters a, b, and τ of equation (1.2.3.9) with $n = 2m$ must satisfy condition (1.2.3.12) for the equation to have periodic solutions $u_{k1}(t) = \sin(\beta_k t)$ and $u_{k2}(t) = \cos(\beta_k t)$, where $\beta_k = \pi k/\tau$.

2°. For equations of an odd order, with $n = 2m+1$ ($m = 0, 1, \ldots$), we separate the real and imaginary parts of (1.2.3.10) to obtain

$$a + b\cos(\beta_k \tau) = 0, \quad (-1)^m \beta_k^{2m+1} = -b\sin(\beta_k \tau). \tag{1.2.3.13}$$

It follows from the first relation of (1.2.3.13) that for $|a| > |b|$, the odd-order equation (1.2.3.9) does not have periodic solutions. Equation (1.2.3.13) implies that the lines in the parametric plane a, b whose points correspond to periodic solutions of odd-order equations (1.2.3.9) can be represented in parametric form as

$$a = \frac{(-1)^m}{\tau^{2m+1}} \frac{\xi^{2m+1} \cos \xi}{\sin \xi}, \quad b = \frac{(-1)^{m+1}}{\tau^{2m+1}} \frac{\xi^{2m+1}}{\sin \xi} \quad (\xi = \tau\beta_k > 0). \tag{1.2.3.14}$$

The ranges of the parameter ξ, $\pi s < \xi < \pi(s+1)$, where $s = 0, 1, \ldots$, determine different branches in the plane a, b for given $\tau > 0$. ◀

The quasi-polynomial

$$\Phi(z) \equiv z^n + \sum_{i=0}^{n-1} \sum_{j=0}^{m} a_{ij} z^i e^{-\tau_j z}, \tag{1.2.3.15}$$

obtained from (1.2.3.5) with the substitution of z for λ, is an entire analytic function of the complex variable $z = x + iy$. If $\Phi(z)$ does not degenerate into a polynomial, implying that equation (1.2.3.3) involves at least one delay, then $\Phi(z)$ has infinitely many zeros, with infinity being the only limit point. All the roots z_k of the quasi-polynomial $\Phi(z)$ lie in the left half-plane, $\operatorname{Re} z_k \leq x_*$ [144].

Let us show that solutions of the nth-order linear homogeneous ODE with constant delay

$$u_t^{(n)}(t) = bu(t - \tau) \tag{1.2.3.16}$$

can be expressed in terms of the Lambert W function (1.1.3.4).

Substituting (1.2.3.4) into (1.2.3.16) and rearranging, we arrive at the following transcendental equation for the exponent λ:

$$\lambda^n e^{\tau\lambda} - b = 0. \tag{1.2.3.17}$$

For any $b > 0$, equation (1.2.3.17) has a real positive root expressible in terms of the Lambert W function as

$$\lambda_{\mathrm{p}} = \frac{n}{\tau} W_{\mathrm{p}}\left(\frac{\tau b^{1/n}}{n}\right).$$

In general, the transcendental equation (1.2.3.17) can be reduced with the substitution $\zeta = \lambda e^{\tau\lambda/n}$ to the algebraic equation $\zeta^n - b = 0$, which has n complex roots [402]:

$$\zeta_k = \begin{cases} b^{1/n}\left(\cos\frac{2(k-1)\pi}{n} + i\sin\frac{2(k-1)\pi}{n}\right) & \text{for } b > 0, \\ |b|^{1/n}\left(\cos\frac{(2k-1)\pi}{n} + i\sin\frac{(2k-1)\pi}{n}\right) & \text{for } b < 0, \end{cases} \quad (1.2.3.18)$$

where $k = 1, \ldots, n$ and $i^2 = -1$. Hence, the difference $\zeta^n - b$ can be factorized and represented as the product $\prod_{k=1}^{n}(\zeta - \zeta_k) = 0$, where $\zeta = \lambda e^{\tau\lambda/n}$. The transcendental equation (1.2.3.17) then breaks into n simpler independent equations

$$\lambda e^{\tau\lambda/n} - \zeta_k = 0, \quad k = 1, \ldots, n. \quad (1.2.3.19)$$

The solutions of these equations are expressed via the Lambert W function of a complex argument as

$$\lambda_k = \frac{n}{\tau} W\left(\frac{\tau\zeta_k}{n}\right), \quad k = 1, \ldots, n, \quad (1.2.3.20)$$

where the numbers ζ_k (generally complex) are defined in (1.2.3.18), and $W(z)$ is understood as the set of all branches of the Lambert W function.

Linear nonhomogeneous ODEs with constant coefficients and constants delays. Linear nonhomogeneous ordinary differential equations of the nth order with constant coefficients and m constant delays have the form

$$u_t^{(n)}(t) + \sum_{i=0}^{n-1}\sum_{j=0}^{m} a_{ij} u_t^{(i)}(t - \tau_j) = f(t), \quad (1.2.3.21)$$

$$\tau_0 = 0, \quad 0 < \tau_1 < \tau_2 < \cdots < \tau_m,$$

where a_{ij} and τ_j are some real constants, $f(t)$ is a continuous function, and $t > t_0$.

The general solution to equation (1.2.3.21) is the sum of the general solution to the respective homogeneous equation (1.2.3.3) and any particular solution to the nonhomogeneous equation.

Table 1.3 describes the structure of particular solutions for some functions on the right-hand side of the linear nonhomogeneous equation (1.2.3.21).

1.2.4. Linear Systems of First- and Second-Order ODEs with Delay. The Cauchy Problem. Exact Solutions

Linear systems of first-order ODEs with delay. A linear homogeneous system of first-order ODEs with constant coefficients and a constant delay with n unknown

Table 1.3. The structure of particular solutions of the linear nonhomogeneous equation with constant delays (1.2.3.21) for some special forms of the function $f(x)$.

Form of function $f(t)$	Roots of characteristic equation $\lambda^n + \sum_{i=0}^{n-1}\sum_{j=0}^{m} a_{ij}\lambda^i e^{-\lambda\tau_j} = 0$	Form of particular solution $u = \widetilde{u}(t)$
$P_m(t)$	zero is not a root of characteristic equation	$\widetilde{P}_m(t)$
	zero is a root of characteristic equation (multiplicity r)	$t^r \widetilde{P}_m(t)$
$P_m(t)e^{\alpha t}$ (α is a real number)	α is not a root of characteristic equation	$\widetilde{P}_m(t)e^{\alpha t}$
	α is a root of characteristic equation (multiplicity r)	$t^r \widetilde{P}_m(t)e^{\alpha t}$
$P_m(t)\cos\beta t$ $+ Q_n(t)\sin\beta t$	$i\beta$ is not a root of characteristic equation	$\widetilde{P}_\nu(t)\cos\beta t$ $+ \widetilde{Q}_\nu(t)\sin\beta t$
	$i\beta$ is a root of characteristic equation (multiplicity r)	$t^r[\widetilde{P}_\nu(t)\cos\beta t$ $+ \widetilde{Q}_\nu(t)\sin\beta t]$
$[P_m(t)\cos\beta t$ $+ Q_n(t)\sin\beta t]e^{\alpha t}$	$\alpha + i\beta$ is not a root of characteristic equation	$[\widetilde{P}_\nu(t)\cos\beta t$ $+ \widetilde{Q}_\nu(t)\sin\beta t]e^{\alpha t}$
	$\alpha + i\beta$ is a root of characteristic equation (multiplicity r)	$t^r[\widetilde{P}_\nu(t)\cos\beta t$ $+ \widetilde{Q}_\nu(t)\sin\beta t]e^{\alpha t}$

Notation: P_m and Q_n are polynomials of degree m and n with prescribed coefficients; $\widetilde{P}_m$, $\widetilde{P}_\nu$, and $\widetilde{Q}_\nu$ are polynomials of degree m and ν whose coefficients are determined by substituting the particular solution into the original equation; $\nu = \max(m, n)$; α and β are real numbers, and $i^2 = -1$.

functions can be written in a matrix form as

$$\mathbf{u}'_t(t) = \mathbb{A}\mathbf{u}(t) + \mathbb{B}\mathbf{u}(t - \tau), \quad t > 0, \qquad (1.2.4.1)$$

where $\mathbf{u} = (u_1, \ldots, u_n)^T$ is a column vector (the superscript T indicates the transpose), $\mathbb{A}$ and $\mathbb{B}$ are $n \times n$ square matrices with constant coefficients that satisfy the commutative condition $\mathbb{A}\mathbb{B} = \mathbb{B}\mathbb{A}$.

The Cauchy problem is stated as follows: find a solution to the system of equations (1.2.4.1) that satisfies the initial condition

$$\mathbf{u} = \boldsymbol{\varphi}(t) \quad \text{at} \quad -\tau \leq t \leq 0, \qquad (1.2.4.2)$$

where $\boldsymbol{\varphi}(t) = (\varphi_1(t), \ldots, \varphi_n(t))^T$ is a given continuous vector function.

A solution of the Cauchy problem (1.2.4.1)–(1.2.4.2) can be represented using two matrix functions as described below.

The exponential of a square matrix $\mathbb{A}t$ is defined by the series

$$\exp(\mathbb{A}t) = \mathbb{E} + \mathbb{A}t + \mathbb{A}^2 \frac{t^2}{2!} + \mathbb{A}^3 \frac{t^3}{3!} + \cdots = \mathbb{E} + \sum_{k=1}^{\infty} \mathbb{A}^k \frac{t^k}{k!},$$

where $\mathbb{E}$ is an identity matrix with elements $e_{ij} = \delta_{ij}$, where δ_{ij} is the Kronecker delta ($\delta_{ij} = 1$ if $i = j$, $\delta_{ij} = 0$ if $i \neq j$). The delayed exponential function of a square matrix $\mathbb{A}t$ was introduced in [267] and is defined by the formulas

$$\exp_d(\mathbb{A}t, \mathbb{A}\tau) = \begin{cases} \Theta, & t < -\tau, \\ \mathbb{E}, & -\tau \leq t < 0, \\ \mathbb{E} + \mathbb{A}\frac{t}{1!} + \cdots + \mathbb{A}^k \frac{[t-(k-1)\tau]^k}{k!}, & (k-1)\tau \leq t < k\tau, \\ & k = 1, 2, \ldots, \end{cases} \quad (1.2.4.3)$$

where Θ is a square zero matrix whose elements are all equal to zero.

It was proved in the studies [267, 268, 439] that the solution to the Cauchy problem (1.2.4.1)–(1.2.4.2) can be represented as

$$\mathbf{u}(t) = \exp(\mathbb{A}(t+\tau))\exp_d(\widetilde{\mathbb{B}}t, \widetilde{\mathbb{B}}\tau)$$
$$+ \int_{-\tau}^{0} \exp(\mathbb{A}(t-\tau-s))\exp_d(\widetilde{\mathbb{B}}(t-\tau-s), \widetilde{\mathbb{B}}\tau)\exp(\mathbb{A}\tau)[\boldsymbol{\varphi}'_s(s) - \mathbb{A}\boldsymbol{\varphi}(s)]\,ds,$$
$$(1.2.4.4)$$

where $\widetilde{\mathbb{B}} = \exp(-\mathbb{A}\tau)\mathbb{B}$. This formula was derived under the assumption that all components of the vector function $\boldsymbol{\varphi}(t)$ are continuously differentiable on the interval $-\tau \leq t \leq 0$.

Linear systems of second-order ODEs with a single delay. Consider a special linear nonhomogeneous system of second-order ODEs with constant coefficients and a single delay which is written in a matrix form as

$$\mathbf{u}''_{tt}(t) = -\mathbb{B}^2 \mathbf{u}(t-\tau) + \mathbf{f}(t), \quad t > 0, \quad (1.2.4.5)$$
$$\mathbf{u} = \boldsymbol{\varphi}(t), \quad \mathbf{u}'_t = \boldsymbol{\varphi}'_t(t) \quad \text{at} \quad -\tau \leq t < 0, \quad (1.2.4.6)$$

where $\mathbf{u} = (u_1, \ldots, u_n)^T$ is a column vector of the unknowns, $\mathbb{B}$ is an $n \times n$ square nondegenerate matrix with constant coefficients, and $\mathbf{f}(t) = (f_1(t), \ldots, f_n(t))^T$ and $\boldsymbol{\varphi}(t) = (\varphi_1(t), \ldots, \varphi_n(t))^T$ are given continuous vector functions.

The solution to the Cauchy problem (1.2.4.5)–(1.2.4.6) can be represented using two matrix functions as described below.

1.2. Second- and Higher-Order Delay ODEs. Systems of Delay ODEs

The matrix delayed cosine and sine are defined as [269]:

$$\cos_d(\mathbb{B}t, \mathbb{B}\tau) = \begin{cases} \Theta, & t < -\tau, \\ \mathbb{E}, & -\tau \leq t < 0, \\ \mathbb{E} - \mathbb{B}^2 \dfrac{t^2}{2!} + \cdots + (-1)^k \mathbb{B}^{2k} \dfrac{[t-(k-1)\tau]^{2k}}{(2k)!}, & (k-1)\tau \leq t < k\tau, \end{cases}$$

(1.2.4.7)

$$\sin_d(\mathbb{B}t, \mathbb{B}\tau) = \begin{cases} \Theta, & t < -\tau, \\ \mathbb{B}(t+\tau), & -\tau \leq t < 0, \\ (t+\tau)\mathbb{B} - \mathbb{B}^3 \dfrac{t^3}{3!} + \cdots + (-1)^k \mathbb{B}^{2k+1} \dfrac{[t-(k-1)\tau]^{2k+1}}{(2k+1)!}, \\ \hspace{5cm} (k-1)\tau \leq t < k\tau, \end{cases}$$

(1.2.4.8)

where $k = 1, 2, \ldots$ The following relations for derivatives hold:

$$\begin{aligned}
[\cos_d(\mathbb{B}t, \mathbb{B}\tau)]'_t &= -\mathbb{B} \sin_d(\mathbb{B}(t-\tau), \mathbb{B}\tau), \\
[\sin_d(\mathbb{B}t, \mathbb{B}\tau)]'_t &= \mathbb{B} \cos_d(\mathbb{B}t, \mathbb{B}\tau), \\
[\cos_d(\mathbb{B}t, \mathbb{B}\tau)]''_{tt} &= -\mathbb{B}^2 \cos_d(\mathbb{B}(t-\tau), \mathbb{B}\tau), \\
[\sin_d(\mathbb{B}t, \mathbb{B}\tau)]''_{tt} &= -\mathbb{B}^2 \sin_d(\mathbb{B}(t-\tau), \mathbb{B}\tau).
\end{aligned}$$

(1.2.4.9)

It was shown in [129] that the solution to the Cauchy problem (1.2.4.5)–(1.2.4.6) can be represented as

$$\begin{aligned}
\mathbf{u}(t) = {} & \varphi(0) \cos_d(\mathbb{B}(t-\tau), \mathbb{B}\tau) + \mathbb{B}^{-1} \varphi'_t(0) \sin_d(\mathbb{B}(t-\tau), \mathbb{B}\tau) \\
& - \mathbb{B} \int_{-\tau}^{0} \sin_d(\mathbb{B}(t-2\tau-s), \mathbb{B}\tau) \varphi(s) \, ds \\
& + \mathbb{B}^{-1} \int_{0}^{t} \sin_d(\mathbb{B}(t-\tau-s), \mathbb{B}\tau) \mathbf{f}(s) \, ds.
\end{aligned}$$

(1.2.4.10)

An alternative but less convenient representation of the solution to the Cauchy problem (1.2.4.5)–(1.2.4.6) can be found in [269].

Remark 1.14. *Consider the Cauchy problem for the system of second-order delay ODEs*

$$\mathbf{u}''_{tt}(t) = -\mathbb{A}\mathbf{u}(t-\tau) + \mathbf{f}(t), \quad t > 0,$$

(1.2.4.11)

subjected to the initial conditions (1.2.4.6). Here, $\mathbb{A}$ is an $n \times n$ positive definite matrix.

The problem is solved in two stages. First, one finds the matrix $\mathbb{B}$ from the equation $\mathbb{B}^2 = \mathbb{A}$. The matrix $\mathbb{B}$ is called the square root of the matrix $\mathbb{A}$ *and denoted $\mathbb{B} = \mathbb{A}^{1/2}$. A positive definite matrix always has exactly one positive definite square root, called the* arithmetic square root. *By using eigenvalue decomposition, one can represent a positive definite matrix as $\mathbb{A} = \mathbb{V}\mathbb{D}\mathbb{V}^{-1}$, where $\mathbb{D}$ is a diagonal matrix consisting of eigenvalues $\lambda_i > 0$. Then the positive definite square root of $\mathbb{A}$ is determined by the formula $\mathbb{B} = \mathbb{A}^{1/2} = \mathbb{V}\mathbb{D}^{1/2}\mathbb{V}^{-1}$, where $\mathbb{D}^{1/2}$ is a diagonal matrix consisting of eigenvalues $\sqrt{\lambda_i}$ [178].*

In the second stage, by setting $\mathbb{A} = \mathbb{B}^2$ in equation (1.2.4.11), one reduces the problem in question to problem (1.2.4.5)–(1.2.4.6).

1.3. Stability (Instability) of Solutions to Delay ODEs

1.3.1. Basic Concepts. General Remarks on Stability of Solutions to Linear Delay ODEs

Some definitions. Let us look at the Cauchy problem for the delay ODE

$$u'_t = f(t, u, w_1, \ldots, w_m), \quad w_i = u(t - \tau_i(t)), \quad i = 1, \ldots, m, \qquad (1.3.1.1)$$

with the initial data

$$u(t) = \varphi_1(t) \quad \text{on} \quad E_{t_0}, \qquad (1.3.1.2)$$

where E_{t_0} is some initial set.

A solution $u_1(t)$ to problem (1.3.1.1), (1.3.1.2) is called *stable* if for any $\varepsilon > 0$, there exists a $\delta(\varepsilon) > 0$ such that if the inequality $|\varphi_1(t) - \varphi_2(t)| < \delta(\varepsilon)$ holds on the initial set, then $|u_1(t) - u_2(t)| < \varepsilon$ for all $t \geq t_0$, where $u_2(t)$ is a solution to equation (1.3.1.1) under the initial condition $u(t) = \varphi_2(t)$ on E_{t_0}.

Solutions that do not possess the above property are called *unstable*.

A stable solution u_1 is called *asymptotically stable* if for any initial function $\varphi_2(t)$ that satisfies the condition $|\varphi_1(t) - \varphi_2(t)| < \delta_1$ for sufficiently small $\delta_1 > 0$, the limit relation $\lim_{t \to \infty} |u_1(t) - u_2(t)| = 0$ holds.

An asymptotically stable solution u_1 is called *globally asymptotically stable* if any other solution of the system in question tends to it as $t \to \infty$ regardless of the initial data. An asymptotically stable solution that is not globally asymptotically stable is called a *locally asymptotically stable* solution.

When carrying out a stability analysis of a solution u_0 to problem (1.3.1.1), (1.3.1.2), one can use the change of variable $v(t) = u(t) - u_0(t)$ to reduce the solution u_0 to the trivial one, $v(t) \equiv 0$. Therefore, in what follows, we will only be performing stability analyses of trivial solutions.

General remarks on solution stability for linear delay ODEs. Just as in solutions of linear ODEs without delays, all solutions of linear ODE with delays (with a fixed initial point t_0) are either simultaneously stable or simultaneously unstable. In particular, all solutions of a linear homogeneous equation behave, in the sense of stability, as the trivial (zero) solution of the same equation.

Solutions to linear homogeneous ODEs with constant coefficients and constant delays,

$$u_t^{(n)}(t) + \sum_{k=0}^{n-1} a_k u_t^{(k)}(t) + \sum_{k=0}^{n-1} \sum_{j=1}^{m} b_{kj} u_t^{(k)}(t - \tau_j) = 0, \qquad (1.3.1.3)$$

are the easiest to analyze for stability; here $\tau_j > 0$ and $t > t_0$. The stability or instability of the trivial solution to this equation is determined by the position of the roots of the respective characteristic equation

$$\lambda^n + \sum_{k=0}^{n-1} a_k \lambda^k + \sum_{k=0}^{n-1} \sum_{j=1}^{m} b_{kj} \lambda^k e^{-\tau_j \lambda} = 0, \qquad (1.3.1.4)$$

which results from substituting the exponential function $u = e^{\lambda t}$ into (1.3.1.3).

If all roots of the characteristic equation (1.3.1.4) have negative real parts, then the zero solution to the linear homogeneous ODE with delays (1.3.1.3) is asymptotically stable. If at least one root of the characteristic equation (1.3.1.4) has a positive real part, then the solution to the linear homogeneous ODE with delays (1.3.1.3) will be unstable.

Therefore, the stability analysis for solutions to linear delay ODEs of the form (1.3.1.3) reduces to a position analysis of the roots of the characteristic equation (1.3.1.4) (this equation is transcendental and has infinitely many complex roots).

Subsection 1.3.2 below will discuss the issues of stability and instability of solutions to specific linear delay ODEs that frequently arise in applications.

The case of small delays. If maximum delay, $\tau_{\max} = \max\limits_{1 \leq j \leq m} \tau_j$, in the linear homogeneous ODE with m delays (1.3.1.3) is sufficiently small, then one would naturally anticipate that many properties of solutions to equation (1.3.1.3) are close to those of solutions to the simpler ODE without delays

$$u_t^{(n)}(t) + \sum_{k=0}^{n-1} a_k u_t^{(k)}(t) + \sum_{k=0}^{n-1} \sum_{j=1}^{m} b_{kj} u_t^{(k)}(t) = 0, \qquad (1.3.1.5)$$

which is formally obtained from (1.3.1.3) by setting $\tau_j = 0$ for all $j = 1, \ldots, m$.

In particular, the following statements are true [144]:

1°. If the real parts of all roots of the characteristic equation for the ODE without delays (1.3.1.5) are negative and, hence, the solutions to equation (1.3.1.5) are asymptotically stable, then the solutions to the ODE with delays (1.3.1.3) are also asymptotically stable for sufficiently small $\tau_{\max}$.

2°. If the characteristic equation for the ODE without delays (1.3.1.5) has at least one root with a positive real part and, hence, the solutions to equation (1.3.1.5) are unstable, then the solutions to the ODE with delays (1.3.1.3) are also unstable for sufficiently small $\tau_{\max}$.

3°. If the characteristic equation for the ODE without delays (1.3.1.5) has a simple root $\lambda = 0$ and the other roots have negative real parts, then the solution to the ODE with delays (1.3.1.3) is stable for sufficiently small $\tau_{\max}$.

1.3.2. Stability of Solutions to Linear ODEs with a Single Constant Delay

First-order linear ODEs with constant delay. Let us revisit the first-order linear ODE with constant coefficients and a constant delay discussed previously in Subsection (1.1.3) (see equation (1.1.3.1)):

$$u_t' = au + bw, \quad w = u(t - \tau). \qquad (1.3.2.1)$$

We look for particular solutions of equation (1.3.2.1) in the form $u = e^{\lambda t}$. As a result, we arrive at the characteristic equation for the parameter λ:

$$\lambda - a - be^{-\lambda \tau} = 0. \qquad (1.3.2.2)$$

The linear delay ODE (1.3.2.1) will be asymptotically stable if all roots of the characteristic equation (1.3.2.2) have negative real parts. The following theorem holds.

Theorem (Hayes) [37]. *All roots of the characteristic equation (1.3.2.2) with real coefficients a and b ($\tau > 0$) have a negative real part ($\mathrm{Re}\,\lambda < 0$) if and only if the following three inequalities hold simultaneously:*

$$\begin{aligned}(i) &\quad a\tau < 1,\\ (ii) &\quad a + b < 0,\\ (iii) &\quad b\tau + \sqrt{(a\tau)^2 + \mu^2} > 0,\end{aligned} \qquad (1.3.2.3)$$

where μ is a root of the transcendental equation $\mu = a\tau \tan \mu$ satisfying the condition $0 < \mu < \pi$. If $a = 0$, one should set $\mu = \pi/2$.

Figure 1.11 displays in white color the region of the plane (A, B), where $A = a\tau$ and $B = b\tau$, in which all roots of the transcendental equation (1.3.2.2) have negative real parts ($\mathrm{Re}\,\lambda < 0$). In this region, the trivial (zero) solution of equation (1.3.2.1) is asymptotically stable. The region of instability, where at least one root of the transcendental equation (1.3.2.2) has a positive real part, is shaded in grey color.

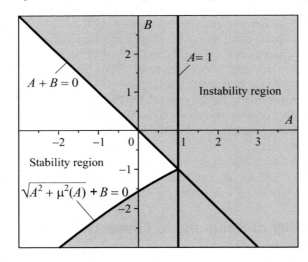

Figure 1.11. Regions of stability and instability for the trivial solution to the first-order ODE with constant delay (1.3.2.1).

▶ **Example 1.14.** Let us look at how the region of stability changes as the delay increases for the simplest two-term linear ODE with constant delay

$$u'_t = bw, \quad w = u(t - \tau), \qquad (1.3.2.4)$$

which corresponds to $a = 0$ in (1.3.2.1).

If there is no delay, $\tau = 0$, the region of stability of solutions to equation (1.3.2.4) is the entire negative semi-axis $-\infty < b < 0$. For $\tau > 0$, we substitute $a = 0$ into conditions (1.3.2.3) to obtain the region of stability of solutions to equation (1.3.2.4) in the form of a finite interval

$$-\frac{\pi}{2\tau} < b < 0. \tag{1.3.2.5}$$

It is apparent that the region of stability shrinks as τ increases, with the size of the region vanishing as $\tau \to \infty$. In other words, the presence of a delay is a destabilizing factor, and the increase of τ can lead to the instability of solutions of the equation. Qualitatively, the same occurs in the overwhelming majority of ODEs and systems of ODEs with constant delay. ◀

Remark 1.15. *The trivial solution of the delay ODE*

$$u'_t = au + b(t)w, \quad w = u(t - \tau),$$

where $b(t)$ is a continuous function, is asymptotically stable if $|b(t)| < -a$ [144].

Second-order linear ODEs with constant delay. Consider the second-order linear ODE with constant coefficients and a constant delay

$$u''_{tt}(t) + a_1 u'_t(t) + b_1 u'_t(t - \tau) + a_0 u(t) + b_0 u(t - \tau) = 0. \tag{1.3.2.6}$$

The associated characteristic equation is

$$\lambda^2 + a_1 \lambda + a_0 + (b_1 \lambda + b_0)e^{-\tau \lambda} = 0. \tag{1.3.2.7}$$

Note that for $\tau = 0$, the trivial solution of equation (1.3.2.6) is asymptotically stable if and only if $a_1 + b_1 > 0$ and $a_0 + b_0 > 0$.

An extensive literature is devoted to the analysis of stability and instability of solutions to equation (1.3.2.6) and related delay ODEs (e.g., see [66, 79, 80, 104, 112, 182, 205, 208, 225, 283, 333, 500, 574, 587]).

For equation (1.3.2.6), the number of different pure imaginary roots (the roots that only differ in sign are treated as the same) of the characteristic equation (1.3.2.7) under the conditions $a_1 + b_1 \neq 0$ and $a_0 + b_0 \neq 0$ can be zero, one, or two. The following three situations are possible [112]:

1. There are no imaginary roots. The stability of the trivial solution does not alter as τ increases from zero to infinity.

2. There is one imaginary root. The trivial solution that is unstable at $\tau = 0$ will never become stable. If the trivial solution is stable at $\tau = 0$, it will become unstable at the least τ for which there is an imaginary root and will remain unstable as τ further increases.

3. There are two imaginary roots. As τ increases, the stability of the trivial solution can alter finitely many times, and it will ultimately become unstable for sufficiently large τ.

Below are the simplest stability and instability conditions.

1°. An equilibrium is unstable for all $\tau \geq 0$ if

(a) it is a saddle point at $\tau = 0$ (i.e., for $a_0 + b_0 < 0$),

(b) it is an unstable node or unstable focus at $\tau = 0$ (i.e., for $a_1 + b_1 > 0$ and $a_0 + b_0 > 0$) and $|a_0| < |b_0|$.

2°. An equilibrium is stable for all $\tau \geq 0$, if it is stable at $\tau = 0$ (for $a_1 + b_1 < 0$ and $a_0 + b_0 > 0$) and

(a) $(a_1^2 - 2a_0 - b_1^2)^2 < 4(a_0^2 - b_0^2)$ or

(b) $a_1^2 > 2a_0 + b_1^2$ and $|a_0| > |b_0|$.

Remark 1.16. *The study [333] conducted a detailed and rather comprehensive position analysis of the roots of the six-parameter transcendental equation*

$$\lambda^2 + a_1\lambda + a_0 + (b_2\lambda^2 + b_1\lambda + b_0)e^{\lambda\tau} = 0$$

in the complex plane $\lambda = \operatorname{Re}\lambda + i\operatorname{Im}\lambda$. *In the special case* $b_2 = 0$, *this equation becomes the characteristic equation (1.3.2.7).*

Higher-order linear ODEs with constant delay. Consider the nth-order linear homogeneous ODE with real constant coefficients and a single constant delay

$$u_t^{(n)}(t) + \sum_{j=0}^{n-1} a_j u_t^{(j)}(t) + \sum_{j=0}^{m} b_j u_t^{(j)}(t-\tau) = 0 \qquad (1.3.2.8)$$

in more detail; here $n > m$, $\tau > 0$, and $t > t_0$.

The characteristic equation associated with (1.3.2.8) reads

$$\Phi(z) \equiv P(z) + Q(z)e^{-\tau z} = 0,$$

$$P(z) \equiv z^n + \sum_{j=0}^{n-1} a_j z^j, \quad Q(z) \equiv \sum_{j=0}^{m} b_j z^j, \qquad (1.3.2.9)$$

where $z = x + iy$ ($i^2 = -1$). The function $\Phi(z)$ in (1.3.2.9) is customarily called a *quasi-polynomial*.

Following [67, 113], we will call the quasi-polynomial $\Phi(z)$ *stable* if all roots of the transcendental equation $\Phi(z) = 0$ have a negative real part, $\operatorname{Re} z < 0$. The quasi-polynomial $\Phi(z)$ will be called *unstable* if the equation $\Phi(z) = 0$ has at least one root with a positive real part, $\operatorname{Re} z > 0$.

Below we describe two methods for investigating roots of quasi-polynomials.

The D-partition method [144]. For fixed τ, the zeros of the quasi-polynomial $\Phi(z)$ are continuous functions of its coefficients. Let us divide the space of coefficients, using hypersurfaces, into regions whose points correspond to pure imaginary zeros of the quasi-polynomial, $z = iy$ (inclusive of the degenerate case $z = 0$). This process is called a *D-partition*.

At the points of each region of a D-partition, the quasi-polynomial has equal numbers of zeros with positive reals parts (the number of zeros is here understood as the sum of their multiplicities). This is because a change in the number of zeros with positive real parts can only occur, as the coefficients vary continuously, when a zero crosses the imaginary axis, or when it crosses the boundary of a region in the D-partition. The regions that have no roots with a positive real part determine regions of asymptotic stability of solutions to the linear delay ODEs concerned.

The stability analysis conducted in the space of the model parameters by the D-partition method suggests finding regions D_k where there are no roots with positive real parts. To single out a region D_k, if it is connected, it suffices to make sure that at least one of its points corresponds to a quasi-polynomial whose roots all have a negative real part. In order to figure out how the number of roots with a positive real part changes when crossing some boundary of the D-partition, one computes the differential of the real part of the root and checks its sign to judge whether the number of roots with a positive real part decreases or increases.

Suppose that the linear delay ODE in question depends on m free parameters a_m (τ is assumed fixed) and the corresponding characteristic equation has the form $\Phi(z, a_1, \ldots, a_m) = 0$, where $z = x + iy$. In view of the inequality $d\Phi = \Phi_z\, dz + \sum_{s=1}^{m} \frac{\partial \Phi}{\partial a_s}\, da_s = 0$, we get

$$dx = -\operatorname{Re}\left(\frac{1}{\Phi_z} \sum_{s=1}^{m} \frac{\partial \Phi}{\partial a_s}\, da_s\right). \quad (1.3.2.10)$$

As a rule, one computes the differential dx at some boundary of the D-partition (at $z = iy$) between two regions where only one parameter varies to ensure that the boundary is crossed. If $dx > 0$, then as we move from one region of the D-partition to the other, the number of roots of the characteristic polynomial that have a positive real part increases by one. If $dx < 0$, the number or roots decreases by one.

▶ **Example 1.15.** Find the region of stability of the second-order linear ODE with constant delay

$$u''_{tt} = au + bw, \quad w = u(t - \tau), \quad (1.3.2.11)$$

in the space of real parameters a and b for $\tau > 0$.

We write the characteristic equation

$$\Phi(z) = 0, \quad \Phi(z) \equiv z^2 - a - be^{-\tau z}. \quad (1.3.2.12)$$

In the degenerate case of $z = 0$, we have $a + b = 0$ (one of the boundaries of the D-partition). Setting $z = iy$ in (1.3.2.12), where $0 < y < \infty$, we get $-y^2 = a + b[\cos(\tau y) - i\sin(\tau y)]$, which leads to the transcendental equations

$$\begin{aligned} y^2 + a + b\cos(\tau y) &= 0, \\ b\sin(\tau y) &= 0. \end{aligned} \quad (1.3.2.13)$$

Assuming that $b \neq 0$, we start from the second equation in (1.3.2.13) and then proceed to the first one to obtain

$$\begin{aligned} y &= \frac{\pi k}{\tau}, \quad k = 1, 2, \ldots, \\ b &= (-1)^{k+1}\left[a + \left(\frac{\pi k}{\tau}\right)^2\right]. \end{aligned} \quad (1.3.2.14)$$

Here, the second relation determines two sets of parallel straight lines with angles of inclination $\pm\pi/4$ that define boundaries of the D-partition in the (a, b) plane. Moreover, the second relation in (1.3.2.14) is satisfied at $b = 0$ and $-\infty < a < 0$ ($y = \sqrt{-a}$), which corresponds to the negative part of the a-axis. So the negative semiaxis also makes up boundaries of the D-partition. Figure 1.12 displays the D-partition lines in the (A, B) plane, where $A = a\tau^2$ and $B = b\tau^2$. The regions where the characteristic equation (1.3.2.12) has equal numbers of roots with a positive real part are shaded in the same color; the circles indicate the numbers of roots.

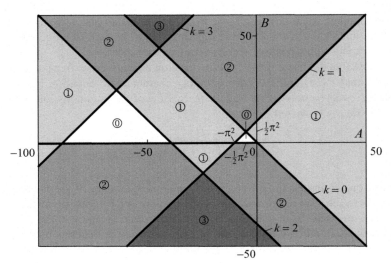

Figure 1.12. The D-partition boundaries for the second-order ODE with constant delay (1.3.2.11). The regions where the characteristic equation (1.3.2.12) has equal numbers of roots with a positive real part are shaded in the same color; the circles indicate the numbers of roots.

We note first that the characteristic equation (1.3.2.12) has a single root with a positive real part at the points of the semiaxis $b = 0$, $a > 0$. Therefore, the characteristic equation also has a single root with a positive real part at all points of the cone containing the semiaxis.

To determine the number of roots with a positive real part in other regions, we will take advantage of formula (1.3.2.10). In view of (1.3.2.12), it becomes

$$dx = -\operatorname{Re}\left(\frac{\Phi_a da + \Phi_b db}{\Phi_z}\right) = \operatorname{Re}\left(\frac{da + e^{-\tau z} db}{2z + b\tau e^{-\tau z}}\right). \tag{1.3.2.15}$$

We will only be interested in the sign of dx as we cross boundaries of regions of the D-partition, which are defined by the linear relations (1.3.2.14). Therefore, we substitute $z = iy$, where $y = \pi k/\tau$ ($k = 1, 2, \ldots$), into the right-hand side of (1.3.2.15) to obtain

$$dx = \operatorname{Re}\frac{da + (-1)^k db}{2iy + (-1)^k b\tau} = \frac{(-1)^k b\tau \, da + b\tau \, db}{4y^2 + (b\tau)^2}. \tag{1.3.2.16}$$

For simplicity, we will further assume that $\tau = 1$. Hence, sign $dx = \text{sign}[(-1)^k b\,da + b\,db]$.

It follows from (1.3.2.16) that for any fixed positive $a = a_* > 0$ ($da = 0$), the differential dx, with sign $dx = \text{sign}(b\,db)$, is positive when b is positive and increases ($db > 0$) or negative and decreases ($db < 0$). This implies that in the regions adjacent with the cone containing the points $a > 0$, the number of roots of the characteristic equation (1.3.2.12) with a positive real part equals two and increases as the vertical line $a = a_*$ crosses further boundaries.

For a sufficiently small fixed positive $b = b_* > 0$ ($db = 0$), we see that $dx < 0$ when we cross the straight line $a + b = 0$ (corresponding to $k = 0$ in (1.3.2.14)) as a decreases ($da < 0$). Therefore, inside the triangle with vertices at $(0,0)$, $(-\pi^2, 0)$, and $(-\frac{1}{2}\pi^2, \frac{1}{2}\pi^2)$, the characteristic equation (1.3.2.12) has no roots with a positive real part, and hence, the trivial solution of the original equation (1.3.2.11) is stable in this region.

When we move left ($da < 0$) along the horizontal line $b = b_* > 0$ ($db = 0$) and cross the straight line (1.3.2.14) with $k = 1$, $b = a + \pi^2$, we have sign $dx = \text{sign}(-b\,da)$ so that $dx > 0$ for sufficiently small b_*. It follows that in the region to the left of the triangle, there is one root of the characteristic equation with a positive real part. Likewise, one can determine the number of roots with a positive real part in other regions of the D-partition. For clarity, in Figure 1.12, the regions where the characteristic equation has equal numbers of roots with a positive real part are shaded in the same color. ◀

The Cooke–van den Driessche method. The regions of stability and instability are separated by pure imaginary roots, $z = iy$, which are zeros of the quasi-polynomial (1.3.2.9) and satisfy the equation $\Phi(iy) = 0$. On representing the quasi-polynomial as $P(iy) = -Q(iy)e^{-i\tau y}$, we take the modulus of both sides and raise to the power of two to obtain the algebraic equation

$$F(y) = 0, \quad F(y) \equiv |P(iy)|^2 - |Q(iy)|^2. \tag{1.3.2.17}$$

Its left-hand side is a polynomial of degree $2n$, since the equation only involves even powers of y. The substitution $\zeta = y^2$ reduces $F(y)$ to a polynomial of degree n, which can be written in the factorized form

$$F(y) = \prod_{j=1}^{n}(\zeta - r_j), \quad \zeta = y^2, \tag{1.3.2.18}$$

where $r_1, \ldots, r_n$ are some numbers, generally complex.

Now assuming that the iy and τ satisfying equations (1.3.2.9) and (1.3.2.17) are known, we will treat the root $z = x + iy$ of equation (1.3.2.9) as a function τ and attempt to determine the direction of motion of z as τ varies. So we need to calculate

$$s = \text{sign}\left\{\text{Re}\left(\frac{dz}{d\tau}\bigg|_{z=iy}\right)\right\} = \text{sign}\left\{\frac{d}{d\tau}(\text{Re}\,z)\bigg|_{z=iy}\right\}. \tag{1.3.2.19}$$

It was shown in [113] that for any simple root iy with $y > 0$ at which the imaginary axis is crossed with a 'positive velocity' ($s \neq 0$), the direction of crossing can be found from the formula

$$s = \operatorname{sign} F'(y), \qquad (1.3.2.20)$$

with the function $F(y)$ defined in (1.3.2.17).

Suppose that the polynomials $P(z)$ and $Q(z)$, defined in (1.3.2.9), do not have common roots and the inequality $P(0) + Q(0) \neq 0$ holds, which means that $z = 0$ is not a root of the quasi-polynomial $\Phi(z)$. Then the following propositions hold [113]:

1°. Suppose the equation $F(y) = 0$, defined in (1.3.2.17), does not have positive roots. Then if the quasi-polynomial $\Phi(z)$ is stable at $\tau = 0$, it will remain stable for all $\tau \geq 0$, and if it is unstable at $\tau = 0$, it will remain unstable for all $\tau \geq 0$.

2°. Let the equation $F(y) = 0$ have at least one positive root and let all positive roots be simple. Then as τ increases, a stable quasi-polynomial can become unstable and vice versa. There is a positive τ_* such that the quasi-polynomial $\Phi(z)$ from (1.3.2.9) is unstable for all $\tau > \tau_*$. For $0 < \tau < \tau_*$, the stability or instability can alter no more than finitely many times.

3°. If iy $(y > 0)$ and τ satisfy relation (1.3.2.9) and if iy is a simple root such that $s \neq 0$, then y is a simple root of the equation $F(y) = 0$, and the root $z(\tau)$ of equation (1.3.2.9) intersects the imaginary axis (as τ increases) in the direction determined by $\operatorname{sign} F'(y)$.

4°. Suppose that all positive roots, $r_1, \ldots, r_p$, appearing on the right-hand side of (1.3.2.18) are different and $r_1 > \cdots > r_p > 0$. Then $\pm i y_k = \pm i \sqrt{r_k}$ $(k = 1, \ldots, p)$ are possible roots of equation (1.3.2.9) on the imaginary axis. Let all these roots be simple. Then the direction of crossing of the imaginary axis at $i y_k$ is determined by

$$s_k = \operatorname{sign} \prod_{j=1, j \neq k}^{p} (r_k - r_j).$$

It follows that the imaginary axis is always crossed from left to right for the largest root r_1, from right to left for the root r_2, and so on. If there is only one positive root, r_1, it is clear that the imaginary axis is crossed from left to right. If there are two positive roots, the imaginary axis is first crossed from left to right (at r_1) and then from right to left (at r_2).

1.3.3. Stability of Solutions to Linear ODEs with Several Constant Delays

First-order linear ODEs with several constant delays. The analysis of stability of a first-order linear equation with constant coefficients and constant delays,

$$u'_t(t) = -\sum_{k=1}^{n} a_k u(t - \tau_k), \qquad (1.3.3.1)$$

reduces to finding the locations of the roots of the characteristic equation

$$\lambda + \sum_{k=1}^{n} a_k e^{-\lambda \tau_k} = 0. \qquad (1.3.3.2)$$

The study [522] proved the following two theorems for equation (1.3.3.1).

Theorem 1. *Let $a_k \in [0, \infty)$ and $\tau_k \in [0, \infty)$ for all $k = 1, \ldots, n$ and let $\sum_{k=1}^{n} a_k > 0$. Then, if the inequality $\sum_{k=1}^{n} a_k \tau_k < \pi/2$ holds, all roots of the transcendental equation (1.3.3.2) have negative real parts, and the trivial solution of equation (1.3.3.1) is asymptotically stable.*

Theorem 2. *If $n = 2$, $a_1 = a_2 > 0$, and $a_1 \tau_1 < 1$, then for any positive τ_2, all roots of the transcendental equation (1.3.3.2) have negative real parts.*

Notably, if $\sum_{k=1}^{n} a_k < 0$ for any nonnegative τ_k, the trivial solution of equation (1.3.3.1) is unstable.

Second-order linear ODEs with several constant delays. Consider the second-order linear ODE with constant coefficients and several constant delays

$$u''_{tt}(t) + au'_t(t) + bu(t) = \sum_{k=1}^{n} c_k u(t - \tau_k), \qquad (1.3.3.3)$$

where $a > 0$ and $b > 0$, and c_k are real numbers satisfying the condition $b > \sum_{k=1}^{n} |c_k|$.

Theorem. *For the trivial solution of equation (1.3.3.3) to be asymptotically stable for any $\tau_k \geq 0$, it suffices that either of the following inequalities holds* [500]:

(a) $\quad a > \sum_{k=1}^{n} |c_k| \Big/ \Big(b - \sum_{k=1}^{n} |c_k|\Big)^{1/2},$

(b) $\quad a > \sum_{k=1}^{n} |c_k| \tau_k.$

The characteristic equation of the delay ODE (1.3.3.3) is $f(\lambda) \equiv \lambda^2 + a\lambda + b - \sum_{k=1}^{n} c_k e^{-\lambda \tau_k} = 0$. Suppose the inequality

$$f(0) = b - \sum_{k=1}^{n} c_k < 0$$

holds. In this case, since $f(+\infty) = +\infty$, the function $f(\lambda)$ has at least one positive root, $\lambda = \lambda_+ > 0$, which generates an indefinitely growing solution, $u_+ = e^{\lambda_+ t}$, of equation (1.3.3.3). Therefore, the trivial solution of equation (1.3.3.3) will be unstable.

Higher-order linear ODEs with several constant delays. Now let us look at the nth-order linear homogeneous ODE with real constant coefficients and several constant delays

$$u_t^{(n)}(t) + \sum_{k=0}^{n-1} a_k u_t^{(k)}(t) + \sum_{k=0}^{n-1} \sum_{j=1}^{m} b_{kj} u_t^{(k)}(t - \tau_j) = 0, \qquad (1.3.3.4)$$

where $\tau_j > 0$ and $t > t_0$.

The characteristic equation associated with the delay ODE (1.3.3.4) is

$$P(\lambda)+\sum_{j=1}^{m}b_{kj}Q_j(\lambda)e^{-\tau_j\lambda}=0, \quad P(\lambda)\equiv\lambda^n+\sum_{k=0}^{n-1}a_k\lambda^k, \quad Q_j(\lambda)\equiv\sum_{k=0}^{n-1}b_{kj}\lambda^k. \tag{1.3.3.5}$$

Theorem. *Let the coefficients of equation (1.3.3.4) satisfy the condition*

$$\sum_{j=1}^{m}|b_{0j}|<|a_0|.$$

Then, solutions of equation (1.3.3.4) are asymptotically stable if and only if the following two conditions hold [144, 609]:

$1°$. *The real parts of all roots of the polynomial $P(\lambda)$ must be negative.*
$2°$. *For any $y>0$, the inequality*

$$\sum_{j=1}^{m}|Q_j(iy)|<|P(iy)|, \quad i^2=-1,$$

must hold.

1.3.4. Stability Analysis of Solutions to Nonlinear Delay ODEs by the First Approximation

Solution stability and instability theorems. In studying the stability of the trivial solution to the nonlinear equation (1.3.1.1) under the assumption that the right-hand side is differentiable with respect to all arguments, starting from the second, in a neighborhood of the zero values of these arguments for $t>t_0$, it is often reasonable to single out the linear part and rewrite the equation in the form

$$\begin{aligned} u'_t &= a_0(t)u + a_1(t)w_1 + \cdots + a_m(t)w_m + N(t,u,w_1,\ldots,w_m), \\ w_i &= u(t-\tau_i(t)), \quad i=1,\ldots,m. \end{aligned} \tag{1.3.4.1}$$

The nonlinear function N has a higher-than-first order in the ensemble of all arguments starting from the second.

In many cases, a stability analysis of the trivial solution to ODE (1.3.4.1) is equivalent to that of the trivial solution to the simpler, linear ODE

$$\begin{aligned} u'_t &= a_0(t)u + a_1(t)w_1 + \cdots + a_m(t)w_m, \\ w_i &= u(t-\tau_i(t)), \quad i=1,\ldots,m, \end{aligned} \tag{1.3.4.2}$$

which is called the first approximation for ODE (1.3.4.1).

In what follows, we will be looking at the constant delay ODE (1.3.4.1) assuming that $a_i = \text{const}$, $\tau_i = \text{const}$, and $N(t,0,0,\ldots,0)=0$. The following two theorems [144], analogous to Lyapunov's theorems for regular ODEs without delay, hold.

1.3. Stability (Instability) of Solutions to Delay ODEs

Theorem 1. *The trivial solution $u = 0$ of the nonlinear ODE with delays*

$$u'_t = a_0 u + a_1 w_1 + \cdots + a_m w_m + N(t, u, w_1, \ldots, w_m),$$
$$w_i = u(t - \tau_i), \quad i = 1, \ldots, m, \tag{1.3.4.3}$$

is asymptotically stable if all roots of the characteristic equation

$$\lambda = a_0 + a_1 e^{-\tau_1 \lambda} + \cdots + a_m e^{-\tau_m \lambda}, \tag{1.3.4.4}$$

associated with the truncated first-approximation equation, with $N \equiv 0$, have negative real parts and the inequality

$$|N(t, u, w_1, \ldots, w_m)| \leq k \sum_{i=0}^{m} |w_i|, \quad w_0 = u, \tag{1.3.4.5}$$

holds. Here, k is a sufficiently small positive constant, all w_i are sufficiently small (such that $|w_i| < \sigma$), and $t \geq t_0$.

Theorem 2. *If at least one root of the characteristic equation (1.3.4.4) has a positive real part and condition (1.3.4.5) holds, then the trivial solution of equation (1.3.4.3) is unstable.*

Remark 1.17. *Similar theorems for systems of nonlinear ODEs with constant delays are stated in [144].*

Stability conditions for solutions to Hutchinson type equations. Below are a few examples of applying the above theorems to the stability and instability analysis of nonlinear Hutchinson type equations.

▶ Example 1.16. Consider the nonlinear ODE with constant delay

$$u'_t = cu(1 - w^k), \quad w = u(t - \tau). \tag{1.3.4.6}$$

At $k = 1$ and $c = b$, it becomes Hutchinson's equation (see Subsection 6.1.1, equation (6.1.1.5)). We call equation (1.3.4.6) a *generalized Hutchinson equation*.

Equation (1.3.4.6) has two stationary solutions: $u = 0$ and $u = 1$. Using Theorems 1 and 2, we will investigate their stability for $c > 0$ and $k > 0$.

1. The stationary solution $u = 0$ is unstable, because the first-approximation characteristic equation is degenerate and it has a single root, $\lambda = c > 0$; in addition, inequality (1.3.4.5) holds.

2. To analyze the stationary solution $u = 1$, we substitute $u = 1 - \bar{u}$ in equation (1.3.4.6) to arrive at the delay ODE

$$\bar{u}'_t = -c(1 - \bar{u})[1 - (1 - \bar{w})^k], \quad \bar{w} = \bar{u}(t - \tau). \tag{1.3.4.7}$$

Considering that the stationary solution $u = 1$ of the original equation (1.3.4.6) has become the trivial solution $\bar{u} = 0$ of the reduced equation (1.3.4.7), we expand the right-hand side of (1.3.4.6) in a Taylor series assuming that $\bar{u}$ and $\bar{w}$ are small. Since $(1 - \bar{w})^k = 1 - k\bar{w} + o(\bar{w})$, we get

$$-c(1 - \bar{u})[1 - (1 - \bar{w})^k] = -ck\bar{w} + ck\bar{u}\bar{w} + o(\bar{w}). \tag{1.3.4.8}$$

Therefore, the linearized first-approximation equation for (1.3.4.7) is

$$\bar{u}'_t = -ck\bar{w}. \tag{1.3.4.9}$$

Equation (1.3.4.9) is a special case of equation (1.3.2.1) where one should set $a = 0$ and $b = -ck$. Consequently, for the stability analysis of the trivial solution of equation (1.3.4.9), we can use the Hayes theorem (see Subsection 1.3.2). The first two conditions of (1.3.2.3) with $a = 0$ and $b < 0$ are satisfied automatically, and one should set $\mu = \pi/2$ in the third condition. Taking into account that the nonlinear part of the function (1.3.4.8) can be neglected for small $|\bar{u}|$ and $|\bar{w}|$, we obtain the following stability and instability conditions for the stationary solution $u = 1$ of the generalized Hutchinson equation (1.3.4.6):

$$\begin{array}{l} \text{solution } u = 1 \text{ is asymptotically stable if } ck\tau < \pi/2; \\ \text{solution } u = 1 \text{ is unstable if } ck\tau > \pi/2. \end{array} \tag{1.3.4.10}$$

◀

▶ **Example 1.17.** The nonlinear delay ODE

$$u'_t = cu[1 - f(w)], \quad w = u(t - \tau), \tag{1.3.4.11}$$

is a further generalization of Hutchinson's equation. In (1.3.4.11), $c > 0$, and $f(w)$ is any monotonic smooth function that satisfies the conditions

$$f(0) = 0, \quad f(1) = 1, \quad f'(1) > 0. \tag{1.3.4.12}$$

Just as (1.3.4.6), equation (1.3.4.11) has the stationary solutions $u = 0$ and $u = 1$. By reasoning like in Example 1.16, we can show that the trivial solution $u = 0$ is unstable, and the stability and instability regions of the stationary solution $u = 1$ are determined by conditions (1.3.4.10), where $k = f'(1) > 0$. ◀

Stability conditions for stationary solutions of other nonlinear delay ODEs. We now consider the nonlinear constant delay ODE of a reasonably general form

$$u'_t = f(u, w), \quad w = u(t - \tau). \tag{1.3.4.13}$$

We assume that equation (1.3.4.13) has a stationary solution $u = u_0$, implying that $f(u_0, u_0) = 0$, and the function $f(u, w)$ has continuous partial derivatives at the point $u = u_0$, $w = u_0$.

For the stability analysis of the stationary solution $u = u_0$, we first linearize the right-hand side of equation (1.3.4.13) about u_0 to obtain

$$f(u, w) = f_u^\circ(u - u_0) + f_w^\circ(w - u_0) + o(|u - u_0| + |w - u_0|),$$
$$f_u^\circ = \frac{\partial f}{\partial u}\bigg|_{u=u_0, w=u_0}, \quad f_w^\circ = \frac{\partial f}{\partial w}\bigg|_{u=u_0, w=u_0}. \tag{1.3.4.14}$$

Then we make the change of variable $\bar{u} = u - u_0$, which takes the stationary solution $u = u_0$ of the original equation to the trivial solution $\bar{u} = 0$ of the reduced equation.

As a result, after discarding the nonlinear part, which has the order $o(|\bar{u}| + |\bar{w}|)$, we arrive at the linear equation

$$\bar{u}'_t = f_u^\circ \bar{u} + f_w^\circ \bar{w}, \quad \bar{w} = \bar{u}(t - \tau). \tag{1.3.4.15}$$

Up to obvious renaming, it coincides with equation (1.3.2.1).

The stability conditions for the trivial solution of equation (1.3.4.15), which coincide with those for the stationary solution $u = u_0$ of the original equation (1.3.4.13), are determined by inequalities (1.3.2.3) with $a = f_u^\circ$ and $b = f_w^\circ$.

1.4. Exact and Approximate Analytical Solution Methods for Delay ODEs

1.4.1. Using Integral Transforms for Solving Linear Problems

To solve linear problems for delay ODEs, one can use the Laplace and Mellin integral transforms [37, 55, 78, 144, 525, 526]. In what follows, we will need some concepts and formulas from the theory of residues.

Residues. A *residue* of a function $f(z)$ holomorphic in a punctured neighborhood of a point $z = a$ of the complex plane z, where a is an isolated singular point of f, is the number

$$\operatorname*{Res}_{z=a} f(z) = \frac{1}{2\pi i} \int_{C_\varepsilon} f(z)\, dz, \quad i^2 = -1,$$

where C_ε denotes a circle of a sufficiently small radius ε described by the equation $|z - a| = \varepsilon$.

For a simple pole, when $f(z) \simeq \mathrm{const}/(z - a)$ as $z \to a$, the following formula holds:

$$\operatorname*{Res}_{z=a} f(z) = \lim_{z \to a} \big[(z - a) f(z)\big].$$

If $f(z) = \dfrac{\varphi(z)}{\psi(z)}$, where $\varphi(a) \neq 0$ and $z = a$ is a simple zero of the function $\psi(z)$, implying that $\psi(a) = 0$ and $\psi'_z(a) \neq 0$, we get

$$\operatorname*{Res}_{z=a} f(z) = \frac{\varphi(a)}{\psi'_z(a)}. \tag{1.4.1.1}$$

If $z = a$ is an nth-order pole* of the function $f(z)$, then

$$\operatorname*{Res}_{z=a} f(z) = \frac{1}{(n-1)!} \lim_{z \to a} \frac{d^{n-1}}{dx^{n-1}} \big[(z - a)^n f(z)\big]. \tag{1.4.1.2}$$

---

*This means that $f(z) \approx \mathrm{const}\,(z - a)^{-n}$ in a neighborhood of the point $z = a$, where n is a positive integer.

The Laplace transform. For an arbitrary complex-valued function $f(t)$ of a real variable t ($t \geq 0$), the *Laplace transform* is defined as

$$\tilde{f}(s) = \int_0^\infty f(t)\, e^{-st}\, dt, \qquad (1.4.1.3)$$

where $s = a + ib$ is a complex variable.

The original function $f(t)$ is piecewise continuous in its entire domain, so it can only have jump discontinuities, with each finite interval containing no more than finitely many discontinuities. Moreover, it can only have a bounded exponential growth, implying that there are numbers $M > 0$ and σ_0 such that $|f(t)| \leq M e^{\sigma_0 t}$ for $t > 0$. Furthermore, we assume that σ_0 is the least possible number, called the *growth constant* of $f(t)$.

The function $\tilde{f}(s)$ is the *Laplace transform* or just the *transform* of $f(t)$. For every $f(t)$, its transform is defined in the half-plane $\operatorname{Re} s > \sigma_0$, where it is analytic.

Formula (1.4.1.3) will be written for short as

$$\tilde{f}(s) = \mathfrak{L}\{f(t)\} \quad \text{or} \quad \tilde{f}(s) = \mathfrak{L}\{f(t), s\}.$$

The inverse Laplace transform. Given a Laplace transform $\tilde{f}(s)$, the original function can be found using the *inverse Laplace transform*

$$f(t) = \frac{1}{2\pi i} \int_{c-i\infty}^{c+i\infty} \tilde{f}(s) e^{sx}\, ds, \qquad i^2 = -1, \qquad (1.4.1.4)$$

where the integration path goes parallel to the imaginary axis of the complex s-plane to the right of all singular points of $\tilde{f}(s)$, which implies that $c > \sigma_0$.

The integral in (1.4.1.4) is understood in the sense of the Cauchy principal value:

$$\int_{c-i\infty}^{c+i\infty} \tilde{f}(s) e^{st}\, ds = \lim_{\omega \to \infty} \int_{c-i\omega}^{c+i\omega} \tilde{f}(s) e^{st}\, ds.$$

In the region $t < 0$, formula (1.4.1.4) gives $f(t) \equiv 0$.

Formula (1.4.1.4) holds for continuous functions. If $f(t)$ has a jump discontinuity at a point $t = t_0 > 0$, then the right-hand side of formula (1.4.1.4) given the value $\frac{1}{2}[f(t_0 - 0) + f(t_0 + 0)]$ at this point (if $t_0 = 0$, the first term in the square brackets must be omitted).

The inversion formula (1.4.1.4) will be written for short as follows:

$$f(t) = \mathfrak{L}^{-1}\{\tilde{f}(s)\} \quad \text{or} \quad f(t) = \mathfrak{L}^{-1}\{\tilde{f}(s), t\}.$$

There are detailed tables of the Laplace transform and inverse Laplace transform, which are handy in solving linear differential equations (e.g., see [35, 130, 132, 145, 365, 403, 440, 441]).

Table 1.4. Basic properties of the Laplace transform.

No.	Function	Laplace transform	Transformation
1	$af_1(t) + bf_2(t)$	$a\tilde{f}_1(s) + b\tilde{f}_2(s)$	Linear superposition
2	$f(t/a)$, $a > 0$	$a\tilde{f}(as)$	Scaling
3	$f(t-a)$, $f(\xi) \equiv 0$ for $\xi < 0$	$e^{-as}\tilde{f}(s)$	Translation in argument
4	$t^n f(t)$; $n = 1, 2, \ldots$	$(-1)^n \tilde{f}_s^{(n)}(s)$	Differentiation of transform
5	$\dfrac{1}{t} f(t)$	$\displaystyle\int_s^\infty \tilde{f}(q)\,dq$	Integration of transform
6	$e^{at} f(t)$	$\tilde{f}(s-a)$	Translation in complex plane
7	$f'_t(t)$	$s\tilde{f}(s) - f(+0)$	Differentiation
8	$f_t^{(n)}(t)$	$s^n \tilde{f}(s) - \displaystyle\sum_{k=1}^{n} s^{n-k} f_t^{(k-1)}(+0)$	Differentiation
9	$t^m f_t^{(n)}(t)$, $m = 1, 2, \ldots$	$(-1)^m \left[s^n \tilde{f}(s) - \displaystyle\sum_{k=1}^{n} s^{n-k} f_t^{(k-1)}(+0) \right]_s^{(m)}$	Differentiation
10	$\left[t^m f(t) \right]_t^{(n)}$, $m \geq n$	$(-1)^m \left[\tilde{f}(s) \right]_s^{(m)}$	Differentiation
11	$\displaystyle\int_0^t f(\xi)\,d\xi$	$\dfrac{\tilde{f}(s)}{s}$	Integration
12	$\displaystyle\int_0^t f_1(\xi) f_2(t-\xi)\,d\xi$	$\tilde{f}_1(s)\tilde{f}_2(s)$	Convolution

Basic properties of the Laplace transform.

$1°$. Table 1.4 lists main formulas of original functions and their Laplace transforms.

$2°$. The Laplace transforms of some functions are listed in Table 1.5; for more detailed tables, see [130, 365, 440].

The inverse transform of rational functions. Let us look at an important case of transforming rational functions of the form

$$\tilde{f}(s) = \frac{R(s)}{Q(s)}, \qquad (1.4.1.5)$$

where $Q(s)$ and $R(s)$ are some polynomials, with the degree of the polynomial $Q(s)$ being higher than that of polynomial $R(s)$.

Let the zeros of the denominator be simple, meaning that

$$Q(s) \equiv \text{const}\,(s - s_1) \ldots (s - s_n).$$

Then the inverse transform can be expressed as

$$f(t) = \sum_{k=1}^{n} \frac{R(s_k)}{Q'(s_k)} \exp(s_k t), \qquad (1.4.1.6)$$

Table 1.5. Laplace transforms of some functions.

No.	Function, $f(t)$	Laplace transform, $\tilde{f}(s)$	Remarks
1	1	$1/s$	
2	t^n	$\dfrac{n!}{s^{n+1}}$	$n = 1, 2, \ldots$
3	t^a	$\Gamma(a+1)s^{-a-1}$	$a > -1$
4	e^{-at}	$(s+a)^{-1}$	
5	$t^a e^{-bt}$	$\Gamma(a+1)(s+b)^{-a-1}$	$a > -1$
6	$\sinh(at)$	$\dfrac{a}{s^2 - a^2}$	
7	$\cosh(at)$	$\dfrac{s}{s^2 - a^2}$	
8	$\ln t$	$-\dfrac{1}{s}(\ln s + \mathcal{C})$	$\mathcal{C}$ is Euler's constant, $\mathcal{C} \approx 0.5772156649$
9	$\sin(at)$	$\dfrac{a}{s^2 + a^2}$	
10	$\cos(at)$	$\dfrac{s}{s^2 + a^2}$	
11	$\operatorname{erfc}\left(\dfrac{a}{2\sqrt{t}}\right)$	$\dfrac{1}{s}\exp(-a\sqrt{s})$	$a \geq 0$
12	$J_0(at)$	$\dfrac{1}{\sqrt{s^2 + a^2}}$	$J_0(t)$ is the Bessel function

where the prime stands for a derivative.

If $Q(s)$ has m zeros with multiplicities $r_1, \ldots, r_m$, or

$$Q(s) \equiv \text{const}\,(s - s_1)^{r_1} \ldots (s - s_m)^{r_m},$$

then

$$f(t) = \sum_{k=1}^{m} \frac{1}{(r_k - 1)!} \lim_{s \to s_k} \frac{d^{r_k - 1}}{ds^{r_k - 1}}\left[(s - s_k)^{r_k} \tilde{f}(s) e^{st}\right]. \quad (1.4.1.7)$$

Inversion of functions with finitely many singular points. If the function $\tilde{f}(s)$ has finitely many singular points $s_1, \ldots, s_n$ and $\tilde{f}(s)$ tends to zero as $s \to \infty$, then the inverse transform can be obtained using the formula

$$f(t) = \sum_{k=1}^{n} \operatorname*{Res}_{s=s_k}\left[\tilde{f}(s) e^{st}\right]. \quad (1.4.1.8)$$

Inversion of functions with infinitely many singular points. Formula (1.4.1.8) can be extended to functions $\tilde{f}(s)$ with infinitely many singular points. In this case, the inverse transform $f(t)$ is expressed as an infinite series.

Theorem (decomposition) (e.g., see [130, 292]). Let the function of complex variable $\tilde{f}(s)$ meet the following conditions:

1.4. Exact and Approximate Analytical Solution Methods for Delay ODEs

1. The function $\tilde{f}(s)$ is meromorphic (i.e., defined on the entire complex plane and does not have, in a finite portion of the plane, singular points other than poles) and differentiable in a half-plane $\operatorname{Re} s > s_0$.

2. There is a nested circle system

$$C_n: \quad |s| = R_n, \quad R_1 < R_2 < \cdots, \quad R_n \to \infty,$$

on which $\tilde{f}(s)$ tends to zero uniformly with respect to $\arg s$.

3. For any $a > s_0$, the integral $\int_{a-i\infty}^{a+i\infty} \tilde{f}(s)\,ds$ is absolutely convergent. Then

$$f(t) = \sum_{k=1}^{\infty} \operatorname*{Res}_{s=s_k} [\tilde{f}(s) e^{st}], \tag{1.4.1.9}$$

where the sum of residues is taken over all singular points s_k of $\tilde{f}(s)$ in decreasing order of their moduli.

If all singular points of $\tilde{f}(s)$ are simple poles, the residues in (1.4.1.9) can be computed using formula (1.4.1.1) by representing $\tilde{f}(s)$ as the ratio of two functions with the roots of the denominator being simple and coinciding with the singular points of $\tilde{f}(s)$.

▶ **Example 1.18.** Consider the linear delay ODE [144, p. 79]:

$$u'_t(t) = au(t) + bu(t-\tau) \tag{1.4.1.10}$$

subjected to the initial condition

$$u(t) = \varphi(t), \quad -\tau \le t \le 0. \tag{1.4.1.11}$$

Applying the Laplace transform (1.4.1.3) to (1.4.1.10) gives

$$\int_0^\infty u'_t(t) e^{-st}\, dt = \int_0^\infty [au(t) + bu(t-\tau)] e^{-st}\, dt.$$

Using rule 7 from Table 1.4 for the left-hand side of the equation, making the change of variable $t - \tau = t_1$ in the second term of the right-hand side, and breaking up the resulting integral $\int_{-\tau}^{\infty}$ into two integrals, $\int_{-\tau}^{0}$ and $\int_{0}^{\infty}$, we get

$$s\tilde{u}(s) - \varphi(0) = a\tilde{u}(s) + b\tilde{u}(s) e^{-s\tau} + b\int_{-\tau}^{0} \varphi(t) e^{-s(t+\tau)}\, dt.$$

As a result, we obtain

$$\tilde{u}(s) = \frac{\varphi(0) + b\int_{-\tau}^{0} \varphi(t) e^{-s(t+\tau)}\, dt}{s - a - be^{-s\tau}}. \tag{1.4.1.12}$$

Suppose that the inequality $\ln(-b\tau) - a\tau + 1 \ne 0$ holds. Then the quasi-polynomial

$$Q(s) = s - a - be^{-s\tau}, \tag{1.4.1.13}$$

which is the denominator of the fraction (1.4.1.12), only has simple zeros, s_k. Using the inversion formula (1.4.1.9), we represent the desired solution as the series

$$u(t) = \sum_{k=0}^{\infty} \frac{e^{s_k t}}{1 + \tau b e^{-s_k \tau}} \left[\varphi(0) + b \int_{-\tau}^{0} \varphi(t) e^{-s_k(t+\tau)} dt \right].$$

The roots of the quasi-polynomial (1.4.1.13) can be expressed in terms of the Lambert W function as (1.1.3.5) by replacing λ with s_k. The symbol W on the right-hand side of formula (1.1.3.5) is understood as the set of all real and complex branches of the Lambert W function.

The article [21] provides a description of how solutions to problem (1.4.1.10)–(1.4.1.11) and other linear systems of delay ODEs can be constructed using the Lambert W function. ◀

Remark 1.18. *The Laplace transform can be used to construct asymptotic solutions to linear ODEs with proportional delay and to study solution stability of such equations* [523].

The Mellin transform. Let the function $f(t)$ be defined on the real semiaxis $t > 0$ and satisfy the conditions [402, 403]:

$$\int_0^1 |f(t)| t^{a_1 - 1} dt < \infty, \quad \int_1^{\infty} |f(t)| t^{a_2 - 1} dt < \infty,$$

where a_1 and a_2 are some real numbers such that $a_1 < a_2$.

The *Mellin transform* of $f(t)$ is defined as follows:

$$\hat{f}(s) = \int_0^{\infty} f(t) t^{s-1} dt, \qquad (1.4.1.14)$$

where $s = a + ib$ is a complex variable ($a_1 < a < a_2$).

Formula (1.4.1.14) will be written for short as

$$\hat{f}(s) = \mathfrak{M}\{f(t)\} \quad \text{or} \quad \hat{f}(s) = \mathfrak{M}\{f(t), s\}.$$

The inverse transform $f(t)$ of the function $\hat{f}(s)$ can be obtained using the inverse Mellin transform

$$f(t) = \frac{1}{2\pi i} \int_{a-i\infty}^{a+i\infty} \hat{f}(s) t^{-s} ds \quad (a_1 < a < a_2). \qquad (1.4.1.15)$$

The integration path goes parallel to the imaginary axis of the complex plane s, and the integral is understood in the sense of the Cauchy principal value.

Formula (1.4.1.15) holds for continuous functions. If the function $f(t)$ has a jump discontinuity at a point $t = t_0 > 0$, then the right-hand side of (1.4.1.15) gives $\frac{1}{2}[f(t_0 - 0) + f(t_0 + 0)]$ at this point (if $t_0 = 0$, the first term in square brackets must be omitted) [130].

The inversion formula of the Mellin transform (1.4.1.15) will be written for short as

$$f(t) = \mathfrak{M}^{-1}\{\hat{f}(s)\} \quad \text{or} \quad f(t) = \mathfrak{M}^{-1}\{\hat{f}(s), t\}.$$

Table 1.6 lists basic formulas giving the correspondence between some functions and their Mellin transforms. Table 1.7 lists the Mellin transforms of some functions. There are detailed tables of the direct and inverse Mellin transforms (see [35, 130, 145, 403]) which are handy in solving linear differential equations and linear differential equations with proportional argument.

Table 1.6. Basic properties of the Mellin transform.

No.	Function	Mellin transform	Operation
1	$af_1(t) + bf_2(t)$	$a\hat{f}_1(s) + b\hat{f}_2(s)$	Linear superposition
2	$f(at)$, $a > 0$	$a^{-s}\hat{f}(s)$	Scaling
3	$t^a f(t)$	$\hat{f}(s+a)$	Translation of argument
4	$f(t^2)$	$\frac{1}{2}\hat{f}(\frac{1}{2}s)$	Argument squaring
5	$f(1/t)$	$\hat{f}(-s)$	Argument inversion
6	$t^\lambda f(at^\beta)$, $a > 0, \beta \neq 0$	$\frac{1}{\beta} a^{-\frac{s+\lambda}{\beta}} \hat{f}\left(\frac{s+\lambda}{\beta}\right)$	Power-law transformation
7	$f'_t(t)$	$-(s-1)\hat{f}(s-1)$	Differentiation
8	$tf'_t(t)$	$-s\hat{f}(s)$	Differentiation
9	$f_t^{(n)}(t)$	$(-1)^n \dfrac{\Gamma(s)}{\Gamma(s-n)} \hat{f}(s-n)$	Differentiation of nth order
10	$\left(t\dfrac{d}{dt}\right)^n f(t)$	$(-1)^n s^n \hat{f}(s)$	Differentiation of nth order
11	$t^\alpha \int_0^\infty \xi^\beta f_1(t\xi) f_2(\xi)\, d\xi$	$\hat{f}_1(s+\alpha)\hat{f}_2(1-s-\alpha+\beta)$	Integration
12	$t^\alpha \int_0^\infty \xi^\beta f_1\left(\dfrac{t}{\xi}\right) f_2(\xi)\, d\xi$	$\hat{f}_1(s+\alpha)\hat{f}_2(s+\alpha+\beta+1)$	Integration

The Mellin transform is related to the Laplace transform by

$$\mathfrak{M}\{f(t), s\} = \mathfrak{L}\{f(e^t), -s\} + \mathfrak{L}\{f(e^{-t}), s\}. \tag{1.4.1.16}$$

Formula (1.4.1.16) allows one to take advantage of the more common tables of the direct and inverse Laplace transforms.

▶ **Example 1.19.** Following [525, 526], consider the ODE with proportional argument

$$u'_t(t) + au(t) = bu(pt), \quad p > 1, \quad a > 0, \tag{1.4.1.17}$$

subjected to homogeneous boundary conditions on a semi-infinite interval,

$$\lim_{t \to 0^+} u(t) = 0, \quad \lim_{t \to \infty} u(t) = 0, \tag{1.4.1.18}$$

Table 1.7. Mellin transforms of some functions.

No.	Function, $f(t)$	Mellin transform, $\tilde{f}(s)$	Remarks
1	$H(t-a)$	$-s^{-1}a^s$	$a>0$, $s<0$
2	$H(a-t)$	$s^{-1}a^s$	$a>0$, $s>0$
3	$t^n H(t-a)$	$-(n+s)^{-1}a^{n+s}$	$a>0$, $\operatorname{Re}(n+s)<0$
4	$t^n H(a-t)$	$(n+s)^{-1}a^{n+s}$	$a>0$, $\operatorname{Re}(n+s)>0$
5	e^{-ct^b}	$b^{-1}c^{-s/b}\Gamma(s/b)$	$b>0$, $\operatorname{Re}c>0$, $\operatorname{Re}s>0$
6	$e^{-ct^{-b}}$	$b^{-1}c^{s/b}\Gamma(-s/b)$	$b>0$, $\operatorname{Re}c>0$, $\operatorname{Re}s<0$
7	$\ln(t)H(a-t)$	$s^{-1}a^{-s}(\ln a - s^{-1})$	$a>0$, $\operatorname{Re}s>0$
8	$\ln(1+at)$	$\dfrac{\pi}{sa^s\sin(\pi s)}$	$\lvert\arg a\rvert<\pi$, $-1<\operatorname{Re}s<0$
9	$\sin(at)$	$a^{-s}\Gamma(s)\sin(\tfrac{1}{2}\pi s)$	$a>0$, $-1<\operatorname{Re}s<1$
10	$\cos(at)$	$a^{-s}\Gamma(s)\cos(\tfrac{1}{2}\pi s)$	$a>0$, $0<\operatorname{Re}s<1$
11	$\operatorname{erfc}(t)$	$\pi^{-1/2}s^{-1}\Gamma[\tfrac{1}{2}(s+1)]$	$\operatorname{Re}s>0$
12	$J_0(at)$	$\dfrac{2^{s-1}\Gamma(s/2)}{a^s\Gamma(1-s/2)}$	$a>0$, $0<\operatorname{Re}s<3/2$, $J_0(t)$ is the Bessel function

Notation: $H(t) = \{1 \text{ if } t \geq 0,\ 0 \text{ if } t < 0\}$ is the Heaviside unit-step function.

and the normalization condition

$$\int_0^\infty u(t)\,dt = 1. \tag{1.4.1.19}$$

Problems where additional conditions like (1.4.1.18) and (1.4.1.19) define the unknown function $u(t)$ as a function of probability arise, for example, in biological models of cell growth [206, 207].

Applying the Mellin transform to (1.4.1.17) and taking into account conditions (1.4.1.18) (the solution is assumed to decay rapidly for large t), we obtain the difference equation

$$-(s-1)\hat{u}(s-1) + a\hat{u}(s) = bp^{-s}\hat{u}(s). \tag{1.4.1.20}$$

In view of the Mellin transform definition (1.4.1.14), condition (1.4.1.19) becomes

$$\hat{u}(1) = 1. \tag{1.4.1.21}$$

We will now consider the auxiliary linear homogeneous ODE

$$v'(t) + av(t) = 0, \tag{1.4.1.22}$$

1.4. Exact and Approximate Analytical Solution Methods for Delay ODEs

obtained by dropping the last term in (1.4.1.17). Applying the Mellin transform to (1.4.1.22) gives

$$-(s-1)F(s-1) + aF(s) = 0, \qquad (1.4.1.23)$$

where $F(s) = \mathfrak{M}\{v(t), s\}$. Equation (1.4.1.22) admits an exact solution $v = e^{-at}$. Considering the above and using transformation 5 (with $b = 1$ and $c = a$) from Table 1.7, we get

$$F(s) = a^{-s}\Gamma(s), \qquad (1.4.1.24)$$

where $\Gamma(s)$ is the gamma function.

A solution to the difference problem (1.4.1.20)–(1.4.1.21) will be sought in the form

$$\hat{u}(s) = F(s)Q(s). \qquad (1.4.1.25)$$

On substituting (1.4.1.25) into (1.4.1.20) and taking into account (1.4.1.23), we arrive at the equation

$$Q(s-1) = (1 - ba^{-1}p^{-s})Q(s). \qquad (1.4.1.26)$$

Its solution can be represented as an infinite product (this can be proved by a direct verification):

$$Q(s) = C \prod_{k=0}^{\infty} (1 - ba^{-1}p^{-k-s-1}), \qquad (1.4.1.27)$$

where C is some constant. Hence, the solution of equation (1.4.1.20) can be written as

$$\hat{u}(s) = Ca^{-s}\Gamma(s) \prod_{k=0}^{\infty} (1 - ba^{-1}p^{-k-s-1}). \qquad (1.4.1.28)$$

The constant C is found from the normalization condition (1.4.1.21):

$$C = a \prod_{k=0}^{\infty} (1 - ba^{-1}p^{-k-2})^{-1}. \qquad (1.4.1.29)$$

To inverse the function (1.4.1.28), one has to convert the infinite product to an infinite series. To this end, we take advantage of Euler's formula [16]:

$$\prod_{k=0}^{\infty} (1 + rq^k) = 1 + \sum_{k=1}^{\infty} \frac{r^k q^{k(k-1)/2}}{\prod_{j=1}^{k} (1 - q^j)},$$

where $|q| < 1$ and r is any complex number. Taking into account that $q = p^{-1}$ and $r = -ba^{-1}p^{-s-1}$, we rewrite solution (1.4.1.28) as

$$\hat{u}(s) = Ca^{-s}\Gamma(s)\left(1 + \sum_{k=1}^{\infty} \beta_k p^{-ks}\right), \qquad (1.4.1.30)$$

where the coefficients β_k are defined by

$$\beta_k = \frac{(-1)^k b^k}{a^k p^{k(k+1)/2} \prod_{j=1}^{k}(1 - p^{-j})}, \quad k = 1, 2, \ldots \qquad (1.4.1.31)$$

The series (1.4.1.30) is uniformly convergent in the right half-plane. Therefore, the function $\hat{u}(s)$ can be transformed back, term by term, using property 5 from Table 1.7 with $b = 1$ and property 2 from Table 1.6. As a result, we obtain a solution of the original problem (1.4.1.17)–(1.4.1.19) in the form

$$u(t) = C \sum_{k=0}^{\infty} \beta_k e^{-ap^k t},$$

where $\beta_0 = 1$; the constants C and β_k are defined by formulas (1.4.1.29) and (1.4.1.31). ◀

1.4.2. Representation of Solutions as Power Series in the Independent Variable

Representation of Solutions as Power Series. Approximate solutions to ODEs with proportional delay can be sought as polynomials (or infinite power series) in the independent variable [46, 199, 471, 474]:

$$u(t) = \sum_{n=0}^{k} \gamma_n t^n, \qquad (1.4.2.1)$$

where k is a given positive integer and γ_n are constants that are sequentially determined during the analysis.

▶ Example 1.20. Consider the Cauchy problem for the linear ODE with constant coefficients and two proportional delays

$$u'_t = au + bw_1 + cw_2, \quad w_1 = u(pt), \quad w_2 = u(qt); \quad u(0) = 1. \qquad (1.4.2.2)$$

We look for its solution as an infinite series

$$u = 1 + \sum_{n=1}^{\infty} \gamma_n t^n. \qquad (1.4.2.3)$$

Substituting (1.4.2.3) into (1.4.2.2) and taking into account the expression of the derivative

$$u'_t = \sum_{n=1}^{\infty} n\gamma_n t^{n-1} = \gamma_1 + \sum_{n=1}^{\infty} (n+1)\gamma_{n+1} t^n,$$

we obtain

$$\gamma_1 + \sum_{n=1}^{\infty} (n+1)\gamma_{n+1} t^n = a + b + c + \sum_{n=1}^{\infty} \gamma_n (a + bp^n + cq^n) t^n.$$

By matching up the coefficients of like powers t^n, we arrive at the following linear algebraic system of equations for γ_n:

$$\gamma_1 = a + b + c, \quad (n+1)\gamma_{n+1} = \gamma_n(a + bp^n + cq^n), \quad n = 1, 2, \dots$$

1.4. Exact and Approximate Analytical Solution Methods for Delay ODEs

We solve this system by starting from $n = 1$ and increasing n sequentially. As a result, we obtain the desired power series solution of the Cauchy problem (1.4.2.2):

$$u = 1 + \sum_{n=1}^{\infty} \gamma_n t^n, \quad \gamma_n = \frac{1}{n!} \prod_{k=0}^{n-1}(a + bp^k + cq^k). \tag{1.4.2.4}$$

For $0 < p < 1$ and $0 < q < 1$, the series (1.4.2.4) has an infinite radius of convergence, in which case we get $a^n < \gamma_n < (a+b+c)^n$ and $e^{at} < u < e^{(a+b+c)t}$, provided that $a \geq 0$, $b \geq 0$, and $c \geq 0$.

Notably, solution (1.4.2.4) was obtained previously in [375] using a different approach. Also, in the special case $c = 0$, the solution becomes the one presented in [166]. ◀

▶ **Example 1.21.** Consider the Cauchy problem for the nonlinear equation with proportional delay

$$u'_t(t) = a - bu^2(pt); \quad u(0) = 0, \tag{1.4.2.5}$$

where $0 < p < 1$.

The right-hand side of equation (1.4.2.5) remains unchanged under the substitution of $-u$ for u. Consequently, u'_t is an even function, and u is the sum of an odd function and a constant. In view of the above and the zero initial condition, we will look for a solution to problem (1.4.2.5) in the form

$$u = \alpha t + \beta t^3 + \gamma t^5 + \delta t^7 + \cdots. \tag{1.4.2.6}$$

Substituting (1.4.2.6) into (1.4.2.5) and collecting the coefficients of like powers of t, we get

$$\alpha - a + (3\beta + bp^2\alpha^2)t^2 + (5\gamma + 2bp^4\alpha\beta)t^4 + [7\delta + bp^6(\beta^2 + 2\alpha\gamma)]t^6 + \cdots = 0.$$

Equating the coefficients of the different powers t^{2n} with zero, we obtain the following algebraic equations for the coefficients of the series (1.4.2.6):

$$\begin{aligned}
\alpha - a &= 0 & \text{(for } t^0\text{)}, \\
3\beta + bp^2\alpha^2 &= 0 & \text{(for } t^2\text{)}, \\
5\gamma + 2bp^4\alpha\beta &= 0 & \text{(for } t^4\text{)}, \\
7\delta + bp^6(\beta^2 + 2\alpha\gamma) &= 0 & \text{(for } t^6\text{)}.
\end{aligned}$$

On solving these equations, we ultimately arrive at an approximate solution to problem (1.4.2.5) in the form of a polynomial:

$$u = at - \tfrac{1}{3}a^2 bp^2 t^3 + \tfrac{2}{15}a^3 b^2 p^6 t^5 - \tfrac{1}{21}a^4 b^3 p^{10}\left(\tfrac{1}{3} + \tfrac{4}{5}p^2\right)t^7. \tag{1.4.2.7}$$

For small t, its error is of the order of $O(t^9)$.

Figure 1.13 displays approximate solutions computed using formula (1.4.2.7) with $p = 0$, 0.25, 0.5, 0.75, and 1.0. It is apparent that for $ab > 0$ and $0 < p \leq 1$, the curves are non-monotonic. They first increase from zero and reach a positive

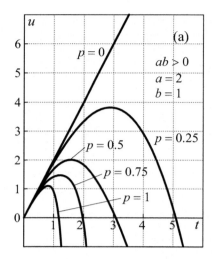 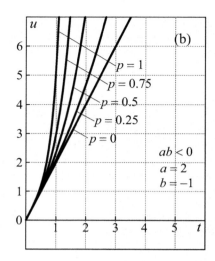

Figure 1.13. Approximate solutions of problem (1.4.2.5) computed using formula (1.4.2.7) with $p = 0$, 0.25, 0.5, 0.75, and 1.0 for two pairs of the determining parameters: (a) $a = 2$ and $b = 1$, (b) $a = 2$ and $b = -1$.

extremum and then decrease, cross the axis $u = 0$ (the smaller the p, the further away the intersection point), and become negative. For $ab < 0$, all curves increase monotonically, and they grow faster as p increases from zero to one.

Formula (1.4.2.7) gives an exact result at $p = 0$. Therefore, one can expect that its error should decrease as the parameter p decreases from one to zero. This is corroborated by the subsequent comparison of formula (1.4.2.7) with exact solutions to problem (1.4.2.5).

For arbitrary nonzero a and b, the nonlinear problem (1.4.2.5) admits, besides the case $p = 0$, two other exact solutions, with $p = \frac{1}{2}$ and $p = 1$. These are given below.

1°. Solution to problem (1.4.2.5) where $p = \frac{1}{2}$:

$$u(t) = \sqrt{\frac{2a}{b}} \sin\left(b\sqrt{\frac{a}{2b}}\,t\right) \qquad \text{for} \quad ab > 0,$$
$$u(t) = -\sqrt{-\frac{2a}{b}} \sinh\left(b\sqrt{-\frac{a}{2b}}\,t\right) \qquad \text{for} \quad ab < 0. \tag{1.4.2.8}$$

2°. Solution to problem (1.4.2.5) with $p = 1$:

$$u(t) = \sqrt{\frac{a}{b}} \tanh\left(b\sqrt{\frac{a}{b}}\,t\right) \qquad \text{for} \quad ab > 0,$$
$$u(t) = -\sqrt{-\frac{a}{b}} \tan\left(b\sqrt{-\frac{a}{b}}\,t\right) \qquad \text{for} \quad ab < 0. \tag{1.4.2.9}$$

Figure 1.14 a shows in solid lines the exact solutions to problem (1.4.2.5) with $a = 2$ and $b = 1$ computed using formulas (1.4.2.8) and (1.4.2.9) for $p = \frac{1}{2}$ and $p = 1$.

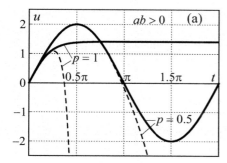 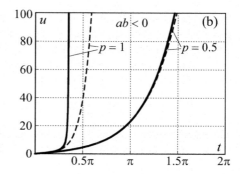

Figure 1.14. Exact solutions to problem (1.4.2.5) computed by formulas (1.4.2.8) and (1.4.2.9) with $p = \frac{1}{2}$ and $p = 1$ for two pairs of the determining parameters: (a) $a = 2$ and $b = 1$, (b) $a = 2$ and $b = -1$.

The dashed lines indicate the approximate solutions represented by the polynomial (1.4.2.7) with the same values of the determining parameters. Figure 1.14 b shows in solid and dashed lines the respective exact and approximate solutions to problem (1.4.2.5) with $a = 2$ and $b = -1$ for $p = \frac{1}{2}$ and $p = 1$. It is apparent that the approximate formula (1.4.2.7) works well at $p = \frac{1}{2}$ on a reasonably large time interval ($0 \leq t \leq \pi$). At $p = 1$, the error of formula (1.4.2.7) is much larger, and it can only be used on the interval $0 \leq t \leq 0.8$. ◀

Remark 1.19. *In problems for proportional delay ODEs where the equation is valid in a region $t \geq t_0$, approximate solutions can be sought in the form $u(t) = \sum_{n=0}^{k} \gamma_n (t - t_0)^n$.*

▶ **Example 1.22.** Consider the following mixed boundary value problem for a second-order linear ODE with proportional delay:
$$u''_{xx}(x) + (a + bx^2)u(px) + c = 0; \tag{1.4.2.10}$$
$$u'_x(0) = 0, \quad u(1) = 0, \tag{1.4.2.11}$$
where $0 < p < 1$.

We look for an approximate solution to problem (1.4.2.10), (1.4.2.11) in the form of a power series:
$$u = \lambda + \alpha x + \beta x^2 + \gamma x^3 + \delta x^4 + \cdots, \tag{1.4.2.12}$$
where the constant $\lambda, \alpha, \beta, \gamma, \delta, \ldots$ are to be determined. On substituting (1.4.2.12) into the first boundary condition (1.4.2.11), we get
$$\alpha = 0. \tag{1.4.2.13}$$
Taking into account (1.4.2.13), we substitute (1.4.2.12) into equation (1.4.2.10) and collect the coefficients of like powers of x to obtain
$$A + Bx + Cx^2 + \cdots = 0, \tag{1.4.2.14}$$
$$A = 2\beta + a\lambda + c, \quad B = 6\gamma, \quad C = 12\delta + b\lambda + ap^2\beta.$$

Equating the coefficients of the different powers of x in (1.4.2.14) with zero, we arrive at a system of linear algebraic equations: $A = 0$, $B = 0$, $C = 0$, ... On solving this system, we express β, γ, δ, ... through λ and substitute (1.4.2.13) and the resulting expressions into (1.4.2.12). Retaining the leading terms up to x^4, we obtain the approximating polynomial

$$u = \lambda - \tfrac{1}{2}(a\lambda + c)x^2 + \tfrac{1}{24}(a^2p^2\lambda + acp^2 - 2b\lambda)x^4. \tag{1.4.2.15}$$

The approximate value of λ results from substituting the polynomial (1.4.2.15) into the second boundary condition (1.4.2.11). This gives

$$\lambda = \frac{c(12 - ap^2)}{a^2p^2 - 2b - 12a + 24}. \tag{1.4.2.16}$$

On substituting (1.4.2.16) into (1.4.2.15), we eventually arrive at an approximate solution to problem (1.4.2.10), (1.4.2.11) in the form

$$u = \frac{c[12 - ap^2 + (b - 12)x^2 + (ap^2 - b)x^4]}{a^2p^2 - 2b - 12a + 24}. \tag{1.4.2.17}$$

For any a, b, and c such that $6a + b \neq 12$, the approximate formula (1.4.2.17) provides the exact result at $p = 0$. Therefore, one should expect that its error decreases as the parameter p decreases from one to zero.

Figure 1.15 shows in solid lines the approximate solutions computed using formula (1.4.2.17) with $a = b = c = 1$ for $p = 0$, 0.5, and 1. A numerical solution obtained by Heun's method for $p = 0.5$ (see Subsections 5.1.4 and 5.1.5) is shown by open circles. The dashed line indicates a numerical solution obtained by the shooting method for $p = 1$ (see Subsection 5.1.6).

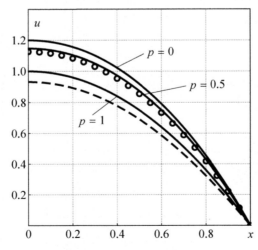

Figure 1.15. Solutions to problem (1.4.2.10), (1.4.2.11) with $a = b = c = 1$ obtained using the approximate formula (1.4.2.17) and by numerical integration for $p = 0$, 0.5, and 1.

It is apparent from Figure 1.15 that the curves representing the approximate and numerical solutions at $p = 0.5$ are very close to each other; the maximum discrepancy between them is 0.02. At $p = 1$, the maximum discrepancy between approximate and numerical solutions is much larger, 0.068. ◂

Padé approximant. The partial sum (1.4.2.1) well approximates the solution for sufficiently small t. However, it usually works poorly for intermediate and large t, since a power series can be slowly converging or can have a small radius of convergence. This also results from the facts that the approximate solution (1.4.2.1) grows indefinitely as $t \to \infty$, while the exact solution is often bounded.

Therefore, it is sometimes reasonable to employ *Padé approximants* rather than power expansions (1.4.2.1). A Padé approximant of order $[N/M]$, $P_M^N(t)$, is the ratio of two polynomials of degree N and M [30]:

$$P_M^N(t) = \frac{A_0 + A_1 t + \cdots + A_N t^N}{1 + B_1 t + \cdots + B_M t^M}, \quad \text{where} \quad N + M = k. \tag{1.4.2.18}$$

The coefficients $A_1, \ldots, A_N$ and $B_1, \ldots, B_M$ are usually selected to ensure that the first $k + 1$ leading terms of the Taylor expansion of (1.4.2.18) coincide with the respective terms of the expansion (1.4.2.1). In other words, the expansions (1.4.2.1) and (1.4.2.18) must be asymptotically equivalent as $t \to 0$.

In practice, one usually chooses a diagonal sequence, with $N = M$. It often turns out that formula (1.4.2.18) approximates the exact solution quite well over the entire range of t (for sufficiently large N).

For examples of using Padé approximants to construct approximate solutions to problems described by ODEs without or with delay, see [30, 423, 613] and [13, 46], respectively.

1.4.3. Method of Regular Expansion in a Small Parameter

The method of regular expansion in a small parameter [255, 421, 423, 613] is used to solve nonlinear ODEs or PDEs. It is also suitable for solving differential equations with proportional delay.

For simplicity, we restrict ourselves to the description of the method of regular expansion in a small parameter ε for a first-order nonlinear ODE with proportional delay:

$$u'_t = f(t, u, w, \varepsilon), \quad w = u(pt). \tag{1.4.3.1}$$

Let the function f be representable as a series in powers of ε:

$$f(t, u, w, \varepsilon) = \sum_{n=0}^{\infty} \varepsilon^n f_n(t, u, w). \tag{1.4.3.2}$$

We will seek a solution of the Cauchy problem for equation (1.4.3.1) subjected to the initial condition

$$u = a \quad \text{at} \quad t = 0 \tag{1.4.3.3}$$

in the form of a regular expansion in powers of the small parameter ($\varepsilon \to 0$):

$$u = \sum_{n=0}^{\infty} \varepsilon^n u_n(t). \qquad (1.4.3.4)$$

Substituting (1.4.3.2) and (1.4.3.4) into equation (1.4.3.1), expanding the functions f_n into series in ε, and equating the coefficients of like powers of ε, we arrive at the following system of equations for $u_n(t)$:

$$u_0' = f_0(t, u_0, w_0), \quad w_0 = u_0(pt); \qquad (1.4.3.5)$$
$$u_1' = f_1(t, u_0, w_0) + g_1(t, u_0, w_0)u_1 + g_2(t, u_0, w_0)w_1, \quad w_1 = u_1(pt), \quad (1.4.3.6)$$
$$g_1(t, u, w) = \frac{\partial f_0}{\partial u}, \quad g_2(t, u, w) = \frac{\partial f_0}{\partial w}.$$

Here we have only written out the first two equations. The primes denote differentiation with respect to t. The initial conditions for u_n can be derived from condition (1.4.3.3) taking into account the expansion (1.4.3.4):

$$u_0(0) = a, \quad u_1(0) = u_2(0) = \cdots = 0.$$

Whether the application of this method is successful depends mainly on the possibility of constructing a solution to equation (1.4.3.5) for the leading term of the expansion, u_0. Importantly, the other terms of the expansion, u_n with $n \geq 1$, are described by linear equations with homogeneous initial conditions.

▶ **Example 1.23.** Consider the Cauchy problem for a nonlinear ODE with proportional delay

$$u_t' + bu = \varepsilon w^2, \quad w = u(pt); \quad u(0) = a, \qquad (1.4.3.7)$$

where ε is a small parameter.

We will look for a solution in the form (1.4.3.4) while retaining three leading terms of the expansion:

$$u = u_0 + \varepsilon u_1 + \varepsilon^2 u_2 + o(\varepsilon^2), \quad u_n = u_n(t). \qquad (1.4.3.8)$$

Substituting (1.4.3.8) into equation (1.4.3.7) and collecting the terms with like powers of ε, we obtain

$$u_0' + bu_0 + \varepsilon(u_1' + bu_1 - w_0^2) + \varepsilon^2(u_2' + bu_2 - 2w_0 w_1) + o(\varepsilon^2) = 0. \qquad (1.4.3.9)$$

Likewise, substituting (1.4.3.8) into the initial condition (1.4.3.7) yields

$$u_0(0) - a + \varepsilon u_1(0) + \varepsilon^2 u_2(0) + o(\varepsilon^2) = 0. \qquad (1.4.3.10)$$

Now equating the coefficients of like powers of ε in (1.4.3.9) and (1.4.3.10) with zero, we arrive at the following sequence of simple linear problems for ODEs without delays:

$$u_0' + bu_0 = 0, \qquad u_0(0) = a;$$
$$u_1' + bu_1 = w_0^2, \qquad u_1(0) = 0;$$
$$u_2' + bu_2 = 2w_0 w_1, \qquad u_2(0) = 0.$$

Integrating these equations one by one beginning with the first and assuming that $p \neq 1/2$, we obtain

$$u_0 = ae^{-bt},$$

$$u_1 = \frac{a^2}{b(2p-1)}\left(e^{-bt} - e^{-2bpt}\right),$$

$$u_2 = \frac{2a^3}{b^2(p+1)(2p-1)^2}\left[pe^{-bt} - (p+1)e^{-2pbt} + e^{-p(2p+1)bt}\right].$$

Substituting these functions into (1.4.3.8), we find the desired solution in the form

$$u = e^{-bt} + \frac{\varepsilon a^2}{b(2p-1)}\left(e^{-bt} - e^{-2bpt}\right)$$
$$+ \frac{2\varepsilon^2 a^3}{b^2(p+1)(2p-1)^2}\left[pe^{-bt} - (p+1)e^{-2pbt} + e^{-p(2p+1)bt}\right] + o(\varepsilon^2).$$

In a similar manner, we obtain an asymptotic solution to problem (1.4.3.7) with $p = 1/2$:

$$u = e^{-bt} + \varepsilon a^2 t e^{-bt} + \tfrac{1}{2}\varepsilon^2 a^3 t^2 e^{-bt} + o(\varepsilon^2). \qquad \blacktriangleleft$$

1.4.4. Method of Matched Asymptotic Expansions. Singular Perturbation Problems with a Boundary Layer

Below we will illustrate the characteristic features of using the method of matched asymptotic expansions. As an example, we will consider a two-point boundary value problem for a quasilinear second-order ODE with proportional argument where the highest derivative is multiplied by a small parameter:

$$\varepsilon u''_{xx} + f(x)u'_x + g(x, u, w) = 0, \quad w = u(px) \quad (0 < x < 1); \qquad (1.4.4.1)$$

$$u(0) = a, \quad u(1) = b, \qquad (1.4.4.2)$$

where $0 < \varepsilon \ll 1$ and $0 < p < 1$.

In general, problem (1.4.4.1)–(1.4.4.2) cannot be solved in a closed analytical form, even though the equation did not involve the term with proportional delay w. However, an approximate solution to the problem for small ε can be obtained using the method of matched asymptotic expansions [289, 354, 355, 407, 423]. We will show this below.

Importantly, if $\varepsilon = 0$, the second-order equation (1.4.4.1) degenerates into a first-order ODE, which has no solutions that can meet both boundary conditions (1.4.4.2) simultaneously. Problems that involve a small parameter and degenerate at $\varepsilon = 0$ are called *singular perturbation* problems.

In what follows, we assume that $f(x) > 0$. In this case, a boundary layer is formed in a small neighborhood of the point $x = 0$ where the solution has a large gradient if ε is small. In the language of the method of matched asymptotic expansions, the boundary layer region $\Omega_i = \{0 \leq x \leq O(\varepsilon)\}$ is customarily called the *inner region*, while the remaining (larger) portion of the interval $0 \leq x \leq 1$ is called the *outer region* and denoted $\Omega_o = \{O(\varepsilon) < x \leq 1\}$.

In the inner region, one uses an extended, boundary layer variable

$$z = x/\varepsilon \tag{1.4.4.3}$$

and looks for an asymptotic solution in the form

$$u = u_i(z) + O(\varepsilon), \quad z = O(1). \tag{1.4.4.4}$$

Substituting (1.4.4.4) into (1.4.4.1), considering (1.4.4.3), and taking into account that $f(x) = f(\varepsilon z) \simeq f(0)$ as $\varepsilon \to 0$ and $z = O(1)$, we obtain

$$\varepsilon^{-1}[(u_i)''_{zz} + f(0)(u_i)'_z] + O(1) = 0.$$

Equating the functional factor of ε^{-1} with zero, we arrive at an equation for the leading term of the asymptotic expansion in the boundary layer region:

$$(u_i)''_{zz} + f(0)(u_i)'_z = 0.$$

This linear ODE is easy to integrate. Then, by satisfying the first boundary condition (1.4.4.2), we get

$$u_i = c(1 - e^{-f_0 z}) + a, \quad z = x/\varepsilon, \quad 0 \leq z \leq O(1), \tag{1.4.4.5}$$

where $f_0 = f(0)$ and c is a constant determined in the subsequent analysis.

The solution in the outer region $\Omega_o = \{O(\varepsilon) < x \leq 1\}$ is sought in the form $u = u_o(x) + O(\varepsilon)$. The leading term of the asymptotic solution to problem (1.4.4.1)–(1.4.4.2) in Ω_o is determined from a truncated equation (the small terms with the second derivative is neglected) and the second boundary condition:

$$f(x)(u_o)'_x + g(x, u_o, w_o) = 0; \quad u_o(1) = b. \tag{1.4.4.6}$$

Importantly, the first-order ODE with proportional argument (1.4.4.6) is significantly simpler than the original second-order equation (1.4.4.1). Let the function

$$u_o = u_o(x) \tag{1.4.4.7}$$

be a solution to problem (1.4.4.6).

The inner and outer solutions (1.4.4.5) and (1.4.4.7) must agree with each other, or satisfy the *matching condition*

$$u_i(z \to \infty) = u_o(x \to 0). \tag{1.4.4.8}$$

It allows us to determine the constant c appearing in (1.4.4.5):

$$c = u_o(0) - a. \tag{1.4.4.9}$$

1.4. Exact and Approximate Analytical Solution Methods for Delay ODEs 73

A *composite asymptotic solution* to problem (1.4.4.1)–(1.4.4.2), uniformly valid in the entire region $0 \leq x \leq 1$, is expressed as

$$u = [a - u_\text{o}(0)]e^{-f_0 z} + u_\text{o}(x) = [a - u_\text{o}(0)]e^{-(f_0/\varepsilon)x} + u_\text{o}(x), \qquad (1.4.4.10)$$

where $f_0 = f(0)$.

Differentiating formula (1.4.4.5) twice with respect to x, we find the derivatives in the inner region:

$$u'_x = \frac{cf_0}{\varepsilon}e^{-(f_0/\varepsilon)x}, \quad u''_{xx} = -\frac{cf_0^2}{\varepsilon^2}e^{-(f_0/\varepsilon)x}. \qquad (1.4.4.11)$$

One can see that for $\varepsilon \to 0$, both derivatives in the boundary layer region $0 \leq x \leq O(\varepsilon)$ are large.

▶ **Example 1.24.** Consider the following nonlinear problem for a second-order ODE with a proportional argument and a small parameter ε multiplying the highest derivative:

$$\varepsilon u''_{xx} + u'_x + ku - sw^2 = 0, \quad w = u(\tfrac{1}{2}x) \quad (0 < x < 1); \qquad (1.4.4.12)$$
$$u(0) = a, \quad u(1) = b. \qquad (1.4.4.13)$$

Problem (1.4.4.12)–(1.4.4.13) is a special case of problem (1.4.4.1)–(1.4.4.2) with $f(x) = 1$, $g(x, u, w) = ku - sw^2$, and $p = 1/2$.

The inner solution is given by formula (1.4.4.5) with $f_0 = 1$:

$$u_\text{i} = c(1 - e^{-z}) + a, \quad z = x/\varepsilon, \quad 0 \leq z \leq O(1). \qquad (1.4.4.14)$$

The constant c will be determined below.

The outer solution is described by the following equation and boundary condition:

$$(u_\text{o})'_x + ku_\text{o} - sw_\text{o}^2 = 0, \quad w_\text{o} = u_\text{o}(\tfrac{1}{2}x); \quad u_\text{o}(1) = b. \qquad (1.4.4.15)$$

The exact solution to problem (1.4.4.15) is

$$u_\text{o} = A\exp[(As - k)x], \qquad (1.4.4.16)$$

where A is a real root of the transcendental equation

$$b = A\exp(As - k). \qquad (1.4.4.17)$$

For $s \neq 0$, this root can be expressed via the Lambert W function: $A = s^{-1}W(bse^k)$.

The constant c appearing in the asymptotic solution (1.4.4.14) can be found using relation (1.4.4.9). We get

$$c = A - a. \qquad (1.4.4.18)$$

Using formula (1.4.4.10) and solutions (1.4.4.14) and (1.4.4.16), we obtain the composite asymptotic solution to problem (1.4.4.12)–(1.4.4.13):

$$u = (a - A)\exp(-\varepsilon^{-1}x) + A\exp[(As - k)x]. \qquad (1.4.4.19)$$

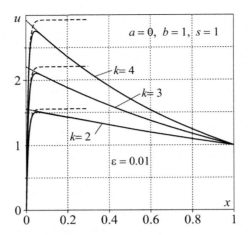

Figure 1.16. Approximate solutions to problem (1.4.4.12)–(1.4.4.13) computed using the composite formula (1.4.4.19) with $\varepsilon = 0.01$ and $a = 0$, $b = 1$, $s = 1$ for $k = 2, 3, 4$ (solid lines). The dashed lines indicate the asymptotic solutions in the boundary layer computed by formulas (1.4.4.14) and (1.4.4.18). The dotted lines indicate the outer solutions obtained with formula (1.4.4.16).

Figure 1.16 displays three approximate solutions to problem (1.4.4.12)–(1.4.4.13) computed using the composite formula (1.4.4.19) with $\varepsilon = 0.01$ for $a = 0, b = 1, s = 1$. These solutions are shown in solid lines and correspond to $k = 2, 3, 4$. The dashed lines indicate the asymptotic solutions in the boundary layer computed by formulas (1.4.4.14) and (1.4.4.18) using the same values of the determining parameters. The dotted lines indicate the outer solutions (1.4.4.16). ◀

One can see that the solutions grow rapidly in a narrow region near the left boundary and then decrease gradually and slowly. Notably, large gradients of solutions severely limit the applicability of standard numerical methods for integrating similar boundary layer problems (e.g., see the introduction in the article [407] and references therein).

1.4.5. Method of Successive Approximations and Other Iterative Methods

Method of successive approximations. Let us look at the nonlinear equation

$$u = F[u], \qquad (1.4.5.1)$$

where $F[u]$ is a nonlinear operator.

An approximate solution of equation (1.4.5.1) is sought using a recurrence relation:

$$u_0 = \varphi, \quad u_n = F[u_{n-1}], \quad n = 1, 2, \ldots. \qquad (1.4.5.2)$$

The initial function φ can be chosen from various considerations. If the unknown function depends on t alone, one usually sets $\varphi = F[u]|_{t=0}$. Under certain restrictions on the operator F, the sequence of function u_n converges to a solution of equation (1.4.5.2) as $n \to \infty$.

▶ **Example 1.25.** Consider the Cauchy problem for a first-order nonlinear ODE with proportional delay

$$u'_t = f(t, u, w), \quad w = u(pt); \quad u(0) = a, \tag{1.4.5.3}$$

where $0 < p < 1$.

Integrating the equation from 0 to t and taking into account the initial condition, we obtain the integral equation

$$u(t) = a + \int_0^t f(\xi, u(\xi), u(p\xi))\, d\xi. \tag{1.4.5.4}$$

An approximate solution of this equation, which is a special case of equation (1.4.5.1), is sought using the recurrence relation

$$u_0 = a, \quad u_n(t) = a + \int_0^t f(\xi, u_{n-1}(\xi), u_{n-1}(p\xi))\, d\xi, \quad n = 1, 2, \ldots.$$

◀

A method based on the expansion of the nonlinear operator. We outline below the iterative method proposed in [120] (see also [56, 375]). It is based on the expansion of the nonlinear operator and can be used for the approximate solution of various linear or nonlinear mathematical equations including delay ODEs.

A solution to equation (1.4.5.1) is sought in the form

$$u = \sum_{n=0}^{\infty} u_n. \tag{1.4.5.5}$$

Assuming that this series is absolutely convergent, we write the identity

$$F\left[\sum_{n=0}^{\infty} u_n\right] = F[u_0] + \sum_{n=1}^{\infty} \left\{ F\left[\sum_{j=0}^{n} u_j\right] - F\left[\sum_{j=0}^{n-1} u_j\right] \right\}. \tag{1.4.5.6}$$

On substituting (1.4.5.5) and (1.4.5.6) into equation (1.4.5.1), we get

$$\sum_{n=0}^{\infty} u_n = F[u_0] + \sum_{n=1}^{\infty} \left\{ F\left[\sum_{j=0}^{n} u_j\right] - F\left[\sum_{j=0}^{n-1} u_j\right] \right\}. \tag{1.4.5.7}$$

This relation can be satisfied by setting [120]:

$$\begin{aligned} u_0 &= \varphi, \\ u_1 &= F[u_0], \\ u_{n+1} &= F[u_0 + \cdots + u_n] - F[u_0 + \cdots + u_{n-1}], \quad n = 1, 2, \ldots \end{aligned} \tag{1.4.5.8}$$

As in the method of successive approximations, the initial function φ can be chosen from various considerations. If the unknown function only depends on t, we set $\varphi = F[u]|_{t=0}$.

Assuming that $1 \leq n \leq m$, we get finitely many recurrence relations (1.4.5.8), which allow one to find an approximate solution to the nonlinear equation (1.4.5.1). The studies [56, 120, 375] found conditions under which the infinite series (1.4.5.5), corresponding to $m = \infty$, is convergent.

▶ **Example 1.26.** Consider the Cauchy problem for a nonlinear ODE with two proportional delays

$$u'_t = f(t, u, w_1, w_2), \quad w_1 = u(pt), \quad w_2 = u(qt), \quad 0 < p < 1, \quad 0 < q < 1; \quad (1.4.5.9)$$
$$u(0) = a.$$

Integrating equation (1.4.5.9) with respect to t, from zero to t, we arrive at the integral equation

$$u(t) = a + \int_0^t f(\xi, u(\xi), u(p\xi), u(q\xi)) \, d\xi, \qquad (1.4.5.10)$$

which is a special case of equation (1.4.5.1) with

$$F[u] = a + \int_0^t f(\xi, u(\xi), u(p\xi), u(q\xi)) \, d\xi.$$

In our case, $F[u]|_{t=0} = a$. Therefore, we choose the initial function in (1.4.5.8) as $\varphi = a$. The studies [57, 375] specify sufficient conditions under which the above iterative method results in a solution to equation (1.4.5.10) as a convergent series (1.4.5.5). ◀

▶ **Example 1.27.** The article [375] shows that applying the iterative method to the Cauchy problem for the linear equation with proportional delays (1.4.2.2) leads to solution (1.4.2.4), obtained in Subsection 1.4.2 using a different approach. ◀

Adomian decomposition method. For clarity, we will outline the main ideas of the Adomian decomposition method (e.g., see [5, 7]) by studying an example of the Cauchy problem for a first-order nonlinear ODE with proportional delay (1.4.5.3). Integrating this equation from 0 to t and employing the initial condition, we arrive at the integral equation (1.4.5.4).

We look for a solution to the integral equation (1.4.5.4) as the series

$$u(t) = \sum_{n=0}^{\infty} \varepsilon^n u_n(t), \qquad (1.4.5.11)$$

where $0 \leq \varepsilon \leq 1$ is an auxiliary parameter. Substituting (1.4.5.11) into the right-hand side of equation (1.4.5.3) and then expanding into a Maclaurin series in ε, we obtain

$$f(t, u(t), u(pt)) = f\left(t, \sum_{n=0}^{\infty} \varepsilon^n u_n(t), \sum_{n=0}^{\infty} \varepsilon^n u_n(pt)\right) = \sum_{n=0}^{\infty} \varepsilon^n A_n, \qquad (1.4.5.12)$$

where A_n are functions that are called *Adomian polynomials* and defined as

$$A_n = \frac{1}{n!}\left\{\frac{\partial^n}{\partial \varepsilon^n} f\left(t, \sum_{k=0}^{\infty} \varepsilon^k u_k(t), \sum_{k=0}^{\infty} \varepsilon^k u_k(pt)\right)\right\}_{\varepsilon=0}. \qquad (1.4.5.13)$$

Substituting (1.4.5.11) and (1.4.5.12) into equation (1.4.5.4) yields

$$\sum_{n=0}^{\infty} \varepsilon^n u_n(t) = a + \int_0^t \sum_{n=0}^{\infty} \varepsilon^n A_n \, dt. \qquad (1.4.5.14)$$

Assuming that the series (1.4.5.11) and (1.4.5.12) are convergent for $0 \leq \varepsilon \leq 1$, we set $\varepsilon = 1$ in (1.4.5.14) to obtain

$$u_0(t) + u_1(t) + u_2(t) + u_3(t) + \cdots = a + \int_0^t A_0 \, dt + \int_0^t A_1 \, dt + \int_0^t A_2 \, dt + \cdots.$$

Equating the terms on the left with the respective terms on the right, we arrive at the Adomian recurrence relations

$$\begin{aligned} u_0(t) &= a; \\ u_n(t) &= \int_0^t A_{n-1} \, dt, \quad n = 1, 2, \ldots \end{aligned} \qquad (1.4.5.15)$$

In general, each polynomial A_n only depends on time t and the components u_j and w_j, where $j \leq n$, so that $A_0 = A_0(t, u_0, w_0)$, $A_1 = A_1(t, u_0, w_0, u_1, w_1)$, and so on. Therefore, the recurrence relations (1.4.5.15) allow one to determine the functions u_n successively.

Substituting the function u_n obtained with the recurrence relations (1.4.5.15) into (1.4.5.11) followed by setting $\varepsilon = 1$, we find the desired solution of the original problem.

Remark 1.20. *If the right-hand side of equation (1.4.5.3) is linear in the unknown function, that is, $f(t, u, w) = g(t)u + h(t)w$, then the Adomian polynomials are defined simply as $A_n = g(t)u_n(t) + h(t)u_n(pt)$.*

▶ **Example 1.28.** Consider the Cauchy problem for a pantograph-type first-order linear ODE with variable coefficients

$$u'_t = g(t)u + h(t)w, \quad w = u(pt); \quad u(0) = a. \qquad (1.4.5.16)$$

We look for a solution to this problem by the Adomian decomposition method as the series

$$u(t) = \sum_{n=0}^{\infty} u_n(t), \qquad (1.4.5.17)$$

whose terms are defined by the recurrence relations

$$u_0(t) = a, \quad u_n(t) = \int_0^t \left[g(\xi)u_{n-1}(\xi) + h(\xi)u_{n-1}(p\xi)\right] d\xi, \quad n = 1, 2, \ldots$$

$$(1.4.5.18)$$

In the special case [143]:

$$a = 1, \quad g(t) = \tfrac{1}{2}, \quad h(t) = \tfrac{1}{2}e^{t/2}, \qquad (1.4.5.19)$$

problem (1.4.5.16) admits the exact solution $u(t) = e^t$.

For the Cauchy problem (1.4.5.16) with functions (1.4.5.19), the study [149] computed 13 terms of the series by formulas (1.4.5.18). The maximum error of the respective approximate solution, expressed as the sum of these terms, was less than 5×10^{-15} for $0 \leq t \leq 1$. ◂

Now we will give a general description of the Adomian decomposition method. Consider the following differential equation with proportional delay written in a short form:

$$L[u] = N[u, w], \qquad (1.4.5.20)$$

where $u = u(t)$ is the unknown function, $w = u(pt)$, $L[u]$ is a linear differential operator containing the highest derivative, and $N[u, w]$ is a nonlinear (in special cases, linear) differential operator or a function of two arguments, u and w. The coefficients of the operators can depend on the independent variable t. To formulate a problem for equation (1.4.5.20), one should use suitable initial or boundary conditions.

In the first stage, one seeks a solution to an auxiliary, simpler problem for the truncated linear equation

$$L[u_0] = 0, \qquad (1.4.5.21)$$

obtained from (1.4.5.20) by dropping the right-hand side. The initial or boundary conditions remain the same. Then the nonlinear term $N[u, w]$ is represented as the series

$$N[u, w] = \sum_{n=0}^{\infty} A_n, \quad A_n = \frac{1}{n!}\left\{\frac{\partial^n}{\partial \varepsilon^n} N\left[\sum_{k=0}^{\infty} \varepsilon^k u_k(t), \sum_{k=0}^{\infty} \varepsilon^k u_k(pt)\right]\right\}_{\varepsilon=0}.$$

One looks for a solution in the form of the series (1.4.5.17), where u_0 is a solution to the above problem for equation (1.4.5.21), while the other functions $u_n(t)$ are determined by solving the linear ODEs

$$L[u_n] = A_{n-1}, \quad n = 1, 2, \ldots, \qquad (1.4.5.22)$$

subjected to homogeneous initial or boundary conditions. Importantly, equations (1.4.5.22) do not involve terms with proportional delay.

For further details and numerous examples of employing the Adomian decomposition method, see, for example, [5–7, 125, 149, 447].

Homotopy analysis method. The homotopy analysis method (e.g., see [298, 299, 307]) is a semi-analytical procedure for solving nonlinear ODEs and PDEs, which can also be employed for solving differential equations with proportional delay. Below we outline the characteristic features of the method.

Let us revisit the nonlinear ODE with proportional delay (1.4.5.20) subjected to some initial or boundary conditions. The key idea of the homotopy method is to replace the analysis of equation (1.4.5.20) with an analysis of an auxiliary family of differential equations (homotopies):

$$(1 - \varepsilon)\{L[u] - L[u_0]\} + \varepsilon h\{L[u] - N[u, w]\} = 0. \qquad (1.4.5.23)$$

This family depends on the decomposition parameter ε ($0 \leq \varepsilon \leq 1$) as well as the *convergence-control parameter* h. As ε increases from 0 to 1, the solution of equation (1.4.5.23) changes from $u_0 = u_0(t)$ to the solution of equation (1.4.5.20).

Notably, if $u = u_0$ is a solution of the original equation (1.4.5.20), then this function will also be a solution of the family of differential equations (1.4.5.23) for any values of the parameters ε and h.

We seek an approximate solution of equation (1.4.5.23) as the finite sum

$$u_*(t) = \sum_{n=0}^{m} \varepsilon^n u_n(t), \qquad (1.4.5.24)$$

with the functions u_n to be determined in the subsequent analysis.

We substitute expression (1.4.5.24) into (1.4.5.23) and collect the terms proportional to the different powers of ε. Then we equate the resulting functional coefficients of ε^n with zero to obtain a system of differential equations for u_n. On solving this system sequentially, we find the terms u_n of the sum (1.4.5.24). Now setting $\varepsilon = 1$ in (1.4.5.24) and assuming m to be sufficiently large, we obtain an approximate analytical solution of the original equation (1.4.5.20). Whether this solution is suitable or not depends on the selection of the initial approximation $u_0 = u_0(t)$, the value of the convergence-control parameter h, and the number of terms m.

One possible way of choosing h is as follows. Take the value $h = h_{\min}$, for which the discrepancy $\|L[u_*] - N[u_*, w_*]\|$ attains a minimum (the choice of the norm $\| \ldots \|$ is down to the researcher).

For further details and numerous examples of employing the homotopy analysis method, see [299, 300]; the drawbacks of the method are discussed in [307].

Other methods based on perturbation-iteration algorithms. We will describe below a general scheme of iterative methods based on introducing a small parameter. To be specific, we will consider second-order nonlinear ODEs with proportional delay of the form

$$u''_{tt} = f(t, u, u'_t, w, w'_t), \quad w = u(pt). \qquad (1.4.5.25)$$

For what follows, we introduce the more general, auxiliary equation with parameter ε

$$u''_{tt} = F(t, u, u'_t, w, w'_t, \varepsilon), \quad w = u(pt). \qquad (1.4.5.26)$$

Its right-hand side must satisfy the matching condition

$$F(t, u, u'_t, w, w'_t, 1) \equiv f(t, u, u'_t, w, w'_t). \qquad (1.4.5.27)$$

The initial or boundary conditions for (1.4.5.26) are exactly the same as for equation (1.4.5.25).

By construction, a solution to the auxiliary problem for equation (1.4.5.26) with $\varepsilon = 1$ is also a solution to the original problem for equation (1.4.5.25).

An approximate solution to the problem for equation (1.4.5.25) is constructed as follows. Assuming ε in (1.4.5.26) to be a small parameter, one looks for a solution to the auxiliary problem as a regular decomposition in powers of ε: $u = \sum_{n=0}^{\infty} \varepsilon^n u_n(t)$. Computing the sum of the first few terms followed by setting $\varepsilon = 1$, one obtains an approximate solution to the original problem for equation (1.4.5.25). The accuracy and applicability of this iterative approach depends on how lucky the choice of the auxiliary function F in (1.4.5.26) was.

For equations with proportional delays

$$u''_{tt} = f(t, u, u'_t, w), \quad w = u(pt), \tag{1.4.5.28}$$

it is advisable to choose the auxiliary equation with parameter ε in the form

$$u''_{tt} = f(t, u, u'_t, \varepsilon w), \quad w = u(pt). \tag{1.4.5.29}$$

At $\varepsilon = 1$, this equation coincides with (1.4.5.28). The decomposition of the solution to equation (1.4.5.29) into a series in powers of ε leads to simpler ODEs without delay for all terms $u_n(t)$.

For examples of employing the above approach for constructing approximate solutions to equations of the form (1.4.5.28), see the article [27].

Remark 1.21. *Approximate analytical methods of iterative type with a large number of iterations are often counted among numerical-analytical or numerical methods.*

1.4.6. Galerkin-Type Projection Methods. Collocation Method

Preliminary remarks. Galerkin-type methods are widely employed to construct approximate solutions to linear and nonlinear boundary value problems for second- and higher-order ODEs and PDEs without delay (e.g., see [161, 251, 280, 282, 342, 402, 423]). These methods are also effective for solving more complicated problems described by differential equations with proportional argument.

The current subsection outlines a few Galerkin-type methods for solving boundary value problems that are described by ODEs with proportional argument. For clarity, we will use x to denote the independent variable (rather than t) and restrict ourselves to studying second-order equations, which, along with the unknown function $u = u(x)$, also involve $w = u(px)$, where $0 < p \leq 1$.

The representation of approximate solutions as linear combinations of basis functions. We will look at a boundary value problem for the equation

$$\mathfrak{F}[u] - f(x) = 0 \tag{1.4.6.1}$$

subjected to linear homogeneous boundary conditions* at points $x = x_1$ and $x = x_2$ (either of the two variants is possible: $x_1 = 0$, $x_2 = L$ or $x_1 = -L$, $x_2 = L$).

---

*For second-order ODEs, nonhomogeneous boundary conditions can be converted to homogeneous ones with the change of variable $z = b_2 x^2 + b_1 x + b_0 + y$ with the constants b_2, b_1, and b_0 selected using the method of undetermined coefficients.

In equation (1.4.6.1), $\mathfrak{F}[u] \equiv \mathfrak{F}(x, u, u_x, u_{xx}, w, w_x)$ is a second-order linear or nonlinear differential operator, $u = u(x)$ is the unknown function, $w = u(px)$, and $f = f(x)$ is a given function.

We chose a sequence of linear independent functions, called *basis functions*,

$$\varphi = \varphi_n(x) \qquad (n = 1, \ldots, N), \tag{1.4.6.2}$$

that satisfy the same boundary conditions as $u = u(x)$. In all the methods discussed below, we seek an approximate solution to equation (1.4.6.1) as the linear combination

$$u_N = \sum_{n=1}^{N} A_n \varphi_n(x), \tag{1.4.6.3}$$

with the coefficients A_n to be determined in the subsequent analysis.

The finite sum (1.4.6.3) is called the *approximating function*. The amount of error (residual), R_N, is determined after substituting the finite sum into the left-hand side of equation (1.4.6.1):

$$R_N = \mathfrak{F}[u_N] - f(x). \tag{1.4.6.4}$$

If the residual R_N is identically zero, then the function u_N is an exact solution of equation (1.4.6.1). In the generic case, $R_N \not\equiv 0$.

Remark 1.22. *The set of basis functions $\varphi_n(x)$ in the approximating function (1.4.6.3) is most commonly chosen to be a sequence of polynomials or trigonometric functions.*

Remark 1.23. *Rather than the approximating function (1.4.6.3), linear in the unknown coefficients A_n, one can use the more general form*

$$u_N = \Phi(x, A_1, \ldots, A_N)$$

for an approximate solution. The function $\Phi(x, A_1, \ldots, A_N)$ is given in advance. It is selected from theoretical considerations while taking into account the characteristic features of the problem or from relevant experimental data. It must satisfy the boundary conditions for any values of the coefficients $A_1, \ldots, A_N$.

The general scheme of applying the Galerkin method. To find the coefficients A_n in (1.4.6.3), we will consider a different sequence of linear independent functions:

$$\psi = \psi_k(x) \qquad (k = 1, 2, \ldots, N). \tag{1.4.6.5}$$

Let us multiply (1.4.6.4) by ψ_k and integrate over the domain $V = \{x_1 \leq x \leq x_2\}$, in which the desired solution to equation (1.4.6.1) is sought. On equating the resulting integral with zero (such integrals are zero for exact solutions), we arrive at the following system of algebraic equations for the unknown coefficients A_n:

$$\int_{x_1}^{x_2} \psi_k R_N \, dx = 0 \qquad (k = 1, 2, \ldots, N), \tag{1.4.6.6}$$

with the residual R_N defined by (1.4.6.4).

Relation (1.4.6.6) implies that the approximating function (1.4.6.3) satisfies equation (1.4.6.1) 'on average' (in the integral sense) with weight functions ψ_k. If we introduce the inner product of arbitrary functions g and h, $\langle g, h \rangle = \int_{x_1}^{x_2} gh\, dx$, we can treat the equations (1.4.6.6) as orthogonality conditions for the residual R_N to all ψ_k.

The Galerkin method is applicable not only to boundary value problems but also to eigenvalue problems, in which case one sets $f = \lambda u$ and looks for eigenfunctions y_n in conjunction with eigenvalues λ_n.

Below we outline a few special techniques that are particular cases of the Galerkin method.

Bubnov–Galerkin method. The sequences of functions (1.4.6.2) and (1.4.6.5) in the Galerkin method can be chosen arbitrarily. The modification in which the sets of functions are the same,

$$\varphi_k(x) = \psi_k(x) \qquad (k = 1, 2, \ldots, N), \tag{1.4.6.7}$$

is called the *Bubnov–Galerkin method*.

Method of moments. The *method of moments* is a Galerkin method with the weight functions (1.4.6.5) chosen as powers of x:

$$\psi_k = x^k. \tag{1.4.6.8}$$

Least squares method. Sometimes the weight functions ψ_k are expressed via φ_k as

$$\psi_k = \mathfrak{F}[\varphi_k] \qquad (k = 1, 2, \ldots),$$

where $\mathfrak{F}$ is the differential operator of equation (1.4.6.1). This modification of the Galerkin method is known as the *least squares method*.

Collocation method. In the *collocation method*, one selects a set of points x_k, $k = 1, \ldots, N$, and imposes the condition that the residual (1.4.6.4) is zero at these points:

$$R_N = 0 \quad \text{at} \quad x = x_k \quad (k = 1, \ldots, N). \tag{1.4.6.9}$$

When solving a specific problem, the points x_k, at which the residual R_N vanishes, are considered most essential. The number of collocation points, N, is taken equal to the number of terms in the sum (1.4.6.3). This results in a complete system of algebraic equations for the unknown coefficients A_n; for linear boundary value problems, this algebraic system is linear.

For polynomial basis functions (1.4.6.2), the collocation points x_k in (1.4.6.9) are reasonable to take as *Chebyshev nodes*, defined on the interval $x \in (-1, 1)$ as

$$x_k = \cos\left(\frac{2k-1}{2N}\pi\right), \quad k = 1, \ldots, N.$$

Other suitable nodes can be used such as roots of some orthogonal polynomials with a weight function. If all roots are equally spaced, the method usually works worse and can even result in divergent solutions as $N \to \infty$.

Notably, the collocation method is a special case of the Galerkin method in which the weight functions (1.4.6.5) are Dirac delta functions:

$$\psi_k = \delta(x - x_k).$$

1.4. Exact and Approximate Analytical Solution Methods for Delay ODEs

With the collocation method, one does not have to calculate integrals, which immensely simplifies the solution of nonlinear problems. This method, however, usually leads to less accurate results than other modifications of the Galerkin method.

▶ **Example 1.29.** Consider the boundary value problem for a second-order linear ordinary differential equation with variable coefficients and proportional delay

$$u''_{xx} + g(x)w - f(x) = 0, \quad w = u(px), \tag{1.4.6.10}$$

subjected to the boundary conditions of the first kind

$$u(-1) = u(1) = 0. \tag{1.4.6.11}$$

We assume the coefficients of equation (1.4.6.10) to be smooth even functions such that $f(x) = f(-x)$ and $g(x) = g(-x)$. We will look for an approximate solution to problem (1.4.6.10)–(1.4.6.11) by employing the collocation method.

1°. We select the polynomials

$$u_n(x) = x^{2n-2}(1 - x^2), \quad n = 1, 2, \dots, N,$$

as basis functions. These satisfy the boundary conditions (1.4.6.11): $u_n(\pm 1) = 0$.

We use three collocation points

$$x_1 = -\sigma, \quad x_2 = 0, \quad x_3 = \sigma \quad (0 < \sigma < 1) \tag{1.4.6.12}$$

and restrict ourselves to two basis functions ($N = 2$). Then the approximating function is

$$\begin{aligned} u &= A_1(1 - x^2) + A_2 x^2(1 - x^2), \\ w &= A_1(1 - p^2 x^2) + A_2 p^2 x^2 (1 - p^2 x^2), \end{aligned} \tag{1.4.6.13}$$

where A_1 and A_2 are unknown coefficients. Substituting (1.4.6.13) into the left-hand side of equation (1.4.6.10) results in the residual

$$R(x) = A_1\big[-2 + (1 - p^2 x^2)g(x)\big] + A_2\big[2 - 12x^2 + p^2 x^2(1 - p^2 x^2)g(x)\big] - f(x).$$

The residual $R(x)$ must vanish at the collocation points (1.4.6.12). Considering that $f(\sigma) = f(-\sigma)$ and $g(\sigma) = g(-\sigma)$, we obtain two linear algebraic equations for A_1 and A_2:

$$\begin{aligned} &A_1\big[-2 + g(0)\big] + 2A_2 - f(0) = 0, \\ &A_1\big[-2 + (1 - p^2\sigma^2)g(\sigma)\big] \\ &\quad + A_2\big[2 - 12\sigma^2 + p^2\sigma^2(1 - p^2\sigma^2)g(\sigma)\big] - f(\sigma) = 0. \end{aligned} \tag{1.4.6.14}$$

2°. To be specific, we choose the following functions in equation (1.4.6.10):

$$f(x) = -1, \quad g(x) = 1 + x^2. \tag{1.4.6.15}$$

On solving the respective system of algebraic equations (1.4.6.14), we find the coefficients

$$A_1 = \frac{12 - p^2(1+\sigma^2)(1-p^2\sigma^2)}{10 + p^2(1+\sigma^2)(1+p^2\sigma^2)}, \quad A_2 = \frac{1 - p^2(1+\sigma^2)}{10 + p^2(1+\sigma^2)(1+p^2\sigma^2)}. \tag{1.4.6.16}$$

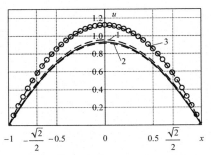

Figure 1.17. Comparison of the approximate analytical solutions (1.4.6.13) and (1.4.6.16) to equation (1.4.6.10) with conditions (1.4.6.11) and (1.4.6.15), obtained by the collocation method, with numerical solutions.

Figure 1.17 depicts in dashed lines, 1 and 2, the approximate solutions to problem (1.4.6.10), (1.4.6.11), (1.4.6.15) with $p = 1$ obtained by the collocation method using formulas (1.4.6.13) and (1.4.6.16) with $\sigma = 1/2$ (the collocation points are equally spaced) and $\sigma = \sqrt{2}/2$ (the collocation points are Chebyshev nodes). The solid line indicates a numerical solution to this problem with $p = 1$ obtained by the shooting method (see Subsection 5.1.6). It is apparent that both approximate solutions are in good agreement with the numerical solution (using Chebyshev nodes gives a more accurate result). For $p = 1/2$, the approximate solutions obtained by formulas (1.4.6.13), (1.4.6.16) for $\sigma = 1/2$ and $\sigma = \sqrt{2}/2$ practically coincide; these are shown in Figure 1.17 by dashed line 3. The circles indicate a numerical solution with $p = 1/2$ obtained using a combination of the shooting method and Heun's method (see Subsections 5.1.4 and 5.1.5). The approximate solutions are in good agreement with the numerical solution; the maximum discrepancy between them is 0.009 at $p = 1/2$ and $\sigma = 1/2$ and 0.006 at $p = 1/2$ and $\sigma = \sqrt{2}/2$. ◂

Method of minimization of the root mean square error. Sometimes, the coefficients A_n of the approximating function (1.4.6.3) are sought using a method based on minimizing the functional

$$\Phi = \int_0^L R_N^2 \, dx \to \min. \tag{1.4.6.17}$$

Given the functions φ_n in the sum (1.4.6.3), the integral Φ is a function of the coefficients A_n. The necessary conditions for the minimization of the functional (1.4.6.17) are

$$\frac{\partial \Phi}{\partial A_n} = 0 \quad (n = 1, \ldots, N). \tag{1.4.6.18}$$

Relations (1.4.6.18) represent a system of algebraic (or transcendental) equations for the desired coefficients A_n.

Remark 1.24. *If the number of terms of the approximating function (1.4.6.3) is large, the Galerkin-type approximate analytical methods (primarily, the collocation method) are often referred to as numerical-analytical or numerical methods.*

Remark 1.25. *Numerical methods for the integration of delay ODEs are discussed in Section 5.1.*

2. Linear Partial Differential Equations with Delay

2.1. Properties and Specific Features of Linear Equations and Problems with Constant Delay

2.1.1. Properties of Solutions to Linear Delay Equations

Examples of linear partial differential equations with delay. Second-order linear partial differential equations of the parabolic or hyperbolic type with constant delay are not uncommon in the literature and applications. Such equations with n spatial variables are expressed as

$$u_t - L_1[u] - L_2[w] = \Phi(\mathbf{x}, t), \qquad (2.1.1.1)$$
$$u_{tt} - L_1[u] - L_2[w] = \Phi(\mathbf{x}, t), \qquad (2.1.1.2)$$

where

$$L_1[u] \equiv \sum_{i,j=1}^{n} a_{ij}^{(1)}(\mathbf{x}, t) \frac{\partial^2 u}{\partial x_i \partial x_j} + \sum_{i=1}^{n} b_i^{(1)}(\mathbf{x}, t) \frac{\partial u}{\partial x_i} + c_1(\mathbf{x}, t) u,$$
$$L_2[w] \equiv \sum_{i,j=1}^{n} a_{ij}^{(2)}(\mathbf{x}, t) \frac{\partial^2 w}{\partial x_i \partial x_j} + \sum_{i=1}^{n} b_2^{(2)}(\mathbf{x}, t) \frac{\partial w}{\partial x_i} + c_2(\mathbf{x}, t) w, \qquad (2.1.1.3)$$
$$w = u(\mathbf{x}, t - \tau), \quad \mathbf{x} = (x_1, \ldots, x_n), \quad \tau > 0.$$

Equations (2.1.1.1) and (2.1.1.2) are said to be homogeneous if $\Phi(\mathbf{x}, t) \equiv 0$.

For variable delays, $\tau = \tau(t)$, one should set $w = u(\mathbf{x}, t - \tau(t))$ in the linear partial differential equations (2.1.1.1)–(2.1.1.3). In particular, for equations with proportional delay, we have $w = u(\mathbf{x}, pt)$, where $0 < p < 1$.

Properties of linear partial differential equations with constant delay. In what follows, for brevity, we will write a linear homogeneous PDE with constant delay as

$$\mathcal{L}[u] = 0. \qquad (2.1.1.4)$$

For parabolic and hyperbolic equations, the linear differential operator $\mathcal{L}[u]$ in (2.1.1.4) is defined by the left-hand side of equations (2.1.1.1) and (2.1.1.2), respectively. In general, equation (2.1.1.4) can be an arbitrary linear homogeneous PDE of any order in the variables t, x_1, ..., x_n with sufficiently smooth coefficients and a constant delay in time.

Linear operator $\mathcal{L}$ has the properties

$$\mathcal{L}[u_1 + u_2] = \mathcal{L}[u_1] + \mathcal{L}[u_2],$$
$$\mathcal{L}[Au] = A\mathcal{L}[u], \quad A = \text{const}.$$

An arbitrary linear homogeneous equation (2.1.1.4) has a trivial solution, $u \equiv 0$.

A function u is called a *classical solution* of equation (2.1.1.4) if, with u substituted into the equation, it becomes an identity and if all partial derivatives of u in (2.1.1.4) are continuous. The concept of a classical solution is directly related to the domain of the independent variables. In what follows, for brevity, we will usually write 'solution' to mean 'classical solution'.

Application of particular solutions to construct other solutions. Below we list the main properties of particular solutions to linear homogeneous PDEs with delay analogous to those of simpler linear homogeneous PDEs without delay [404].

$1°$. Let $u_1 = u_1(\mathbf{x}, t)$, $u_2 = u_2(\mathbf{x}, t)$, ..., $u_k = u_k(\mathbf{x}, t)$ be any particular solutions to the homogeneous equation (2.1.1.4). Then the linear combination of these solutions

$$u = A_1 u_1 + A_2 u_2 + \cdots + A_k u_k, \tag{2.1.1.5}$$

where $A_1, A_2, \ldots, A_k$ are arbitrary constants, is also a solution to equation (2.1.1.4). In physics, this property is known as the *principle of linear superposition*.

We assume that $\{u_k\}$ is an infinite sequence of solutions to equation (2.1.1.4). Then, regardless of whether the series $\sum_{k=1}^{\infty} u_k$ is convergent or not, it is called a *formal solution* to equation (2.1.1.4). If all solutions u_k are classical and, in addition, the series $\sum_{k=1}^{\infty} u_k$ and its derivatives involved in the equation are uniformly convergent, then the series determines a classical solution to equation (2.1.1.4).

$2°$. Let the coefficients of the linear differential operator $\mathcal{L}$ be independent of time t. If equation (2.1.1.4) has a particular solution $\tilde{u} = \tilde{u}(\mathbf{x}, t)$, then the function $\tilde{u}(\mathbf{x}, t + a)$, where a is an arbitrary constant, is also a solution to the equation.

If the coefficients of the operator $\mathcal{L}$ are independent of only one space variable, x_k, and equation (2.1.1.4) has a particular solution $\tilde{u} = \tilde{u}(\mathbf{x}, t)$, then the function $\tilde{u}(\mathbf{x}, t)|_{x_k \Rightarrow x_k + b}$, where b is an arbitrary constant, is also a solution to the equation.

$3°$. Suppose the coefficients of the linear differential operator $\mathcal{L}$ are independent of t. Then if equation (2.1.1.4) has a particular solution $\tilde{u} = \tilde{u}(\mathbf{x}, t)$, the partial derivatives of $\tilde{u}$ with respect to time*

$$\frac{\partial \tilde{u}}{\partial t}, \quad \frac{\partial^2 \tilde{u}}{\partial t^2}, \quad \ldots, \quad \frac{\partial^k \tilde{u}}{\partial t^k}, \quad \ldots$$

are also solutions to equation (2.1.1.4).

---

*Here and henceforth, we assume that the particular solution $\tilde{u}$ is differentiable sufficiently many times with respect to t and $x_1, \ldots, x_n$ (or the parameters).

4°. Let the coefficients of the linear differential operator $\mathcal{L}$ be independent of the space variables $x_1, \ldots, x_n$. Then if equation (2.1.1.4) has a particular solution $\tilde{u} = \tilde{u}(\mathbf{x}, t)$, the partial derivatives $\tilde{u}$ with respect to these variables

$$\frac{\partial \tilde{u}}{\partial x_1}, \quad \frac{\partial \tilde{u}}{\partial x_2}, \quad \frac{\partial \tilde{u}}{\partial x_3}, \quad \ldots, \quad \frac{\partial^2 \tilde{u}}{\partial x_1^2}, \quad \frac{\partial^2 \tilde{u}}{\partial x_1 \partial x_2}, \quad \ldots, \quad \frac{\partial^{k+m} \tilde{u}}{\partial x_2^k \partial x_3^m}, \quad \ldots$$

are also solutions to equation (2.1.1.4).

If the coefficients of the operator $\mathcal{L}$ are independent of only one space variable, x_1, and equation (2.1.1.4) has a particular solution $\tilde{u} = \tilde{u}(\mathbf{x}, t)$, then the partial derivatives

$$\frac{\partial \tilde{u}}{\partial x_1}, \quad \frac{\partial^2 \tilde{u}}{\partial x_1^2}, \quad \ldots, \quad \frac{\partial^k \tilde{u}}{\partial x_1^k}, \quad \ldots$$

are also solutions to equation (2.1.1.4).

5°. Suppose that the coefficients of the linear differential operator $\mathcal{L}$ are all constant and equation (2.1.1.4) has a particular solution $\tilde{u} = \tilde{u}(\mathbf{x}, t)$. Then any partial derivatives of $\tilde{u}$ with respect to time and the spatial variables (inclusive of mixed derivatives)

$$\frac{\partial \tilde{u}}{\partial t}, \quad \frac{\partial \tilde{u}}{\partial x_1}, \quad \ldots, \quad \frac{\partial^2 \tilde{u}}{\partial x_2^2}, \quad \frac{\partial^2 \tilde{u}}{\partial t \partial x_1}, \quad \ldots, \quad \frac{\partial^k \tilde{u}}{\partial x_3^k}, \quad \ldots$$

are all solutions to equation (2.1.1.4).

6°. Let equation (2.1.1.4) have a particular solution $\tilde{u} = \tilde{u}(\mathbf{x}, t; \mu)$ dependent on the free parameter μ and let the coefficients of the linear differential operator $\mathcal{L}$ be independent of μ (but can depend on time and spatial variables). Then, the differentiation of $\tilde{u}$ with respect to μ gives other solutions to equation (2.1.1.4):

$$\frac{\partial \tilde{u}}{\partial \mu}, \quad \frac{\partial^2 \tilde{u}}{\partial \mu^2}, \quad \ldots, \quad \frac{\partial^k \tilde{u}}{\partial \mu^k}, \quad \ldots$$

Let some constants $\mu_1, \ldots, \mu_k$ belong to the domain of the parameter μ. Then the sum

$$u = A_1 \tilde{u}(\mathbf{x}, t; \mu_1) + \cdots + A_k \tilde{u}(\mathbf{x}, t; \mu_k), \tag{2.1.1.6}$$

where $A_1, \ldots, A_k$ are arbitrary constants, is also a solution to the linear homogeneous equation (2.1.1.4). The number of terms in the sum (2.1.1.6) can be finite or infinite.

7°. Further solutions may also be constructed in the following manner. Suppose $\tilde{u}(\mathbf{x}, t; \mu)$ is a particular solution dependent on the parameters μ (just as in Item 6°, the coefficients of the linear operator $\mathcal{L}$ are assumed to be independent of μ). We first multiply this solution by an arbitrary function $\varphi(\mu)$ and then integrate with respect to μ on the interval $[\alpha, \beta]$ to obtain the function

$$\int_\alpha^\beta \varphi(\mu) \tilde{u}(\mathbf{x}, t; \mu) \, d\mu$$

that is also a solution to the original linear homogeneous equation (2.1.1.4).

8°. In case we deal with a linear partial differential equation with delay, the following proposition [9] can also be used to construct complicated solutions out of simpler ones.

Proposition. Suppose that a linear homogeneous PDE with delay has a one-parameter particular solution of the form $u = \tilde{u}(\mathbf{x}, t; \mu)$, where μ is a free parameter not involved in the original equation. Then the equation also has the two-parameter solutions
$$u_1 = \operatorname{Re} \tilde{u}(\mathbf{x}, t, \alpha + i\beta), \qquad u_2 = \operatorname{Im} \tilde{u}(\mathbf{x}, t, \alpha + i\beta),$$
where α and β are arbitrary real constants, $i^2 = -1$, and $\operatorname{Re} z$ and $\operatorname{Im} z$ are the real and imaginary parts of the complex variable z.

Corollary 1. Let a linear homogeneous PDE with constant delay whose coefficients are independent of time t have a particular solution $u = \tilde{u}(\mathbf{x}, t)$. Then this equation also admits the one-parameter family of solutions
$$u_1 = \operatorname{Re} \tilde{u}(\mathbf{x}, t + i\alpha), \qquad u_2 = \operatorname{Im} \tilde{u}(\mathbf{x}, t + i\alpha),$$
where α is an arbitrary real constant and $i^2 = -1$.

Corollary 2. Let a linear homogeneous PDE with constant delay whose coefficients are independent of a space variable x_k have a particular solution $u = \tilde{u}(\mathbf{x}, t)$. Then this equation also admits the one-parameter family of solutions
$$u_1 = \operatorname{Re} \tilde{u}(\mathbf{x}, t)\big|_{x_k \Rightarrow x_k + i\alpha}, \qquad u_2 = \operatorname{Im} \tilde{u}(\mathbf{x}, t)\big|_{x_k \Rightarrow x_k + i\alpha},$$
where α is an arbitrary real constant.

The properties specified in Items 1° to 8° allow one to use known particular solutions to construct other particular solutions to linear homogeneous PDEs with constant delay.

▶ Example 2.1. The linear thermal conduction PDE with constant delay
$$u_t = u_{xx} - aw, \qquad w = u(x, t - \tau),$$
has a particular solution
$$\tilde{u}(x, t) = \exp\big(\sqrt{\mu + ae^{-\mu\tau}}\, x + \mu t\big),$$
where μ is an arbitrary constant. Differentiating this expression with respect to μ (see Item 6°), we arrive at the more complex solution
$$\tilde{u}_1(x, t) = \left(\frac{1 - a\tau e^{-\mu\tau}}{2\sqrt{\mu + ae^{-\mu\tau}}} x + t \right) \exp\big(\sqrt{\mu + ae^{-\mu\tau}}\, x + \mu t\big).$$
◀

▶ Example 2.2. The wave-type linear PDE with delay
$$u_{tt} = w_{xx}, \qquad w = u(x, t - \tau),$$

has a particular solution
$$\tilde{u}(x,t) = \exp\bigl(\mu e^{\mu\tau/2}x + \mu t\bigr),$$
where μ is an arbitrary constant. Setting $\mu = i\alpha$ in this solution (see Item 8°) and then extracting the real and imaginary parts, we obtain two more complicated solutions
$$u_1 = \exp\bigl[-\alpha\sin(\tfrac{1}{2}\alpha\tau)x\bigr]\cos\bigl[\alpha\cos(\tfrac{1}{2}\alpha\tau)x + \alpha t\bigr],$$
$$u_2 = \exp\bigl[-\alpha\sin(\tfrac{1}{2}\alpha\tau)x\bigr]\sin\bigl[\alpha\cos(\tfrac{1}{2}\alpha\tau)x + \alpha t\bigr].$$ ◀

Separable solutions in the form of the product or sum of functions with different arguments.

1°. Many linear homogeneous partial differential equations have solutions that can be represented as the product of two or more functions dependent on different arguments. Such solutions are referred to as *multiplicative separable solutions*; they are also often called *separable solutions* for short.

Table 2.1 lists frequent types of linear homogeneous differential equations with several independent variables and a constant delay that admit exact separable solutions. Linear combinations of particular solutions associated with distinct values of the separation parameters, λ, $\beta_1, \ldots, \beta_n$, are also solutions to these equations. The last row of Table 2.1 uses short notations: L_t is a linear differential operator whose coefficients are only dependent on time t, and $L_\mathbf{x}$ is a linear differential operator with coefficients on the space variables alone; it is assumed that $L_t[C] = 0$ and $L_\mathbf{x}[C] = 0$, where C is an arbitrary constant.

Table 2.1. Some linear nonhomogeneous PDEs with constant delay admitting multiplicative separable solutions.

No.	Form of equation (2.1.1.4)	Form of particular solutions
1	Equation coefficients are constant	$u(\mathbf{x},t) = A\exp(\lambda t + \beta_1 x_1 + \cdots + \beta_n x_n)$, $\lambda, \beta_1, \ldots, \beta_n$ are related by an algebraic (transcendental) equation
2	Equation coefficients are independent of t	$u(\mathbf{x},t) = e^{\lambda t}\psi(\mathbf{x})$, where λ is an arbitrary constant
3	Equation coefficients are independent of $x_1, \ldots, x_n$	$u(\mathbf{x},t) = \exp(\beta_1 x_1 + \cdots + \beta_n x_n)\psi(t)$, where $\beta_1, \ldots, \beta_n$ are arbitrary constants
4	Equation coefficients are independent of $x_1, \ldots, x_k$ $(k < n)$	$u(\mathbf{x},t) = \exp(\beta_1 x_1 + \cdots + \beta_k x_k)\psi(t, x_{k+1}, \ldots, x_n)$, where $\beta_1, \ldots, \beta_k$ are arbitrary constants
5	Equation can be represented as $L_t[u] + L_\mathbf{x}[w] = 0$	$u(\mathbf{x},t) = \varphi(t)\psi(\mathbf{x})$, $\varphi(t)$ satisfies equation $L_t[\varphi] + \lambda\varphi(t-\tau) = 0$, $\psi(\mathbf{x})$ satisfies equation $L_\mathbf{x}[\psi] - \lambda\psi = 0$

For constant coefficient equations (see row 1 of Table 2.1), the separation parameters must satisfy a single algebraic (or transcendental) equation
$$D(\lambda, \beta_1, \ldots, \beta_n) = 0, \qquad (2.1.1.7)$$

which results from substituting the solution into equation (2.1.1.4). In physical applications, equation (2.1.1.7) is usually referred to as a *dispersion equation*. Any n out of $n + 1$ separation parameters in (2.1.1.7) can be treated as arbitrary.

▶ **Example 2.3.** Consider the linear telegraph equation with constant delay

$$u_{tt} + ku_t = au_{xx} + bu_x + c_1 u + c_2 w, \quad w = u(x, t - \tau). \tag{2.1.1.8}$$

We look for a (traveling wave) particular solution to this equation in the exponential form

$$u = A \exp(\beta x + \lambda t),$$

where A is an arbitrary constant, to arrive at the dispersion equation

$$\lambda^2 + k\lambda = a\beta^2 + b\beta + c_1 + c_2 e^{-\lambda \tau}.$$

Either parameter β or λ can be treated as fixed, and the other is determined from this equation.

Notably, the constant coefficient equation (2.1.1.8) also admits more complicated solutions (see the second and third rows of Table 2.1, last column). ◀

▶ **Example 2.4.** The linear heat equation with constant delay

$$u_t = aw_{xx}, \quad w = u(x, t - \tau),$$

admits multiplicative separable solutions

$$\begin{aligned} u &= [A\cos(kx) + B\sin(kx)]e^{-\lambda t}, & k &= (\lambda e^{-\lambda \tau}/a)^{1/2} & (\lambda > 0); \\ u &= [A\exp(kx) + B\exp(-kx)]e^{-\lambda t}, & k &= (-\lambda e^{-\lambda \tau}/a)^{1/2} & (\lambda < 0), \end{aligned} \tag{2.1.1.9}$$

where A, B, and λ are arbitrary constants.

It is noteworthy that solutions (2.1.1.9) are special cases of a multiplicative separable solution $u = \varphi(x)\psi(t)$. ◀

2°. Linear delay PDEs with two independent variables, x and t, of the form

$$L_t^{(1)}[u] + L_t^{(2)}[w] + M_x^{(1)}[u] + M_x^{(2)}[w] + c_1 u + c_2 w = f(x) + g(t),$$

where

$$L_t^{(j)}[v] \equiv \sum_{i=1}^{K_j} a_i^{(j)}(t) \frac{\partial^i v}{\partial t^i}, \quad M_x^{(j)}[v] \equiv \sum_{i=1}^{N_j} b_i^{(j)}(x) \frac{\partial^i v}{\partial x^i}, \quad j = 1, 2,$$

have exact solutions representable as the sum of functions with different arguments:

$$u = \varphi(x) + \psi(t). \tag{2.1.1.10}$$

Such solutions are known as *additive separable solutions*.

▶ **Example 2.5.** The linear delay PDE (2.1.1.8) has an additive separable solution of the form (2.1.1.10) with the functions $\varphi = \varphi(x)$ and $\psi = \psi(t)$ satisfying the ODE and delay ODE

$$\begin{aligned} a\varphi_{xx}'' + b\varphi_x' + (c_1 + c_2)\varphi &= 0, \\ \psi_{tt}'' + k\psi_t' &= c_1 \psi + c_2 \bar\psi, \quad \bar\psi = \psi(t - \tau). \end{aligned}$$

◀

2.1.2. General Properties and Qualitative Features of Delay Problems

Many properties of linear partial differential equations with delay are similar to those of respective partial differential equations without delay. Boundary conditions in initial-boundary value problems for partial differential equations with delay are stated in exactly the same way as for partial differential equations without delay.

Below we note three main qualitative features that distinguish problems for delay PDEs from those for PDEs without delay.

First, the initial conditions for PDEs in problems with a constant delay, $\tau > 0$, are set on a closed interval, $t_0 - \tau \leq t \leq t_0$, rather than at a point, $t = t_0$, as in problems without delay. Moreover, one looks for a solution continuous at $t = t_0$. Most commonly, the initial point is chosen as $t_0 = 0$ or, sometimes, $t_0 = \tau$.

Secondly, even though the initial data have continuous derivatives with respect to t up to an arbitrary order, solutions to Cauchy problems and initial-boundary value problems will generally have jump-discontinuous partial derivatives of order k and above ($k = 1, 2, \ldots$) at the points $t = t_0 + (k-1)\tau$. However, the lower-order derivatives will be continuous at these points. This means that jump discontinuities propagate throughout derivatives with gradual solution smoothing (as in the delay ODEs; see Section 1.1.2).

Thirdly, initial-boundary value problems for parabolic and hyperbolic equations with delay can, under certain conditions, be ill-posed in the sense of Hadamard, meaning that solutions of such problems are unstable with respect to small perturbations of the initial data.

Linear problems described by partial differential equations with delay can be solved using separation of variables or methods of integral transforms in the same ways as those for linear partial differential equations without delay [404, 514].

2.2. Linear Initial-Boundary Value Problems with Constant Delay

2.2.1. First Initial-Boundary Value Problem for One-Dimensional Parabolic Equations with Constant Delay

Statement of the problem. Let us look at the first initial-boundary value problem for one-dimensional parabolic equations with constant coefficients and a constant delay

$$\frac{\partial u}{\partial t} = a_1 \frac{\partial^2 u}{\partial x^2} + a_2 \frac{\partial^2 w}{\partial x^2} + c_1 u + c_2 w + f(x,t), \quad w = u(x, t-\tau), \quad (2.2.1.1)$$

where $a_1 > a_2 \geq 0$, defined in the domain $\Omega = \{0 < x < h, \; t > 0\}$. We supplement equation (2.2.1.1) with the nonhomogeneous boundary conditions of the first kind (Dirichlet conditions)

$$u = g_1(t) \quad \text{at} \quad x = 0, \; t > -\tau; \quad u = g_2(t) \quad \text{at} \quad x = h, \; t > -\tau, \quad (2.2.1.2)$$

and the general initial condition

$$u = \varphi(x,t) \quad \text{at} \quad 0 < x < h, \ -\tau \leq t \leq 0. \tag{2.2.1.3}$$

We assume the functions f and φ, appearing in equation (2.2.1.1) and the initial condition (2.2.1.3), to be continuous and the functions g_1 and g_2, involved in the boundary conditions (2.2.1.2), to be continuously differentiable with respect to t. In addition, we assume that the boundary and initial conditions (2.2.1.2) and (2.2.1.3) are consistent, implying that

$$\varphi(0,t) = g_1(t), \quad \varphi(h,t) = g_2(t).$$

The studies [270–272, 335, 452] employed the method of separation of variables to solve one-dimensional problems described by parabolic equations with constant delay (2.2.1.1) and related equations.

Representation of solutions as the sum of several functions. Following [270, 271], we look for solutions to problem (2.2.1.1)–(2.2.1.3) in the form

$$u = u_0(x,t) + u_1(x,t) + u_2(x,t), \tag{2.2.1.4}$$

where the function

$$u_0(x,t) = g_1(t) + \frac{x}{h}[g_2(t) - g_1(t)] \tag{2.2.1.5}$$

satisfies the nonhomogeneous boundary conditions (2.2.1.2). The functions $u_1 = u_1(x,t)$ and $u_2 = u_2(x,t)$ are determined by solving the simpler initial-boundary value problems with homogeneous (zero) boundary conditions described below.

Problem 1. The function u_1 satisfies the linear homogeneous PDE with constant delay

$$\frac{\partial u_1}{\partial t} = a_1 \frac{\partial^2 u_1}{\partial x^2} + a_2 \frac{\partial^2 w_1}{\partial x^2} + c_1 u_1 + c_2 w_1, \quad w_1 = u_1(x, t-\tau), \tag{2.2.1.6}$$

homogeneous boundary conditions

$$u_1 = 0 \quad \text{at} \quad x = 0, \ t > -\tau; \quad u_1 = 0 \quad \text{at} \quad x = h, \ t > -\tau, \tag{2.2.1.7}$$

and nonhomogeneous initial condition

$$u_1 = \Phi(x,t) \quad \text{at} \quad 0 < x < h, \ -\tau \leq t \leq 0, \tag{2.2.1.8}$$

where

$$\Phi(x,t) = \varphi(x,t) - g_1(t) - \frac{x}{h}[g_2(t) - g_1(t)]. \tag{2.2.1.9}$$

Problem 2. The function u_2 satisfies the linear nonhomogeneous PDE with constant delay

$$\frac{\partial u_2}{\partial t} = a_1 \frac{\partial^2 u_2}{\partial x^2} + a_2 \frac{\partial^2 w_2}{\partial x^2} + c_1 u_2 + c_2 w_2 + F(x,t), \quad w_2 = u_2(x, t-\tau), \tag{2.2.1.10}$$

2.2. Linear Initial-Boundary Value Problems with Constant Delay

where

$$F(x,t) = f(x,t) - \frac{\partial}{\partial t}\left\{g_1(t) + \frac{x}{h}[g_2(t) - g_1(t)]\right\}$$
$$+ c_1\left\{g_1(t) + \frac{x}{h}[g_2(t) - g_1(t)]\right\} \qquad (2.2.1.11)$$
$$+ c_2\left\{g_1(t-\tau) + \frac{x}{h}[g_2(t-\tau) - g_1(t-\tau)]\right\},$$

and the zero (homogeneous) boundary and initial conditions

$$u_2 = 0 \quad \text{at} \quad x = 0,\ t > -\tau; \quad u_2 = 0 \quad \text{at} \quad x = h,\ t > -\tau; \qquad (2.2.1.12)$$
$$u_2 = 0 \quad \text{at} \quad 0 < x < h,\ -\tau \leq t \leq 0. \qquad (2.2.1.13)$$

Solution of problem 1. Consider the linear homogeneous PDE with delay (2.2.1.6) subjected to the boundary and initial conditions (2.2.1.7) and (2.2.1.8). We first look for particular solutions to equation (2.2.1.6) as the product of functions with different arguments

$$u_{1p} = X(x)T(t). \qquad (2.2.1.14)$$

On substituting (2.2.1.14) into (2.2.1.6) and rearranging, we get

$$X(x)[T'(t) - c_1 T(t) - c_2 T(t-\tau)] = X''(x)[a_1 T(t) + a_2 T(t-\tau)]. \qquad (2.2.1.15)$$

Separating the variables in this equation, we arrive at the second-order linear ODE without delay and first-order ODE with constant delay

$$X''(x) = -\lambda^2 X(x), \qquad (2.2.1.16)$$
$$T'(t) = (c_1 - a_1\lambda^2)T(t) + (c_2 - a_2\lambda^2)T(t-\tau). \qquad (2.2.1.17)$$

Considering that the function u_1 must satisfy the homogeneous boundary conditions (2.2.1.7) and using (2.2.1.14), we obtain homogeneous boundary conditions for X:

$$X(0) = X(h) = 0. \qquad (2.2.1.18)$$

The linear homogeneous problem (2.2.1.16), (2.2.1.18), which is a special case of a *Sturm–Liouville problem* (also known as an *eigenvalue problem*), has nontrivial solutions at only the following discrete values of the parameter λ:

$$\lambda_n = \frac{\pi n}{h}, \quad n = 1, 2, \ldots. \qquad (2.2.1.19)$$

The respective eigenfunctions are

$$X_n(x) = \sin\left(\frac{\pi n x}{h}\right). \qquad (2.2.1.20)$$

Notably, two eigenfunctions $X_n(x)$ and $X_m(x)$ are orthogonal in the sense that

$$\int_0^h X_n(x) X_m(x)\, dx = 0 \quad \text{if} \quad n \neq m. \qquad (2.2.1.21)$$

On substituting the *eigenvalues* (2.2.1.19) into (2.2.1.17), we get a delay ODE for $T = T_n(t)$:

$$T'_n(t) = \left[c_1 - a_1\left(\frac{\pi n}{h}\right)^2\right]T_n(t) + \left[c_2 - a_2\left(\frac{\pi n}{h}\right)^2\right]T_n(t-\tau). \quad (2.2.1.22)$$

We seek a solution to the linear initial-boundary value problem (2.2.1.6)–(2.2.1.7) as the series

$$u_1(x,t) = \sum_{n=1}^{\infty} X_n(x)T_n(t), \quad (2.2.1.23)$$

where the functions $u_{1n}(x,t) = X_n(x)T_n(t)$ are particular solutions to equation (2.2.1.6) satisfying the homogeneous boundary conditions (2.2.1.7). By the linear superposition principle, the series (2.2.1.23) is also a formal solution to the original partial differential equation with delay (2.2.1.6) and satisfies the boundary conditions (2.2.1.7).

To find the initial conditions for the delay ODE (2.2.1.22), we represent the initial condition (2.2.1.8) as an expansion in the eigenfunctions (2.2.1.20):

$$\Phi(x,t) = \sum_{n=1}^{\infty} \Phi_n(t)X_n(x) = \sum_{n=1}^{\infty} \Phi_n(t)\sin\frac{\pi n x}{h}, \quad 0 \leq x \leq h, \quad -\tau \leq t \leq 0. \quad (2.2.1.24)$$

Multiplying (2.2.1.24) by $X_m(x) = \sin\frac{\pi m x}{h}$ ($m = 1, 2, \ldots$), integrating with respect to the space variable x from 0 to h, and taking into account (2.2.1.21), we obtain

$$\Phi_n(t) = \frac{2}{h}\int_0^h \Phi(\xi,t)\sin\left(\frac{\pi n \xi}{h}\right)d\xi, \quad -\tau \leq t \leq 0, \quad (2.2.1.25)$$

where the function $\Phi(\xi,t)$ is defined by (2.2.1.9).

From relations (2.2.1.23) and (2.2.1.24) we find the initial conditions for the delay ODE (2.2.1.17):

$$T_n(t) = \Phi_n(t), \quad -\tau \leq t \leq 0. \quad (2.2.1.26)$$

The functions $\Phi_n(t)$ are defined by (2.2.1.25).

Up to notation, problem (2.2.1.22), (2.2.1.26) coincides with problem (1.1.5.5), (1.1.5.6) discussed in Subsection 1.1.5. Introducing the notation

$$\alpha_n = c_1 - a_1\left(\frac{\pi n}{h}\right)^2, \quad \beta_n = c_2 - a_2\left(\frac{\pi n}{h}\right)^2, \quad \sigma_n = e^{-\alpha_n \tau}\beta_n \quad (2.2.1.27)$$

and using formulas (1.1.5.7) and (1.1.5.3), we represent the solution to problem (2.2.1.22), (2.2.1.26) as

$$T_n(t) = e^{\alpha_n(t+\tau)}\exp_\mathrm{d}(\sigma_n t, \sigma_n \tau)\Phi_n(-\tau)$$
$$+ \int_{-\tau}^0 e^{\alpha_n(t-s)}\exp_\mathrm{d}(\sigma_n(t-\tau-s), \sigma_n \tau)[\Phi'_n(s) - \alpha_n \Phi_n(s)]\,ds, \quad (2.2.1.28)$$

2.2. Linear Initial-Boundary Value Problems with Constant Delay

where $\exp_d(t, \tau)$ is the delayed exponential function, which is defined as

$$\exp_d(t, \tau) \equiv \sum_{k=0}^{[t/\tau]+1} \frac{[t-(k-1)\tau]^k}{k!}. \qquad (2.2.1.29)$$

The symbol $[A]$ stands for the integer part of the number A.

On substituting (2.2.1.20) and (2.2.1.28) into (2.2.1.23), we can write the solution to problem (2.2.1.6)–(2.2.1.7) as [271]:

$$u_1(x,t) = \sum_{n=1}^{\infty} \sin\left(\frac{\pi n x}{h}\right) \Big\{ e^{\alpha_n(t+\tau)} \exp_d(\sigma_n t, \sigma_n \tau) \Phi_n(-\tau)$$

$$+ \int_{-\tau}^{0} e^{\alpha_n(t-s)} \exp_d(\sigma_n(t-\tau-s), \sigma_n \tau) [\Phi'_n(s) - \alpha_n \Phi_n(s)] \, ds \Big\}, \quad (2.2.1.30)$$

where

$$\Phi_n(t) = \frac{2}{h} \int_0^h \Big\{ \varphi(\xi, t) - g_1(t) - \frac{\xi}{h}[g_2(t) - g_1(t)] \Big\} \sin\left(\frac{\pi n \xi}{h}\right) d\xi. \quad (2.2.1.31)$$

▶ **Example 2.6.** For homogeneous boundary conditions, $g_1(t) = g_2(t) \equiv 0$, on a unit-length interval, $h = 1$, and with stationary initial data as a portion of a parabola, $\varphi(x,t) = 4x(1-x)$, the Fourier coefficients (2.2.1.31) are

$$\Phi_n = \begin{cases} \frac{32}{\pi^3 n^3} & \text{for odd } n, \\ 0 & \text{for even } n. \end{cases}$$

◀

Solution of problem 2. We now consider the linear nonhomogeneous delay PDE (2.2.1.10)–(2.2.1.11) with homogeneous boundary and initial conditions (2.2.1.12) and (2.2.1.13).

We first expand the nonhomogeneous component of equation (2.2.1.10) into a series in the eigenfunctions (2.2.1.20):

$$F(x,t) = \sum_{n=1}^{\infty} F_n(t) \sin\frac{\pi n x}{h}, \quad F_n(t) = \frac{2}{h} \int_0^h F(\xi, t) \sin\left(\frac{\pi n \xi}{h}\right) d\xi, \quad (2.2.1.32)$$

where the function $F(x,t)$ is defined by (2.2.1.11).

We seek a solution to problem (2.2.1.10)–(2.2.1.13) as the series

$$u_2(x,t) = \sum_{n=1}^{\infty} U_n(t) \sin\frac{\pi n x}{h}, \qquad (2.2.1.33)$$

which satisfies the homogeneous boundary conditions (2.2.1.12). On substituting (2.2.1.33) into (2.2.1.10), we obtain a linear nonhomogeneous delay ODE for $U_n(t)$:

$$U'_n(t) = \Big[c_1 - a_1\Big(\frac{\pi n}{h}\Big)^2\Big] U_n(t) + \Big[c_2 - a_2\Big(\frac{\pi n}{h}\Big)^2\Big] U_n(t-\tau) + F_n(t), \quad (2.2.1.34)$$

where the functions $F_n(t)$ are determined with the second formula in (2.2.1.32). To complete the statement of the problem, we supplement equation (2.2.1.34) with the homogeneous initial conditions

$$U_n(t) = 0, \quad -\tau \le t \le 0, \qquad (2.2.1.35)$$

which follow from (2.2.1.13) and (2.2.1.33).

Up to notation, problem (2.2.1.34)–(2.2.1.35) coincides with problem (1.1.5.8)–(1.1.5.9) discussed in Subsection 1.1.5. Therefore, the solution to problem (2.2.1.12)–(2.2.1.35) for $t \ge 0$ can be represented as the integral

$$U_n(t) = \int_0^t e^{\alpha_n(t-s)} \exp_{\mathrm{d}}(\sigma_n(t-s), \sigma_n \tau) F_n(s)\, ds, \quad \sigma_n = e^{-\alpha_n \tau} \beta_n, \qquad (2.2.1.36)$$

with the parameters α_n and β_n defined in (2.2.1.27). Substituting (2.2.1.36) into (2.2.1.33) yields the following solution to problem (2.2.1.10)–(2.2.1.13) [271]:

$$u_2(x,t) = \sum_{n=1}^{\infty} \left[\int_0^t e^{\alpha_n(t-s)} \exp_{\mathrm{d}}(\sigma_n(t-s), \sigma_n \tau) F_n(s)\, ds \right] \sin \frac{\pi n x}{h}. \qquad (2.2.1.37)$$

On substituting the functions (2.2.1.5), (2.2.1.30), and (2.2.1.37) into (2.2.1.4), one arrives at the solution to the original problem (2.2.1.1)–(2.2.1.3).

Remark 2.1. *The studies [270, 271] describe sufficient conditions for the convergence of the series (2.2.1.30) and (2.2.1.37), which define the solution to problem (2.2.1.1)–(2.2.1.3). The solution stability and instability conditions for this problem are discussed in Section 2.2.5, which shows that if the inequality $a_2 > a_1$ holds, the initial-boundary value problem (2.2.1.1)–(2.2.1.3) is ill-posed in the sense of Hadamard.*

2.2.2. Other Problems for a One-Dimensional Parabolic Equation with Constant Delay

Representation of solutions to initial-boundary value problems as the sum of solutions to simpler problems. Below we outline the procedure of constructing solutions by separation of variables for other initial-boundary value problems described by the one-dimensional linear partial differential equation with constant delay (2.2.1.1). For brevity, we will write the equation as

$$\mathcal{L}[u,w] = f(x,t), \quad t > 0, \qquad (2.2.2.1)$$

where $\mathcal{L}[u,w] \equiv u_t - a_1 u_{xx} - a_2 w_{xx} - c_1 u - c_2 w$, $w = u(x, t-\tau)$, and $a_1 > a_2 \ge 0$.

We will study equation (2.2.2.1) while supplying it with different linear nonhomogeneous boundary conditions, which we will write in the general form

$$\begin{aligned} \Gamma_1[u] &= g_1(t) \quad \text{at} \quad x = 0, \ t > -\tau; \\ \Gamma_2[u] &= g_2(t) \quad \text{at} \quad x = h, \ t > -\tau, \end{aligned} \qquad (2.2.2.2)$$

and the general initial condition

$$u = \varphi(x,t) \quad \text{at} \quad 0 < x < h, \ -\tau \leq t \leq 0. \tag{2.2.2.3}$$

We assume that the linear operators $\Gamma_{1,2}[u]$ involved in the boundary conditions (2.2.2.2) are explicitly independent of t. The most common boundary conditions are shown in the third row of Table 2.2.

We look for a solution to problem (2.2.2.1)–(2.2.2.3) as the sum (2.2.1.4), where

$$u_0 = u_0(x,t) \tag{2.2.2.4}$$

is any twice continuously differentiable function that satisfies the boundary conditions (2.2.2.2), implying that

$$\Gamma_1[u_0] = g_1(t) \quad \text{at} \quad x = 0, \quad \Gamma_2[u_0] = g_2(t) \quad \text{at} \quad x = h. \tag{2.2.2.5}$$

The definition of u_0 is unrelated to the solution of the differential equation. This function can be sought by the method of undetermined coefficients in the form, for example, of a quadratic polynomial in x: $u_0 = \alpha_0(t) + \alpha_1(t)x + \alpha_2(t)x^2$ (in most cases, setting $\alpha_2 \equiv 0$ would suffice). The functional coefficients $\alpha_k(t)$ are determined by substituting the polynomial into the boundary conditions (2.2.2.5).

Table 2.2 lists the simplest functions $u_0 = u_0(x,t)$ that satisfy the most common nonhomogeneous boundary conditions in initial-boundary value problems for parabolic or hyperbolic equations with a single space variable. In the boundary conditions of the third kind, it is assumed that $k_1 > 0$ and $k_2 > 0$.

Table 2.2. Simplest functions $u_0 = u_0(x,t)$ that satisfy the most common nonhomogeneous boundary conditions at the endpoints of the interval $0 \leq x \leq h$.

No.	Initial-boundary value problem	Boundary conditions	Function $u_0 = u_0(x,t)$ satisfying the boundary conditions
1	First	$u = g_1(t)$ at $x=0$ $u = g_2(t)$ at $x=h$	$u_0 = g_1(t) + \dfrac{x}{h}\big[g_2(t) - g_1(t)\big]$
2	Second	$u_x = g_1(t)$ at $x=0$ $u_x = g_2(t)$ at $x=h$	$u_0 = xg_1(t) + \dfrac{x^2}{2h}\big[g_2(t) - g_1(t)\big]$
3	Third	$u_x - k_1 u = g_1(t)$ at $x=0$ $u_x + k_2 u = g_2(t)$ at $x=h$	$u_0 = \dfrac{(k_2 x - 1 - k_2 h)g_1(t) + (1 + k_1 x)g_2(t)}{k_2 + k_1 + k_1 k_2 h}$
4	Mixed	$u = g_1(t)$ at $x=0$ $u_x = g_2(t)$ at $x=h$	$u_0 = g_1(t) + xg_2(t)$
5	Mixed	$u_x = g_1(t)$ at $x=0$ $u = g_2(t)$ at $x=h$	$u_0 = (x - l)g_1(t) + g_2(t)$

Remark 2.2. *The boundary conditions of the first, second, and third initial-boundary value of problems are also known as Dirichlet, Neumann, and Robin boundary conditions, respectively.*

The other two functions, $u_1 = u_1(x,t)$ and $u_2 = u_2(x,t)$, appearing in (2.2.1.4) are determined by solving the simpler initial-boundary value problems with homogeneous (zero) boundary conditions stated below.

Problem 1. The function u_1 satisfies the linear homogeneous PDE with constant delay
$$\mathcal{L}[u_1, w_1] = 0, \quad w_1 = u_1(x, t - \tau), \qquad (2.2.2.6)$$

homogeneous boundary conditions
$$\Gamma_1[u_1] = 0 \quad \text{at} \quad x = 0, \, t > -\tau; \quad \Gamma_2[u_1] = 0 \quad \text{at} \quad x = h, \, t > -\tau, \qquad (2.2.2.7)$$

and nonhomogeneous initial condition
$$u_1 = \Phi(x, t) \quad \text{at} \quad 0 < x < h, \, -\tau \leq t \leq 0, \qquad (2.2.2.8)$$

where
$$\Phi(x, t) = \varphi(x, t) - u_0(x, t). \qquad (2.2.2.9)$$

Problem 2. The function u_2 satisfies the linear nonhomogeneous PDE with constant delay
$$\mathcal{L}[u_2, w_2] = F(x, t), \quad w_2 = u_2(x, t - \tau), \qquad (2.2.2.10)$$

where
$$F(x, t) = f(x, t) - \mathcal{L}[u_0, w_0], \quad w_0 = u_0(x, t - \tau), \qquad (2.2.2.11)$$

and the zero boundary and initial conditions
$$\Gamma_1[u_2] = 0 \quad \text{at} \quad x = 0, \, t > -\tau; \quad \Gamma_2[u_2] = 0 \quad \text{at} \quad x = h, \, t > -\tau; \qquad (2.2.2.12)$$
$$u_2 = 0 \quad \text{at} \quad 0 < x < h, \, -\tau \leq t \leq 0. \qquad (2.2.2.13)$$

Solution of problem 1. Let us look at the linear homogeneous PDE with delay (2.2.1.6) (or (2.2.2.6)) subjected to the boundary and initial conditions (2.2.2.7) and (2.2.2.8). Just as previously, we first look for particular solutions to equation (2.2.1.6) as the product of functions with different arguments (2.2.1.14): $u_{1p} = X(x)T(t)$. On separating the variables in the resulting equation, we arrive at the linear ODE and linear delay ODE
$$X''(x) = -\lambda^2 X(x), \qquad (2.2.2.14)$$
$$T'(t) = (c_1 - a_1 \lambda^2) T(t) + (c_2 - a_2 \lambda^2) T(t - \tau), \qquad (2.2.2.15)$$

which coincide with equations (2.2.1.16) and (2.2.1.17). Requiring that the function $u_{1p} = X(x)T(t)$ must satisfy the homogeneous boundary conditions (2.2.1.7), we obtain the following homogeneous boundary conditions for X:
$$\Gamma_1[X] = 0 \quad \text{at} \quad x = 0, \quad \Gamma_2[X] = 0 \quad \text{at} \quad x = h. \qquad (2.2.2.16)$$

The linear homogeneous eigenvalue problem (2.2.2.14), (2.2.2.16) has nontrivial solutions $X = X_n(x)$ for only a discrete set of values of λ:
$$\lambda = \lambda_n, \quad X = X_n(x), \quad n = 1, 2, \ldots \qquad (2.2.2.17)$$

Table 2.3 lists eigenvalues and eigenfunctions of homogeneous linear boundary value problems described by ODE (2.2.2.14) subjected to the five most common boundary conditions.

Table 2.3. Eigenvalues and eigenfunctions of eigenvalue problems described by the homogeneous ODE $X''_{xx} = -\lambda^2 X$ subjected to the most common homogeneous boundary conditions at the endpoints of the interval $0 \leq x \leq h$.

No.	Initial-boundary value problem	Boundary conditions	Eigenvalues λ_n and eigenfunctions $X_n = X_n(x)$, $n = 1, 2, \ldots$
1	First	$X = 0$ at $x = 0$ $X = 0$ at $x = h$	$\lambda_n = \dfrac{\pi n}{h}$; $X_n = \sin \dfrac{\pi n x}{h}$
2	Second	$X'_x = 0$ at $x = 0$ $X'_x = 0$ at $x = h$	$\lambda_0 = 0$, $\lambda_n = \dfrac{\pi n}{h}$; $X_0 = 1$, $X_n = \cos \dfrac{\pi n x}{h}$
3	Third	$X'_x - k_1 X = 0$ at $x = 0$ $X'_x + k_2 X = 0$ at $x = h$	λ_n are roots of the transcendental equation $\dfrac{\tan(\lambda h)}{\lambda} = \dfrac{k_1 + k_2}{\lambda^2 - k_1 k_2}$ ($\lambda_n > 0$); $X_n = \cos(\lambda_n x) + \dfrac{k_1}{\lambda_n} \sin(\lambda_n x)$
4	Mixed	$X = 0$ at $x = 0$ $X'_x = 0$ at $x = h$	$\lambda_n = \dfrac{\pi(2n-1)}{2h}$; $X_n = \sin \dfrac{\pi(2n-1)x}{2h}$
5	Mixed	$X'_x = 0$ at $x = 0$ $X = 0$ at $x = h$	$\lambda_n = \dfrac{\pi(2n-1)}{2h}$; $X_n = \cos \dfrac{\pi(2n-1)x}{2h}$

On substituting the eigenvalues $\lambda = \lambda_n$ into (2.2.2.15), we obtain the respective delay ODEs for the functions $T = T_n(t)$.

We look for a solution to the linear initial-boundary value problems (2.2.2.6)–(2.2.2.9) as the series

$$u_1(x, t) = \sum_{n=1}^{\infty} X_n(x) T_n(t), \qquad (2.2.2.18)$$

where the functions $u_{1n}(x, t) = X_n(x) T_n(t)$ are particular solutions to equation (2.2.2.6) satisfying the homogeneous boundary conditions (2.2.2.7).

To determine the initial conditions for the delay ODE (2.2.2.15) at $\lambda = \lambda_n$, we rewrite the initial condition (2.2.2.8) as a series expansion in the eigenfunctions:

$$\Phi(x, t) = \sum_{n=1}^{\infty} \Phi_n(t) X_n(x), \quad 0 \leq x \leq h, \ -\tau \leq t \leq 0. \qquad (2.2.2.19)$$

Multiplying (2.2.2.19) by $X_m(x)$ ($m = 1, 2, \ldots$), integrating with respect to the space variable x from 0 to h, and taking into account that the eigenfunctions $X_n(x)$ and $X_m(x)$ are orthogonal if $n \neq m$, implying that relations (2.2.1.21) hold, we obtain

$$\Phi_n(t) = \frac{1}{\|X_n\|^2} \int_0^h \Phi(\xi, t) X_n(\xi)\, d\xi, \quad -\tau \leq t \leq 0. \tag{2.2.2.20}$$

The function $\Phi(\xi, t)$ is defined by formula (2.2.2.9), and $\|X_n\|^2 = \int_0^h X_n^2(\xi)\, d\xi$. From relations (2.2.2.18) and (2.2.2.19) we obtain the initial conditions for the delay ODE (2.2.2.15) at $\lambda = \lambda_n$:

$$T_n(t) = \Phi_n(t), \quad -\tau \leq t \leq 0, \tag{2.2.2.21}$$

where the functions $\Phi_n(t)$ are given by (2.2.2.20).

Introducing the notations

$$\alpha_n = c_1 - a_1 \lambda_n^2, \quad \beta_n = c_2 - a_2 \lambda_n^2, \quad \sigma_n = e^{-\alpha_n \tau} \beta_n \tag{2.2.2.22}$$

and reasoning as in Subsection 2.2.1, we obtain the solution to problem (2.2.2.15), (2.2.2.21) at $\lambda = \lambda_n$ in the form

$$T_n(t) = e^{\alpha_n(t+\tau)} \exp_\mathrm{d}(\sigma_n t, \sigma_n \tau) \Phi_n(-\tau)$$
$$+ \int_{-\tau}^0 e^{\alpha_n(t-s)} \exp_\mathrm{d}(\sigma_n(t-\tau-s), \sigma_n \tau)[\Phi_n'(s) - \alpha_n \Phi_n(s)]\, ds, \tag{2.2.2.23}$$

where $\exp_\mathrm{d}(t, \tau) \equiv \sum_{k=0}^{[t/\tau]+1} \frac{[t-(k-1)\tau]^k}{k!}$ is the delayed exponential function.

On substituting the functions (2.2.2.23) into formula (2.2.2.18), we find the solution to problem (2.2.2.6)–(2.2.2.9):

$$u_1(x, t) = \sum_{n=1}^\infty X_n(x) \bigg\{ e^{\alpha_n(t+\tau)} \exp_\mathrm{d}(\sigma_n t, \sigma_n \tau) \Phi_n(-\tau)$$
$$+ \int_{-\tau}^0 e^{\alpha_n(t-s)} \exp_\mathrm{d}(\sigma_n(t-\tau-s), \sigma_n \tau)[\Phi_n'(s) - \alpha_n \Phi_n(s)]\, ds \bigg\}, \tag{2.2.2.24}$$

where

$$\Phi_n(t) = \frac{1}{\|X_n\|^2} \int_0^h [\varphi(\xi, t) - u_0(\xi, t)] X_n(\xi)\, d\xi, \quad \|X_n\|^2 = \int_0^h X_n^2(\xi)\, d\xi. \tag{2.2.2.25}$$

For any of the five main initial-boundary value problems, whose boundary conditions are presented in Table 2.2, the eigenvalues λ_n and eigenfunctions $X_n(x)$ in formulas (2.2.2.24) and (2.2.2.25) should be taken from Table 2.3.

Solution of problem 2. Now we consider the linear nonhomogeneous delay PDE (2.2.2.10)–(2.2.2.11) with the homogeneous boundary and initial conditions (2.2.2.12) and (2.2.2.13).

2.2. Linear Initial-Boundary Value Problems with Constant Delay

We first expand the nonhomogeneous component of equation (2.2.2.10) into a series in eigenfunctions (2.2.2.17):

$$F(x,t) = \sum_{n=1}^{\infty} F_n(t) X_n(x), \quad F_n(t) = \frac{1}{\|X_n\|^2} \int_0^h F(\xi, t) X_n(\xi)\, d\xi, \quad (2.2.2.26)$$

where the function $F(x,t)$ is defined by (2.2.2.11), and $\|X_n\|^2 = \int_0^h X_n^2(\xi)\, d\xi$. Then we look for a solution to problem (2.2.2.10)–(2.2.2.13) as the series

$$u_2(x,t) = \sum_{n=1}^{\infty} U_n(t) X_n(x), \quad (2.2.2.27)$$

which satisfies the homogeneous boundary conditions (2.2.2.12). Inserting (2.2.2.27) into (2.2.2.10) and taking into account (2.2.2.26), we obtain linear nonhomogeneous delay ODEs for $U_n(t)$:

$$U_n'(t) = (c_1 - a_1 \lambda_n^2) U_n(t) + (c_2 - a_2 \lambda_n^2) U_n(t - \tau) + F_n(t), \quad (2.2.2.28)$$

where the functions $F_n(t)$ are determined using the second formula in (2.2.2.26). To complete the statement of the problem, we supplement equations (2.2.2.28) with the homogeneous initial conditions

$$U_n(t) = 0, \quad -\tau \leq t \leq 0, \quad (2.2.2.29)$$

which follow from (2.2.2.13) and (2.2.2.27).

Up to notation, problem (2.2.2.28)–(2.2.2.29) coincides with problem (2.2.1.34)–(2.2.1.35). Therefore, its solution for $t \geq 0$ can be represented as

$$U_n(t) = \int_0^t e^{\alpha_n(t-s)} \exp_d(\sigma_n(t-s), \sigma_n \tau) F_n(s)\, ds, \quad \sigma_n = e^{-\alpha_n \tau} \beta_n, \quad (2.2.2.30)$$

where the parameters α_n and β_n are defined in (2.2.2.22). Substituting (2.2.2.30) into (2.2.2.27) yields the solution to problem (2.2.2.10)–(2.2.2.13):

$$u_2(x,t) = \sum_{n=1}^{\infty} \left[\int_0^t e^{\alpha_n(t-s)} \exp_d(\sigma_n(t-s), \sigma_n \tau) F_n(s)\, ds \right] X_n(x). \quad (2.2.2.31)$$

A solution to the initial-boundary value problem (2.2.2.1)–(2.2.2.3) with any boundary conditions presented in Table 2.2 can be obtained by substituting the functions (2.2.2.4), (2.2.2.24), and (2.2.2.31) into (2.2.1.4) and taking the functions $u_0 = u_0(x,t)$ from Table 2.2. The respective eigenvalues λ_n and eigenfunctions $X_n(x)$ should be taken from Table 2.3.

2.2.3. Problems for Linear Parabolic Equations with Several Variables and Constant Delay

Statement of the problem. Consider the m-dimensional linear parabolic equation with constant delay

$$u_t = a_1 L[u] + a_2 L[w] + c_1 u + c_2 w + f(\mathbf{x}, t), \quad w = u(\mathbf{x}, t - \tau), \quad (2.2.3.1)$$
$$a_1 > a_2 \geq 0, \quad \mathbf{x} = (x_1, \ldots, x_m),$$

defined in a domain $\Omega = \{\mathbf{x} \in V, \ t > 0\}$, where V is an open connected bounded domain in $\mathbb{R}^m$ with a smooth boundary $S = \partial V$.

We assume that the coefficients of the second-order linear differential operator with respect to the space variables $L[u]$ appearing in (2.2.3.1) can be dependent on $x_1, \ldots, x_m$ but are independent of time t. In particular, the right-hand side of equation (2.2.3.1) can involve the m-dimensional Laplace operator $L[u] = \Delta u \equiv \sum_{i=1}^{m} \frac{\partial^2 u}{\partial x_i^2}$ or a more complex operator with variable coefficients, $L[u] = \text{div}[p(\mathbf{x})\nabla u]$, where $p(\mathbf{x}) > 0$.

We supplement equation (2.2.3.1) with a nonhomogeneous linear boundary condition that we will write as

$$\Gamma[u] = g(\mathbf{x}, t) \quad \text{at} \quad \mathbf{x} \in S, \ t > -\tau, \quad (2.2.3.2)$$

and the general initial condition

$$u = \varphi(\mathbf{x}, t) \quad \text{at} \quad \mathbf{x} \in V, \ -\tau \leq t \leq 0. \quad (2.2.3.3)$$

The coefficients of the differential operator $\Gamma[u]$ in (2.2.3.2) can depend on the space variable $x_1, \ldots, x_m$ but are independent of time t.

The construction of solutions to problem (2.2.3.1)–(2.2.3.3) consists of a few stages described below.

Representation of solutions to the problem concerned as the sum of solutions to simpler problems. We seek a solution to problem (2.2.3.1)–(2.2.3.3) as the sum

$$u = u_0(\mathbf{x}, t) + u_1(\mathbf{x}, t) + u_2(\mathbf{x}, t), \quad (2.2.3.4)$$

where

$$u_0 = u_0(\mathbf{x}, t) \quad (2.2.3.5)$$

is any twice differentiable function satisfying the boundary condition (2.2.3.2), which implies that

$$\Gamma[u_0] = g(\mathbf{x}, t) \quad \text{at} \quad \mathbf{x} \in S, \ t > -\tau. \quad (2.2.3.6)$$

Remark 2.3. For the Dirichlet boundary condition, one should set $\Gamma[u] \equiv u$ in (2.2.3.2). In this case, any sufficiently smooth function $G(\mathbf{x}, t)$ defined for $t > -\tau$ in the closed domain $V \bigcup S$ and satisfying the condition $G(\mathbf{x}, t)|_{\mathbf{x} \in S} = g(\mathbf{x}, t)$ can be chosen as u_0.

The functions $u_1 = u_1(\mathbf{x}, t)$ and $u_2 = u_2(\mathbf{x}, t)$ in (2.2.3.4) are determined by solving two initial-boundary value of problems, simpler than the original one, with homogeneous (zero) boundary conditions as described below.

Problem 1. The function u_1 satisfies the linear homogeneous PDE with constant delay

$$(u_1)_t = a_1 L[u_1] + a_2 L[w_1] + c_1 u_1 + c_2 w_1, \quad w_1 = u_1(\mathbf{x}, t - \tau), \quad t > 0, \quad (2.2.3.7)$$

homogeneous boundary condition

$$\Gamma[u_1] = 0 \quad \text{at} \quad \mathbf{x} \in S, \ t > -\tau, \quad (2.2.3.8)$$

and nonhomogeneous initial condition

$$u_1 = \Phi(\mathbf{x}, t) \quad \text{at} \quad \mathbf{x} \in V, \ -\tau \leq t \leq 0, \quad (2.2.3.9)$$

where

$$\Phi(\mathbf{x}, t) = \varphi(\mathbf{x}, t) - u_0(\mathbf{x}, t). \quad (2.2.3.10)$$

Problem 2. The function u_2 satisfies the linear nonhomogeneous PDE with constant delay

$$(u_1)_2 = a_1 L[u_2] + a_2 L[w_2] + c_1 u_2 + c_2 w_2 + F(\mathbf{x}, t), \quad w_2 = u_2(\mathbf{x}, t - \tau), \quad (2.2.3.11)$$

where

$$F(\mathbf{x}, t) = f(\mathbf{x}, t) - (u_0)_t + a_1 L[u_0] + a_2 L[w_0] + c_1 u_0 + c_2 w_0, \quad w_0 = u_0(\mathbf{x}, t - \tau), \quad (2.2.3.12)$$

and the zero boundary and initial conditions

$$\Gamma[u_2] = 0 \quad \text{at} \quad \mathbf{x} \in S, \ t > -\tau; \quad (2.2.3.13)$$
$$u_2 = 0 \quad \text{at} \quad \mathbf{x} \in V, \ -\tau \leq t \leq 0. \quad (2.2.3.14)$$

Solution of problem 1. Consider the linear homogeneous delay PDE (2.2.3.7) with the boundary and initial conditions (2.2.3.8) and (2.2.3.9). As previously, we first look for particular solutions to equation (2.2.3.7) as the product of functions with different arguments

$$u_{1p} = X(\mathbf{x}) T(t). \quad (2.2.3.15)$$

Substituting (2.2.3.15) into (2.2.3.7) and separating the variables in the resulting equation, we arrive at a second-order linear partial differential equation and a first-order ODE with constant delay:

$$L[X] = -\mu X, \quad (2.2.3.16)$$
$$T'(t) = (c_1 - a_1 \mu) T(t) + (c_2 - a_2 \mu) T(t - \tau). \quad (2.2.3.17)$$

For $\mu = \lambda^2$, equation (2.2.3.17) coincides with (2.2.2.15). Requiring that the function (2.2.3.15) must satisfy the homogeneous boundary condition (2.2.3.8), we arrive at a homogeneous boundary condition for $X = X(\mathbf{x})$:

$$\Gamma[X] = 0 \quad \text{at} \quad \mathbf{x} \in S. \quad (2.2.3.18)$$

With respect to the linear homogeneous stationary eigenvalue problem (2.2.3.16), (2.2.3.18), we assume that the following three standard conditions hold.

1°. Problem (2.2.3.16), (2.2.3.18) has nontrivial solutions for only a discrete set of eigenvalues, $\mu = \mu_n$, $n = 1, 2, \ldots$, such that

$$0 \leq \mu_1 \leq \mu_2 \leq \cdots \leq \mu_n \leq \mu_{n+1} \leq \cdots, \quad \mu_n \to \infty \text{ as } n \to \infty. \quad (2.2.3.19)$$

Each eigenvalue appears in this ordered sequence as many times as its multiplicity.

2°. The eigenfunctions, $X = X_n(\mathbf{x})$, corresponding to the eigenvalues $\mu = \mu_n$, can be chosen real-valued and orthonormal, so that the relations

$$\int_V X_m(\mathbf{x}) X_n(\mathbf{x}) \, dV = \delta_{mn}, \quad \delta_{mn} = \begin{cases} 1 & \text{if } m = n, \\ 0 & \text{if } m \neq n \end{cases} \quad (2.2.3.20)$$

hold.

3°. Any function $F(\mathbf{x})$ that is twice continuously differentiable in an open domain $V_{+\varepsilon}$ (containing V) and satisfies the boundary condition (2.2.3.18) is expandable in a Fourier series in the orthonormal system of eigenfunctions $X_n(\mathbf{x})$:

$$F(\mathbf{x}) = \sum_{n=1}^{\infty} F_n X_n(\mathbf{x}), \quad F_n = \int_V F(\mathbf{x}) X_n(\mathbf{x}) \, dV. \quad (2.2.3.21)$$

This series is assumed to be regularly convergent in the domain $\bar{V} = V \cup S$.

▶ Example 2.7. The linear homogeneous eigenvalue problem [534] determined by the operator $L[X] \equiv \text{div}[p(\mathbf{x}) \nabla X]$ and described by the equation

$$\text{div}[p(\mathbf{x}) \nabla X] = -\mu X, \quad \mathbf{x} \in V, \quad (2.2.3.22)$$

subjected to the boundary condition

$$\alpha(\mathbf{x}) X + \beta(\mathbf{x}) \frac{\partial X}{\partial \nu} = 0 \quad \text{at} \quad \mathbf{x} \in S \quad (2.2.3.23)$$

satisfies conditions 1°–3° above. The symbol $\partial/\partial \nu$ stands for a derivative with respect to the outward normal to the surface S. The functional coefficients in (2.2.3.22) and (2.2.3.23) satisfy the conditions

$$\begin{aligned} & p(\mathbf{x}) \in C^1(\bar{V}), \quad \alpha(\mathbf{x}) \in C(S), \quad \beta(\mathbf{x}) \in C(S), \\ & p(\mathbf{x}) > 0, \quad \alpha(\mathbf{x}) \geq 0, \quad \beta(\mathbf{x}) \geq 0, \quad \alpha(\mathbf{x}) + \beta(\mathbf{x}) > 0. \end{aligned} \quad (2.2.3.24)$$

The linear homogeneous eigenvalue problems for the *Helmholtz equation*

$$\Delta X = -\mu X, \quad \mathbf{x} \in V \quad (L = \Delta), \quad (2.2.3.25)$$

with homogeneous Dirichlet, Neumann, and Robin boundary conditions are important special cases of problem (2.2.3.22)–(2.2.3.23).

The book [404] lists eigenvalues and eigenfunctions of many boundary value problems for equation (2.2.3.25) with various homogeneous boundary conditions for domains V of different shape. ◀

2.2. Linear Initial-Boundary Value Problems with Constant Delay

In what follows, we assume that the eigenvalues and eigenfunctions of the linear homogeneous boundary value problem (2.2.3.16), (2.2.3.18) are known, and condition 1°–3° above are satisfied. On substituting the eigenvalues $\mu = \mu_n$ into (2.2.3.17), we obtain respective delay ODEs for the functions $T = T_n(t)$.

We will look for solutions to the linear initial-boundary value problem (2.2.3.7)–(2.2.3.10) as the series

$$u_1(\mathbf{x}, t) = \sum_{n=1}^{\infty} X_n(\mathbf{x}) T_n(t), \quad (2.2.3.26)$$

with the functions $u_{1n}(\mathbf{x}, t) = X_n(\mathbf{x}) T_n(t)$ being particular solutions to equation (2.2.3.7) satisfying the homogeneous boundary condition (2.2.3.8).

To determine the initial conditions for the delay ODE (2.2.3.17), we rewrite condition (2.2.3.9) to expand into a series in eigenfunctions:

$$\Phi(\mathbf{x}, t) = \sum_{n=1}^{\infty} \Phi_n(t) X_n(\mathbf{x}), \quad \mathbf{x} \in V, \ -\tau \leq t \leq 0. \quad (2.2.3.27)$$

Multiplying (2.2.3.27) by $X_m(\mathbf{x})$ ($m = 1, 2, \ldots$), integrating over the domain V, and taking into account (2.2.3.20), we find the functions $\Phi_n(t)$:

$$\Phi_n(t) = \int_V \Phi(\mathbf{x}, t) X_n(\mathbf{x}) \, dV, \quad -\tau \leq t \leq 0. \quad (2.2.3.28)$$

The function $\Phi(\mathbf{x}, t)$ is defined by formula (2.2.3.10). From relations (2.2.3.26) and (2.2.3.28) we obtain the following initial conditions for the delay ODE (2.2.3.17) with $\mu = \mu_n$:

$$T_n(t) = \Phi_n(t), \quad -\tau \leq t \leq 0, \quad (2.2.3.29)$$

where the functions $\Phi_n(t)$ are given by (2.2.3.28).

Since problem (2.2.3.17), (2.2.3.29) with $\mu = \mu_n$ coincides, up to notation, with problem (2.2.2.15), (2.2.2.21), its solution can be expressed as

$$T_n(t) = e^{\alpha_n(t+\tau)} \exp_\mathrm{d}(\sigma_n t, \sigma_n \tau) \Phi_n(-\tau)$$
$$+ \int_{-\tau}^{0} e^{\alpha_n(t-s)} \exp_\mathrm{d}(\sigma_n(t - \tau - s), \sigma_n \tau) [\Phi_n'(s) - \alpha_n \Phi_n(s)] \, ds, \quad (2.2.3.30)$$

$$\alpha_n = c_1 - a_1 \mu_n, \quad \beta_n = c_2 - a_2 \mu_n, \quad \sigma_n = e^{-\alpha_n \tau} \beta_n,$$

where $\exp_\mathrm{d}(t, \tau)$ is the delayed exponential function (2.2.1.29).

Substituting the functions (2.2.3.30) into (2.2.3.26) yields the solution to problem (2.2.3.7)–(2.2.3.10):

$$u_1(\mathbf{x}, t) = \sum_{n=1}^{\infty} X_n(\mathbf{x}) \bigg\{ e^{\alpha_n(t+\tau)} \exp_\mathrm{d}(\sigma_n t, \sigma_n \tau) \, \Phi_n(-\tau)$$
$$+ \int_{-\tau}^{0} e^{\alpha_n(t-s)} \exp_\mathrm{d}(\sigma_n(t - \tau - s), \sigma_n \tau) [\Phi_n'(s) - \alpha_n \Phi_n(s)] \, ds \bigg\}, \quad (2.2.3.31)$$

where $\Phi_n(t) = \int_V [\varphi(\mathbf{x}, t) - u_0(\mathbf{x}, t)] X_n(\mathbf{x}) \, dV$.

Solution of problem 2. We now look at the linear nonhomogeneous delay PDE (2.2.3.11)–(2.2.3.12) with the homogeneous boundary and initial conditions (2.2.3.13) and (2.2.3.14).

First, we expand the nonhomogeneous component of equation (2.2.3.11) into a series in eigenfunctions of problem (2.2.3.16), (2.2.3.18):

$$F(\mathbf{x}, t) = \sum_{n=1}^{\infty} F_n(t) X_n(\mathbf{x}), \quad F_n(t) = \int_V F(\mathbf{x}, t) X_n(\mathbf{x}) \, d\xi, \qquad (2.2.3.32)$$

where the function $F(\mathbf{x}, t)$ is defined by formula (2.2.3.12).

Then, we look for a solution to problem (2.2.3.11)–(2.2.3.14) as the series

$$u_2(\mathbf{x}, t) = \sum_{n=1}^{\infty} U_n(t) X_n(\mathbf{x}), \qquad (2.2.3.33)$$

which satisfies the homogeneous boundary conditions (2.2.3.13). On substituting (2.2.3.33) into (2.2.3.11) and taking into account (2.2.3.32), we obtain linear nonhomogeneous ODEs with constant delay for $U_n(t)$:

$$U_n'(t) = (c_1 - a_1 \mu_n) U_n(t) + (c_2 - a_2 \mu_n) U_n(t - \tau) + F_n(t), \qquad (2.2.3.34)$$

where the functions $F_n(t)$ are found from the second formula in (2.2.3.32). To complete the statement of the problem, we supplement equations (2.2.3.34) with the homogeneous initial conditions

$$U_n(t) = 0, \quad -\tau \leq t \leq 0, \qquad (2.2.3.35)$$

which follow from (2.2.3.14) and (2.2.3.33).

Up to notation, problem (2.2.3.34)–(2.2.3.35) coincides with problem (2.2.2.28)–(2.2.2.29). Therefore, its solution for $t \geq 0$ can be represented as

$$U_n(t) = \int_0^t e^{\alpha_n(t-s)} \exp_d(\sigma_n(t-s), \sigma_n \tau) F_n(s) \, ds, \quad \sigma_n = e^{-\alpha_n \tau} \beta_n, \qquad (2.2.3.36)$$

where the parameters α_n and β_n are defined in (2.2.3.30). On substituting (2.2.3.36) into (2.2.3.33), we obtain the solution to problem (2.2.3.11)–(2.2.3.14):

$$u_2(\mathbf{x}, t) = \sum_{n=1}^{\infty} \left[\int_0^t e^{\alpha_n(t-s)} \exp_d(\sigma_n(t-s), \sigma_n \tau) F_n(s) \, ds \right] X_n(\mathbf{x}). \qquad (2.2.3.37)$$

Formulas (2.2.3.4), (2.2.3.31), and (2.2.3.37) form a basis for the construction of analytical solutions to the original problem (2.2.3.1)–(2.2.3.3). The eigenvalues μ_n and eigenfunctions are determined by solving the homogeneous problem (2.2.3.16), (2.2.3.18).

▶ **Example 2.8.** Setting $L[u] = \Delta u$ in (2.2.3.1), where Δ is the Laplace operator, we arrive at an eigenvalue problem for the homogeneous Helmholtz equation

(2.2.3.25). In the three-dimensional case, the homogeneous Dirichlet boundary conditions on the faces of a rectangular parallelepiped with edges a, b, and c are written as

$$X|_{x=0} = X|_{x=a} = X|_{y=0} = X|_{y=b} = X|_{z=0} = X|_{z=c} = 0.$$

Then the eigenvalues and normalized eigenfunctions of the problem are [404]:

$$\mu_{klm} = \pi^2\left(\frac{k^2}{a^2} + \frac{l^2}{b^2} + \frac{m^2}{c^2}\right), \quad k,l,m = 1, 2, 3, \ldots,$$

$$X_{klm}(x,y,z) = \frac{8}{abc}\sin\left(\frac{\pi kx}{a}\right)\sin\left(\frac{\pi ly}{b}\right)\sin\left(\frac{\pi mz}{c}\right).$$

◀

2.2.4. Problems for Linear Hyperbolic Equations with Constant Delay

Statement of the problem. Consider the initial-boundary value problem for the one-dimensional homogeneous linear hyperbolic equation with constant coefficients and a constant delay

$$u_{tt} = a_1 u_{xx} + a_2 w_{xx} + c_1 u + c_2 w, \quad w = u(x, t-\tau), \qquad (2.2.4.1)$$

where $a_1 > a_2 \geq 0$. The equation is defined in the domain $\Omega = \{0 < x < h,\ t > 0\}$. We supplement equation (2.2.4.1) with the homogeneous boundary conditions

$$\begin{aligned}\Gamma_1[u] &= 0 \quad \text{at} \quad x = 0,\ t > -\tau;\\ \Gamma_2[u] &= 0 \quad \text{at} \quad x = h,\ t > -\tau,\end{aligned} \qquad (2.2.4.2)$$

and consistent initial conditions

$$\begin{aligned}u &= \varphi(x,t) \quad \text{at} \quad 0 < x < h,\ -\tau \leq t \leq 0,\\ u_t &= \varphi_t(x,t) \quad \text{at} \quad 0 < x < h,\ -\tau \leq t \leq 0.\end{aligned} \qquad (2.2.4.3)$$

We assume that the linear operators $\Gamma_{1,2}[u]$ involved in the boundary condition (2.2.4.2) are explicitly independent of t. The most common boundary conditions are specified in the third row of Table 2.2, where one should set $g_1(t) = g_2(t) \equiv 0$. In particular, in the case of the boundary Dirichlet boundary conditions, one should set $\Gamma_1[u] = \Gamma_2[u] = u$ in (2.2.4.2).

Construction of solutions to problem (2.2.4.1)–(2.2.4.3). We first look for particular solutions of equation (2.2.4.1) as the product of functions with different arguments: $u_p = X(x)T(t)$. Separating the variables in the resulting equation, we arrive at the following linear ODE and second-order delay ODE:

$$X''(x) = -\lambda^2 X(x), \qquad (2.2.4.4)$$

$$T''(t) = (c_1 - a_1\lambda^2)T(t) + (c_2 - a_2\lambda^2)T(t-\tau). \qquad (2.2.4.5)$$

Requiring that the function $u_p = X(x)T(t)$ must satisfy the homogeneous boundary conditions (2.2.4.2), we obtain the homogeneous boundary conditions for X:

$$\Gamma_1[X] = 0 \quad \text{at} \quad x = 0, \quad \Gamma_2[X] = 0 \quad \text{at} \quad x = h. \tag{2.2.4.6}$$

The linear homogeneous eigenvalue problem (2.2.4.4), (2.2.4.6), which coincides with problem (2.2.2.14), (2.2.2.16) has nontrivial solutions $X = X_n(x)$ for only a discrete set of values of the parameter λ:

$$\lambda = \lambda_n, \quad X = X_n(x), \quad n = 1, 2, \ldots \tag{2.2.4.7}$$

The eigenvalues and eigenfunctions for the homogeneous linear boundary value problems described by ODE (2.2.4.4) with five most common types of boundary conditions are specified in Table 2.3.

Substituting the eigenvalues $\lambda = \lambda_n$ into (2.2.4.5) yields respective second-order delay ODEs for the functions $T = T_n(t)$.

We seek solutions to the linear initial-boundary value problem (2.2.4.1)–(2.2.4.3) as the series

$$u(x,t) = \sum_{n=1}^{\infty} X_n(x)T_n(t), \tag{2.2.4.8}$$

where the functions $u_{pn}(x,t) = X_n(x)T_n(t)$ are particular solutions to equation (2.2.4.1) satisfying the homogeneous boundary conditions (2.2.4.2).

To determine the initial conditions for the delay ODE (2.2.4.5) with $\lambda = \lambda_n$, we rewrite the first initial condition of (2.2.4.3) to expand into a series in the eigenfunctions:

$$\varphi(x,t) = \sum_{n=1}^{\infty} \varphi_n(t) X_n(x), \quad 0 \leq x \leq h, \quad -\tau \leq t \leq 0. \tag{2.2.4.9}$$

Multiplying (2.2.4.9) by $X_m(x)$ ($m = 1, 2, \ldots$), integrating with respect to the space variable x from 0 to h, and taking into account that the eigenfunctions $X_n(x)$ and $X_m(x)$ are orthogonal if $n \neq m$, implying that relations (2.2.1.21) hold, we find the functions $\varphi_n(t)$:

$$\varphi_n(t) = \frac{1}{\|X_n\|^2} \int_0^h \varphi(\xi, t) X_n(\xi) \, d\xi, \quad -\tau \leq t \leq 0, \tag{2.2.4.10}$$

where $\|X_n\|^2 = \int_0^h X_n^2(\xi) \, d\xi$.

From relations (2.2.4.8) and (2.2.4.9) and the initial conditions (2.2.4.3) we find the initial conditions for the delay ODE (2.2.2.15) with $\lambda = \lambda_n$:

$$T_n(t) = \varphi_n(t), \quad T_n'(t) = \varphi_n'(t) \quad \text{at} \quad -\tau \leq t \leq 0, \tag{2.2.4.11}$$

where the functions $\varphi_n(t)$ are given by (2.2.4.10).

To construct analytical solutions of Cauchy-type problems for the second-order delay ODE (2.2.4.5) with $c_1 = 0$ and the initial data (2.2.4.11), one can take advantage of the results obtained in the study [455], which dealt with a similar problem for a constant-delay ODE. The solution for functions $T_n(t)$, obtained by the method of steps, is quite cumbersome; it is presented in Subsection 1.2.2. Furthermore, problem (2.2.4.5), (2.2.4.11) can be solved using the Laplace transform (see Subsection 1.4.1) or numerical methods (see Subsection 5.1).

Once the functions $T_n(t)$ have been found, the solution to the original problem (2.2.4.1)–(2.2.4.3) is determined by the series (2.2.4.8); the respective eigenfunctions $X_n(x)$ (and eigenvalues λ_n) for the most common types of boundary conditions can be taken from Table 2.3.

Notably, if the inequality $a_2 > a_1$ holds, the initial-boundary value problems (2.2.4.1)–(2.2.4.3) are ill-posed in the sense of Hadamard (see Remark 2.7 in Subsection 2.2.5).

Remark 2.4. *The study [455] obtained an analytical solution to the initial-boundary value problem for the one-dimensional hyperbolic linear homogeneous equation with constant delay (2.2.4.1) where $a_2 = c_1 = 0$ and which is subjected to the homogeneous Dirichlet boundary conditions $u|_{x=0} = u|_{x=h} = 0$.*

2.2.5. Stability and Instability Conditions for Solutions to Linear Initial-Boundary Value Problems

An initial-boundary value problem with Dirichlet boundary conditions. Solutions of special form. Let us consider exponential-trigonometric functions of the form

$$u_n = A_n \exp(\varrho_n t) \sin\left(\frac{\pi n x}{h}\right), \quad n = 1, 2, \ldots, \qquad (2.2.5.1)$$

where A_n is a free parameter, and the constant ϱ_n satisfies the transcendental equation

$$\varrho_n = -\left(a_1 + a_2 e^{-\varrho_n \tau}\right)\left(\frac{\pi n}{h}\right)^2 + c_1 + c_2 e^{-\varrho_n \tau}. \qquad (2.2.5.2)$$

Under condition (2.2.5.2), the functions (2.2.5.1) are exact solutions to the homogeneous parabolic equation with constant delay (2.2.1.1) where $f(x,t) \equiv 0$ and satisfy the homogeneous boundary conditions (2.2.1.2) with $g_1(t) = g_2(t) \equiv 0$. It is noteworthy that the transcendental equation (2.2.5.2) can be derived from the delay ODE (2.2.1.22) by seeking its exact solutions in the exponential from $T_n = A_n \exp(\varrho_n t)$.

The solution to the initial-boundary value problems for the homogeneous parabolic equation with constant delay (2.2.1.1), subjected to the homogeneous boundary conditions (2.2.1.2) and a fairly arbitrary initial condition, can be represented as a series whose terms are functions of the form (2.2.5.1)–(2.2.5.2). This solution will tend to a trivial one if all roots of the characteristic equation (2.2.5.2) have a negative real part, in which case the trivial solution is said to be asymptotically stable. On the other hand, if at least one root of the characteristic equation (2.2.5.2) has a positive real part, then the trivial solution will be unstable as well as any other.

Necessary and sufficient stability conditions for solutions. If there is no delay, $\tau = 0$, equation (2.2.5.2) has a single, real root: $\varrho_n = c_1 + c_2 - (a_1 + a_2)(\pi n/h)^2$.

In this case, if the condition
$$c_1 + c_2 - (a_1 + a_2)(\pi/h)^2 < 0$$
holds, all constant ϱ_n will be negative and the respective solutions (2.2.5.1) will tend to zero as $t \to \infty$.

In what follows, we assume that $\tau > 0$, $a_1 > 0$, and $a_2 \geq 0$ (we do not require the condition $a_1 > a_2$ to be satisfied here). Using the notations (2.2.1.27), we rewrite equation (2.2.5.2) in the more compact form
$$\varrho_n - \alpha_n - \beta_n e^{-\varrho_n \tau} = 0,$$
$$\alpha_n = c_1 - a_1 \lambda_n^2, \quad \beta_n = c_2 - a_2 \lambda_n^2, \quad \lambda_n = \pi n/h. \tag{2.2.5.3}$$

Up to obvious renaming, equation (2.2.5.3) coincides with the transcendental equation (1.1.3.3), which was treated in Subsection 1.1.3.

All roots of the characteristic equation (2.2.5.3) have a negative real part if and only if the following three inequalities hold simultaneously (see the Hayes theorem in Subsection 1.3.2):

$$\begin{aligned}&\text{(i)} \quad (c_1 - a_1 \lambda_n^2)\tau < 1, \\ &\text{(ii)} \quad c_1 + c_2 - (a_1 + a_2)\lambda_n^2 < 0, \\ &\text{(iii)} \quad c_2 - a_2 \lambda_n^2 + \sqrt{(c_1 - a_1 \lambda_n^2)^2 + (\mu/\tau)^2} > 0,\end{aligned} \tag{2.2.5.4}$$

where $\lambda_n = \pi n/h$, $n = 1, 2, \ldots$, μ is a root of the transcendental equation $\mu = \tau(c_1 - a_1 \lambda_n^2) \tan \mu$ that satisfies the condition $0 < \mu < \pi$ (in the degenerate case of $c_1 - a_1 \lambda_n^2 = 0$, one should set $\mu = \pi/2$).

If inequalities (2.2.5.4) all hold simultaneously, all solutions of the homogeneous parabolic equation with constant delay (2.2.1.1) satisfying homogeneous boundary conditions of the form (2.2.1.2) with $g_1 = g_2 = 0$ tend to the trivial solution as $t \to \infty$.

Sufficient conditions for asymptotic stability of solutions. The inequalities (2.2.5.4), which involve six continuous parameters, a_1, a_2, c_1, c_2, h, and τ, and a discrete parameter, n, are inconvenient in practical usage. Below are some simpler sufficient conditions under which all roots of the characteristic equation (2.2.5.3) will have a negative real part.

Since $a_1 > 0$, $a_2 \geq 0$, and $\lambda_n \geq \lambda_1$, the first two inequalities in (2.2.5.4) hold for any n if they hold for $n = 1$. Using the inequality $\sqrt{z_1^2 + z_2^2} > -z_1$ with $z_2 \neq 0$, one can easily see that the last condition in (2.2.5.4) holds if the inequality $c_2 - c_1 + (a_1 - a_2)\lambda_n^2 \geq 0$ is true. For $a_1 \geq a_2$, λ_n in this inequality can be replaced with λ_1. As a result, we obtain the following conditions:

$$\begin{aligned}&\text{(i)} \quad (c_1 - a_1 \lambda_1^2)\tau < 1, \\ &\text{(ii)} \quad c_1 + c_2 - (a_1 + a_2)\lambda_1^2 < 0, \\ &\text{(iii)} \quad c_2 - c_1 + (a_1 - a_2)\lambda_1^2 \geq 0, \\ &\text{(iv)} \quad a_1 \geq a_2 \geq 0, \quad a_1 > 0,\end{aligned} \tag{2.2.5.5}$$

where $\lambda_1 = \pi/h$. If inequalities (2.2.5.5) all hold simultaneously, all roots of the characteristic equation (2.2.5.3) will have a negative real part.

▶ **Example 2.9.** For $c_1 = c_2 = 0$ and $a_1 > a_2 \geq 0$, conditions (2.2.5.5) hold, and all solutions to the homogeneous parabolic equation with constant delay (2.2.1.1) that satisfy the homogeneous boundary conditions (2.2.1.2) tend to the trivial solution as $t \to \infty$. ◀

Remark 2.5. *The stability conditions (2.2.5.5) can also be used if the inequalities $a_1 \geq a_2 \geq 0$ in Item (iv) are replaced with $a_2 < 0$ and $a_1 + a_2 > 0$.*

Instability conditions for solutions. Solutions to the homogeneous parabolic equation with constant delay (2.2.1.1) that satisfy the homogeneous boundary conditions (2.2.1.2) will be unstable if at least one of the inequalities (2.2.5.4) is violated. Therefore, for solutions to be unstable it suffices that either condition holds:

$$\begin{aligned}&\text{(i)} \quad (c_1 - a_1\lambda_1^2)\tau > 1, \\ &\text{(ii)} \quad c_1 + c_2 - (a_1 + a_2)\lambda_1^2 > 0,\end{aligned} \qquad (2.2.5.6)$$

where $\lambda_1 = \pi/h$.

It is noteworthy that if $a_1 > a_2$, the characteristic equation (2.2.5.2) can only have finitely many roots with a positive real part.

Now let us look at the left-hand side of the last inequality in (2.2.5.4). For large n, we have $\lambda_n \to \infty$ and

$$c_2 - a_2\lambda_n^2 + \sqrt{(c_1 - a_1\lambda_n^2)^2 + (\mu/\tau)^2} = (a_1 - a_2)\lambda_n^2 + c_2 - c_1 + O(\lambda_n^{-2}).$$

It follows that if the condition

$$a_2 > a_1 \qquad (2.2.5.7)$$

holds for a sufficiently large n, the inequality

$$c_2 - a_2\lambda_n^2 + \sqrt{(c_1 - a_1\lambda_n^2)^2 + (\mu/\tau)^2} < 0$$

will also hold, which indicates the domain of instability.

An asymptotic formula for the real part of the constant ϱ_n at large n. Solutions to the transcendental equation (2.2.5.3) can be expressed via the Lambert W function $W = W(z)$ as (cf. formula (1.1.3.5)):

$$\varrho_n = \alpha_n + \frac{1}{\tau}W(z), \quad z = \beta_n\tau e^{-\alpha_n\tau}, \qquad (2.2.5.8)$$

where W is understood as all branches of the function.

For $a_1 > 0$ and $a_2 > 0$ and sufficiently large n, the coefficients α_n and β_n are negative, with $\lim_{n\to\infty} \alpha_n = -\infty$ and $\lim_{n\to\infty} \beta_n = -\infty$. Considering the above and using the two leading terms of the asymptotic representation of the Lambert W function (1.1.3.10), we obtain the following expression of the real part of the constant ϱ_n for $a_1/a_2 = O(1)$ [410]:

$$\operatorname{Re}\varrho_n = \frac{1}{\tau}\ln\frac{a_2}{a_1} \quad (\text{as } n \to \infty). \qquad (2.2.5.9)$$

To sum up, for sufficiently large n, two qualitatively different situations can occur:

$$\begin{aligned} \operatorname{Re} \varrho_n < 0 \quad &\text{at} \quad a_2 < a_1, \\ \operatorname{Re} \varrho_n > 0 \quad &\text{at} \quad a_2 > a_1. \end{aligned} \qquad (2.2.5.10)$$

In the first case of (2.2.5.10), the trivial solution can be stable or unstable depending on whether inequalities (2.2.5.4), or the simpler conditions (2.2.5.5) or (2.2.5.6), hold or not. In the second case of (2.2.5.10), for a sufficiently large n, the associated exact solution obtained by extracting the real part of (2.2.5.1), will exponentially grow as time t increases. This means that the trivial solution to the initial-boundary value problem for the homogeneous equation with constant delay (2.2.1.1) subjected to the homogeneous boundary conditions (2.2.1.2) will be unstable.

Remark 2.6. *The special case of $a_1 = 0$ and $c_1 = c_2 = 0$ is discussed in Subsection 2.3.2.*

Instability of the nonhomogeneous initial-boundary value problems with delay. We will show that instability of the trivial solution of a homogeneous equation with homogeneous boundary conditions leads to instability of any solution to the general nonhomogeneous initial-boundary value problems with delay (2.2.1.1)–(2.2.1.3).

Indeed, suppose that the transcendental equation (2.2.5.2) has a root ϱ_n with $\operatorname{Re} \varrho_n > 0$, and $u = u(x,t)$ is an arbitrary solution to the first initial-boundary value problems with delay (2.2.1.1)–(2.2.1.3). Then the function

$$u_s(x,t) = u(x,t) + u_n(x,t), \qquad (2.2.5.11)$$

where $u_n(x,t)$ is an exponentially growing function of the form (2.2.5.1) such that $\operatorname{Re} \varrho_n > 0$, is a solution to equation (2.2.1.1) and satisfies the nonhomogeneous boundary conditions (2.2.1.2). It is easy to verify that for sufficiently large n, the inequality

$$|u_s(x,t) - u(x,t)| \le |A_n| \quad \text{for} \quad 0 < x < h, \ -\tau \le t \le 0, \qquad (2.2.5.12)$$

holds. It is apparent from (2.2.5.12) that for a sufficiently small $|A_n| = \varepsilon$, the initial data for the solutions u and u_s of equation (2.2.1.1) are indefinitely close to each other. However, as $t \to \infty$, the two solutions will diverge indefinitely at the point $x = h/2$ because the real part of ϱ_n is positive for large n in formula (2.2.5.1). So we have

$$\lim_{t \to \infty} |u_s(x,t) - u(x,t)|_{x=h/2} \to \infty \quad \text{as} \quad t \to \infty.$$

This means that any solution to the delay initial-boundary value problems (2.2.1.1)–(2.2.1.3) is unstable with respect to variations in the initial data as long as the transcendental equation (2.2.5.2) has a root with $\operatorname{Re} \varrho_n > 0$.

Stability and instability conditions for solutions to other initial-boundary value problems. Conditions (2.2.5.4), (2.2.5.5), (2.2.5.6), and (2.2.5.7) can be used to analyze the stability and instability of solutions to other initial-boundary value problems. In particular, the solutions to all problems discussed in Subsection 2.2.2 are unstable if inequality (2.2.5.7) holds.

To determine whether a solution to an initial-boundary value problem is stable or unstable, one should insert the least eigenvalue λ_1 from Table 2.3 into the sufficient conditions (2.2.5.5) or (2.2.5.6). For the Neumann boundary conditions (second row in Table 2.3), one has, in addition, to investigate the sign of the real part of the root ϱ_0 of the transcendental equation (2.2.5.2) with $n = 0$.

Remark 2.7. *It can be shown [410] that the initial-boundary value problems for hyperbolic equations with constant delay (2.2.4.1), discussed in Subsection 2.2.4, are also unstable if $a_2 > a_1$.*

2.3. Hyperbolic and Differential-Difference Heat Equations

2.3.1. Derivation of the Hyperbolic and Differential-Difference Heat Equations

The classical heat (diffusion) equation. The classical model of thermal conduction or diffusion is based on *Fourier's law* [325]:

$$\mathbf{q} = -\lambda \nabla \theta, \tag{2.3.1.1}$$

where $\mathbf{q}$ is the heat flux density, λ is thermal conductivity, θ is temperature, and ∇ is the gradient operator.

In the simplest case where thermal sources are absent, the *energy conservation law* is

$$\rho c_p \theta_t = -\operatorname{div} \mathbf{q}, \tag{2.3.1.2}$$

where t is time, ρ is density, and c_p is the specific heat capacity of the substance (medium).

Substituting (2.3.1.1) into (2.3.1.2) gives the *classical heat equation* [88, 325]:

$$\theta_t = a \Delta \theta, \tag{2.3.1.3}$$

where $a = \lambda/(\rho c_p)$ is thermal diffusivity, Δ is the Laplace operator (in the three-dimensional case, we have $\Delta \theta = \theta_{xx} + \theta_{yy} + \theta_{zz}$, where x, y, z are the Cartesian coordinates).

Notably, the diffusion equation is also written in the form (2.3.1.3), where θ is concentration, and a is the diffusion coefficient.

The heat equation (2.3.1.3) is of parabolic type and it possesses a physically paradoxical property of infinite speed of propagation of perturbations. This is not observed in practice, which indicates a limited area of applicability of such equations.

Hyperbolic heat (diffusion) equation. The above drawback of Fourier's model (2.3.1.1) led to the need to develop other thermal conduction (diffusion) models that would give a finite perturbation propagation speed. This resulted in a more complex thermal conduction model based on the *Cattaneo–Vernotte differential law* [91, 92, 531, 532] (see also [165, 257, 325, 489]):

$$\mathbf{q} = -\lambda \nabla \theta - \tau \mathbf{q}_t. \tag{2.3.1.4}$$

Model (2.3.1.4) differs from Fourier's law (2.3.1.1) in the presence of an additional nonstationary term proportional to τ. When $\tau=0$, model (2.3.1.4) becomes (2.3.1.1).

Using model (2.3.1.4) and in view of the conservation law (2.3.1.2), one obtains a hyperbolic heat equation:
$$\tau\theta_{tt} + \theta_t = a\,\Delta\theta, \tag{2.3.1.5}$$
where the *relaxation time* τ is assumed small. If $\tau = 0$, equation (2.3.1.5) becomes (2.3.1.3).

Estimates of the thermal and diffusion relaxation times.

$1°$. *Estimates of the thermal relaxation time.* The relaxation time τ appearing in the model (2.3.1.4) and equation (2.3.1.5) characterizes nonequilibrium properties of the thermal conduction process and takes into account the inertia of the thermal flux. For metals, superconductors, and semiconductors, theoretical estimates give $\tau \approx 10^{-12}\text{--}10^{-6}$ s [176, 370, 374, 529].

Although very small, such values of τ must be considered when analyzing high-intensity nonstationary processes, the duration of which is comparable to the relaxation time when, for example, materials are treated using ultrashort laser pulses and high-speed electronic devices [17, 370, 456]. These also include the processes of high-speed friction heating, local heating during transonic dynamic crack propagation, etc. [244, 285]. For materials and media with an inhomogeneous internal structure (e.g., capillary-porous bodies, pastes, suspensions, powders, gas-liquid multiphase media, biological substances, food products, wood, etc.), the relaxation time can be much larger [47, 121, 126, 325, 507]. For example, the studies [253, 345] state that the thermal relaxation time of meat products and some bulk media can give τ-values of the order of ten or more seconds.

$2°$. *Thermal and diffusion speeds of perturbation propagation. Diffusion relaxation time.* In simple systems such as ideal gas mixtures, the characteristic diffusion relaxation time τ_D, which is the time of establishing the local equilibrium concentration of the diffusing component, coincides with the characteristic thermal relaxation time τ_T, the time of establishing local equilibrium temperature. (The subscript 'T' is added here for clarity.) However, in systems with a more complex structure [489, 490], such as molten metals, we have $\tau_D \gg \tau_T$. In such systems, thermal equilibrium establishes prior to diffusion equilibrium. Either stage of establishing local equilibrium has its own characteristic speed (determined from the hyperbolic heat equation (2.3.1.5)): diffusion speed, $V_D = (D/\tau_D)^{1/2}$, and thermal wave speed, $V_T = (a/\tau_T)^{1/2}$. For homogeneous gaseous and liquid media, we can roughly assume that the thermal wave speed V_T is approximately equal to the speed of sound. For molten metals, $V_D \approx 1\text{--}10$ m/s and $V_T \approx 10^3\text{--}10^4$ m/s, or $V_D \ll V_T$. The speed of heat propagation in the air is approximately equal to the speed of sound $V_T \approx 330$ m/s. The diffusion speed V_D in capillary-porous media is less than V_T by about $10^6\text{--}10^7$ times, so it must be taken into account in mass transfer equations [325, p. 455]. For diffusion in polymers, the relaxation time is a few seconds [250]. The above examples testify that the thermal and diffusion relaxation times can vary within very wide limits and must be taken into account when solving many heat and mass transfer problems.

2.3. Hyperbolic and Differential-Difference Heat Equations

The differential-difference heat equation with a finite relaxation time. To substantiate the Cattaneo–Vernotte differential model (2.3.1.4) theoretically, one often (but not always) uses the following differential-difference relation for the thermal flux [165, 390, 427, 489, 521]:

$$\mathbf{q}|_{t+\tau} = -\lambda \nabla \theta, \qquad (2.3.1.6)$$

in which the left-hand side is evaluated at time $t + \tau$, while the right-hand side is evaluated at time t. The physical meaning of relation (2.3.1.6) is as follows: unlike the classical local equilibrium case, transfer processes in local non-equilibrium media exhibit inertial properties: the system reacts to a thermal action (or the thermal flux responds to a change in the temperature gradient) a relaxation time τ later than the current time t.

As a result, we arrive at the linear differential-difference heat equation with a finite relaxation time

$$\theta_t|_{t+\tau} = a \Delta \theta, \qquad (2.3.1.7)$$

where $\theta|_{t+\tau} = \theta(\mathbf{x}, t + \tau)$.

Remark 2.8. *The fact that the heat equation with a time delay can happen to be a more adequate model was first noted by Maxwell [337].*

If we formally expand the left-hand side of equation (2.3.1.7) into a Taylor series in the small τ and retain the two leading terms, we will get the hyperbolic heat equation (2.3.1.5). However, as will be shown in the next section, this formal reasoning frequently used in the literature is not justified in any way since it significantly changes the properties of the equation.

Denoting $u = \theta(\mathbf{x}, t + \tau)$ in (2.3.1.7), we obtain the delay PDE

$$u_t = a \Delta w, \quad w = u(\mathbf{x}, t - \tau), \qquad (2.3.1.8)$$

which is a special case of equation (2.1.1.1).

2.3.2. Stokes Problem and Initial-Boundary Value Problems for the Differential-Difference Heat Equation

Exact solutions to the one-dimensional differential-difference heat equation. In the one-dimensional case, the differential-difference heat equation (2.3.1.7) simplifies to become

$$\theta_t|_{t+\tau} = a\theta_{xx}. \qquad (2.3.2.1)$$

Here, as previously, the left-hand side is evaluated at time $t + \tau$, while the right-hand side is evaluated at time t.

Below we list a number of exact solutions to equation (2.3.2.1).

1°. Multiplicative separable solutions:

$$\theta = [A\cos(kx) + B\sin(kx)]e^{-\lambda t}, \quad ak^2 = \lambda e^{-\lambda \tau} \quad (\lambda > 0); \qquad (2.3.2.2)$$

$$\theta = [A\exp(kx) + B\exp(-kx)]e^{-\lambda t}, \quad ak^2 = -\lambda e^{-\lambda \tau} \quad (\lambda < 0), \qquad (2.3.2.3)$$

where A, B, and λ are arbitrary constants.

Solution (2.3.2.2) is periodic in the space coordinate x and decays as $t \to \infty$. For $0 < \lambda < \infty$ and $\tau > 0$, the range of the parameter k is limited: $0 < k \leq k_{\max} = (ea\tau)^{-1/2}$. For a given k such that $0 < k < k_{\max}$, equation (2.3.2.1) admits two real-valued solutions of the form (2.3.2.2) associated with two positive roots, λ_1 and λ_2, of the transcendental equation $\lambda e^{-\lambda \tau} = ak^2$.

Solutions (2.3.2.2) and (2.3.2.3) are special cases of the multiplicative separable solution

$$\theta = \varphi(x)\psi(t), \qquad (2.3.2.4)$$

where the functions $\varphi(x)$ and $\psi(t)$ satisfy the linear constant-coefficient equations

$$\varphi''_{xx} + c\varphi = 0, \qquad (2.3.2.5)$$
$$\psi'_t(t+\tau) + ac\psi(t) = 0. \qquad (2.3.2.6)$$

The former is an ordinary differential equation, while the latter is a differential-difference equation. The change of variable $\bar{t} = t + \tau$ reduces the latter equation to a standard delay ODE (for its solution, see Subsection 1.1.3).

2°. Solution periodic in t:

$$\theta = e^{-\gamma x}[A\cos(\omega t - \beta x) + B\sin(\omega t - \beta x)] + C, \qquad (2.3.2.7)$$

$$\beta = \left(\frac{\omega}{2a}\right)^{1/2}[1+\sin(\tau\omega)]^{1/2}, \quad \gamma = \left(\frac{\omega}{2a}\right)^{1/2}\frac{\cos(\tau\omega)}{[1+\sin(\tau\omega)]^{1/2}},$$

where A, B, C, and ω are arbitrary constants.

Solution (2.3.2.7) decays as $x \to \infty$ if $C = 0$ and $\tau\omega < \frac{1}{2}\pi$.

3°. Polynomial solutions:

$$\theta = Ax + B,$$
$$\theta = A(x^2 + 2at) + B,$$
$$\theta = A(x^3 + 6atx) + B,$$
$$\theta = A[x^4 + 12a(t-\tau)x^2 + 12a^2(t-2\tau)^2] + B,$$
$$\theta = A[x^5 + 20a(t-\tau)x^3 + 60a^2(t-2\tau)^2 x] + B,$$
$$\theta = x^{2n} + \sum_{k=1}^{n}\frac{(2n)(2n-1)\ldots(2n-2k+1)}{k!}a^k(t-k\tau)^k x^{2n-2k},$$
$$\theta = x^{2n+1} + \sum_{k=1}^{n}\frac{(2n+1)(2n)\ldots(2n-2k+2)}{k!}a^k(t-k\tau)^k x^{2n-2k+1},$$

where A and B are arbitrary constants, and n is a positive integer. The first three solutions are independent of the relaxation time τ.

Stokes problem with a periodic boundary condition ($0 \leq x < \infty$). Below we consider the Stokes problem for the one-dimensional differential-difference heat equation (2.3.2.1) without initial data and with special boundary conditions

$$\theta = A\cos(\omega t) \quad \text{at} \quad x = 0, \qquad \theta \to 0 \quad \text{at} \quad x \to \infty, \qquad (2.3.2.8)$$

2.3. Hyperbolic and Differential-Difference Heat Equations 117

where A and ω are arbitrary constants. We will look for a solution to equation (2.3.2.1) periodic in time t and satisfying the boundary conditions (2.3.2.8).

Problem (2.3.2.1), (2.3.2.8) has an exact solution that is a special case of solution (2.3.2.7) (see [390, 427]):

$$\theta = Ae^{-\gamma x}\cos(\omega t - \beta x), \qquad (2.3.2.9)$$

where

$$\beta = \left(\frac{\omega}{2a}\right)^{1/2}[1 + \sin(\tau\omega)]^{1/2}, \quad \gamma = \left(\frac{\omega}{2a}\right)^{1/2}\frac{\cos(\tau\omega)}{[1 + \sin(\tau\omega)]^{1/2}}, \qquad (2.3.2.10)$$

$$\omega \neq \frac{1}{\tau}\left(\frac{3\pi}{2} + 2\pi k\right), \quad k = 0, 1, 2, \ldots$$

At $\tau = 0$, solution (2.3.2.9), (2.3.2.10) becomes the solution to the related Stokes problem without initial conditions for the classical parabolic heat equation, which is given by formula (2.3.2.9) with

$$\beta = \gamma = \left(\frac{\omega}{2a}\right)^{1/2}. \qquad (2.3.2.11)$$

The solution to the problem without initial conditions for the one-dimensional hyperbolic heat equation (2.3.1.5) (with $\Delta\theta = \theta_{xx}$) resulting from the Cattaneo–Vernotte differential model (2.3.1.4) is described by formula (2.3.2.9) in which

$$\begin{aligned}\beta &= \left(\frac{\omega}{2a}\right)^{1/2}\left[\tau\omega + (1 + \tau^2\omega^2)^{1/2}\right]^{1/2}, \\ \gamma &= \left(\frac{\omega}{2a}\right)^{1/2}\left[\tau\omega + (1 + \tau^2\omega^2)^{1/2}\right]^{-1/2}.\end{aligned} \qquad (2.3.2.12)$$

A comparison of formulas (2.3.2.9) and (2.3.2.10) with (2.3.2.9) and (2.3.2.11) shows that for small $\omega\tau$ (to be precise, for $0 < \omega\tau < \pi/2$), the decay constant γ for the differential-difference model is less than that for the classical model (described by the parabolic equation). However, the coefficient β for the differential-difference model is greater than that for the classical model.

The two leading terms in the expansions of formulas (2.3.2.10) and (2.3.2.12) into a series in small τ (with $\omega\tau \ll 1$) coincide. For small $\tau > 0$ and large frequencies $\omega \gg \tau^{-1}$, the coefficients (2.3.2.12) have the following asymptotics:

$$\beta = \omega\sqrt{\frac{\tau}{a}}, \quad \gamma = \frac{1}{2\sqrt{a\tau}}. \qquad (2.3.2.13)$$

This means that for large frequencies, the decay constant γ is independent of ω, which is a qualitative difference from the respective solution of the parabolic heat equation (2.3.2.11). Both determining parameters in (2.3.2.13) depend on the perturbation parameter τ significantly. For large values of $\omega\tau$, solutions (2.3.2.9), (2.3.2.10) and (2.3.2.9), (2.3.2.12) differ qualitatively. Specifically, the decay constant γ in the differential-difference model depends on the frequency ω significantly:

it does not tend to a constant quantity as in the Cattaneo–Vernotte model (see the asymptotics (2.3.2.13)).

Exact solution of the Stokes problems with a first-order volume reaction. The one-dimensional heat and mass transfer linear differential-difference equation with a source is expressed as

$$\theta_t|_{t+\tau} = a\theta_{xx} - k\theta|_{t+\tau}, \qquad (2.3.2.14)$$

where $\theta|_{t+\tau} = \theta(x, t + \tau)$. In mass transfer problems, the last term in equation (2.3.2.14) with $k > 0$ describes a first-order chemical reaction [401].

Consider the Stokes problem for the one-dimensional differential-difference heat equation with a source (2.3.2.14) without initial data and with the special periodic boundary conditions (2.3.2.8).

Problem (2.3.2.14), (2.3.2.8) has the following exact solution:

$$\theta = Ae^{-\gamma x} \cos(\omega t - \beta x),$$

$$\beta = \frac{1}{\sqrt{2a}} \left[\sqrt{\omega^2 + k^2} + \omega \sin(\tau\omega) - k \cos(\tau\omega) \right]^{1/2}, \qquad (2.3.2.15)$$

$$\gamma = \frac{\omega \cos(\tau\omega) + k \sin(\tau\omega)}{\sqrt{2a}\left[\sqrt{\omega^2 + k^2} + \omega \sin(\tau\omega) - k \cos(\tau\omega)\right]^{1/2}}.$$

In the limiting case $k = 0$, formulas (2.3.2.15) become (2.3.2.9) and (2.3.2.10).

Remark 2.9. *The change of variable*

$$\theta(x, t) = e^{-kt} \eta(x, t)$$

converts equation (2.3.2.14) to a simpler, sourceless equation of the form (2.3.2.1):

$$\eta_t|_{t+\tau} = ae^{k\tau} \eta_{xx}.$$

An initial-boundary value problem for the differential-difference heat equation. The first initial-boundary value problem for the one-dimensional differential-difference heat equation (2.3.2.1) can be written as

$$u_t = aw_{xx}, \quad w = u(x, t - \tau);$$
$$u = g_1(t) \quad \text{at} \quad x = 0, \ t > -\tau; \quad u = g_2(t) \quad \text{at} \quad x = h, \ t > -\tau; \qquad (2.3.2.16)$$
$$u = \varphi(x, t) \quad \text{at} \quad 0 < x < h, \ -\tau \leq t \leq 0,$$

where the notation $u(x, t) = \theta(x, t + \tau)$ is used.

Problem (2.3.2.16) is a special case of problem (2.2.1.1)–(2.2.1.3) with $a_1 = 0$, $a_2 = a$, $c_1 = c_2 = 0$, and $f(x, t) \equiv 0$. It follows from the results presented in Subsection 2.2.5 that the homogeneous problem (2.3.2.16) with $g_1(t) = g_2(t) \equiv 0$ and $\varphi(x, t) \equiv 0$ admits the exact solutions (2.2.5.1), where the coefficient λ_n in the exponential is expressed via the Lambert W function (see formula (2.2.5.8) with $\alpha_n = 0$):

$$\lambda_n = \frac{1}{\tau} W(z), \quad z = -a\tau \left(\frac{\pi n}{h}\right)^2. \qquad (2.3.2.17)$$

Using the asymptotic representation of the Lambert W function (1.1.3.10), we can express the real part of λ_n for sufficiently large n as [410]:

$$\operatorname{Re}\lambda_n = \frac{1}{\tau}\bigl(2\ln n - \ln\ln n + O(1)\bigr). \qquad (2.3.2.18)$$

Hence, $\lambda_n \to \infty$ as $n \to \infty$. Therefore, for large n, the associated exact solution obtained by extracting the real part of (2.2.5.1) will exponentially grow with time t; moreover, the initial-boundary value problem with delay (2.3.2.16) is unstable with respect to small perturbations in the initial data (this fact can be proved using a similar reasoning as in Subsection 2.2.5). Apparently, this fact was first established in [243], where equation (2.3.2.1) was treated using first-kind homogeneous boundary conditions and a special initial condition (see also [234, 390]).

Consequently, attempting to generalize the classical thermal conduction model (2.3.1.1) by employing the differential-difference model (2.3.1.6) with a finite relaxation time τ (delay) leads to problem (2.3.2.16) whose solutions are unstable for any $\tau > 0$. This means that the differential-difference model (2.3.1.6) is unsuitable for the description of thermal (and diffusion) processes.

Notably, the Cattaneo–Vernotte differential model (2.3.1.4) results in a stable trivial solution to the initial-boundary value problem for the hyperbolic heat equation (2.3.1.5), which is because both roots of the associated characteristic equation are either negative or have a negative real part. The derivation of the hyperbolic heat equation (2.3.1.5) from the differential-difference equation (2.3.1.7) through the expansion with respect to small τ is fallacious for $t \sim \tau$ (it is assumed in the expansion that $\tau \ll t$). It is at small $t \sim \tau$ that using the hyperbolic heat equation (2.3.1.5) allows one to remove the above drawback of the parabolic heat equation (2.3.1.3).

2.4. Linear Initial-Boundary Value Problems with Proportional Delay

2.4.1. Preliminary Remarks

There are relatively few publications devoted to analyzing and solving partial differential equations with proportional delay. The study [591] uses a first-order linear PDE with proportional delay to simulate the growth and division of cells distributed by size. It looks for a solution to this equation as a series whose terms are determined by solving simpler PDEs without delay. The article [141] explores a more complex linear reaction-diffusion equation with a proportional space argument (obtained by adding a diffusion term to the equation dealt with in [591]).

The study [308] briefly outlines the following initial-boundary value problems for linear heat and wave equations with proportional delay in two arguments,

$$u_t(\alpha^2 x, t) = u_{xx}(x, \beta t) \quad \text{and} \quad u_{tt}(\alpha^2 x, t) = u_{xx}(x, \beta^2 t),$$

subjected to homogeneous Dirichlet boundary conditions and a general initial condition. The authors employ the method of separation of variables to construct solutions.

A complicating factor in these problems is the non-orthogonality of the system of eigenfunctions $X_n(x)$.

The articles [458, 459, 484] investigate the issues of unique solvability and smoothness of linear boundary value problems for elliptic PDEs with dilated or contracted arguments in the highest-order derivatives of the unknown (see also [364]).

The studies [1, 9, 194, 416] discuss analytical solution methods for some linear and nonlinear PDEs with proportional delays. The article [492] constructs a finite-difference scheme of numerical integration of first-order constant delay PDEs with respect to t and proportional delay PDEs with respect to x. The studies [36, 461, 509] are devoted to numerical solution methods for PDEs with proportional delays [36, 461] and more complex, variable delays [509].

2.4.2. First Initial-Boundary Value Problem for a Parabolic Equation with Proportional Delay

Statement of the problem. Consider the first initial-boundary value problem for a one-dimensional parabolic linear homogeneous equation with constant coefficients and a proportional delay:

$$u_t = a_1 u_{xx} + a_2 w_{xx} + c_1 u + c_2 w, \qquad w = u(x, pt). \qquad (2.4.2.1)$$

The equation is defined in the domain $\Omega = \{0 < x < h,\ t > 0\}$, and it is assumed that $a_1 > a_2 \geq 0$ and $0 < p < 1$. We supplement equation (2.4.2.1) with first-kind homogeneous boundary conditions (Dirichlet conditions)

$$u = 0 \quad \text{at} \quad x = 0, \qquad u = 0 \quad \text{at} \quad x = h \qquad (2.4.2.2)$$

and a general initial condition

$$u = \varphi(x) \quad \text{at} \quad t = 0. \qquad (2.4.2.3)$$

By reasoning in the same way as for the linear homogeneous equation with constant delay (2.2.1.1) with $f = 0$, we can show by separation of variables that the proportional delay equation (2.4.2.1) admits exact solutions as the product of two functions with different arguments:

$$u_n(x, t) = T_n(t) \sin\left(\frac{\pi n x}{h}\right), \qquad n = 1, 2, \ldots . \qquad (2.4.2.4)$$

The functions $T_n(t)$ satisfy the first-order linear ODEs with proportional delay

$$T_n'(t) = \left[c_1 - a_1\left(\frac{\pi n}{h}\right)^2\right]T_n(t) + \left[c_2 - a_2\left(\frac{\pi n}{h}\right)^2\right]T_n(pt), \qquad (2.4.2.5)$$

which one can obtain from (2.2.1.22) by formally replacing $T(t-\tau)$ with $T(pt)$. The particular solutions (2.4.2.4) satisfy the homogeneous boundary conditions (2.4.2.2).

2.4. Linear Initial-Boundary Value Problems with Proportional Delay

Using the linear superposition principle, we look for a solution to the initial-boundary value problem (2.4.2.1)–(2.4.2.3) as the infinite series

$$u(x,t) = \sum_{n=1}^{\infty} T_n(t) \sin\left(\frac{\pi n x}{h}\right) \qquad (2.4.2.6)$$

that satisfies equation (2.4.2.1) and the homogeneous boundary conditions (2.4.2.2).

To determine the initial condition for the ODE with proportional delay (2.4.2.5), we rewrite the function $\varphi(x)$ involved in the initial condition (2.4.2.3) to expand into a series in the eigenfunctions

$$\varphi(x) = \sum_{n=1}^{\infty} A_n \sin\frac{\pi n x}{h}. \qquad (2.4.2.7)$$

Multiplying (2.4.2.7) by $X_m(x) = \sin\frac{\pi m x}{h}$ ($m = 1, 2, \ldots$) and integrating with respect to the space variable x from 0 to h, we find the coefficients A_n:

$$A_n = \frac{2}{h} \int_0^h \varphi(\xi) \sin\left(\frac{\pi n \xi}{h}\right) d\xi. \qquad (2.4.2.8)$$

From relations (2.4.2.6) and (2.4.2.7) we obtain the initial condition for the proportional delay ODE (2.4.2.5):

$$T_n(0) = A_n, \qquad (2.4.2.9)$$

where the coefficients A_n are determined by formula (2.4.2.8).

Up to notation, the linear problem (2.4.2.5), (2.4.2.9) with the normalized initial condition $A_n = 1$ coincides with problem (1.4.2.2) where $c = 0$, which was treated in Subsection 1.4.2. Introducing the notations

$$\alpha_n = c_1 - a_1\left(\frac{\pi n}{h}\right)^2, \quad \beta_n = c_2 - a_2\left(\frac{\pi n}{h}\right)^2 \qquad (2.4.2.10)$$

and using formulas (1.4.2.4), we represent the solution to problem (2.4.2.5), (2.4.2.9) as the power series

$$T_n(t) = A_n\left(1 + \sum_{m=1}^{\infty} \gamma_{mn} t^m\right), \quad \gamma_{mn} = \frac{1}{m!} \prod_{k=0}^{m-1} (\alpha_n + \beta_n p^k). \qquad (2.4.2.11)$$

For $0 < p < 1$, the series (2.4.2.11) has an infinite radius of convergence.

Substituting expressions (2.4.2.11) into formula (2.4.2.6) yields the solution to problem (2.4.2.1)–(2.4.2.3):

$$u(x,t) = \sum_{n=1}^{\infty} A_n\left(1 + \sum_{m=1}^{\infty} \gamma_{mn} t^m\right) \sin\left(\frac{\pi n x}{h}\right),$$

$$\gamma_{mn} = \frac{1}{m!} \prod_{k=0}^{m-1} (\alpha_n + \beta_n p^k), \qquad (2.4.2.12)$$

where the coefficients A_n, α_n, β_n are defined by formulas (2.4.2.8) and (2.4.2.10).

2.4.3. Other Initial-Boundary Value Problems for a Parabolic Equation with Proportional Delay

Representation of solutions to initial-boundary value problems as the sum of solutions to simpler problems. Below we describe a procedure for constructing solutions, by the method of separation of variables, to other initial-boundary value problems for the one-dimensional parabolic linear homogeneous PDE with proportional delay (2.4.2.1). For brevity, we will write this equation as

$$\mathcal{L}[u,w] = 0, \quad t > 0, \qquad (2.4.3.1)$$

where $\mathcal{L}[u,w] \equiv u_t - a_1 u_{xx} - a_2 w_{xx} - c_1 u - c_2 w$ and $w = u(x, pt)$ with $0 < p < 1$.

We will supplement equation (2.4.3.1) with different linear homogeneous boundary conditions

$$\Gamma_1[u] = 0 \quad \text{at} \quad x = 0, \qquad \Gamma_2[u] = 0 \quad \text{at} \quad x = h, \qquad (2.4.3.2)$$

and the general initial condition

$$u = \varphi(x) \quad \text{at} \quad t = 0. \qquad (2.4.3.3)$$

The third column of Table 2.2 with $g_1(t) = g_2(t) \equiv 0$ shows the most common homogeneous boundary conditions, which define the form of the operators (functions) $\Gamma_{1,2}[u]$.

As previously, we first look for particular solutions to the linear homogeneous equation (2.4.3.1) as the product of two functions with different arguments: $u_1 = X(x)T(t)$. Separating the variables in the resulting equation, we arrive at the linear ODE and ODE with proportional delay

$$X''(x) = -\lambda^2 X(x), \qquad (2.4.3.4)$$

$$T'(t) = (c_1 - a_1 \lambda^2) T(t) + (c_2 - a_2 \lambda^2) T(pt). \qquad (2.4.3.5)$$

Requiring that the function $u_1 = X(x)T(t)$ must satisfy the homogeneous boundary conditions (2.4.3.2), we obtain homogeneous boundary conditions for X:

$$\Gamma_1[X] = 0 \quad \text{at} \quad x = 0, \qquad \Gamma_2[X] = 0 \quad \text{at} \quad x = h. \qquad (2.4.3.6)$$

The nontrivial solutions $X = X_n(x)$ of the linear homogeneous eigenvalue problem (2.4.3.4), (2.4.3.6) exist for only a discrete set of values of λ:

$$\lambda = \lambda_n, \quad X = X_n(x), \quad n = 1, 2, \ldots \qquad (2.4.3.7)$$

Table 2.3 presents the eigenvalues and eigenfunctions of homogeneous linear boundary value problems for ODE (2.4.3.4) subjected to the five most common types of boundary conditions.

Using the linear superposition principle, we look for a solution to the initial-boundary value problem (2.4.3.1)–(2.4.3.1) as the infinite series

$$u(x,t) = \sum_{n=1}^{\infty} T_n(t) X_n(x), \qquad (2.4.3.8)$$

2.4. Linear Initial-Boundary Value Problems with Proportional Delay

with the functions $T_n(t)$ described by equation (2.4.3.5) at $\lambda = \lambda_n$. By construction, the series (2.4.3.8) automatically satisfies equation (2.4.3.1) and the homogeneous boundary conditions (2.4.3.2).

To determine the initial conditions for the ODE with proportional delay (2.4.3.5) at $\lambda = \lambda_n$, we represent the function $\varphi(x)$ appearing in the initial condition (2.4.3.3) as a series in the eigenfunctions:

$$\varphi(x) = \sum_{n=1}^{\infty} A_n X_n(x). \qquad (2.4.3.9)$$

Multiplying (2.4.3.9) by $X_m(x)$ ($m = 1, 2, \ldots$), integrating with respect to the space variable x from 0 to h, and taking into account that any two eigenfunctions $X_n(x)$ and $X_m(x)$ with $n \neq m$ are orthogonal, meaning that relations (2.2.1.21) hold, we find the coefficients A_n:

$$A_n = \frac{1}{\|X_n\|^2} \int_0^h \varphi(\xi) X_n(\xi)\, d\xi, \quad \|X_n\|^2 = \int_0^h X_n^2(\xi)\, d\xi. \qquad (2.4.3.10)$$

From relations (2.4.3.8) and (2.4.3.9) we obtain the initial conditions for the ODE with proportional delay (2.4.3.5) where $\lambda = \lambda_n$:

$$T_n(0) = A_n. \qquad (2.4.3.11)$$

The coefficients A_n are determined by formula (2.4.3.10).

Up to notation, the linear problem with proportional delay (2.4.3.5), (2.4.3.11) where $\lambda = \lambda_n$ coincides with problem (2.4.2.5), (2.4.2.9) discussed in Subsection 2.4.2. Consequently, the solution to problem (2.4.2.5), (2.4.2.9) can be represented as the power series

$$T_n(t) = A_n \left(1 + \sum_{m=1}^{\infty} \gamma_{mn} t^m \right), \quad \gamma_{mn} = \frac{1}{m!} \prod_{k=0}^{m-1} (\alpha_n + \beta_n p^k),$$
$$\alpha_n = c_1 - a_1 \lambda_n^2, \quad \beta_n = c_2 - a_2 \lambda_n^2. \qquad (2.4.3.12)$$

Substituting expressions (2.4.3.12) into formula (2.4.3.8) yields the solution to problem (2.4.3.1)–(2.4.3.3):

$$u(x,t) = \sum_{n=1}^{\infty} A_n \left(1 + \sum_{m=1}^{\infty} \gamma_{mn} t^m \right) X_n(x), \qquad (2.4.3.13)$$

where the coefficients A_n and γ_{mn} are evaluated using expressions (2.4.3.10) and (2.4.3.12).

Solutions to the initial-boundary value problems (2.4.3.1)–(2.4.3.3) with any type of boundary conditions (see Table 2.2) can be obtained with formulas (2.4.3.10), (2.4.3.12), and (2.4.3.13) by taking the respective eigenvalues λ_n and eigenfunctions $X_n(x)$ from Table 2.3.

Remark 2.10. *Solutions to more complicated initial-boundary value problems described by n-dimensional homogeneous PDEs with proportional delay subjected to homogeneous boundary conditions can be constructed in a similar way to those in Subsection 2.2.3 for equations with constant delay.*

A self-similar problem for a linear PDE with two proportional arguments. We now consider the parabolic-type equation with two proportional arguments

$$u_t = a_1 u_{xx} + a_2 w_{xx}, \quad w = u(px, qt) \quad (x > 0, \ t > 0), \tag{2.4.3.14}$$

where $p > 0$ and $q > 0$ are scaling factors.

We supplement equation (2.4.3.14) with a special initial and a special boundary condition

$$u = A \quad \text{at} \quad t = 0, \quad u = B \quad \text{at} \quad x = 0, \tag{2.4.3.15}$$

where A and B are arbitrary constants.

The solution to problem (2.4.3.14)–(2.4.3.15) is *self-similar* and it can be represented as

$$u = U(z), \quad z = xt^{-1/2}, \tag{2.4.3.16}$$

where the function $U(z)$ satisfies the following boundary value problem for an ODE with proportional argument:

$$-\tfrac{1}{2} z U'_z = a_1 U''_{zz} + a_2 W''_{zz}, \quad W = U(\sigma z), \quad \sigma = pq^{-1/2}; \tag{2.4.3.17}$$

$$U(0) = B, \quad U(\infty) = A. \tag{2.4.3.18}$$

Suppose that the scaling factors are related by the parabolic formula $q = p^2$. Then we get $\sigma = 1$ and $U = W$. In this special case, equation (2.4.3.17) is easy to integrate, and the solution to the original problem (2.4.3.14)–(2.4.3.15) is expressed as

$$u = B + (A - B)\,\text{erf}\!\left(\frac{x}{2\sqrt{at}}\right), \quad a = a_1 + a_2, \tag{2.4.3.19}$$

where $\text{erf}\,\zeta = \frac{2}{\sqrt{\pi}} \int_0^\zeta \exp(-\xi^2)\,d\xi$ is the error function.

2.4.4. Initial-Boundary Value Problem for a Linear Hyperbolic Equation with Proportional Delay

Statement of the problem. Let us look at the first initial-boundary value problem for a one-dimensional hyperbolic linear homogeneous equation with constant coefficients and a proportional delay

$$u_{tt} = a_1 u_{xx} + a_2 w_{xx} + c_1 u + c_2 w, \quad w = u(x, pt). \tag{2.4.4.1}$$

The equation is defined in the domain $\Omega = \{0 < x < h, \ t > 0\}$ and it is assumed that $a_1 > a_2 \geq 0$ and $0 < p < 1$. We supplement equation (2.4.4.1) with the first-kind (Dirichlet) homogeneous boundary conditions

$$u = 0 \quad \text{at} \quad x = 0, \quad u = 0 \quad \text{at} \quad x = h, \tag{2.4.4.2}$$

2.4. Linear Initial-Boundary Value Problems with Proportional Delay

and general initial conditions

$$u = \varphi(x) \quad \text{at} \quad t = 0, \qquad u_t = \psi(x) \quad \text{at} \quad t = 0. \tag{2.4.4.3}$$

Reasoning in the same way as for the parabolic equation (2.4.2.1), we can employ the method of separation of variables to show that the hyperbolic equation with proportional delay (2.4.4.1) admits exact solutions as the product of two functions with different arguments

$$u_n(x,t) = T_n(t) \sin\left(\frac{\pi n x}{h}\right), \quad n = 1, 2, \ldots, \tag{2.4.4.4}$$

with the functions $T_n(t)$ satisfying the second-order linear ODEs with proportional delay

$$T_n''(t) = \left[c_1 - a_1\left(\frac{\pi n}{h}\right)^2\right] T_n(t) + \left[c_2 - a_2\left(\frac{\pi n}{h}\right)^2\right] T_n(pt). \tag{2.4.4.5}$$

These equations can be obtained from (2.4.2.5) by formally replacing the first derivative with the second derivative. The particular solutions (2.4.4.4) satisfy the homogeneous boundary conditions (2.4.4.2).

Using the linear superposition principle, we look for a solution to the initial-boundary value problem (2.4.4.1)–(2.4.4.3) as the infinite series

$$u(x,t) = \sum_{n=1}^{\infty} T_n(t) \sin\left(\frac{\pi n x}{h}\right). \tag{2.4.4.6}$$

It satisfies equation (2.4.4.1) and the homogeneous boundary conditions (2.4.4.2).

To determine the initial conditions for the ODE with proportional delay (2.4.4.5), we rewrite the functions $\varphi(x)$ and $\psi(x)$ appearing in the initial conditions (2.4.4.3) to expand into series in the eigenfunctions:

$$\varphi(x) = \sum_{n=1}^{\infty} A_n \sin\frac{\pi n x}{h}, \quad \psi(x) = \sum_{n=1}^{\infty} B_n \sin\frac{\pi n x}{h}. \tag{2.4.4.7}$$

Multiplying (2.4.4.7) by $X_m(x) = \sin\frac{\pi m x}{h}$ ($m = 1, 2, \ldots$) and integrating with respect to the space variable x from 0 to h, we find the coefficients A_n and B_n:

$$A_n = \frac{2}{h}\int_0^h \varphi(\xi)\sin\left(\frac{\pi n \xi}{h}\right)d\xi, \quad B_n = \frac{2}{h}\int_0^h \psi(\xi)\sin\left(\frac{\pi n \xi}{h}\right)d\xi. \tag{2.4.4.8}$$

From relations (2.4.4.6) and (2.4.4.7) we obtain the initial conditions for the ODE with proportional delay (2.4.4.5):

$$T_n(0) = A_n, \quad T_n'(0) = B_n, \tag{2.4.4.9}$$

where the coefficients A_n and B_n are determined by formulas (2.4.4.8).

Up to notation, the linear problem for the second-order ODE with proportional delay (2.4.4.5), (2.4.4.9) coincides with problem (1.2.2.24), (1.2.2.25) with $c = 0$ discussed in Subsection 1.2.2. Considering the above and using formulas (1.2.2.26) and (1.2.2.26), we can represent the solution to problem (2.4.4.5), (2.4.4.9) as a linear combination of two power series:

$$T_n(t) = A_n T_{n1}(t) + B_n T_{n2}(t), \qquad (2.4.4.10)$$

where

$$T_{n1}(t) = 1 + \sum_{m=1}^{\infty} \gamma_{n,2m} t^{2m}, \quad \gamma_{n,2m} = \frac{1}{(2m)!} \prod_{k=0}^{m-1} (\alpha_n + \beta_n p^{2k});$$

$$T_{n2}(t) = t + \sum_{m=1}^{\infty} \gamma_{n,2m+1} t^{2m+1}, \quad \gamma_{n,2m+1} = \frac{1}{(2m+1)!} \prod_{k=0}^{m-1} (\alpha_n + \beta_n p^{2k+1}).$$
$$(2.4.4.11)$$

The coefficients α_n and β_n are defined by formulas (2.4.2.10). For $0 < p < 1$, both series in (2.4.4.11) have an infinite radius of convergence.

Substituting expressions (2.4.4.10) into (2.4.4.6) yields the solution to the initial-boundary value problem (2.4.4.1)–(2.4.4.3):

$$u(x,t) = \sum_{n=1}^{\infty} [A_n T_{n1}(t) + B_n T_{n2}(t)] \sin\left(\frac{\pi n x}{h}\right),$$

$$T_{n1}(t) = 1 + \sum_{m=1}^{\infty} \gamma_{n,2m} t^{2m}, \quad T_{n2}(t) = t + \sum_{m=1}^{\infty} \gamma_{n,2m+1} t^{2m+1},$$
$$(2.4.4.12)$$

where the coefficients A_n, B_n, $\gamma_{n,2m}$, $\gamma_{n,2m+1}$ are evaluated by formulas (2.4.4.8) and (2.4.4.11).

3. Analytical Methods and Exact Solutions to Nonlinear Delay PDEs. Part I

3.1. Remarks and Definitions. Traveling Wave Solutions

3.1.1. Preliminary Remarks. Terminology. Classes of Equations Concerned

Preliminary Remarks. Nonlinear second- and higher-order partial differential equations with delay (nonlinear equations of mathematical physics with delay) arise in various areas of applied mathematics, physics, mechanics, biology, medicine, chemistry, and numerous applications. General closed-form solutions of such equations are impossible to find even in the simplest cases. Therefore, researchers have to restrict themselves to seeking and analyzing particular solutions which are frequently referred to as *exact solutions*.

Exact solutions of differential equations have always played and continue to play a huge role in shaping the correct understanding of the qualitative features of many phenomena and processes in various fields of natural science. Exact solutions of nonlinear equations clearly demonstrate and contribute to the better understanding of the mechanisms of such complicated nonlinear effects as the spatial localization of transfer processes, the multiplicity or absence of stationary states under certain conditions, the existence of blow-up modes, the possible non-smoothness or discontinuity of the unknown quantities, and many more.

Even those particular exact solutions of differential equations that do not have a clear physical meaning can fit for designing test problems to check the correctness and evaluate the accuracy of various numerical, asymptotic, and approximate analytical methods. In addition, model equations and problems that admit exact solutions serve as a basis for developing new numerical, asymptotic, and approximate methods, making it possible to study more complex problems that do not have an exact analytical solution. Finally, exact methods and solutions are also necessary for developing and improving the relevant sections of computer software intended for symbolic calculations (computer algebra systems such as Mathematica, Maple, Maxima, and more).

Factors leading to the need to consider the delay. The studies [182, 213, 283, 346, 570] mention various factors that lead to the need to introduce delay into mathematical models described by reaction-diffusion type equations and other nonlinear PDEs. In particular, in biology and biomechanics, delays are associated with limited transmission rates of nerve and muscle reactions in living tissues. In medical problems of the spread of infectious diseases, the delay time is determined by the incubation period (the time interval between initial contact with an infectious agent and appearance of the first signs or symptoms of the disease). In population dynamics, a delay arises because individuals participate in reproduction only after reaching a certain age. In control theory, delays occur due to limited speeds of signal propagation and limited rates of technological processes.

Delay reaction-diffusion equations. In natural science and numerous applications, nonlinear reaction-diffusion equations with constant delay are very common to model phenomena and processes with aftereffects (e.g., see [420, 569, 572]):

$$u_t = au_{xx} + F(u,w), \quad w = u(x, t - \tau), \tag{3.1.1.1}$$

where $a > 0$ is the transfer (diffusion) coefficient, $F(u, w)$ is the kinetic function, and τ is the delay time.

The special case $F(u, w) = f(w)$ in (3.1.1.1) allows a simple physical interpretation: the transfer of a substance in a local-nonequilibrium medium exhibits inertial properties, meaning that the system does not respond to an action instantaneously, as in the classical local-equilibrium case, but a delay time τ later.

Nonlinear delay reaction-diffusion equations of the form (3.1.1.1) and related more complicated equations and systems of such equations arise in various applications in such disciplines as biology, biophysics, biochemistry, chemistry, medicine, ecology, economics, control theory, the theory of climate models, and many others (e.g., see the studies [152, 189, 224, 315, 339, 377, 420, 483, 519, 569, 572] and references therein). Notably, similar equations occur in the mathematical theory of artificial neural networks, whose results are used in signal and image processing and pattern recognition problems [19, 85, 86, 319, 320, 493, 542, 607].

The models described by nonlinear delay reaction-diffusion equations of the form (3.1.1.1) are usually obtained from generalizing simpler models. To this end, the following two techniques are most frequently used:

(i) in models described by first-order ordinary differential equations with independent variable t and constant delay τ, one adds a diffusion term au_{xx};

(ii) in models described by reaction-diffusion equations without delay, one replaces the kinetic function $f(u)$ with a more complicated kinetic function with delay, $F(u, w)$, that satisfies the condition $F(u, u) = f(u)$.

A further generalization of reaction-diffusion models would be to use nonlinear equations with a variable transfer coefficient and a constant delay,

$$u_t = [g(u)u_x]_x + F(u,w), \quad w = u(x, t - \tau), \tag{3.1.1.2}$$

or even more complex related equations with variable delay.

Remark 3.1. *For exact solutions to various reaction-diffusion type equations (3.1.1.1) and (3.1.1.2) with $\tau = 0$ and related nonlinear PDEs without delay, see, for example, [99, 100, 134, 137, 148, 171, 175, 229, 394, 396, 397, 400, 422, 425, 430, 437, 442, 445, 603].*

Nonlinear delay Klein–Gordon type wave equations. Apart from reaction-diffusion equations with constant delay (3.1.1.1) and (3.1.1.2), the present book also deals with nonlinear Klein–Gordon type wave equations with constant delay [429]:

$$u_{tt} = au_{xx} + F(u, w), \quad w = u(x, t - \tau). \tag{3.1.1.3}$$

In addition, it also treats some more complicated related nonlinear wave type equations with delay, including telegraph and hyperbolic reaction-diffusion equations with delay [419].

Remark 3.2. Klein–Gordon type wave equations occur in various areas of theoretical physics, including relativistic quantum mechanics and field theory. For exact solutions to various nonlinear equations of the form (3.1.1.3) with $\tau = 0$ and related nonlinear PDEs without delay, see, for example, [12, 15, 62, 108, 147, 195, 218, 220–223, 229, 371, 399, 422, 425, 430, 443, 494, 593, 608, 611].

Remark 3.3. For example, see [117, 118, 304, 538] for the investigation of oscillatory properties of solutions to some nonlinear hyperbolic equations with delay.

Terminology: which solutions are called exact. In the present book, the following solutions are understood as exact solutions to nonlinear partial differential equations with constant or variable delay [416, 428, 432]:

(*a*) Solutions expressible in terms of elementary functions, the functions appearing in the equation (in case the equation involves arbitrary or special functions), and indefinite or/and definite integrals.

(*b*) Solutions expressible in terms of solutions to ordinary differential equations or systems of such equations.

(*c*) Solutions expressible in terms of solutions to ordinary differential equations with constant or variable delay or systems of such equations.

Various combinations of cases (*a*)–(*c*) are also allowed. In case (*a*), exact solutions can be represented explicitly, implicitly, or parametrically.

Exact methods for solving nonlinear PDEs (including delay PDEs) are methods that allow one to obtain exact solutions.

Following [422, 425, 430], we will further be using a simple and clear classification of most common solutions by their appearance (see Table 3.1) unrelated to the type or form of the equations concerned.

Difficulties arising in using standard analytical methods. The presence of a constant or variable delay in nonlinear equations of mathematical physics dramatically complicates the analysis. Many exact methods that are effective in finding exact solutions for nonlinear PDEs without delay, such as the nonclassical method of symmetry reductions (Bluman–Cole method) [20, 60, 105, 107, 363], direct method of symmetry reductions (Clarkson–Kruskal method) [20, 106, 107, 363, 422, 425, 430], method of differential constraints [171, 256, 422, 425, 430, 480], method of inverse scattering problem [3, 82, 362, 383], and method of truncated Painlevé expansions [239, 286, 422, 425, 550, 551] are inapplicable for constructing exact solutions to nonlinear PDEs with constant or variable delay. The classical method of symmetry reductions (Lie group analysis) [61, 229, 368, 372] shows very limited capabilities

Table 3.1. Most common types of exact solutions to equations of mathematical physics with two independent variables, x and t, and the unknown function u.

No.	Type of solution	General structure of solution (x and t can be swapped)
1	Traveling wave solution	$u = U(z)$, $z = kx + \lambda t$, $k\lambda \neq 0$
2	Additive separable solution	$u = \varphi(x) + \psi(t)$
3	Multiplicative separable solution	$u = \varphi(x)\psi(t)$
4	Self-similar solution	$u = t^\alpha F(z)$, $z = xt^\beta$
5	Generalized self-similar solution	$u = \varphi(t)F(z)$, $z = \psi(t)x$
6	Generalized traveling wave solution	$u = U(z)$, $z = \varphi(t)x + \psi(t)$
7	Generalized separable solution	$u = \varphi_1(x)\psi_1(t) + \cdots + \varphi_n(x)\psi_n(t)$
8	Functional separable solution (special case)	$u = U(z)$, $z = \varphi(x) + \psi(t)$
9	Functional separable solution	$u = U(z)$, $z = \varphi_1(x)\psi_1(t) + \cdots + \varphi_n(x)\psi_n(t)$

as only relatively few exact solutions to isolated nonlinear PDEs with constant delay have been reported so far [316, 341, 510]. Notably, equations of mathematical physics with two independent variables and a delay have the following essential qualitative features: (i) PDEs with constant delay do not admit self-similar solutions, unlike PDEs without delay, many of which do, and (ii) PDEs with proportional delay in either independent variable do not have traveling wave solutions, unlike simpler PDEs without delay, which often have.

In subsequent sections, we will describe quite effective methods, developed in recent years, for constructing exact solutions to nonlinear PDEs with constant or variable delay. We will give many examples of constructing exact solutions for specific equations. When selecting suitable material, we paid the greatest attention to nonlinear reaction-diffusion type equations with delay that often arise in applications and nonlinear equations of general form with one or more arbitrary functions. Exact solutions of such equations are of greatest interest for testing numerical and approximate analytical methods. Apart from equations with constant delay (3.1.1.1) and (3.1.1.3), we will also consider more complex equations, with a proportional delay $\tau = pt$ and a variable delay of general form $\tau = \tau(t)$.

3.1.2. States of Equilibrium. Traveling Wave Solutions. Exact Solutions in Closed Form

States of equilibrium. A constant

$$u = u_*, \quad u_* = \text{const}, \tag{3.1.2.1}$$

that solves an equation of mathematical physics is called a *state of equilibrium*, also known as a *stationary point*, a *rest point*, a *fixed point*, or simply an *equilibrium*. In

3.1. Remarks and Definitions. Traveling Wave Solutions

the case of a PDE with constant delay independent of the variables x and t,

$$\Phi(u, u_x, u_t, u_{xx}, \ldots; w, w_x, w_t, w_{xx}, \ldots) = 0, \quad w = u(x, t - \tau), \quad (3.1.2.2)$$

the states of equilibrium are determined from the algebraic (or transcendental) equation

$$\Phi(u_*, 0, 0, 0, \ldots; u_*, 0, 0, 0, \ldots) = 0.$$

The equilibrium states (3.1.2.1) of the delay reaction-diffusion equation (3.1.1.1) are determined from the algebraic (transcendental) equation $F(u_*, u_*) = 0$ and, hence, are zeros of the kinetic function.

Traveling wave solutions. Non-constant solutions of the form

$$u = U(z), \quad z = kx + \lambda t, \quad (3.1.2.3)$$

where k and λ are nonzero constants, are called *traveling wave solutions*. As a rule, PDEs with constant delay of the form (3.1.2.2) have traveling wave solutions (3.1.2.3), where the function $U(z)$ is described by the ordinary differential equation with constant delay

$$\Phi(U, kU'_z, \lambda U'_z, k^2 U''_{zz}, \ldots; W, kW'_z, \lambda W'_z, k^2 W''_{zz}, \ldots) = 0, \quad W = U(z - \lambda\tau). \quad (3.1.2.4)$$

Remark 3.4. There are very rare occasions when equations of the form (3.1.2.2) do not have traveling wave solutions (3.1.2.3). In such cases, the left-hand side of equation (3.1.2.4) is nonzero for any k and λ.

▶ **Example 3.1.** The nonlinear Monge–Ampére type equation with constant delay

$$u_{xt}^2 - u_{xx} u_{tt} + w = 0, \quad w = u(x, t - \tau),$$

which is an equation of the form (3.1.2.2), does not have traveling wave solutions, because inserting (3.1.2.3) results in the false equality $W = 0$ (since $U \neq \text{const}$). ◀

Traveling wave solutions of reaction-diffusion equations with constant delay. Substituting (3.1.2.3) into (3.1.1.1) yields the nonlinear ordinary differential equation with constant delay

$$ak^2 U''_{zz} - \lambda U'_z + F(U, W) = 0, \quad (3.1.2.5)$$

where $W = U(z - \sigma)$, $\sigma = \lambda\tau$.

The questions of existence and stability of traveling wave solutions to delay reaction-diffusion equations have been addressed in many studies (e.g., see [152, 224, 339, 483, 519, 572] and references therein).

Below we specify several reduced nonlinear delay ODEs of the form (3.1.2.5) that admit traveling wave solutions expressible in terms of elementary functions. These derive from respective original delay reaction-diffusion equations (3.1.1.1) whose kinetic function contains functional arbitrariness. The solutions listed below as well as some other solutions were obtained in [412]) (see also [495, 497]). The situations $\sigma > 0$ (delay ODEs) and $\sigma < 0$ (advanced ODEs) are treated simultaneously.

Equation 1. Consider the nonlinear delay ODE

$$ak^2 U''_{zz} - \lambda U'_z + U f(W/U) = 0, \qquad (3.1.2.6)$$

where $f(\zeta)$ is an arbitrary function, and $W = U(z - \sigma)$.

1°. Equation (3.1.2.6) admits exponential exact solutions of the form

$$U = C \exp(\beta z),$$

where C is an arbitrary constant, and β is a root of the algebraic (transcendental) equation

$$ak^2 \beta^2 - \lambda \beta + f(e^{-\sigma \beta}) = 0.$$

The parameters k and λ in (3.1.2.6) can be any.

2°. Equation (3.1.2.6) admits exact exponential-trigonometric solutions of the form

$$U = e^{\mu z}\left[A_n \cos(\beta_n z) + B_n \sin(\beta_n z)\right], \qquad \beta_n = \frac{\pi n}{\sigma}, \quad n = \pm 1, \pm 2, \ldots,$$

where A_n, B_n, and μ are arbitrary constants. The equation parameters λ and k are expressed as

$$\lambda = 2ak^2 \mu, \qquad k = \pm \left[\frac{f\bigl((-1)^n e^{-\mu \sigma}\bigr)}{a(\beta_n^2 + \mu^2)}\right]^{1/2}.$$

Equation 2. The nonlinear delay ODE

$$ak^2 U''_{zz} - \lambda U'_z + U f(U - cW) + W g(U - cW) + h(U - cW) = 0, \quad c > 0, \quad (3.1.2.7)$$

where $f(\zeta)$, $g(\zeta)$, and $h(\zeta)$ are arbitrary functions, and $W = U(z - \sigma)$, admits exponential-trigonometric solutions of the form

$$U = e^{\mu z}\left[A_n \cos(\beta_n z) + B_n \sin(\beta_n z)\right] + D,$$
$$\mu = \frac{1}{\sigma} \ln c, \quad \beta_n = \frac{2\pi n}{\sigma}, \quad n = \pm 1, \pm 2, \ldots,$$

where A_n and B_n are arbitrary constants, and D is a root of the algebraic (transcendental) equation

$$D\bigl[f(\xi) + g(\xi)\bigr] + h(\xi) = 0, \qquad \xi = (1 - c)D.$$

The equation parameters λ and k are expressed as

$$\lambda = 2ak^2 \mu, \qquad k = \pm \left[\frac{cf(\xi) + g(\xi)}{ac(\beta_n^2 + \mu^2)}\right]^{1/2}.$$

Equation 3. The nonlinear delay ODE

$$ak^2 U''_{zz} - \lambda U'_z + U f(U+cW) + W g(U+cW) + h(U+cW) = 0, \quad c > 0, \quad (3.1.2.8)$$

where $f(\zeta)$, $g(\zeta)$, and $h(\zeta)$ are arbitrary functions, and $W = U(z - \sigma)$, admits exponential-trigonometric solutions

$$U = e^{\mu z}\left[A_n \cos(\beta_n z) + B_n \sin(\beta_n z)\right] + D,$$

$$\mu = \frac{1}{\sigma} \ln c, \quad \beta_n = \frac{(2n-1)\pi}{\sigma}, \quad n = 0, \pm 1, \pm 2, \ldots,$$

where A_n and B_n are arbitrary constants, and D is a root of the algebraic (or transcendental) equation

$$D\left[f(\xi) + g(\xi)\right] + h(\xi) = 0, \quad \xi = (1+c)D.$$

The equation parameters λ and k are expressed as

$$\lambda = 2ak^2\mu, \quad k = \pm\left[\frac{cf(\xi) - g(\xi)}{ac(\beta_n^2 + \mu^2)}\right]^{1/2}.$$

Equation 4. The nonlinear delay ODE

$$ak^2 U''_{zz} - \lambda U'_z + U f(U^2 + W^2) + W g(U^2 + W^2) = 0, \quad (3.1.2.9)$$

where $f(\zeta)$ and $g(\zeta)$ are arbitrary functions, and $W = U(z - \sigma)$, admits exact trigonometric solutions

$$U = A_n \cos(\beta_n z) + B_n \sin(\beta_n z),$$
$$\beta_n = \frac{\pi(2n+1)}{2\sigma}, \quad n = 0, \pm 1, \pm 2, \ldots \quad (3.1.2.10)$$

The coefficients A_n and B_n in (3.1.2.10) are determined from the system of algebraic (transcendental) equations

$$-ak^2\beta_n^2 A_n - \lambda \beta_n B_n + A_n f(A_n^2 + B_n^2) + (-1)^{n+1} B_n g(A_n^2 + B_n^2) = 0,$$
$$-ak^2\beta_n^2 B_n + \lambda \beta_n A_n + B_n f(A_n^2 + B_n^2) + (-1)^n A_n g(A_n^2 + B_n^2) = 0. \quad (3.1.2.11)$$

If the constants A_n and B_n are treated as arbitrary, then it follows from (3.1.2.11) that the parameters λ and k are expressed as

$$\lambda = \frac{(-1)^{n+1} g(A_n^2 + B_n^2)}{\beta_n}, \quad k = \pm\left[\frac{f(A_n^2 + B_n^2)}{a\beta_n^2}\right]^{1/2}.$$

Equation 5. Now consider the nonlinear delay ODE

$$ak^2 U''_{zz} - \lambda U'_z + \frac{1}{\varphi'_U} f\bigl(\varphi(U) - \varphi(W)\bigr) + \frac{\varphi''_{UU}}{(\varphi'_U)^3} g\bigl(\varphi(U) - \varphi(W)\bigr) = 0, \quad (3.1.2.12)$$

where $f(\zeta)$, $g(\zeta)$, and $\varphi(U)$ are arbitrary functions. It admits exact solutions that can be represented in implicit form:

$$\varphi(U) = Az + B, \qquad (3.1.2.13)$$

where A and B are arbitrary constants. The parameters k and λ are expressed as

$$k = \pm\left[\frac{g(A\sigma)}{aA^2}\right]^{1/2}, \quad \lambda = \frac{f(A\sigma)}{A}. \qquad (3.1.2.14)$$

▶ **Example 3.2.** Assuming that $\varphi(U) = U^k$ in (3.1.2.12) and (3.1.2.13), we arrive at the equation

$$ak^2 U''_{zz} - \lambda U'_z + U^{1-k}\hat{f}(U^k - W^k) + U^{1-2k}\hat{g}(U^k - W^k) = 0,$$

which has an exact solution $U = (Az + B)^{1/k}$. The new arbitrary functions are related to the original ones as: $\hat{f}(\zeta) = \frac{1}{k}f(\zeta)$ and $\hat{g}(\zeta) = \frac{k-1}{k^2}g(\zeta)$. ◀

▶ **Example 3.3.** Setting $\varphi(U) = e^{\beta U}$ in (3.1.2.12) and (3.1.2.13), we get the equation

$$ak^2 U''_{zz} - \lambda U'_z + e^{-\beta U}\hat{f}(e^{\beta U} - e^{\beta W}) + e^{-2\beta U}\hat{g}(e^{\beta U} - e^{\beta W}) = 0,$$

which has an exact solution $U = \frac{1}{\beta}\ln(Az + B)$. The new arbitrary functions are related to the original ones as: $\hat{f}(\zeta) = \frac{1}{\beta}f(\zeta)$ and $\hat{g}(\zeta) = \frac{1}{\beta}g(\zeta)$. ◀

Reaction-diffusion equations with proportional delay. We note right away that the partial differential equation with proportional delay in one independent variable does not admit traveling wave solutions.

However, partial differential equations explicitly independent of x and t with identical proportional delays in two independent variables can have traveling wave solutions. Such equations involve the unknown functions $u = u(x,t)$ and $w = u(px, qt)$ with $q = p$. In particular, reaction-diffusion equations with proportional delay [416]:

$$u_t = [g(u)u_x]_x + F(u,w), \quad w = u(px, pt),$$

admit traveling wave solutions (3.1.2.3), where the function $U(z)$ is described by the ODE with proportional delay

$$\lambda U'_z = k^2[g(U)U'_z]'_z + F(U,W), \quad W = U(pz).$$

3.1.3. Traveling Wave Front Solutions to Nonlinear Reaction-Diffusion Type Equations

Traveling wave front solutions. We will consider traveling wave solutions

$$u = U(z), \quad z = x + \lambda t, \qquad (3.1.3.1)$$

with $\lambda > 0$. Substituting (3.1.3.1) into the PDE with constant delay (3.1.1.1) yields the following second-order delay ODE for $U(z)$:

$$aU''_{zz} - \lambda U'_z + f(U, W) = 0, \quad W = U(z - \lambda\tau). \tag{3.1.3.2}$$

Bounded solutions allowing a physical interpretation are of greatest interest. In what follows, we assume that equation (3.1.1.1) has simple stationary solutions $u = u_1$ and $u = u_2$, where u_1 and u_2 are some constants. This implies that the kinetic function $f(u, w)$ vanishes at these points: $f(u_1, u_1) = f(u_2, u_2) = 0$.

Applications pay special attention to traveling wave solutions (3.1.3.1) in which the function $U(z)$ satisfies not only the delay ODE (3.1.3.2) but also additional boundary conditions to match the stationary solutions:

$$U(z) \to u_1 \quad \text{as} \quad z \to -\infty, \quad U(z) \to u_2 \quad \text{as} \quad z \to \infty. \tag{3.1.3.3}$$

The constants u_1 and u_2 can be swapped. Bounded monotonic solutions (3.1.3.1) of equation (3.1.1.1) that satisfy conditions (3.1.3.3) are called *traveling wave front solutions* or *traveling front solutions* for short.

Below we describe the qualitative features of traveling wave front solutions to some PDEs with constant delay occurring in applications.

Delay diffusive logistic equation. The *reaction-diffusion logistic equation with constant delay* has the form

$$u_t = au_{xx} + bu(1 - cw), \quad w = u(x, t - \tau), \tag{3.1.3.4}$$

where $a > 0$, $b > 0$, and $c > 0$. It describes the dynamics of populations (collections of individuals of the same species) considering the period of maturation, when individuals are not capable of reproduction (see Subsection 6.3.2 for details). Sometimes, equation (3.1.3.4) is also called the *delay Fisher equation*; with $\tau = 0$, it was discussed in [162, 278].

Equation (3.1.3.4) has two simple stationary solutions, $u = 0$ and $u = 1/c$, and admits a traveling wave solution (3.1.3.1), where the function $U(z)$ is described by the delay ODE

$$aU''(z) - \lambda U'(z) + bU(z)[1 - cU(z - \lambda\tau)] = 0. \tag{3.1.3.5}$$

The monotonic function $U(z)$ satisfying equation (3.1.3.5) and the asymptotic boundary conditions matching the stationary conditions

$$U(z) \to 0 \quad \text{as} \quad z \to -\infty, \quad U(z) \to 1/c \quad \text{as} \quad z \to \infty \tag{3.1.3.6}$$

will define a traveling wave front solution.

The following theorems hold true.

Theorem 1 [160, 353]. *If $\tau = 0$, the boundary value problem (3.1.3.5)–(3.1.3.6) has a monotonic solution if and only if $\lambda \geq 2\sqrt{ab}$. In other words, if there is no delay, $\lambda = 2\sqrt{ab}$ is the minimum allowed traveling wave speed for the reaction-diffusion equation (3.1.3.4).*

Theorem 2 *[224, 572].* *For any $\lambda > 2\sqrt{ab}$, there is a $\tau_*(\lambda) > 0$ such that for $\tau \le \tau_*(\lambda)$, equation (3.1.3.4) has a traveling wave front solution with the wave front moving at a speed λ.*

Let us explain why a bounded traveling wave speed arises in Theorems 1 and 2. To this end, we linearize equation (3.1.3.5) for large negative z while assuming that $|U| \ll 1$. As a result, we arrive at the approximate ODE without delay

$$aU''(z) - \lambda U'(z) + bU(z) = 0.$$

We look for its particular solutions in the exponential form $U = \exp(\beta z)$. The constant β must satisfy the quadratic equation

$$a\beta^2 - \lambda\beta + b = 0,$$

whose roots are

$$\beta_{1,2} = \frac{\lambda \pm \sqrt{\lambda^2 - 4ab}}{2a}.$$

It is apparent that for $\lambda > 2\sqrt{ab}$, both roots are positive real numbers. The associated exponential solutions tend to zero monotonically as $z \to -\infty$. For $\lambda < 2\sqrt{ab}$, both roots are complex numbers. Although the associated solutions rapidly decay as $z \to -\infty$, they also oscillate and, hence, are not monotonic.

The theorem below refines Theorem 1.

Theorem 3 *[181] (see also [288]).* *The delay reaction-diffusion logistic equation (3.1.3.4) with $a = b = c = 1$ has a positive monotonic traveling wave front solution of the form (3.1.3.1) that connects the stationary solutions 0 and 1 if and only if either of the following conditions holds:*

 (i) $0 \le \tau \le 1/e = 0.367879441\ldots$ *and* $2 \le \lambda < +\infty$;
 (ii) $1/e < \tau \le \tau_1 = 0.560771160\ldots$ *and* $2 \le \lambda \le \lambda_*(\tau) = 1/\sqrt{\phi(\tau)}$.

The constant τ_1 is a root of the transcendental equation

$$2\tau^2 \exp\bigl(1 + \sqrt{1 + 4\tau^2} - 2\tau\bigr) = 1 + \sqrt{1 + 4\tau^2},$$

and the function $\phi(\tau)$ is defined parametrically as

$$\phi = \xi h(\xi), \quad \tau = h(\xi) \equiv \bigl(2\xi + \sqrt{1 + 4\xi^2}\bigr) \exp\left(-1 - \frac{2\xi}{\sqrt{1 + 4\xi^2}}\right),$$

where $0 \le \xi \le 0.445\ldots$.

If $\tau > \tau_1 = 0.560771160\ldots$, the delay diffusive logistic equation (3.1.3.4) with $a = b = c = 1$ does not have a monotonic traveling wave front solution.

Delay diffusive logistic equation under limited food conditions. The diffusive logistic equation with constant delay and limited food supply, which generalizes equation (3.1.3.4), has the form

$$u_t = au_{xx} + bu\frac{1 - cw}{1 + \gamma w}, \quad w = u(x, t - \tau), \qquad (3.1.3.7)$$

where $\gamma > 0$. In the special case $\gamma = 0$, it becomes equation (3.1.3.4).

Just as (3.1.3.4), equation (3.1.3.7) has two simple stationary solutions, $u = 0$ and $u = 1/c$, and admits a traveling wave solution of the form (3.1.3.1).

The study [187] (see also [224]) showed that equation (3.1.3.7) has a traveling wave front solution under the conditions of Theorem 2 stated above.

Remark 3.5. *A similar statement holds true for the more complex delay equation obtained from (3.1.3.7) by replacing the denominator* $1 + \gamma w$ *with* $1 + \gamma_1 u + \gamma_2 w$, *where* $\gamma_1 \geq 0$ *and* $\gamma_2 \geq 0$.

Nicholson's blowflies delay model. The nonlinear reaction-diffusion equation with delay

$$u_t = u_{xx} - \delta u + pwe^{-\kappa w}, \quad w = u(x, t - \tau), \qquad (3.1.3.8)$$

where $p > 0$, $\delta > 0$, and $\kappa > 0$, is known as Nicholson's blowflies delay model.

For $p/\delta > 1$, equation (3.1.3.8) has two simple stationary solutions, $u = 0$ and $u = (1/\kappa) \ln(p/\delta)$, and admits an exact traveling wave solution (3.1.3.1) with the function $U(z)$ described by the delay ODE

$$U''(z) - \lambda U'(z) - \delta U(z) + pU(z - \lambda\tau)e^{-\kappa U(z-\lambda\tau)} = 0. \qquad (3.1.3.9)$$

The monotonic function $U(z)$ that solves equation (3.1.3.9) and satisfies the boundary conditions matching the stationary solutions

$$U(z) \to 0 \quad \text{as} \quad z \to -\infty, \quad U(z) \to (1/\kappa)\ln(p/\delta) \quad \text{as} \quad z \to \infty \qquad (3.1.3.10)$$

defines a traveling front solution.

The study [488] proved the following theorem.

Theorem 4. *Let* $1 < p/\delta \leq e$. *Then there exists a* $\lambda_* > 0$ *such that for any* $\lambda > \lambda_*$, *equation (3.1.3.9) has a traveling front solution (3.1.3.1) with the wave front moving at a speed* λ.

Belousov–Zhabotinsky delay reaction-diffusion model. The Belousov–Zhabotinsky delay reaction-diffusion model is described by the quasilinear system of equations [572]:

$$\begin{aligned} u_t &= u_{xx} + u(1 - u - a\bar{v}), \quad \bar{v} = v(x, t - \tau), \\ v_t &= v_{xx} - buv, \end{aligned} \qquad (3.1.3.11)$$

where u and v are the bromous acid and bromide ion concentrations, while a and b are positive constants. Equations (3.1.3.11) and their generalizations can serve to describe more complicated biochemical and biological processes. The simpler system without delay, with $\tau = 0$ in (3.1.3.11), was studied in [351].

System (3.1.3.11) has two simple stationary solutions: $u = 0$, $v = \text{const}$ and $u = 1, v = 0$.

System (3.1.3.11) admits traveling wave solutions

$$u = U(z), \quad v = V(z), \quad z = x + \lambda t, \qquad (3.1.3.12)$$

where $\lambda > 0$; the functions $U(z)$ and $V(z)$ are described by the system of delay ODEs

$$U''(z) - \lambda U'(z) + U(z)[1 - U(z) - aV(z - \lambda\tau)] = 0,$$
$$V''(z) - \lambda V'(z) - bU(z)V(z) = 0.$$
(3.1.3.13)

We supplement equation (3.1.3.13) with asymptotic boundary conditions matching the stationary solutions:

$$\begin{aligned}U(z) \to 0 \quad &\text{as} \quad z \to -\infty, \quad U(z) \to 1 \quad \text{as} \quad z \to \infty,\\ V(z) \to 1 \quad &\text{as} \quad z \to -\infty, \quad V(z) \to 0 \quad \text{as} \quad z \to \infty.\end{aligned}$$
(3.1.3.14)

The functions $U(z)$ and $V(z)$ that solve the system of delay ODEs (3.1.3.13) and satisfy the boundary conditions (3.1.3.14) define a traveling front solution for the original system of delay reaction-diffusion equations (3.1.3.11).

The study [572] proved the following theorem.

Theorem 5. *Depending on the values of the determining parameters a and b, this theorem consists of two items:*

$1°$. *Let $0 < b \leq 1 - a$. Then for any $\lambda \geq 2\sqrt{1-a}$ and $\tau > 0$, system (3.1.3.11) has a traveling front solution with speed λ.*

$2°$. *Let $1 - a < b$. Then for any $\lambda \geq 2\sqrt{b}$ and $\tau > 0$, system (3.1.3.11) has a traveling front solution with speed λ.*

Lotka–Volterra type diffusive model with several delays. The Lotka–Volterra type reaction-diffusion model with cooperative interactions and four delay times is described by the system of equations [224]:

$$\begin{aligned}\frac{\partial u(x,t)}{\partial t} &= a_1 \frac{\partial^2 u(x,t)}{\partial x^2} + b_1 u(x,t)[1 - c_1 u(x, t - \tau_1) + d_1 v(x, t - \tau_2)],\\ \frac{\partial v(x,t)}{\partial t} &= a_2 \frac{\partial^2 v(x,t)}{\partial x^2} + b_2 v(x,t)[1 + d_2 u(x, t - \tau_3) - c_2 v(x, t - \tau_4)],\end{aligned}$$
(3.1.3.15)

where a_i, b_i, c_i, d_i, and τ_j ($i = 1, 2$; $j = 1, 2, 3, 4$) are positive constants.

Let $c_1 c_2 - d_1 d_2 > 0$. Then system (3.1.3.15) has four states of equilibrium: $(0,0)$, $(1/c_1, 0)$, $(0, 1/c_2)$, (k_1, k_2), where

$$k_1 = \frac{d_1 + c_2}{c_1 c_2 - d_1 d_2}, \quad k_2 = \frac{c_1 + d_2}{c_1 c_2 - d_1 d_2}.$$
(3.1.3.16)

By changing in (3.1.3.15) to the traveling wave variables (3.1.3.12), we obtain the system of delay ODEs

$$\begin{aligned}a_1 U''(z) - \lambda U'_z(z) + b_1 U(z)[1 - c_1 U(z - \lambda\tau_1) + d_1 V(z - \lambda\tau_2)] &= 0,\\ a_2 V''(z) - \lambda V'_z(z) + b_2 V(z)[1 + d_2 U(z - \lambda\tau_3) - c_2 V(z - \lambda\tau_4)] &= 0.\end{aligned}$$
(3.1.3.17)

We supplement equation (3.1.3.17) with asymptotic boundary conditions matching the stationary solutions:

$$\begin{aligned}U(z) \to 0 \quad &\text{as} \quad z \to -\infty, \quad U(z) \to k_1 \quad \text{as} \quad z \to \infty,\\ V(z) \to 0 \quad &\text{as} \quad z \to -\infty, \quad V(z) \to k_2 \quad \text{as} \quad z \to \infty,\end{aligned}$$
(3.1.3.18)

where the constants k_1 and k_2 are defined in (3.1.3.16).

The study [224] proved the following theorem.

Theorem 6. *Let* $c_1 c_2 - d_1 d_2 > 0$. *Then for any*
$$\lambda > \max\left[2\sqrt{a_1 b_1 c_1 k_1},\ 2\sqrt{a_2 b_2 c_2 k_2}\right]$$
and sufficiently small τ_1 *and* τ_4, *the Lotka–Volterra type delay reaction-diffusion system (3.1.3.15) has a traveling front solution (3.1.3.12) with the wave front moving at a speed* λ; *this solution is asymptotically related to the stationary solutions* $(0,0)$ *and* (k_1, k_2).

3.2. Multiplicative and Additive Separable Solutions

3.2.1. Preliminary Remarks. Terminology. Examples

Preliminary remarks and definitions. The method of separation of variables is the most common approach to solving linear equations of mathematical physics without delay [89, 280, 402, 404, 514, 613]. For equations in two independent variables, x and t, and one unknown function, $u = u(x,t)$, this method is based on seeking exact solutions as the product of two functions with different arguments:

$$u = \varphi(x)\psi(t), \qquad (3.2.1.1)$$

with the functions $\varphi = \varphi(x)$ and $\psi = \psi(t)$ described by linear ordinary differential equations and determined in a subsequent analysis.

Integrating certain classes of nonlinear first-order PDEs without delay implies seeking exact solutions as the sum of two functions with different arguments [254, 402, 424]:

$$u = \varphi(x) + \psi(t). \qquad (3.2.1.2)$$

Some nonlinear second- or higher-order equations of mathematical physics without or with delay also have exact solutions of the form (3.2.1.1) or (3.2.1.2). These kinds of solution will be called a *multiplicative separable solution* or *additive separable solution*, respectively [422, 425, 430]. Either kind of exact solution will sometimes be referred to as a *separable solution*.

Examples of nonlinear delay PDEs admitting separable solutions. In the simplest cases, the separation of variables in nonlinear partial differential equations with two independent variables and a constant time delay is carried out following the same procedure as in linear equations without delay. One looks for an exact solution as the product or sum of two functions with different arguments. Substituting (3.2.1.1) or (3.2.1.2) into the equation concerned and performing simple algebraic rearrangements, one arrives at an equality of two expressions (for equations in two variables) dependent on different arguments. This situation is only possible when both expressions are equal to the same constant quantity. As a result, one obtains an ODE without delay for $\varphi = \varphi(x)$ and a delay ODE for $\psi = \psi(t)$.

Let us illustrate the aforesaid by simple specific examples.

▶ **Example 3.4.** We will show that the reaction-diffusion equation with a constant delay and a power-law nonlinearity

$$u_t = a(u^k u_x)_x + bw, \quad w = u(x, t - \tau), \tag{3.2.1.3}$$

has an exact solution as the product of two functions with different arguments. Indeed, substituting (3.2.1.1) into equation (3.2.1.3) yields the relation

$$\varphi \psi'_t = a \psi^{k+1} (\varphi^k \varphi'_x)'_x + b \varphi \bar{\psi}, \quad \bar{\psi} = \psi(t - \tau). \tag{3.2.1.4}$$

Moving the term $b\varphi\bar{\psi}$ across to the left-hand side of (3.2.1.4) and dividing by $\varphi \psi^{k+1}$, we obtain

$$\frac{\psi'_t - b\bar{\psi}}{\psi^{k+1}} = \frac{a(\varphi^k \varphi'_x)'_x}{\varphi}.$$

The left-hand side of this equation only depends on the variable t, while the right-hand side depends on x alone. This is possible only if

$$\frac{\psi'_t - b\bar{\psi}}{\psi^{k+1}} = C, \quad \frac{a(\varphi^k \varphi'_x)'_x}{\varphi} = C, \tag{3.2.1.5}$$

where C is an arbitrary constant. The first-order ODE with constant delay for $\psi = \psi(t)$ in (3.2.1.5) can be solved using the method of steps (see Subsection 1.1.5). The second-order ODE without delay for $\varphi = \varphi(x)$ in (3.2.1.5) admits order reduction (since it is explicitly independent of x), and its general solution can be represented in implicit form.

The procedure for constructing a separable solution of the form (3.2.1.1) to the nonlinear PDE (3.2.1.3) is completely analogous to that for solving the simpler linear PDE with constant delay at $k = 0$. The fundamental difference between linear and nonlinear differential equations is that the principle of superposition is inapplicable to nonlinear equations. This means that solutions of the form (3.2.1.1) for the nonlinear equation (3.2.1.3) with $k \neq 0$ obtained by integrating equations (3.2.1.5) for various values of C cannot be added up. ◀

Remark 3.6. *The constant delay time τ in equations (3.2.1.3) and (3.2.1.5) can be replaced with an arbitrary variable delay $\tau = \tau(t)$. In particular, a proportional delay $\tau = (1 - p)t$, implying that $t - \tau = pt$, can be used.*

▶ **Example 3.5.** The reaction-diffusion type equation with an exponential nonlinearity and constant delay

$$u_t = au_{xx} + be^{\lambda(u-w)}, \quad w = u(x, t - \tau), \tag{3.2.1.6}$$

has an additive separable solution expressed as the sum of two functions with different arguments. Substituting (3.2.1.2) into equation (3.2.1.6) and rearranging yields the equation

$$\psi'_t - be^{\lambda(\psi - \bar{\psi})} = a\varphi''_{xx}, \quad \bar{\psi} = \psi(t - \tau). \tag{3.2.1.7}$$

Its left-hand side is only dependent on t, while the right-hand side depends on x alone. Equating the left- and right-hand sides of (3.2.1.7) with the same constant, we obtain
$$\psi'_t - be^{\lambda(\psi-\bar\psi)} = C, \quad a\varphi''_{xx} = C. \qquad (3.2.1.8)$$
The nonlinear first-order ODE with constant delay for $\psi = \psi(t)$ in (3.2.1.8) reduces to a linear delay ODE with the substitution $\theta = e^{-\lambda\psi}$. Integrating the second-order ODE for $\varphi = \varphi(x)$ in (3.2.1.8) twice, we get $\varphi = \frac{C}{2a}x^2 + C_1 x + C_2$. ◂

Remark 3.7. *The constant delay τ in equations (3.2.1.6) and (3.2.1.8) can be replaced with an arbitrary variable delay $\tau = \tau(t)$. In particular, a proportional delay $\tau = (1-p)t$, implying that $t - \tau = pt$, can be used.*

▶ **Example 3.6.** We will show that the reaction-diffusion equation with constant delay and a logarithmic source
$$u_t = au_{xx} + bu\ln w, \quad w = u(x, t-\tau), \qquad (3.2.1.9)$$
has a multiplicative separable solution expressed as the sum of two functions with different arguments
$$u = \varphi(x)\psi(t). \qquad (3.2.1.10)$$
To this end, we substitute expression (3.2.1.10) into equation (3.2.1.9) and divide by $\varphi\psi$. As a result, after moving one term from the right- to the left-hand side, we get
$$\frac{\psi'_t}{\psi} - b\ln\bar\psi = a\frac{\varphi''_{xx}}{\varphi} + b\ln\varphi, \quad \bar\psi = \psi(t-\tau).$$
The left-hand side of this equations is only dependent on t, while the right-hand side depend on x alone. Equating them with the same constant, we obtain a delay ODE for $\psi(t)$ and ODE without delay for $\varphi(x)$:
$$\frac{\psi'_t}{\psi} - b\ln\bar\psi = C, \quad a\frac{\varphi''_{xx}}{\varphi} + b\ln\varphi = C. \qquad (3.2.1.11)$$
With the change of variable $\psi = e^\theta$, the nonlinear first-order ODE with constant delay for $\psi = \psi(t)$ in (3.2.1.11) reduces to a linear delay ODE for θ. ◂

Remark 3.8. *The constant delay τ in equations (3.2.1.9) and (3.2.1.11) can be replaced with an arbitrary variable delay $\tau = \tau(t)$. In particular, a proportional delay $\tau = (1-p)t$, implying that $t - \tau = pt$, can be used.*

Below we describe some nonlinear delay PDEs of the form (3.1.1.1) that involve arbitrary functions, dependent on combinations of u and w, and admit additive or multiplicative separable solutions.

3.2.2. Delay Reaction-Diffusion Equations Admitting Separable Solutions

Reaction-diffusion equations with constant delay involving an arbitrary function. Below we list a number of nonlinear reaction-diffusion equations with constant delay that involve one arbitrary continuous function $f(\ldots)$ and admit separable exact solutions.

Equation 1. Consider the nonlinear reaction-diffusion equation with constant delay
$$u_t = au_{xx} + uf(w/u), \quad w = u(x, t - \tau). \tag{3.2.2.1}$$

1°. It has a multiplicative separable solution periodic in the space coordinate x:
$$u = [C_1 \cos(\beta x) + C_2 \sin(\beta x)]\psi(t), \tag{3.2.2.2}$$

where C_1, C_2, and β are arbitrary constants, and $\psi(t)$ is a function satisfying the first-order ODE with constant delay
$$\psi'_t(t) = -a\beta^2 \psi(t) + \psi(t) f\bigl(\psi(t-\tau)/\psi(t)\bigr). \tag{3.2.2.3}$$

2°. Equation (3.2.2.1) has another multiplicative separable solution:
$$u = [C_1 \exp(-\beta x) + C_2 \exp(\beta x)]\psi(t), \tag{3.2.2.4}$$

where C_1, C_2, and β are arbitrary constants, and $\psi(t)$ is a function satisfying the first-order ODE with constant delay
$$\psi'_t(t) = a\beta^2 \psi(t) + \psi(t) f\bigl(\psi(t-\tau)/\psi(t)\bigr). \tag{3.2.2.5}$$

3°. Equation (3.2.2.1) has a degenerate multiplicative separable solution:
$$u = (C_1 x + C_2)\psi(t), \tag{3.2.2.6}$$

where C_1 and C_2 are arbitrary constants; the function $\psi(t)$ satisfies the delay ODE (3.2.2.3) with $\beta = 0$.

4°. Equation (3.2.2.1) also has a mixed multiplicative separable solution:
$$u = e^{\alpha x + \beta t}\theta(z), \quad z = \lambda x + \gamma t, \tag{3.2.2.7}$$

where α, β, γ, and λ are arbitrary constants; the function $\theta(z)$ satisfies the second-order ODE with constant delay
$$a\lambda^2 \theta''_{zz}(z) + (2a\alpha\lambda - \gamma)\theta'_z(z) + (a\alpha^2 - \beta)\theta(z)$$
$$+ \theta(z) f\bigl(e^{-\beta\tau}\theta(z-\sigma)/\theta(z)\bigr) = 0, \quad \sigma = \gamma\tau.$$

Solution (3.2.2.7) can be treated as a nonlinear superposition of two traveling waves.

Remark 3.9. *The delay ODEs (3.2.2.3) and (3.2.2.5) admit particular solutions of the exponential form:*
$$\psi(t) = Ae^{\lambda_n t}, \quad n = 1, 2,$$
where A is an arbitrary constant, and λ_1 and λ_2 are roots of the transcendental equations

$$\lambda_1 = -a\beta^2 + f(e^{-\lambda_1 \tau}) \qquad \text{for equation (3.2.2.3),}$$
$$\lambda_2 = a\beta^2 + f(e^{-\lambda_2 \tau}) \qquad \text{for equation (3.2.2.5).}$$

Equation 2. The nonlinear reaction-diffusion equation with constant delay

$$u_t = au_{xx} + bu\ln u + uf(w/u) \tag{3.2.2.8}$$

admits a multiplicative separable solution

$$u = \varphi(x)\psi(t). \tag{3.2.2.9}$$

The functions $\varphi(x)$ and $\psi(t)$ are described, respectively, by the ODE without delay and ODE with constant delay

$$a\varphi''_{xx} = C_1\varphi - b\varphi\ln\varphi, \tag{3.2.2.10}$$
$$\psi'_t(t) = C_1\psi(t) + \psi(t)f(\psi(t-\tau)/\psi(t)) + b\psi(t)\ln\psi(t), \tag{3.2.2.11}$$

where C_1 is an arbitrary constant.

Remark 3.10. The second-order ODE (3.2.2.10) is explicitly independent of x, and so its general solution can be expressed in implicit form. The equation has a particular one-parameter solution

$$\varphi = \exp\left[-\frac{b}{4a}(x+C_2)^2 + \frac{C_1}{b} + \frac{1}{2}\right],$$

where C_2 is an arbitrary constant.

Equation 3. Consider the nonlinear reaction-diffusion equation with constant delay

$$u_t = au_{xx} + f(u-w). \tag{3.2.2.12}$$

1°. It has an exact additive separable solution quadratic in x:

$$u = C_2 x^2 + C_1 x + \psi(t), \tag{3.2.2.13}$$

where C_1 and C_2 are arbitrary constants; the function $\psi(t)$ is described by first-order ODE with constant delay

$$\psi'_t(t) = 2C_2 a + f(\psi(t) - \psi(t-\tau)). \tag{3.2.2.14}$$

2°. Equation (3.2.2.12) also has a more general solution than (3.2.2.13):

$$u = C_1 x^2 + C_2 x + C_3 t + \theta(z), \quad z = \beta x + \gamma t, \tag{3.2.2.15}$$

where C_1, C_2, C_3, β, and γ are arbitrary constants; the function $\theta(z)$ is described by the second-order ODE with constant delay

$$a\beta^2\theta''_{zz}(z) - \gamma\theta'_z(z) + 2C_1 a - C_3 + f(\theta(z) - \theta(z-\sigma) + C_3\tau) = 0, \quad \sigma = \gamma\tau.$$

With $C_1 = C_2 = C_3 = 0$, relation (3.2.2.15) represents a traveling wave solution.

Remark 3.11. The delay ODE (3.2.2.14) has a particular solution $\psi(t) = \lambda t + C_3$ linear in t, where C_3 is an arbitrary constant and λ is a root of the algebraic (transcendental) equation $2C_2 a - \lambda + f(\tau\lambda) = 0$.

Equation 4. Let us look at the nonlinear reaction-diffusion equation with constant delay
$$u_t = au_{xx} + bu + f(u - w), \tag{3.2.2.16}$$
which becomes equation (3.2.2.12) at $b = 0$.

$1°$. For $ab > 0$, equation (3.2.2.16) has an additive separable solution periodic in the space variable x:
$$u = C_1 \cos(\lambda x) + C_2 \sin(\lambda x) + \psi(t), \quad \lambda = \sqrt{b/a}, \tag{3.2.2.17}$$
where C_1 and C_2 are arbitrary constants, and the function $\psi(t)$ satisfies the delay ODE
$$\psi'_t(t) = b\psi(t) + f(\psi(t) - \psi(t - \tau)). \tag{3.2.2.18}$$

$2°$. For $ab < 0$, equation (3.2.2.16) has another additive separable solution
$$u = C_1 \exp(-\lambda x) + C_2 \exp(\lambda x) + \psi(t), \quad \lambda = \sqrt{-b/a}, \tag{3.2.2.19}$$
where C_1 and C_2 are arbitrary constants, and the function $\psi(t)$ satisfies the delay ODE (3.2.2.18).

$3°$. For $b = 0$, equation (3.2.2.16) has a degenerate additive separable solution
$$u = C_1 x + C_2 + \psi(t),$$
where the function $\psi(t)$ satisfies the delay ODE (3.2.2.18) with $b = 0$.

$4°$. For $ab > 0$, equation (3.2.2.16) also has a more general solution
$$u = C_1 \cos(\lambda x) + C_2 \sin(\lambda x) + \theta(z), \quad z = \beta x + \gamma t, \quad \lambda = \sqrt{b/a}, \tag{3.2.2.20}$$
than (3.2.2.17). Here, C_1, C_2, β, and γ are arbitrary constants; the function $\theta(z)$ is described by the delay ODE
$$\gamma \theta'_z(z) = a\beta^2 \theta''_{zz}(z) + b\theta(z) + f(\theta(z) - \theta(z - \sigma)), \quad \sigma = \gamma \tau. \tag{3.2.2.21}$$
Unlike (3.2.2.17), solution (3.2.2.20) is not periodic in the space variable x; it describes a nonlinear interaction of a periodic standing wave with a traveling wave.

$5°$. For $ab < 0$, equation (3.2.2.16) also has a more general solution
$$u = C_1 \exp(-\lambda x) + C_2 \exp(\lambda x) + \theta(z), \tag{3.2.2.22}$$
$$z = \beta x + \gamma t, \quad \lambda = \sqrt{-b/a},$$
than (3.2.2.19). Here, C_1, C_2, β, and γ are arbitrary constants, and $\theta(z)$ is a function satisfying the delay ODE (3.2.2.21).

Table 3.2 collects the above and some other nonlinear reaction-diffusion type equations with constant delay that admit additive or multiplicative separable solutions (according to [426, 432, 435]). Eleven equations involve one or two arbitrary continuous functions of a single argument, $f(z)$ and $g(z)$, where $z = u - w$ or $z = u/w$, and one equation involves an arbitrary function of two arguments, $f(z_1, z_2)$.

3.2. Multiplicative and Additive Separable Solutions

Table 3.2. Reaction-diffusion equations with constant delay admitting additive or multiplicative separable solutions. Notations: $w = u(x, t - \tau)$, $\bar{\psi} = \psi(t - \tau)$; C_1, C_2, and C_3 are arbitrary constants.

Original equation	Forms of solution	Determining equations or constants
$u_t = au_{xx} + uf(w/u)$	$u = [C_1\cos(\beta x) + C_2\sin(\beta x)]\psi(t)$; $u = [C_1\exp(-\beta x) + C_2\exp(\beta x)]\psi(t)$; $u = (C_1 x + C_2)\psi(t)$	$\psi'_t = -a\beta^2\psi + \psi f(\bar{\psi}/\psi)$; $\psi'_t = a\beta^2\psi + \psi f(\bar{\psi}/\psi)$; $\psi'_t = \psi f(\bar{\psi}/\psi)$
$u_t = au_{xx} + bu\ln u$ $+ uf(w/u)$	$u = \varphi(x)\psi(t)$	$a\varphi''_{xx} = C_1\varphi - b\varphi\ln\varphi$, $\psi'_t = C_1\psi + b\psi\ln\psi$ $+ \psi f(\bar{\psi}/\psi)$
$u_t = au_{xx} + f(u - w)$	$u = C_2 x^2 + C_1 x + \psi(t)$	$\psi'_t = 2C_2 a + f(\psi - \bar{\psi})$
$u_t = au_{xx}$ $+ bu + f(u - w)$	$u = C_1\cos(\lambda x) + C_2\sin(\lambda x) + \psi(t)$, where $\lambda = \sqrt{b/a}$ (for $ab > 0$); $u = C_1\exp(-\lambda x) + C_2\exp(\lambda x) + \psi(t)$, where $\lambda = \sqrt{-b/a}$ (for $ab < 0$)	$\psi'_t = b\psi + f(\psi - \bar{\psi})$; $\psi'_t = b\psi + f(\psi - \bar{\psi})$
$u_t = au_{xx}$ $+ uf(u - kw, w/u)$	$u = e^{ct}[C_1\cos(\lambda x) + C_2\sin(\lambda x)]$, if $b = f(0, 1/k) - c > 0$; $u = e^{ct}[C_1\exp(-\lambda x) + C_2\exp(\lambda x)]$, if $b = f(0, 1/k) - c < 0$	$c = (\ln k)/\tau$, $\lambda = (b/a)^{1/2}$, $k > 0$; $c = (\ln k)/\tau$, $\lambda = \|b/a\|^{1/2}$, $k > 0$
$u_t = a(u^k u_x)_x$ $+ uf(w/u)$	$u = \varphi(x)\psi(t)$	$a(\varphi^k \varphi'_x)'_x = C_1\varphi$, $\psi'_t = C_1\psi^{k+1} + \psi f(\bar{\psi}/\psi)$
$u_t = a(u^k u_x)_x + bu^{k+1}$ $+ uf(w/u)$	$u = [C_1\cos(\beta x) + C_2\sin(\beta x)]^{\frac{1}{k+1}}\psi(t)$, where $\beta = \sqrt{b(k+1)/a}$, $b(k+1) > 0$; $u = (C_1 e^{-\beta x} + C_2 e^{\beta x})^{\frac{1}{k+1}}\psi(t)$, where $\beta = \sqrt{-b(k+1)/a}$, $b(k+1) < 0$; $u = C_1\exp\left(-\frac{b}{2a}x^2 + C_2 x\right)\psi(t)$ at $k = -1$; $u = \varphi(x)\psi(t)$ (generalizes preceding solutions)	$\psi'_t = \psi f(\bar{\psi}/\psi)$; $\psi'_t = \psi f(\bar{\psi}/\psi)$; $\psi'_t = \psi f(\bar{\psi}/\psi)$; $a(\varphi^k \varphi'_x)'_x + b\varphi^{k+1} = C_1\varphi$, $\psi'_t = C_1\psi^{k+1} + \psi f(\bar{\psi}/\psi)$
$u_t = a(e^{\lambda u} u_x)_x + f(u - w)$	$u = \frac{1}{\lambda}\ln(C_1\lambda x^2 + C_2 x + C_3) + \psi(t)$	$\psi'_t = 2aC_1 e^{\lambda\psi} + f(\psi - \bar{\psi})$
$u_t = a(e^{\lambda u} u_x)_x + be^{\lambda u}$ $+ f(u - w)$	$u = \frac{1}{\lambda}\ln[C_1\cos(\beta x) + C_2\sin(\beta x)] + \psi(t)$, where $\beta = \sqrt{b\lambda/a}$, $b\lambda > 0$; $u = \frac{1}{\lambda}\ln(C_1 e^{-\beta x} + C_2 e^{\beta x}) + \psi(t)$, where $\beta = \sqrt{-b\lambda/a}$, $b\lambda < 0$; $u = \varphi(x) + \psi(t)$ (generalizes preceding solutions)	$\psi'_t = f(\psi - \bar{\psi})$; $\psi'_t = f(\psi - \bar{\psi})$; $a(e^{\lambda\varphi}\varphi'_x)'_x + be^{\lambda\varphi} = C_1$, $\psi'_t = C_1 e^{\lambda\psi} + f(\psi - \bar{\psi})$
$u_t = a(u^k u_x)_x$ $+ uf(w/u)$ $+ u^{k+1} g(w/u)$	$u = e^{\lambda t}\varphi(x)$, where λ is a root of transcendental equation $\lambda = f(e^{-\lambda\tau})$	$a(\varphi^k \varphi'_x)'_x$ $+ g(e^{-\lambda\tau})\varphi^{k+1} = 0$, this ODE linearizes with change of variable $\xi = \varphi^{k+1}$
$u_t = a(e^{\lambda u} u_x)_x$ $+ f(u - w)$ $+ e^{\lambda u} g(u - w)$	$u = \beta t + \varphi(x)$, where β is a root of algebraic equation $\beta = f(\beta\tau)$	$a(e^{\lambda\varphi}\varphi'_x)'_x + g(\beta\tau)e^{\lambda\varphi} = 0$, this ODE linearizes with change of variable $\xi = e^{\lambda\varphi}$
$u_t = [(a\ln u + b)u_x]_x$ $- cu\ln u + uf(w/u)$	$u = \exp(\pm\lambda x)\psi(t)$, $\lambda = \sqrt{c/a}$	$\psi'_t = \lambda^2(a + b)\psi$ $+ \psi f(\bar{\psi}/\psi)$

The listed solutions were obtained with the method of functional constraints, which is discussed below in Section 3.4; this method allows one to find even more sophisticated exact solutions.

Reaction-diffusion equations with a general variable delay. Many additive and multiplicative separable solutions obtained previously for nonlinear reaction-diffusion equations with constant delay (see Table 3.2) are extendable to the case of more complicated nonlinear PDEs with general variable delay.

Table 3.3 lists nonlinear reaction-diffusion equations with variable delay that admit exact separable solutions. It is assumed that $\tau = \tau(t)$ is an arbitrary positive continuous function that can vanish at one or more isolated points; in particular, one should set $\tau = (1-p)t$, implying that $t - \tau = pt$, in the case of proportional delay.

Table 3.3. Reaction-diffusion equations with general variable delay that admit additive or multiplicative separable solutions. Notations: $w = u(x, t - \tau(t))$, $\bar{\psi} = \psi(t - \tau(t))$; C_1, C_2, and C_3 are arbitrary constants.

Original equation	Forms of solution	Determining equations
$u_t = au_{xx} + uf(w/u)$	$u = [C_1\cos(\beta x) + C_2\sin(\beta x)]\psi(t)$; $u = [C_1\exp(-\beta x) + C_2\exp(\beta x)]\psi(t)$; $u = (C_1 x + C_2)\psi(t)$	$\psi'_t = -a\beta^2\psi + \psi f(\bar\psi/\psi)$; $\psi'_t = a\beta^2\psi + \psi f(\bar\psi/\psi)$; $\psi'_t = \psi f(\bar\psi/\psi)$
$u_t = au_{xx} + bu\ln u + uf(w/u)$	$u = \varphi(x)\psi(t)$	$a\varphi''_{xx} = C_1\varphi - b\varphi\ln\varphi$, $\psi'_t = C_1\psi + b\psi\ln\psi + \psi f(\bar\psi/\psi)$
$u_t = au_{xx} + f(u-w)$	$u = C_2 x^2 + C_1 x + \psi(t)$	$\psi'_t = 2C_2 a + f(\psi - \bar\psi)$
$u_t = au_{xx} + bu + f(u-w)$	$u = C_1\cos(\lambda x) + C_2\sin(\lambda x) + \psi(t)$, where $\lambda = \sqrt{b/a}$ (for $ab > 0$); $u = C_1\exp(-\lambda x) + C_2\exp(\lambda x) + \psi(t)$, where $\lambda = \sqrt{-b/a}$ (for $ab < 0$)	$\psi'_t = b\psi + f(\psi - \bar\psi)$; $\psi'_t = b\psi + f(\psi - \bar\psi)$
$u_t = a(u^k u_x)_x + uf(w/u)$	$u = \varphi(x)\psi(t)$	$a(\varphi^k\varphi'_x)'_x = C_1\varphi$, $\psi'_t = C_1\psi^{k+1} + \psi f(\bar\psi/\psi)$
$u_t = a(u^k u_x)_x + bu^{k+1} + uf(w/u)$	$u = \varphi(x)\psi(t)$	$a(\varphi^k\varphi'_x)'_x + b\varphi^{k+1} = C_1\varphi$, $\psi'_t = C_1\psi^{k+1} + \psi f(\bar\psi/\psi)$
$u_t = a(e^{\lambda u}u_x)_x + f(u-w)$	$u = \dfrac{1}{\lambda}\ln(C_1\lambda x^2 + C_2 x + C_3) + \psi(t)$	$\psi'_t = 2aC_1 e^{\lambda\psi} + f(\psi - \bar\psi)$
$u_t = a(e^{\lambda u}u_x)_x + be^{\lambda u} + f(u-w)$	$u = \varphi(x) + \psi(t)$	$a(e^{\lambda\varphi}\varphi'_x)'_x + be^{\lambda\varphi} = C_1$, $\psi'_t = C_1 e^{\lambda\psi} + f(\psi - \bar\psi)$

Remark 3.12. All the equations and solutions listed in Table 3.3 can be generalized by replacing the single-argument arbitrary functions $f(w/u)$ and $f(u-w)$ in the original equations with two-argument arbitrary functions $f(t, w/u)$ and $f(t, u-w)$ and also the functions $f(\bar\psi/\psi)$ and $f(\psi - \bar\psi)$ in the determining equations with $f(t, \bar\psi/\psi)$ and $f(t, \psi - \bar\psi)$.

Reaction-diffusion equations with several delays. All the equations and their additive and multiplicative separable solutions specified in Table 3.3 can be generalized to nonlinear reaction-diffusion equations with several variable delays of general form. To this end, the single-argument arbitrary functions in the equations and their solutions should be replaced with multi-argument arbitrary functions following the rules

$$f(w/u) \Rightarrow f(w_1/u, \ldots, w_n/u), \quad f(u-w) \Rightarrow f(u-w_1, \ldots, u-w_n);$$
$$f(\bar\psi/\psi) \Rightarrow f(\bar\psi_1/\psi, \ldots, \bar\psi_n/\psi), \quad f(\psi-\bar\psi) \Rightarrow f(\psi-\bar\psi_1, \ldots, \psi-\bar\psi_n); \quad (3.2.2.23)$$
$$w_k = u(x, t-\tau_k(t)), \quad \bar\psi_k = \psi(t-\tau_k(t)), \quad k=1,\ldots,n.$$

▶ **Example 3.7.** The reaction-diffusion equation with several delays

$$u_t = au_{xx} + uf(w_1/u, \ldots, w_n/u), \quad w_k = u(x, t-\tau_k(t)), \quad k=1,\ldots,n,$$

which generalizes the first equation from Table 3.3, admits a periodic multiplicative separable solution in the space variable x:

$$u = [C_1 \cos(\beta x) + C_2 \sin(\beta x)]\psi(t).$$

The function $\psi = \psi(t)$ is described by the ODE with several variable delays

$$\psi'_t = -a\beta^2 \psi + \psi f(\bar\psi_1/\psi, \ldots, \bar\psi_n/\psi), \quad \bar\psi_k = \psi(t-\tau_k(t)). \quad ◀$$

▶ **Example 3.8.** Another reaction-diffusion equation with several delays

$$u_t = au_{xx} + f(u-w_1, \ldots, u-w_n), \quad w_k = u(x, t-\tau_k(t)), \quad k=1,\ldots,n,$$

which generalizes the third equation from Table 3.3, admits an additive separable solution

$$u = C_2 x^2 + C_1 x + \psi(t),$$

where $\psi = \psi(t)$ is a function satisfying the ODE with several variable delays

$$\psi'_t = 2C_2 a + f(\psi-\bar\psi_1, \ldots, \psi-\bar\psi_n), \quad \bar\psi_k = \psi(t-\tau_k(t)). \quad ◀$$

Reaction-diffusion equations with several space variables and a constant delay. Below we will discuss some generalizations of the previous one-dimensional nonlinear reaction-diffusion equations with constant delay and their separable solutions to the case of more complicated, n-dimensional delay reaction-diffusion equations.

Table 3.4 displays some nonlinear reaction-diffusion equations with several space variables and a constant delay time that admit additive and multiplicative separable solutions. The equations are written using the short notations

$$\mathbf{x} = (x_1, \ldots, x_m), \quad u = u(\mathbf{x}, t), \quad w = u(\mathbf{x}, t-\tau),$$
$$\Delta u = \sum_{j=1}^{m} \frac{\partial^2 u}{\partial x_j^2}, \quad \nabla u = \sum_{j=1}^{m} \mathbf{e}_j \frac{\partial}{\partial x_j}, \quad \operatorname{div}[s(u)\nabla u] = \sum_{j=1}^{m} \frac{\partial}{\partial x_j}\left[s(u)\frac{\partial u}{\partial x_j}\right],$$

where x_j are Cartesian coordinates and $\mathbf{e}_j$ is a unit vector defining the direction of change of the space coordinate x_j. The values $m=2$ and $m=3$ correspond to two- and three-dimensional equations. The separation of variables results in the m-dimensional stationary equation for $\varphi = \varphi(\mathbf{x})$ specified in the last column of Table 3.4. Eight out of the eleven equations for φ are linear or can be linearized. For exact solutions to these PDEs, see, for example, the books [404, 514].

Table 3.4. Reaction-diffusion equations with several space variables and a constant time delay that admit additive and multiplicative separable solutions. Notations: $w = u(\mathbf{x}, t - \tau)$, $\bar\psi = \psi(t - \tau)$, and C is an arbitrary constant.

Original equation	Forms of solution	Determining equations
$u_t = a\Delta u + uf(w/u)$	$u = \varphi(\mathbf{x})\psi(t)$	$\Delta\varphi = C\varphi$, $\psi'_t = aC\psi + \psi f(\bar\psi/\psi)$
$u_t = a\Delta u + bu\ln u$ $+ uf(w/u)$	$u = \varphi(\mathbf{x})\psi(t)$	$a\Delta\varphi = C\varphi - b\varphi\ln\varphi$, $\psi'_t = C\psi + b\psi\ln\psi$ $+ \psi f(\bar\psi/\psi)$
$u_t = a\Delta u + f(u - w)$	$u = \varphi(\mathbf{x}) + \psi(t)$	$\Delta\varphi = C$, $\psi'_t = aC + f(\psi - \bar\psi)$
$u_t = a\Delta u + bu + f(u - w)$	$u = \varphi(\mathbf{x}) + \psi(t)$	$a\Delta\varphi + b\varphi = 0$, $\psi'_t = b\psi + f(\psi - \bar\psi)$
$u_t = a\Delta u$ $+ uf(u - kw, w/u)$	$u = e^{ct}\varphi(\mathbf{x})$, $c = (\ln k)/\tau$, $k > 0$	$a\Delta\varphi + [f(0, 1/k) - c]\varphi = 0$
$u_t = a\,\text{div}(u^k\nabla u)$ $+ uf(w/u)$	$u = \varphi(\mathbf{x})\psi(t)$	$a\,\text{div}(\varphi^k\nabla\varphi) = C\varphi$, $\psi'_t = C\psi^{k+1} + \psi f(\bar\psi/\psi)$
$u_t = a\,\text{div}(u^k\nabla u)$ $+ bu^{k+1} + uf(w/u)$	$u = \varphi(\mathbf{x})\psi(t)$	$a\,\text{div}(\varphi^k\nabla\varphi) + b\varphi^{k+1} = C\varphi$, $\psi'_t = C\psi^{k+1} + \psi f(\bar\psi/\psi)$
$u_t = a\,\text{div}(e^{\lambda u}\nabla u)$ $+ f(u - w)$	$u = \frac{1}{\lambda}\ln\varphi(\mathbf{x}) + \psi(t)$	$\Delta\varphi = C\lambda$, $\psi'_t = aCe^{\lambda\psi} + f(\psi - \bar\psi)$
$u_t = a\,\text{div}(e^{\lambda u}\nabla u)$ $+ be^{\lambda u} + f(u - w)$	$u = \frac{1}{\lambda}\ln\varphi(\mathbf{x}) + \psi(t)$	$(a/\lambda)\Delta\varphi + b\varphi = C$, $\psi'_t = Ce^{\lambda\psi} + f(\psi - \bar\psi)$
$u_t = a\,\text{div}(u^k\nabla u)$ $+ uf(w/u)$ $+ u^{k+1}g(w/u)$	$u = e^{\lambda t}\varphi(\mathbf{x})$, where λ is a root of transcendental equation $\lambda = f(e^{-\lambda\tau})$	$a\,\text{div}(\varphi^k\nabla\varphi) + g(e^{-\lambda\tau})\varphi^{k+1} = 0$, change of variable $\xi = \varphi^{k+1}$ linearizes this PDE
$u_t = a\,\text{div}(e^{\lambda u}\nabla u)$ $+ f(u - w)$ $+ e^{\lambda u}g(u - w)$	$u = \beta t + \varphi(\mathbf{x})$, where β is a root of algebraic equation $\beta = f(\beta\tau)$	$a\,\text{div}(e^{\lambda\varphi}\nabla\varphi) + g(\beta\tau)e^{\lambda\varphi} = 0$, change of variable $\xi = e^{\lambda\varphi}$ linearizes this PDE

3.2.3. Delay Klein–Gordon Type Equations Admitting Separable Solutions

Klein–Gordon type wave equations with a constant delay and arbitrary functions. Nonlinear Klein–Gordon type wave equations with delay differ from delay reaction-diffusion equations by formally replacing the first time derivative u_t with the second time derivative u_{tt}. In many cases, the general structure of additive and multiplicative separable solutions to these different nonlinear equations of mathematical physics with delay is the same; this means that the principle of solution analogy works here [9].

Table 3.5 displays nonlinear Klein–Gordon type wave equations with constant delay admitting additive or multiplicative separable solutions (those to the first five equations were obtained in [429]). Ten equations involve one or two arbitrary continuous functions of a single argument, $f(z)$ and $g(z)$, where $z = u - w$ or $z = u/w$, and one equation involves a two-argument arbitrary function, $f(z_1, z_2)$.

Klein–Gordon type wave equations with a variable delay of general form. Many additive and multiplicative separable solutions obtained previously for nonlinear Klein–Gordon type wave equations with constant delay (see Table 3.6) are extendable to more complicated, nonlinear equations with a variable delay of general form.

Table 3.6 shows nonlinear Klein–Gordon type wave equations with variable delay that admit separable solutions. It is assumed that $\tau = \tau(t)$ is an arbitrary positive continuous function that can vanish at one or more isolated points; in particular, for proportional delay, one should set $\tau = (1 - p)t$, implying that $t - \tau = pt$, in the equations.

All the equations and their additive and multiplicative separable solutions listed in Table 3.6 can be generalized to nonlinear Klein–Gordon type wave equations with several variable delays of general form. To this end, the single-argument arbitrary functions should be replaced with multi-argument arbitrary functions in the equations and their solutions following the rules (3.2.2.23).

Klein–Gordon type wave equations with several space variables and a constant delay. We will now describe some generalizations of the previously discussed nonlinear Klein–Gordon type wave equations with constant delay and their separable solutions to more complex, n-dimensional Klein–Gordon type wave equations with delay.

Table 3.7 displays some nonlinear Klein–Gordon type wave equations with several space variables and a constant delay time that admit additive or multiplicative separable solutions.

3.2.4. Some Generalizations

We will now look at nonlinear delay PDEs of a fairly general form

$$\mathbf{L}[u] = \mathbf{M}[u] + F(t, u, w),$$
$$w = u(\mathbf{x}, t - \tau), \quad \mathbf{x} = (x_1, \ldots, x_m),$$
(3.2.4.1)

where L is a linear differential operator in time t of order n whose coefficients can be time dependent,

$$\mathbf{L}[u] = \sum_{i=1}^{n} c_i(t) \frac{\partial^i u}{\partial t^i},$$
(3.2.4.2)

and M is a linear differential operator of any order in the space variables $x_1, \ldots, x_m$ whose coefficients can be dependent on $x_1, \ldots, x_m$.

Table 3.5. Klein–Gordon type wave equations with constant delay admitting additive or multiplicative separable solutions. Notations: $w = u(x, t - \tau)$, $\bar{\psi} = \psi(t - \tau)$; C_1, C_2, and C_3 are arbitrary constants.

Original equation	Forms of solution	Determining equations or constants
$u_{tt} = au_{xx} + uf(w/u)$	$u = [C_1\cos(\beta x) + C_2\sin(\beta x)]\psi(t)$; $u = [C_1\exp(-\beta x) + C_2\exp(\beta x)]\psi(t)$; $u = (C_1 x + C_2)\psi(t)$	$\psi''_{tt} = -a\beta^2\psi + \psi f(\bar{\psi}/\psi)$; $\psi''_{tt} = a\beta^2\psi + \psi f(\bar{\psi}/\psi)$; $\psi''_{tt} = \psi f(\bar{\psi}/\psi)$
$u_{tt} = au_{xx} + bu\ln u + uf(w/u)$	$u = \varphi(x)\psi(t)$	$a\varphi''_{xx} = C_1\varphi - b\varphi\ln\varphi$, $\psi''_{tt} = C_1\psi + b\psi\ln\psi + \psi f(\bar{\psi}/\psi)$
$u_{tt} = au_{xx} + f(u-w)$	$u = C_2 x^2 + C_1 x + \psi(t)$	$\psi''_{tt} = 2C_2 a + f(\psi - \bar{\psi})$
$u_{tt} = au_{xx} + bu + f(u-w)$	$u = C_1\cos(\lambda x) + C_2\sin(\lambda x) + \psi(t)$, where $\lambda = \sqrt{b/a}$ (for $ab > 0$); $u = C_1\exp(-\lambda x) + C_2\exp(\lambda x) + \psi(t)$, where $\lambda = \sqrt{-b/a}$ (for $ab < 0$)	$\psi''_{tt} = b\psi + f(\psi - \bar{\psi})$; $\psi''_{tt} = b\psi + f(\psi - \bar{\psi})$
$u_{tt} = au_{xx} + uf(u-kw, w/u)$	$u = e^{ct}[C_1\cos(\lambda x) + C_2\sin(\lambda x)]$, if $b = f(0, 1/k) - c^2 > 0$; $u = e^{ct}[C_1\exp(-\lambda x) + C_2\exp(\lambda x)]$, if $b = f(0, 1/k) - c^2 < 0$	$c = (\ln k)/\tau$, $\lambda = (b/a)^{1/2}$, $k > 0$; $c = (\ln k)/\tau$, $\lambda = \|b/a\|^{1/2}$, $k > 0$
$u_{tt} = a(u^k u_x)_x + uf(w/u)$	$u = \varphi(x)\psi(t)$	$a(\varphi^k \varphi'_x)'_x = C_1\varphi$, $\psi''_{tt} = C_1\psi^{k+1} + \psi f(\bar{\psi}/\psi)$
$u_{tt} = a(u^k u_x)_x + bu^{k+1} + uf(w/u)$	$u = [C_1\cos(\beta x) + C_2\sin(\beta x)]^{\frac{1}{k+1}}\psi(t)$, where $\beta = \sqrt{b(k+1)/a}$, $b(k+1) > 0$; $u = (C_1 e^{-\beta x} + C_2 e^{\beta x})^{\frac{1}{k+1}}\psi(t)$, where $\beta = \sqrt{-b(k+1)/a}$, $b(k+1) < 0$; $u = C_1\exp\!\left(-\frac{b}{2a}x^2 + C_2 x\right)\psi(t)$ if $k = -1$; $u = \varphi(x)\psi(t)$ (generalizes preceding solutions)	$\psi''_{tt} = \psi f(\bar{\psi}/\psi)$; $\psi''_{tt} = \psi f(\bar{\psi}/\psi)$; $\psi''_{tt} = \psi f(\bar{\psi}/\psi)$; $a(\varphi^k \varphi'_x)'_x + b\varphi^{k+1} = C_1\varphi$, $\psi''_{tt} = C_1\psi^{k+1} + \psi f(\bar{\psi}/\psi)$
$u_{tt} = a(e^{\lambda u} u_x)_x + f(u-w)$	$u = \frac{1}{\lambda}\ln(C_1\lambda x^2 + C_2 x + C_3) + \psi(t)$	$\psi''_{tt} = 2aC_1 e^{\lambda\psi} + f(\psi - \bar{\psi})$
$u_{tt} = a(e^{\lambda u} u_x)_x + be^{\lambda u} + f(u-w)$	$u = \frac{1}{\lambda}\ln[C_1\cos(\beta x) + C_2\sin(\beta x)] + \psi(t)$, where $\beta = \sqrt{b\lambda/a}$, $b\lambda > 0$; $u = \frac{1}{\lambda}\ln(C_1 e^{-\beta x} + C_2 e^{\beta x}) + \psi(t)$, where $\beta = \sqrt{-b\lambda/a}$, $b\lambda < 0$; $u = \varphi(x) + \psi(t)$ (generalizes preceding solutions)	$\psi''_{tt} = f(\psi - \bar{\psi})$; $\psi''_{tt} = f(\psi - \bar{\psi})$; $a(e^{\lambda\varphi}\varphi'_x)'_x + be^{\lambda\varphi} = C_1$, $\psi''_{tt} = C_1 e^{\lambda\psi} + f(\psi - \bar{\psi})$
$u_{tt} = a(u^k u_x)_x + uf(w/u) + u^{k+1} g(w/u)$	$u = e^{\lambda t}\varphi(x)$, where λ is a root of transcendental equation $\lambda^2 = f(e^{-\lambda\tau})$	$a(\varphi^k \varphi'_x)'_x + g(e^{-\lambda\tau})\varphi^{k+1} = 0$, change of variable $\xi = \varphi^{k+1}$ linearizes this ODE
$u_{tt} = [(a\ln u + b)u_x]_x - cu\ln u + uf(w/u)$	$u = \exp(\pm\lambda x)\psi(t)$, $\lambda = \sqrt{c/a}$	$\psi''_{tt} = \lambda^2(a+b)\psi + \psi f(\bar{\psi}/\psi)$

3.2. Multiplicative and Additive Separable Solutions

Table 3.6. Klein–Gordon type wave equations with a variable delay of general form that admit additive or multiplicative separable solutions. Notations: $w = u(x, t - \tau(t))$, $\bar{\psi} = \psi(t - \tau(t))$; C_1, C_2, and C_3 are arbitrary constants.

Original equation	Forms of solution	Determining equations
$u_{tt} = au_{xx} + uf(w/u)$	$u = [C_1 \cos(\beta x) + C_2 \sin(\beta x)]\psi(t)$; $u = [C_1 \exp(-\beta x) + C_2 \exp(\beta x)]\psi(t)$; $u = (C_1 x + C_2)\psi(t)$	$\psi''_{tt} = -a\beta^2\psi + \psi f(\bar{\psi}/\psi)$; $\psi''_{tt} = a\beta^2\psi + \psi f(\bar{\psi}/\psi)$; $\psi''_{tt} = \psi f(\bar{\psi}/\psi)$
$u_{tt} = au_{xx} + bu \ln u$ $+ uf(w/u)$	$u = \varphi(x)\psi(t)$	$a\varphi''_{xx} = C_1\varphi - b\varphi \ln \varphi$, $\psi''_{tt} = C_1\psi + b\psi \ln \psi$ $+ \psi f(\bar{\psi}/\psi)$
$u_{tt} = au_{xx} + f(u - w)$	$u = C_2 x^2 + C_1 x + \psi(t)$	$\psi''_{tt} = 2C_2 a + f(\psi - \bar{\psi})$
$u_{tt} = au_{xx} + bu$ $+ f(u - w)$	$u = C_1 \cos(\lambda x) + C_2 \sin(\lambda x) + \psi(t)$, where $\lambda = \sqrt{b/a}$ (for $ab > 0$); $u = C_1 \exp(-\lambda x) + C_2 \exp(\lambda x) + \psi(t)$, where $\lambda = \sqrt{-b/a}$ (for $ab < 0$)	$\psi''_{tt} = b\psi + f(\psi - \bar{\psi})$; $\psi''_{tt} = b\psi + f(\psi - \bar{\psi})$
$u_{tt} = a(u^k u_x)_x$ $+ uf(w/u)$	$u = \varphi(x)\psi(t)$	$a(\varphi^k \varphi'_x)'_x = C_1\varphi$, $\psi''_{tt} = C_1 \psi^{k+1} + \psi f(\bar{\psi}/\psi)$
$u_{tt} = a(u^k u_x)_x + bu^{k+1}$ $+ uf(w/u)$	$u = \varphi(x)\psi(t)$	$a(\varphi^k \varphi'_x)'_x + b\varphi^{k+1} = C_1\varphi$, $\psi''_{tt} = C_1 \psi^{k+1} + \psi f(\bar{\psi}/\psi)$
$u_{tt} = a(e^{\lambda u} u_x)_x$ $+ f(u - w)$	$u = \frac{1}{\lambda} \ln(C_1 \lambda x^2 + C_2 x + C_3) + \psi(t)$	$\psi''_{tt} = 2aC_1 e^{\lambda \psi} + f(\psi - \bar{\psi})$
$u_{tt} = a(e^{\lambda u} u_x)_x + be^{\lambda u}$ $+ f(u - w)$	$u = \varphi(x) + \psi(t)$	$a(e^{\lambda\varphi}\varphi'_x)'_x + be^{\lambda\varphi} = C_1$, $\psi''_{tt} = C_1 e^{\lambda\psi} + f(\psi - \bar{\psi})$

In particular, M can be an elliptic operator:

$$\mathbf{M}[u] = \sum_{i,j=1}^{m} \frac{\partial}{\partial x_i}\left(a_{ij}(\mathbf{x}) \frac{\partial u}{\partial x_j}\right) + \sum_{i=1}^{m} b_i(\mathbf{x}) \frac{\partial u}{\partial x_i}. \quad (3.2.4.3)$$

Also, M can be a biharmonic operator of the form

$$\mathbf{M}[u] = a\Delta\Delta u, \qquad \Delta u \equiv \sum_{i=1}^{m} \frac{\partial^2 u}{\partial x_i^2}.$$

Setting $\mathbf{L}[u] = u_t$, $\mathbf{M}[u] = au_{xx}$, $F_t = 0$, and $m = 1$ in (3.2.4.1) yields the nonlinear delay reaction-diffusion equation (3.1.1.1). Setting $\mathbf{L}[u] = u_{tt}$, $\mathbf{M}[u] = au_{xx}$, $F_t = 0$, and $m = 1$ in (3.2.4.1) results in the nonlinear delay Klein–Gordon equation (3.1.1.3).

Listed below are a few multiplicative or additive separable solutions to nonlinear partial differential equations of the form (3.2.4.1) that involve an arbitrary two-argument function $f(t, z)$, where $z = z(u, w)$. The determining equations are derived

Table 3.7. Klein–Gordon type wave equations with several space variables and a constant delay admitting additive or multiplicative separable solutions. Notations: $w = u(\mathbf{x}, t - \tau)$, $\bar{\psi} = \psi(t - \tau)$, and C is an arbitrary constant.

Original equation	Form of solution	Determining equations
$u_{tt} = a\Delta u + uf(w/u)$	$u = \varphi(\mathbf{x})\psi(t)$	$\Delta\varphi = C\varphi$, $\psi''_{tt} = aC\psi + \psi f(\bar{\psi}/\psi)$
$u_{tt} = a\Delta u + bu\ln u + uf(w/u)$	$u = \varphi(\mathbf{x})\psi(t)$	$a\Delta\varphi = C\varphi - b\varphi\ln\varphi$, $\psi''_{tt} = C\psi + b\psi\ln\psi + \psi f(\bar{\psi}/\psi)$
$u_{tt} = a\Delta u + f(u - w)$	$u = \varphi(\mathbf{x}) + \psi(t)$	$\Delta\varphi = C$, $\psi''_{tt} = aC + f(\psi - \bar{\psi})$
$u_{tt} = a\Delta u + bu + f(u - w)$	$u = \varphi(\mathbf{x}) + \psi(t)$	$a\Delta\varphi + b\varphi = 0$, $\psi''_{tt} = b\psi + f(\psi - \bar{\psi})$
$u_{tt} = a\Delta u + uf(u - kw, w/u)$	$u = e^{ct}\varphi(\mathbf{x})$, $c = (\ln k)/\tau$, $k > 0$	$a\Delta\varphi + [f(0, 1/k) - c^2]\varphi = 0$
$u_{tt} = a\,\mathrm{div}(u^k\nabla u) + uf(w/u)$	$u = \varphi(\mathbf{x})\psi(t)$	$a\,\mathrm{div}(\varphi^k\nabla\varphi) = C\varphi$, $\psi''_{tt} = C\psi^{k+1} + \psi f(\bar{\psi}/\psi)$
$u_{tt} = a\,\mathrm{div}(u^k\nabla u)$ $+ bu^{k+1} + uf(w/u)$	$u = \varphi(\mathbf{x})\psi(t)$	$a\,\mathrm{div}(\varphi^k\nabla\varphi) + b\varphi^{k+1} = C\varphi$, $\psi''_{tt} = C\psi^{k+1} + \psi f(\bar{\psi}/\psi)$
$u_{tt} = a\,\mathrm{div}(e^{\lambda u}\nabla u) + f(u - w)$	$u = \dfrac{1}{\lambda}\ln\varphi(\mathbf{x}) + \psi(t)$	$\Delta\varphi = C\lambda$, $\psi''_{tt} = aCe^{\lambda\psi} + f(\psi - \bar{\psi})$
$u_{tt} = a\,\mathrm{div}(e^{\lambda u}\nabla u)$ $+ be^{\lambda u} + f(u - w)$	$u = \dfrac{1}{\lambda}\ln\varphi(\mathbf{x}) + \psi(t)$	$(a/\lambda)\Delta\varphi + b\varphi = C$, $\psi''_{tt} = Ce^{\lambda\psi} + f(\psi - \bar{\psi})$
$u_{tt} = a\,\mathrm{div}(u^k\nabla u)$ $+ uf(w/u)$ $+ u^{k+1}g(w/u)$	$u = e^{\lambda t}\varphi(\mathbf{x})$, where λ is a root of transcendental equation $\lambda^2 = f(e^{-\lambda\tau})$	$a\,\mathrm{div}(\varphi^k\nabla\varphi) + g(e^{-\lambda\tau})\varphi^{k+1} = 0$, change of variable $\xi = \varphi^{k+1}$ linearizes this PDE

by using the following simple properties of the linear operators L and M:

$$\mathrm{L}[\varphi(\mathbf{x})\psi(t)] = \varphi(\mathbf{x})\mathrm{L}[\psi(t)], \quad \mathrm{L}[\varphi(\mathbf{x}) + \psi(t)] = \mathrm{L}[\psi(t)],$$
$$\mathrm{M}[\varphi(\mathbf{x})\psi(t)] = \psi(t)\mathrm{M}[\varphi(\mathbf{x})], \quad \mathrm{M}[\varphi(\mathbf{x}) + \psi(t)] = \mathrm{M}[\varphi(\mathbf{x})].$$

Equation 1. The nonlinear PDE with constant delay

$$\mathrm{L}[u] = \mathrm{M}[u] + uf(t, w/u), \quad w = u(\mathbf{x}, t - \tau), \qquad (3.2.4.4)$$

admits a multiplicative separable solution

$$u = \varphi(\mathbf{x})\psi(t). \qquad (3.2.4.5)$$

The functions $\varphi = \varphi(\mathbf{x})$ and $\psi = \psi(t)$ are described by the following linear stationary PDE without delay and nonlinear ODE with constant delay:

$$\mathrm{M}[\varphi] = C\varphi; \qquad (3.2.4.6)$$
$$\mathrm{L}[\psi] = C\psi + \psi f(t, \bar{\psi}/\psi), \quad \bar{\psi} = \psi(t - \tau), \qquad (3.2.4.7)$$

where C is an arbitrary constant.

Below are two simple cases where particular solutions to equation (3.2.4.6) or (3.2.4.7) can be found.

1°. If M is a linear differential operator with constant coefficients, then equation (3.2.4.6) admits an exponential exact solution of the form $\varphi(\mathbf{x}) = A\exp\bigl(\sum_{i=1}^{m}\beta_i x_i\bigr)$, where A is an arbitrary constant and $\beta_1,\ldots,\beta_m$ are arbitrary constants linked by a single polynomial dispersion relation. For $C = 0$ and $m \geq 2$, equation (3.2.4.6) can also have polynomial particular solutions.

2°. If L is a linear differential operator of the form (3.2.4.2) with constant coefficients ($c_i = \text{const}$) and if the source function is explicitly independent of time t, implying that $f = f(w/u)$, then the constant delay ODE (3.2.4.7) admits exponential solutions $\psi(t) = Be^{\lambda t}$, where B is an arbitrary constant and λ is a root of the algebraic (transcendental) equation

$$\sum_{i=1}^{n} c_i \lambda^i = C + f(e^{-\tau\lambda}).$$

Remark 3.13. *If the linear differential operators L and M have constant coefficients and if the source function f is explicitly independent of time t, then the equation with constant delay (3.2.4.4) admits exact solutions of the form*

$$u = \exp\Bigl(\alpha t + \sum_{i=1}^{m}\beta_i x_i\Bigr)\theta(z), \quad z = \gamma t + \sum_{i=1}^{m}\lambda_i x_i, \qquad (3.2.4.8)$$

where α, β_i, γ, and λ_i are arbitrary constants, and the function $\theta(z)$ is described by a delay ODE.

Equation 2. The more complicated nonlinear PDE with constant delay

$$\mathbf{L}[u] = \mathbf{M}[u] + bu\ln u + uf(t, w/u), \qquad (3.2.4.9)$$

also admits a multiplicative separable solution of the form (3.2.4.5), where the functions $\varphi = \varphi(\mathbf{x})$ and $\psi = \psi(t)$ are described by the nonlinear stationary PDE and nonlinear constant delay ODE

$$\mathbf{M}[\varphi] = C\varphi - b\varphi\ln\varphi;$$
$$\mathbf{L}[\psi] = C\psi + b\psi\ln\psi + \psi f(t, \bar\psi/\psi), \quad \bar\psi = \psi(t-\tau);$$

C is an arbitrary constant.

Equation 3. Another nonlinear partial differential equation with constant delay

$$\mathbf{L}[u] = \mathbf{M}[u] + bu + f(t, u - w) \qquad (3.2.4.10)$$

admits an additive separable solution

$$u = \varphi(\mathbf{x}) + \psi(t). \qquad (3.2.4.11)$$

The functions $\varphi = \varphi(\mathbf{x})$ and $\psi = \psi(t)$ are described by the linear stationary PDE and nonlinear constant delay ODE

$$\mathbf{M}[\varphi] = C - b\varphi;$$
$$\mathbf{L}[\psi] = C + b\psi + f(t, \psi - \bar\psi), \quad \bar\psi = \psi(t-\tau),$$

where C is an arbitrary constant.

Remark 3.14. *For a variable delay of general form $\tau = \tau(t)$, the nonlinear equations (3.2.4.4), (3.2.4.9), and (3.2.4.10) also have multiplicative or additive separable solutions (3.2.4.5) and (3.2.4.11).*

3.3. Generalized and Functional Separable Solutions

3.3.1. Generalized Separable Solutions

Preliminary remarks and definitions. Just as previously, we will consider partial differential equations with two independent variables, x and t, and a constant delay time, τ.

Linear equations of mathematical physics with constant coefficients without delay and many related linear PDEs with variable coefficients have exact solutions as the sum of pairwise products of functions with different arguments (e.g., see [402, 404, 514]):

$$u(x,t) = \varphi_1(x)\psi_1(t) + \varphi_2(x)\psi_2(t) + \cdots + \varphi_k(x)\psi_k(t). \qquad (3.3.1.1)$$

Many nonlinear partial differential equations of mathematical physics with quadratic and power-law nonlinearities, inclusive of some delay PDEs, also have exact solutions of the form (3.3.1.1). Such solutions will be referred to as *generalized separable solutions*. In general, the functions $\varphi_j(x)$ and $\psi_j(t)$ are not known in advance and so have to be determined in a subsequent investigation.

Remark 3.15. *For generalized separable solutions and methods for constructing such solution to nonlinear PDE without delay, see, for example, [169, 170, 172–175, 393, 422, 425, 430, 438, 515–517].*

Remark 3.16. *Expressions of the form (3.3.1.1) are frequently used in applied and computational mathematics for constructing approximate analytical and numerical solutions by projection methods such as Bubnov–Galerkin methods [161, 163, 446].*

In practice, one often deals with generalized separable solutions of special form that involve three unknown functions [169, 170, 422, 425, 430]:

$$u(x,t) = \varphi(t)\theta(x) + \psi(t). \qquad (3.3.1.2)$$

(The independent variables can be swapped in the right-hand side.) In the special case $\psi(t) = 0$, this solution becomes a multiplicative separable solution, and if $\varphi(t) = 1$, it becomes an additive separable solution.

The method based on a priori setting of a system of coordinate functions. To construct exact solutions of PDEs with a quadratic or power-law nonlinearity that is explicitly independent of x, one can employ the following simplified approach. One looks for a solution as the finite sum (3.3.1.1) and assumes that the system of coordinate functions $\varphi_m(x)$ is described by linear ODEs with constant coefficients. The most common solutions to such equations are

$$\varphi_m(x) = x^{\alpha_m}, \quad \varphi_m(x) = e^{\beta_m x}, \quad \varphi_m(x) = \cos(\lambda_m x), \quad \varphi_m(x) = \sin(\lambda_m x). \qquad (3.3.1.3)$$

Finite sets of these functions or various combinations can be used to search for generalized separable solutions of the form (3.3.1.1) with the constants α_m, β_m, and λ_m either set or determined in a subsequent analysis. The other system of functions $\psi_k(t)$ is found by solving the nonlinear delay ODEs obtained by substituting expression (3.3.1.1) with functions (3.3.1.3) into the original delay PDE.

By explicitly setting one system of coordinate functions $\{\varphi_j(x)\}$, one seriously facilitates the construction of exact solutions. However, isolated solutions of the form (3.3.1.1) may be lost with this approach. Notably, the overwhelming majority of generalized separable solutions for PDEs with a quadratic nonlinearity, known so far, are determined by coordinate functions (3.3.1.3), most frequently with $k = 2$ in (3.3.1.1).

Method of invariant subspaces. Consider the nonlinear evolution equation with constant delay [433]:

$$u_t = F[u] + sw, \quad w = u(x, t - \tau), \qquad (3.3.1.4)$$

where $F[u]$ is a nonlinear differential operator in the space variable x of the form

$$F[u] \equiv F(x, u, u_x, \ldots, u_x^{(n)}) \qquad (3.3.1.5)$$

and s is some constant.

Definition [175]. A finite-dimensional linear subspace

$$\mathscr{L}_k = \{\varphi_1(x), \ldots, \varphi_k(x)\} \qquad (3.3.1.6)$$

whose elements are all possible linear combinations of the linearly independent functions $\varphi_1(x), \ldots, \varphi_k(x)$ is said to be *invariant under a differential operator F* if $F[\mathscr{L}_k] \subseteq \mathscr{L}_k$. In this case, there exist functions $f_1, \ldots, f_k$ such that

$$F\left[\sum_{j=1}^{k} C_j \varphi_j(x)\right] = \sum_{j=1}^{k} f_j(C_1, \ldots, C_k) \varphi_j(x) \qquad (3.3.1.7)$$

for arbitrary constants $C_1, \ldots, C_k$. Note that the functions $\varphi_j(x)$ appearing in (3.3.1.7) cannot depend on $C_1, \ldots, C_k$.

Proposition 1. Let the linear subspace (3.3.1.6) be invariant under a differential operator F. Then equation (3.3.1.4) has a generalized separable solution of the form

$$u = \sum_{j=1}^{k} \psi_j(t) \varphi_j(x), \qquad (3.3.1.8)$$

with the functions $\psi_1(t), \ldots, \psi_k(t)$ described by the system of ordinary differential equations with delay [433]:

$$\psi_j' = f_j(\psi_1, \ldots, \psi_k) + s\bar{\psi}_j, \quad \bar{\psi}_j = \psi_j(t - \tau), \quad j = 1, \ldots, k. \qquad (3.3.1.9)$$

Here the prime stands for a derivative with respect to t.

This proposition can be proved as follows. First, one substitutes expression (3.3.1.8) into equation (3.3.1.4) and then uses relation (3.3.1.7) in which the constants C_j are replaced with $\psi_j = \psi_j(t)$. After collecting the terms proportional to $\varphi_j = \varphi_j(x)$, one obtains the equation

$$\sum_{j=1}^{k}[\psi_j' - f_j(\psi_1, \ldots, \psi_k) - s\bar{\psi}_j]\varphi_j(x) = 0.$$

Since the functions φ_j are linearly independent, all expressions in square brackets must be set equal to zero. This results in the system of ODEs (3.3.1.9).

Remark 3.17. *The study [175] formulated Proposition 1 for equation (3.3.1.4) without the delay term, i.e., for $s = 0$.*

Remark 3.18. *The delay in equations (3.3.1.4) and (3.3.1.9) can depend arbitrarily on time, so that $\tau = \tau(t)$.*

Table 3.8 displays some nonlinear differential operators and linear subspaces invariant under these operators (according to [175, 422, 430]). Adding a linear operator $L[u] = \alpha u_{xx} + \beta u_x + \gamma u + \delta$ to the nonlinear operators Nos. 4–8 does not change the invariant subspaces.

Examples of constructing generalized separable solutions for nonlinear delay equations. Below we consider a few examples of applying Proposition 1 and Table 3.8 to construct exact solutions for reaction-diffusion type equations with a quadratic nonlinearity and delay.

▶ **Example 3.9.** Let us look at the delay reaction-diffusion equation

$$u_t = [(a_1 u + a_0) u_x]_x + b_1 u + b_2 w, \quad w = u(x, t - \tau). \quad (3.3.1.10)$$

It follows from row 3 of Table 3.8 with $a = b$ and $c = 0$ that the nonlinear differential operator on the right-hand side of equation (3.3.1.10) with $b_2 = 0$ admits the invariant linear subspace $\mathscr{L}_3 = \{1, x, x^2\}$ (the operator converts the quadratic polynomial $C_1 + C_2 x + C_3 x^2$ to a quadratic polynomial with other coefficients). Considering the above and Proposition 1, one can conclude that the original equation (3.3.1.10) has a polynomial generalized separable solution in the space variable:

$$u = \psi_1(t) + \psi_2(t)x + \psi_3(t)x^2. \quad (3.3.1.11)$$

The functions $\psi_j = \psi_j(t)$ ($j = 1, 2, 3$) are described by the system of delay ODEs

$$\psi_1' = 2a_1 \psi_1 \psi_3 + a_1 \psi_2^2 + 2a_0 \psi_3 + b_1 \psi_1 + b_2 \bar{\psi}_1,$$
$$\psi_2' = 6a_1 \psi_2 \psi_3 + b_1 \psi_2 + b_2 \bar{\psi}_2,$$
$$\psi_3' = 6a_1 \psi_3^2 + b_1 \psi_3 + b_2 \bar{\psi}_3,$$

where $\bar{\psi}_j = \psi_j(t - \tau)$. ◀

▶ **Example 3.10.** Let us look at the more complicated delay reaction-diffusion equation with a quadratic nonlinearity

$$u_t = [(a_1 u + a_0) u_x]_x + ku^2 + b_1 u + b_2 w, \quad w = u(x, t - \tau). \quad (3.3.1.12)$$

At $k = 0$, it becomes equation (3.3.1.10). In what follows, we assume that $k \neq 0$.

3.3. Generalized and Functional Separable Solutions   157

Table 3.8. Some nonlinear differential operators and linear subspaces invariant under these operators (a, b, c, α, β, γ, and δ are free parameters).

No.	Nonlinear operator $F[u]$	Subspaces invariant under $F[u]$
1	$au_{xx} + bu_x^2 + \beta u_x + \gamma u + \delta$	$\mathscr{L}_3 = \{1, x, x^2\}$
2	$au_{xx} + bu_x^2 + cu^2$ $+ \beta u_x + \gamma u + \delta$	$\mathscr{L}_3 = \{1, \sin(x\sqrt{c/b}), \cos(x\sqrt{c/b})\}$ if $bc > 0$, $\mathscr{L}_3 = \{1, \sinh(x\sqrt{|c/b|}), \cosh(x\sqrt{|c/b|})\}$ if $bc < 0$
3	$auu_{xx} + bu_x^2 + cu^2$ $+ \alpha u_{xx} + \beta u_x + \gamma u + \delta$	$\mathscr{L}_3 = \{1, \sin(\lambda x), \cos(\lambda x)\}$ if $c/(a+b) = \lambda^2 > 0$, $\mathscr{L}_3 = \{1, \sinh(\lambda x), \cosh(\lambda x)\}$ if $c/(a+b) = -\lambda^2 < 0$, $\mathscr{L}_3 = \{1, x, x^2\}$ if $c = 0$, $\mathscr{L}_2 = \{x^2, x^\sigma\}$, $\sigma = a/(a+b)$ if $c = \alpha = \beta = \delta = 0$, $a \neq -b$
4	$uu_{xx} - u_x^2$ (special case of operator 3)	$\mathscr{L}_3 = \{1, \sin(\lambda x), \cos(\lambda x)\}$, λ is an arbitrary constant, $\mathscr{L}_3 = \{1, \sinh(\lambda x), \cosh(\lambda x)\}$, λ is an arbitrary constant, $\mathscr{L}_3 = \{1, x, x^2\}$
5	$uu_{xx} - \frac{2}{3} u_x^2$ (special case of operator 3)	$\mathscr{L}_4 = \{1, x, x^2, x^3\}$
6	$uu_{xx} - \frac{3}{4} u_x^2 + au^2$ (special case of operator 3)	$\mathscr{L}_5 = \{1, \cos(kx), \sin(kx), \cos(2kx), \sin(2kx)\}$ if $a = k^2 > 0$, $\mathscr{L}_5 = \{1, \cosh(kx), \sinh(kx), \cosh(2kx), \sinh(2kx)\}$ if $a = -k^2 < 0$, $\mathscr{L}_5 = \{1, x, x^2, x^3, x^4\}$ if $a = 0$
7	$[(au^2 + bu + c)u_x]_x$	$\mathscr{L}_2 = \{1, x\}$
8	$u^2 u_{xx} - \frac{1}{2} uu_x^2 + au^3$	$\mathscr{L}_3 = \{1, \cos(\sqrt{2a}\, x), \sin(\sqrt{2a}\, x)\}$ if $a > 0$, $\mathscr{L}_3 = \{1, \cosh(\sqrt{2|a|}\, x), \sinh(\sqrt{2|a|}\, x)\}$ if $a < 0$, $\mathscr{L}_3 = \{1, x, x^2\}$ if $a = 0$
9	$u_x u_{xx}$	$\mathscr{L}_4 = \{1, x, x^2, x^3\}$, $\mathscr{L}_3 = \{1, x^{3/2}, x^3\}$, $\mathscr{L}_2 = \{1, \varphi(x)\}$, $\varphi'_x \varphi''_{xx} = p_1 + p_2 \varphi$, p_1, p_2 are constants

1°. For $a_1 k < 0$, the nonlinear differential operator on the right-hand side of equation (3.3.1.12) with $b_2 = 0$ admits an invariant linear three-dimensional subspace $\mathscr{L}_3 = \{1, e^{-\lambda x}, e^{\lambda x}\}$, where $\lambda = \sqrt{-k/(2a_1)}$ (see row 3 in Table 3.8 with $a = b$ and $c = k \neq 0$). In this case, it follows from Proposition 1 that equation (3.3.1.12) admits a generalized separable solution of the form

$$u = \psi_1(t) + \psi_2(t)\exp(-\lambda x) + \psi_3(t)\exp(\lambda x), \quad \lambda = \sqrt{-\frac{k}{2a_1}}. \qquad (3.3.1.13)$$

The functions $\psi = \psi_n(t)$ are described by the system of delay ODEs

$$\psi'_1 = k\psi_1^2 + 2k\psi_2\psi_3 + b_1\psi_1 + b_2\bar{\psi}_1,$$
$$\psi'_2 = \left(\tfrac{3}{2}k\psi_1 + a_0\lambda^2 + b_1\right)\psi_2 + b_2\bar{\psi}_2,$$
$$\psi'_3 = \left(\tfrac{3}{2}k\psi_1 + a_0\lambda^2 + b_1\right)\psi_3 + b_2\bar{\psi}_3,$$

where $\bar{\psi}_j = \psi_j(t - \tau)$ ($i = 1, 2, 3$).

2°. For $a_1 k > 0$, it can be shown in a similar way that equation (3.3.1.12) admits a generalized separable solution

$$u = \psi_1(t) + \psi_2(t)\cos(\lambda x) + \psi_3(t)\sin(\lambda x), \quad \lambda = \sqrt{\frac{k}{2a_1}}, \quad (3.3.1.14)$$

where the functions $\psi = \psi_n(t)$ are described by the system of delay ODEs

$$\psi_1' = k\psi_1^2 + \tfrac{1}{2}k(\psi_2^2 + \psi_3^2) + b_1\psi_1 + b_2\bar{\psi}_1,$$
$$\psi_2' = (\tfrac{3}{2}k\psi_1 + b_1 - a_0\lambda^2)\psi_2 + b_2\bar{\psi}_2,$$
$$\psi_3' = (\tfrac{3}{2}k\psi_1 + b_1 - a_0\lambda^2)\psi_3 + b_2\bar{\psi}_3.$$

◀

Some generalizations. Below are two more general propositions that enable one to obtain generalized separable solutions to some nonlinear delay PDEs.

1°. Consider a more complex nonlinear PDE with several delays than (3.3.1.4):

$$u_t = F[u] + \sum_{i=1}^p s_i w_i, \quad w_i = u(x, t - \tau_i), \quad (3.3.1.15)$$

where $F[u]$ is an nth-order nonlinear differential operator in x of the form (3.3.1.5), and τ_i are delays ($i = 1, \ldots, p$), which are assumed to be independent constants.

Proposition 2. Let the linear subspace (3.3.1.6) be invariant under the operator F, implying that relation (3.3.1.7) holds. Then equation (3.3.1.15) has a generalized separable solution of the form (3.3.1.8) with the functions $\psi_1(t), \ldots, \psi_k(t)$ described by the system of ODEs with p delays

$$\psi_j'(t) = f_j(\psi_1(t), \ldots, \psi_k(t)) + \sum_{i=1}^p s_i \psi_j(t - \tau_i), \quad j = 1, \ldots, k. \quad (3.3.1.16)$$

2°. We now look at another nonlinear delay PDE

$$L[u] = F[u; w], \quad w = u(x, t - \tau), \quad (3.3.1.17)$$

where $L[u]$ is an arbitrary linear differential operator in t of the form

$$L[u] \equiv \sum_{j=1}^q a_j(t) u_t^{(j)}, \quad (3.3.1.18)$$

and $F[u; w]$ is a nonlinear differential operator in x involving the functions u and w:

$$F[u; w] \equiv F(u, u_x, u_{xx}, \ldots, u_x^{(m)}; w, w_x, w_{xx}, \ldots, w_x^{(r)}). \quad (3.3.1.19)$$

Suppose that the linearly independent functions $\varphi_1(x), \ldots, \varphi_k(x)$ form a finite-dimensional linear subspace $\mathscr{L}_k$.

3.3. Generalized and Functional Separable Solutions

Proposition 3. Let $C_1, \ldots, C_k$ and $\bar{C}_1, \ldots, \bar{C}_k$ be two sets or arbitrary real constants and let there exist functions $f_1, \ldots, f_k$ such that

$$F\left[\sum_{j=1}^{k} C_j \varphi_j(x); \sum_{j=1}^{k} \bar{C}_j \varphi_j(x)\right] = \sum_{j=1}^{k} f_j(C_1, \ldots, C_n; \bar{C}_1, \ldots, \bar{C}_n) \varphi_j(x). \quad (3.3.1.20)$$

Then equation (3.3.1.17) has generalized separable solutions of the form (3.3.1.8) where the functions $\psi_1(t), \ldots, \psi_k(t)$ are described by the system of delay ODEs

$$L[\psi_j(t)] = f_j\big(\psi_1(t), \ldots, \psi_k(t); \psi_1(t-\tau), \ldots, \psi_k(t-\tau)\big), \quad j = 1, \ldots, k. \quad (3.3.1.21)$$

Proposition 3 can be used to construct generalized separable solutions to nonlinear delay PDEs other than those discussed above, including nonlinear delay Klein–Gordon type wave equations. The delay in equations (3.3.1.17) and (3.3.1.21) can depend on time: $\tau = \tau(t)$.

Remark 3.19. To seek generalized separable solutions, one may find it helpful to employ the method of functional constraints, which is described below in Section 3.4.

3.3.2. Functional Separable Solutions

Preliminary remarks and definitions. Suppose there is a linear equation of mathematical physics for $z = z(x, t)$ that admits a generalized separable solution. Then a nonlinear equation obtained from this linear equation with a change of variable $u = U(z)$ will have an exact solution of the form

$$u(x, t) = U(z), \quad \text{where} \quad z = \sum_{j=1}^{k} \varphi_j(x) \psi_j(t). \quad (3.3.2.1)$$

Many nonlinear PDEs without delay irreducible to linear equations also have exact solutions of the form (3.3.2.1). Such solutions will be referred to as *functional separable solutions*. In general, the functions $\varphi_j(x)$, $\psi_j(t)$, and $U(z)$ in (3.3.2.1) are not known in advance and so have to be determined in a subsequent analysis. The function U will be called the *outer function*, while φ_j and ψ_j will be called *inner functions*. This terminology also applies to nonlinear equations of mathematical physics with delay, which sometimes admit exact solutions of the form (3.3.2.1).

Remark 3.20. Generalized separable solutions (see Subsection 3.3.1) are special functional separable solutions with $U(z) = z$. The presence of the outer function U in (3.3.2.1), which must be found, is a complicating factor in constructing functional separable solutions.

In the narrow sense, the term *functional separable solution* is frequently used for simpler exact solutions of the following form (e.g., see [15, 137, 195, 238, 343, 344, 425, 608]):

$$u = U(z), \quad z = \varphi(x) + \psi(t), \quad (3.3.2.2)$$

where the three functions $U(z)$, $\varphi(x)$, and $\psi(t)$ are unknown. In constructing solutions (3.3.2.2), one assumes that $\varphi \neq \text{const}$ and $\psi \neq \text{const}$.

Remark 3.21. *In functional separation of variables, seeking the simplest solutions* $u = U(\varphi(x) + \psi(t))$ *and* $u = U(\varphi(x)\psi(t))$ *leads to the same results, since the representation* $U(\varphi(x)\psi(t)) = U_1(\varphi_1(x) + \psi_1(t))$, *where* $U_1(z) = U(e^z)$, $\varphi_1(x) = \ln \varphi(x)$, *and* $\psi_1(t) = \ln \psi(t)$, *holds true.*

The method based on transformations of the unknown function. In certain cases, seeking solutions in the form (3.3.2.1) can be carried out in two steps. First, one applies a transformation converting the original equation to a simpler one with a quadratic or power-law nonlinearity. Then, one looks for a generalized separable solution of the resulting equation.

Unfortunately, there are no regular methods for reducing a PDE of a given form to a PDE with a quadratic nonlinearity. Sometimes, equations involving a quadratic nonlinearity can be obtained through a transformation $u = U(z)$ of the unknown functions. The most common transformations are

$$u = z^\lambda \quad \text{(for equations with a power-law nonlinearity)},$$
$$u = \lambda \ln z \quad \text{(for equations with an exponential nonlinearity)},$$
$$u = e^{\lambda z} \quad \text{(for equations with a logarithmic nonlinearity)};$$

where λ is a constant that has to be determined. This approach is equivalent to a priori setting of the form of the outer function $U(z)$ in (3.3.2.1); whether this is successful or not mainly depends on the researcher's experience and intuition.

Remark 3.22. *Many nonlinear equations of mathematical physics without delay reducible to equations with a quadratic nonlinearity using appropriate transformations can be found in [169, 172–175, 422, 425, 430].*

Examples of constructing functional separable solutions to nonlinear delay equations. Below we will give a few examples of transforming the unknown function to construct functional separable solutions for second-order nonlinear PDEs with delay.

▶ **Example 3.11.** Let us look at the six-parameter family of reaction-diffusion equations with power-law nonlinearities and a time delay

$$u_t = a(u^n u_x)_x + bu^{n+1} + cu + ku^{1-n} + mu^{1-n}w^n, \quad w = u(x, t-\tau), \quad (3.3.2.3)$$

where a, b, c, k, n, and m are free parameters. The substitution $z = u^n$ converts (3.3.2.3) to an equation with a quadratic nonlinearity

$$z_t = azz_{xx} + \frac{a}{n}z_x^2 + bnz^2 + cnz + kn + mn\bar{z}, \quad \bar{z} = z(x, t-\tau). \quad (3.3.2.4)$$

This equation admits various generalized separable solutions whose forms depend on the coefficients of the nonlinear terms on the right-hand side of (3.3.2.4). Exact solutions of equation (3.3.2.4) can be found using Table 3.8 (see rows 3–6). In particular, if $ab(n+1) > 0$, it will have solutions with trigonometric functions and if $ab(n+1) < 0$, it will have exponential solutions.

3.3. Generalized and Functional Separable Solutions

This approach allows one to obtain functional separable solutions of the form

$$u = \{\varphi(t)[C_1\cos(\beta x) + C_2\sin(\beta x)] + \psi(t)\}^{1/n} \quad \text{for } ab(n+1) > 0,$$
$$u = \{\varphi(t)[C_1\cosh(\beta x) + C_2\sinh(\beta x)] + \psi(t)\}^{1/n} \quad \text{for } ab(n+1) < 0.$$
(3.3.2.5)

Here C_1 and C_2 are arbitrary constants,

$$\beta = \sqrt{\frac{|b|n^2}{|a(n+1)|}},$$

and the functions $\varphi = \varphi(t)$ and $\psi = \psi(t)$ are described by the system of ordinary differential equations with delay

$$\varphi'_t = \frac{bn(n+2)}{n+1}\varphi\psi + cn\varphi + mn\bar{\varphi}, \quad \bar{\varphi} = \varphi(t-\tau),$$
$$\psi'_t = n(b\psi^2 + c\psi + k) + \frac{bn}{n+1}(C_1^2 \pm C_2^2)\varphi^2 + mn\bar{\psi}, \quad \bar{\psi} = \psi(t-\tau).$$
(3.3.2.6)

The upper sign in the second equation corresponds to the first solution in (3.3.2.5), and the lower sign corresponds to the second solution.

If $C_1 = C_2$, the last equation in (3.3.2.6), with the lower sign, can be satisfied if we set $\psi = \text{const}$, where ψ is a root of quadratic equation $b\psi^2 + (c+m)\psi + k = 0$. In this case, the first equation in (3.3.2.6) is a linear homogeneous delay ODE of the form (1.1.3.1), which was studied in Subsection 1.1.3 in detail. This equation admits particular exponential solutions $\varphi = C_3 e^{\lambda t}$, where C_3 is an arbitrary constant and λ is a root of the transcendental equation

$$\lambda = \frac{bn(n+2)}{n+1}\psi + cn + mne^{-\lambda\tau}.$$

◀

▶ **Example 3.12.** Now let us consider the six-parameter family of reaction-diffusion equations with exponential nonlinearities and a time delay

$$u_t = a(e^{\lambda u}u_x)_x + be^{\lambda u} + c + ke^{-\lambda u} + me^{\lambda(w-u)}, \quad w = u(x, t-\tau). \quad (3.3.2.7)$$

The change of variable $z = e^{\lambda u}$ converts (3.3.2.7) to an equation with a quadratic nonlinearity

$$z_t = azz_{xx} + b\lambda z^2 + c\lambda z + k\lambda + m\lambda\bar{z}, \quad \bar{z} = z(x, t-\tau). \quad (3.3.2.8)$$

Solutions to equation (3.3.2.8) can be obtained using Table 3.8 (see row 3). It is apparent that if $ab\lambda > 0$, equation (3.3.2.8) has a solution with trigonometric functions, if $ab\lambda < 0$, it has a solution with exponential functions, and if $b = 0$, there is a solution in the form of a quadratic polynomial in x.

In particular, the above change of variable leads to functional separable solutions to equation (3.3.2.7) expressible in terms of elementary functions:

$$u = \frac{1}{\lambda}\ln\{e^{\alpha t}[C_1\cos(x\sqrt{\beta}) + C_2\sin(x\sqrt{\beta})] + \gamma\} \quad \text{for } ab\lambda > 0,$$
$$u = \frac{1}{\lambda}\ln\{e^{\alpha t}[C_1\cosh(x\sqrt{-\beta}) + C_2\sinh(x\sqrt{-\beta})] + \gamma\} \quad \text{for } ab\lambda < 0.$$

Here C_1 and C_2 are arbitrary constants, and α is a root of the transcendental equation

$$\alpha = \lambda(b\gamma + c + me^{-\alpha\tau}), \quad \beta = b\lambda/a,$$

where $\gamma = \gamma_{1,2}$ are roots of the quadratic equation $b\gamma^2 + (c+m)\gamma + k = 0$. ◀

▶ **Example 3.13.** The five-parameter family of reaction-diffusion equations with logarithmic nonlinearities and a time delay

$$u_t = au_{xx} + bu\ln^2 u + cu\ln u + ku + mu\ln w, \quad w = u(x, t-\tau), \quad (3.3.2.9)$$

can be reduced with the change of variable $u = e^z$ to an equation with a quadratic nonlinearity

$$z_t = az_{xx} + az_x^2 + bz^2 + cz + k + m\bar{z}, \quad \bar{z} = z(x, t-\tau). \quad (3.3.2.10)$$

One can find solutions to equation (3.3.2.10) using Table 3.8 (see equations 1 and 2). It is apparent that if $ab > 0$, equation (3.3.2.10) has a solution with trigonometric functions and if $ab < 0$, there is an exponential solution. If $b = 0$, equation (3.3.2.10) admits a generalized separable solution as a quadratic polynomial in the space variable, $z = \psi_1(t)x^2 + \psi_2(t)x + \psi_3(t)$, which leads to a functional separable solution for the original equation (3.3.2.9):

$$u = \exp[\psi_1(t)x^2 + \psi_2(t)x + \psi_3(t)].$$

The functions $\psi_k = \psi_k(t)$ are described by the system of delay ODEs

$$\psi_1' = 4a\psi_1^2 + c\psi_1 + m\bar{\psi}_1,$$
$$\psi_2' = 4a\psi_1\psi_2 + c\psi_2 + m\bar{\psi}_2,$$
$$\psi_3' = c\psi_3 + 2a\psi_1 + a\psi_2^2 + k + m\bar{\psi}_3,$$

where $\bar{\psi}_j = \psi_j(t-\tau)$, $j = 1, 2, 3$. The first equation of the system has a stationary particular solution $\psi_1 = -(c+m)/(4a)$. In this case, the second equation is a linear homogeneous delay ODE, which was thoroughly investigated in Subsection 1.1.3 and has an exponential particular solution $\psi_2 = Ce^{\lambda t}$, and the last equation becomes a linear nonhomogeneous delay ODE. ◀

3.3.3. Using Linear Transformations to Construct Generalized and Functional Separable Solutions

Preliminary remarks. In certain cases, to obtain exact solutions of nonlinear delay PDEs, one could first transform the independent variables and then look for generalized or functional separable solutions. The simplest linear transformations include

$$x, t, u \quad \Longrightarrow \quad y, z, u, \quad \text{where} \quad y = k_1 x + \lambda_1 t, \quad z = k_2 x + \lambda_2 t, \quad (3.3.3.1)$$

where k_1, k_2, λ_1, and λ_2 are some constants. Previously, this approach made it possible to obtain, in terms of new variables, multiplicative and additive separable solutions to the nonlinear delay reaction-diffusion equations (3.2.2.1) (see solution (3.2.2.7)), (3.2.2.12) (solution (3.2.2.15) with $C_1 = 0$), and (3.2.2.16) (solution (3.2.2.20)).

Linear transformations for nonlinear delay Klein–Gordon type wave equations. To find exact solutions of nonlinear delay Klein–Gordon equations (3.1.1.3), one may find it helpful to start with the following two linear transformations:

$$x, t, u \implies x, z, u, \quad \text{where} \quad z = t \pm a^{-1/2}x. \tag{3.3.3.2}$$

This results in two delay PDEs [429]:

$$au_{xx} \pm 2a^{1/2}u_{xz} + F(u, w) = 0,$$
$$u = u(x, z), \quad z = t \pm a^{-1/2}x, \quad w = u(x, z - \tau). \tag{3.3.3.3}$$

Notably, the relations $t \pm a^{-1/2}x = C_\pm$, where $C_\pm$ are arbitrary constants, define two different families of characteristics for the linear wave equation (3.1.1.3) with $F \equiv 0$ (e.g., see [404, 514]).

The transformed equations (3.3.3.3) are often more convenient than the original delay PDE (3.1.1.3); these make it possible to find generalized and functional separable solutions with respect to the new arguments x and z.

Exact solutions to nonlinear delay PDEs [429]. To illustrate the effectiveness of employing linear transformations of the form (3.3.3.2), we will list several nonlinear hyperbolic PDEs with delay and their exact solutions.

Equation 1. The nonlinear delay Klein–Gordon type wave equation

$$u_{tt} = au_{xx} + f(w/u) \tag{3.3.3.4}$$

written in terms of the variables (3.3.3.2) admits multiplicative separable solutions

$$u = (x + C)\varphi(z), \quad z = t \pm a^{-1/2}x,$$

where C is an arbitrary constant; the functions $\varphi(z)$ satisfy the nonlinear first-order delay ODE

$$\pm 2a^{1/2}\varphi'(z) + f(\varphi(z-\tau)/\varphi(z)) = 0.$$

Equation 2. The nonlinear delay Klein–Gordon type wave equation

$$u_{tt} = au_{xx} + f(u - w) \tag{3.3.3.5}$$

written in terms of the variables (3.3.3.2) admits a few exact solutions described below.

1°. Generalized separable solutions:

$$u = Cx^2 + \varphi(z)x + \psi(z), \quad z = t \pm a^{-1/2}x,$$

where C is an arbitrary constant, and $\varphi(z)$ and $\psi(z)$ are functions satisfying the difference equations

$$\varphi(z) = \varphi(z - \tau), \tag{3.3.3.6}$$

$$f(\psi(z) - \psi(z - \tau)) = \mp 2a^{1/2}\varphi'(z) - 2Ca. \tag{3.3.3.7}$$

It follows from the linear equation (3.3.3.6) that $\varphi(z)$ is any τ-periodic function that can generally be represented as the convergent series

$$\varphi(z) = A_0 + \sum_{n=1}^{\infty}\left(A_n \cos\frac{2\pi n z}{\tau} + B_n \sin\frac{2\pi n z}{\tau}\right), \qquad (3.3.3.8)$$

where A_n and B_n are arbitrary constants. Substituting (3.3.3.8) into (3.3.3.7) yields an equation that reduces to a linear nonhomogeneous difference equation of the form $\psi(z) - \psi(z-\tau) = g_{\mp}(z)$ with a known right-hand side.

2°. Generalized separable solutions:

$$u = Cxz + \varphi(x) + \psi(z), \quad z = t \pm a^{-1/2} x,$$

where C is an arbitrary constant, and $\varphi(x)$ and $\psi(z)$ are functions satisfying the linear ODE and linear difference equation

$$a\varphi''_{xx} \pm 2Ca^{1/2} + f(C\tau x + B) = 0,$$
$$\psi(z) - \psi(z-\tau) = B,$$

in which B is an arbitrary constant. The functions $\varphi(x)$ and $\psi(z)$ admit a closed-form representation.

Equation 3. The nonlinear delay Klein–Gordon type wave equation

$$u_{tt} = au_{xx} + f(u + kw) \qquad (3.3.3.9)$$

admits generalized separable solutions

$$u = \varphi(z)x + \psi(z), \quad z = t \pm a^{-1/2} x, \qquad (3.3.3.10)$$

where the functions $\varphi(z)$ and $\psi(z)$ are described by the difference equations

$$\varphi(z) + k\varphi(z-\tau) = 0, \qquad (3.3.3.11)$$
$$f\big(\psi(z) + k\psi(z-\tau)\big) = \mp 2a^{1/2}\varphi'(z). \qquad (3.3.3.12)$$

For $k > 0$, the general solution to equation (3.3.3.11) can be represented as

$$\varphi(z) = k^{z/\tau}\sum_{n=1}^{\infty}\left[A_n \cos\frac{(2n-1)\pi z}{\tau} + B_n \sin\frac{(2n-1)\pi z}{\tau}\right], \qquad (3.3.3.13)$$

where A_n and B_n are arbitrary constants such that the series in (3.3.3.13) is convergent. If $k = 1$, we get a general τ-antiperiodic function (3.3.3.13).

For $k < 0$, equation (3.3.3.11) has the general solution

$$\varphi(z) = |k|^{z/\tau}\sum_{n=0}^{\infty}\left(A_n \cos\frac{2\pi n z}{\tau} + B_n \sin\frac{2\pi n z}{\tau}\right). \qquad (3.3.3.14)$$

Substituting (3.3.3.13) or (3.3.3.14) into (3.3.3.12) gives an equation reducible to linear nonhomogeneous difference equations of the form $\psi(z) + k\psi(z-\tau) = g_{\mp}(z)$ with a known right-hand side.

3.3. Generalized and Functional Separable Solutions

Equation 4. The nonlinear delay PDE

$$u_{tt} = au_{xx} + u^{1-2k}f(u^k - w^k), \quad k \neq 1, \tag{3.3.3.15}$$

admits functional separable solutions

$$u = [x + \theta(z)]^{1/k}, \quad z = t \pm a^{-1/2}x,$$

where the function $\theta = \theta(z)$ is described by the nonlinear first-order delay ODE

$$\pm 2a^{1/2}\theta'_z + a + \frac{k^2}{1-k}f(\theta - \bar{\theta}) = 0, \quad \bar{\theta} = \theta(z - \tau).$$

Equation 5. The nonlinear delay PDE

$$u_{tt} = au_{xx} + e^{bu+cw}f(u - w) \tag{3.3.3.16}$$

admits additive separable solutions

$$u = \varphi(x) + \theta(z), \quad z = t \pm a^{-1/2}x,$$

with the function $\varphi = \varphi(x)$ satisfying the nonlinear second-order ODE

$$\varphi''_{xx} = Ke^{(b+c)\varphi},$$

where K is an arbitrary constant and $\theta = \theta(z)$ is a function described by the difference equation

$$aK + e^{b\theta + c\bar{\theta}}f(\theta - \bar{\theta}) = 0, \quad \bar{\theta} = \theta(z - \tau).$$

Notably, the general solution of the ODE for φ is expressed in terms of elementary functions [423].

Equation 6. The nonlinear delay PDE

$$u_{tt} = au_{xx} + e^{-2\beta u}f(be^{\beta u} + ce^{\beta w}) \tag{3.3.3.17}$$

admits functional separable solutions of the form

$$u = \frac{1}{\beta}\ln[\varphi(z)x + \psi(z)], \quad z = t \pm a^{-1/2}x,$$

with the function $\varphi = \varphi(z)$ satisfying the linear difference equation

$$b\varphi + c\bar{\varphi} = 0, \quad \bar{\varphi} = \varphi(z - \tau),$$

and the function $\psi = \psi(z)$ described by the nonlinear first-order delay ODE

$$\pm 2a^{1/2}(\varphi'_z\psi - \varphi\psi'_z) - a\varphi^2 + \beta f(b\psi + c\bar{\psi}) = 0, \quad \bar{\psi} = \psi(z - \tau).$$

3.4. Method of Functional Constraints

3.4.1. General Description of the Method of Functional Constraints

Following [428], we will consider the class of nonlinear delay reaction-diffusion equations

$$u_t = au_{xx} + uf(z) + wg(z) + h(z),$$
$$w = u(x, t - \tau), \quad z = z(u, w),$$
(3.4.1.1)

where $f(z)$, $g(z)$, and $h(z)$ are arbitrary functions, and $z = z(u, w)$ is a given function (in certain cases, it can be unknown and has to be determined). In addition, along the way, we will sometimes be looking at more general equations where the functions f, g, and h can also depend on either independent variable, x or t.

Remark 3.23. *The first derivative u_t on the left-hand side of equation (3.4.1.1) can be replaced with the second derivative u_{tt} or a linear combination of time derivatives, $L[u] = \sum_{m=1}^{n} c_m u_t^{(m)}$, where c_m are arbitrary constants.*

We seek generalized separable solutions of the form

$$u = \sum_{n=1}^{N} \varphi_n(x)\psi_n(t),$$
(3.4.1.2)

where the functions $\varphi_n(x)$ and $\psi_n(t)$ are to be determined in the subsequent analysis. It should be reminded that the sum (3.4.1.2) most frequently contains the coordinate functions specified in (3.3.1.3).

Notably, nonlinear delay PDEs of the form (3.4.1.1), which involve arbitrary functions, cannot be solved with the use of the standard methods of generalized separation of variables described in the books [175, 422, 425, 430] and Section 3.3.

The method of functional constraints developed in [428] is based on seeking generalized separable solutions (3.4.1.2) for equations of the form (3.4.1.1) and related more complex equations by employing one of the following two additional *functional constraints*:

$$z(u, w) = p(x), \quad w = u(x, t - \tau);$$
(3.4.1.3)
$$z(u, w) = q(t), \quad w = u(x, t - \tau).$$
(3.4.1.4)

These represent difference equations in the variable t, where the space variable x serves as a free parameter. The function $z = z(u, w)$ is the argument of the arbitrary functions f, g, and h involved in (3.4.1.1). The functions $p(x)$ and $q(t)$ depend on x and t implicitly (expressed in terms of $\varphi_n(x)$ and $\psi_n(t)$) and are determined in the investigation of equation (3.4.1.3) or (3.4.1.4) taking into account (3.4.1.2). Notably, there is no need to obtain the general solutions of equation (3.4.1.3) or (3.4.1.4); particular solutions will suffice.

In view of (3.4.1.2), a solution to the difference equation (3.4.1.3) (or (3.4.1.4)) defines allowed structures of exact solutions. Their final forms are further determined by substituting these solutions into equation (3.4.1.1) in question. Constraints

(3.4.1.3) and (3.4.1.4) will further be referred to as a *functional constraint of the first kind* and a *functional constraint of the second kind*, respectively.

If either relation (3.4.1.3) or (3.4.1.4) holds, equation (3.4.1.1) becomes 'linear', which then enables one to apply the procedure of separation of variables.

Remark 3.24. *To avoid ambiguity, in Section 3.4, we will treat the simplest (degenerate) functional constraint $z(u, w) = \text{const}$ as a functional constraint of the first kind (3.4.1.3).*

Remark 3.25. *The term 'functional constraint' is introduced by analogy with 'differential constraint' used in the method of differential constraints (proposed by N. N. Yanenko in [579]) to seek exact solutions of nonlinear PDEs and systems of such equations. For the description of this method and examples of its application, see, for example, [422, 425, 430, 480].*

Remark 3.26. *In certain cases, the functional equations (3.4.1.3) and (3.4.1.4) allow one to obtain more complex functional separable solutions than (3.4.1.2); for examples of such solutions, see Subsection 3.4.3.*

Below we give a number of examples illustrating the application of the method of functional constraints to construct generalized separable solutions for some nonlinear delay equations of the form (3.4.1.1) and related more complex equations.

3.4.2. Exact Solutions to Quasilinear Delay Reaction-Diffusion Equations

This subsection will deal with quasilinear delay reaction-diffusion equations of the form (3.4.1.1), which are linear in both derivatives. The exact solutions listed below were obtained in [428].

Equations involving a single arbitrary function dependent on w/u.

Equation 1. Consider the nonlinear reaction-diffusion equation with constant delay that involves a single arbitrary function dependent on the ratio w/u:

$$u_t = au_{xx} + uf(w/u), \quad w = u(x, t - \tau). \tag{3.4.2.1}$$

It is a special case of equation (3.4.1.1) with $g = h = 0$ and $z = w/u$.

1°. In this case, the functional constraint of the second kind (3.4.1.4) is

$$w/u = q(t), \quad w = u(x, t - \tau). \tag{3.4.2.2}$$

It is clear that the difference equation (3.4.2.2) can be satisfied if we take the simple multiplicative separable solution

$$u = \varphi(x)\psi(t), \tag{3.4.2.3}$$

which gives $q(t) = \psi(t-\tau)/\psi(t)$. Substituting (3.4.2.3) into (3.4.2.1) and separating the variables, we get the equations for $\varphi = \varphi(x)$ and $\psi = \psi(t)$:

$$\varphi''_{xx} = k\varphi, \tag{3.4.2.4}$$

$$\psi'_t = ak\psi + \psi f(\bar\psi/\psi), \quad \bar\psi = \psi(t - \tau), \tag{3.4.2.5}$$

where k is an arbitrary constant.

The general solution to ODE (3.4.2.4) is given by

$$\varphi(x) = \begin{cases} C_1 \cos(\sqrt{|k|}\,x) + C_2 \sin(\sqrt{|k|}\,x) & \text{if } k < 0; \\ C_1 \exp(-\sqrt{k}\,x) + C_2 \exp(\sqrt{k}\,x) & \text{if } k > 0; \\ C_1 x + C_2 & \text{if } k = 0, \end{cases} \qquad (3.4.2.6)$$

where C_1 and C_2 are arbitrary constants. The delay ODE (3.4.2.5) admits particular solutions in the exponential form

$$\psi(t) = C_3 e^{\lambda t},$$

where C_3 is an arbitrary constant, and λ is a root of the transcendental equation

$$\lambda = ak + f(e^{-\lambda \tau}).$$

2°. The functional constraint of the first kind (3.4.1.3) for equation (3.4.2.1) is written in the simplest case $p(x) = p_0 = \text{const}$ as

$$w/u = p_0, \qquad w = u(x, t - \tau). \qquad (3.4.2.7)$$

We look for a solution to the difference equation (3.4.2.7) with $p_0 > 0$ in the form

$$u = e^{ct} v(x, t), \qquad v(x, t) = v(x, t - \tau), \qquad (3.4.2.8)$$

where c is an arbitrary constant, and $v(x, t)$ is a τ-periodic function that has to be determined. In this case, we have $w/u = p_0 = e^{-c\tau}$.

Substituting (3.4.2.8) into equation (3.4.2.1) yields a linear problem for v:

$$v_t = a v_{xx} + bv, \qquad v(x, t) = v(x, t - \tau), \qquad (3.4.2.9)$$

where $b = f(e^{-c\tau}) - c$.

For convenience, we denote the general solution of problem (3.4.2.9) by $v = V_1(x, t; b)$. It is expressed as

$$V_1(x, t; b) = \sum_{n=0}^{\infty} \exp(-\lambda_n x)\bigl[A_n \cos(\beta_n t - \gamma_n x) + B_n \sin(\beta_n t - \gamma_n x)\bigr]$$

$$+ \sum_{n=1}^{\infty} \exp(\lambda_n x)\bigl[C_n \cos(\beta_n t + \gamma_n x) + D_n \sin(\beta_n t + \gamma_n x)\bigr],$$

(3.4.2.10)

$$\beta_n = \frac{2\pi n}{\tau}, \qquad \lambda_n = \left(\frac{\sqrt{b^2 + \beta_n^2} - b}{2a}\right)^{1/2}, \qquad \gamma_n = \left(\frac{\sqrt{b^2 + \beta_n^2} + b}{2a}\right)^{1/2},$$

(3.4.2.11)

where A_n, B_n, C_n, and D_n are arbitrary constants such that the series (3.4.2.10)–(3.4.2.11) and its derivatives $(V_1)_t$ and $(V_1)_{xx}$ are all convergent; for example, one can ensure the convergence by setting $A_n = B_n = C_n = D_n = 0$ with $n > N$, where N is an arbitrary positive integer.

We can single out the following special cases:

(i) formulas (3.4.2.10)–(3.4.2.11) with $A_0 = B_0 = 0$, $C_n = D_n = 0$, $n = 1, 2, \ldots$, define time τ-periodic solutions to problem (3.4.2.9) that decay as $x \to \infty$;

(ii) formulas (3.4.2.10)–(3.4.2.11) with $C_n = D_n = 0$, $n = 1, 2, \ldots$, define time τ-periodic solutions to problem (3.4.2.9) that are bounded as $x \to \infty$;

(iii) formulas (3.4.2.10)–(3.4.2.11) with $A_n = B_n = C_n = D_n = 0$, $n = 1, 2, \ldots$, define a stationary solution.

To sum up, equation (3.4.2.1) has the exact solution

$$u = e^{ct} V_1(x, t; b), \quad b = f(e^{-c\tau}) - c, \tag{3.4.2.12}$$

where c is an arbitrary constant, and $V_1(x, t; b)$ is a τ-periodic function defined by (3.4.2.10)–(3.4.2.11).

$3°$. We look for a solution to the difference equation (3.4.2.7) with $p_0 < 0$ in the form

$$u = e^{ct} v(x, t), \quad v(x, t) = -v(x, t - \tau), \tag{3.4.2.13}$$

where c is an arbitrary constant, and $v(x, t)$ is a τ-antiperiodic function. In our case, $w/u = p_0 = -e^{-c\tau}$.

Substituting (3.4.2.13) into equation (3.4.2.1) yields a linear problem for v:

$$v_t = a v_{xx} + bv, \quad v(x, t) = -v(x, t - \tau), \tag{3.4.2.14}$$

where $b = f(-e^{-c\tau}) - c$.

The general solution of problem (3.4.2.14), which we denote by $v = V_2(x, t; b)$ for convenience, is

$$V_2(x, t; b) = \sum_{n=1}^{\infty} \exp(-\lambda_n x) \big[A_n \cos(\beta_n t - \gamma_n x) + B_n \sin(\beta_n t - \gamma_n x) \big]$$

$$+ \sum_{n=1}^{\infty} \exp(\lambda_n x) \big[C_n \cos(\beta_n t + \gamma_n x) + D_n \sin(\beta_n t + \gamma_n x) \big],$$

(3.4.2.15)

$$\beta_n = \frac{\pi(2n-1)}{\tau}, \quad \lambda_n = \left(\frac{\sqrt{b^2 + \beta_n^2} - b}{2a} \right)^{1/2}, \quad \gamma_n = \left(\frac{\sqrt{b^2 + \beta_n^2} + b}{2a} \right)^{1/2},$$

(3.4.2.16)

where A_n, B_n, C_n, and D_n are arbitrary constants such that the series (3.4.2.15)–(3.4.2.16) and respective derivatives $(V_2)_t$ and $(V_2)_{xx}$ are convergent. Formulas (3.4.2.15)–(3.4.2.16) with $C_n = D_n = 0$, $n = 1, 2, \ldots$, define τ-antiperiodic solutions, in time t, of problem (3.4.2.14) that decay as $x \to \infty$.

To sum up, equation (3.4.2.1) has the exact solution

$$u = e^{ct} V_2(x, t; b), \quad b = f(-e^{-c\tau}) - c, \tag{3.4.2.17}$$

where c is an arbitrary constant, and $V_2(x, t; b)$ is a τ-antiperiodic function defined by formulas (3.4.2.15)–(3.4.2.16).

Remark 3.27. Solutions (3.4.2.10)–(3.4.2.11) and (3.4.2.15)–(3.4.2.16) are very similar in appearance. However, the first sum starts with $n = 0$ in the first solution and with $n = 1$ in the second solution; the values of β_n are also different.

Equation 2. Consider the nonlinear reaction-diffusion equation with constant delay that involves a single arbitrary function

$$u_t = au_{xx} + bu \ln u + uf(w/u), \quad w = u(x, t - \tau). \qquad (3.4.2.18)$$

At $b = 0$, it becomes equation (3.4.2.1). The exact solution to equation (3.4.2.18) determined by the functional constraint of the second kind (3.4.2.2) has the form (3.4.2.3), or

$$u = \varphi(x)\psi(t),$$

where the functions $\varphi = \varphi(x)$ and $\psi = \psi(t)$ are described by the ODE and delay ODE

$$a\varphi''_{xx} + b\varphi \ln \varphi = C\varphi,$$
$$\psi'_t = b\psi \ln \psi + C\psi + \psi f(\bar{\psi}/\psi), \quad \bar{\psi} = \psi(t - \tau),$$

where C is an arbitrary constant.

Equations involving a single arbitrary function dependent on a linear combination of u and w.

Equation 3. Consider the nonlinear reaction-diffusion equation with constant delay that involves a single arbitrary function dependent on the difference $u - w$:

$$u_t = au_{xx} + bu + f(u - w), \quad w = u(x, t - \tau). \qquad (3.4.2.19)$$

It is a special case of equation (3.4.1.1) with $f(z) = b$, $g = 0$, and $z = u - w$; for convenience, the function h has been renamed f.

1°. In this case, the functional constraint of the second kind (3.4.1.4) has the form

$$u - w = q(t), \quad w = u(x, t - \tau). \qquad (3.4.2.20)$$

It is clear that the linear difference equation (3.4.2.20) can be satisfied if we take the additive separable solution

$$u = \varphi(x) + \psi(t), \qquad (3.4.2.21)$$

which gives $q(t) = \psi(t) - \psi(t - \tau)$. Substituting (3.4.2.21) into (3.4.2.19) and separating the variables, we obtain equations for $\varphi = \varphi(x)$ and $\psi = \psi(t)$:

$$a\varphi''_{xx} + b\varphi = k, \qquad (3.4.2.22)$$
$$\psi'_t = b\psi + k + f(\psi - \bar{\psi}), \quad \bar{\psi} = \psi(t - \tau), \qquad (3.4.2.23)$$

where k is an arbitrary constant.

The general solution of equation (3.4.2.22) with $b \neq 0$ and $k = 0$ is

$$\varphi(x) = \begin{cases} C_1 \cos(\beta x) + C_2 \sin(\beta x), & \beta = \sqrt{b/a} \quad \text{for } b > 0; \\ C_1 \exp(-\beta x) + C_2 \exp(\beta x), & \beta = \sqrt{-b/a} \quad \text{for } b < 0, \end{cases} \qquad (3.4.2.24)$$

where C_1 and C_2 are arbitrary constants. For $b > 0$, solution (3.4.2.21), (3.4.2.24) is periodic in the space variable x.

The general solution of equation (3.4.2.22) with $b = 0$ and $k \neq 0$ is expressed as

$$\varphi(x) = \frac{k}{2a} x^2 + C_1 x + C_2. \qquad (3.4.2.25)$$

2°. The functional constraint of the first kind (3.4.1.3) for equation (3.4.2.19) is

$$u - w = p(x), \qquad w = u(x, t - \tau). \qquad (3.4.2.26)$$

The difference equation (3.4.2.26) can be satisfied if we take, for example, the generalized separable solution

$$u = t\varphi(x) + \psi(x), \qquad (3.4.2.27)$$

which gives $p(x) = \tau\varphi(x)$.

Substituting (3.4.2.27) into (3.4.2.19) gives ordinary differential equations for $\varphi = \varphi(x)$ and $\psi = \psi(x)$:

$$a\varphi''_{xx} + b\varphi = 0, \qquad (3.4.2.28)$$
$$a\psi''_{xx} + b\psi + f(\tau\varphi) - \varphi = 0. \qquad (3.4.2.29)$$

Equation (3.4.2.28) coincides with equation (3.4.2.22) at $k = 0$ and its solution is given by formulas (3.4.2.24). The linear nonhomogeneous ODE with constant coefficients (3.4.2.29) is easy to integrate.

One can obtain more complicated exact solutions to equation (3.4.2.19), involving any number of arbitrary parameters by utilizing the above solutions (3.4.2.21) and (3.4.2.27) and the following theorem.

Theorem 1 (*on nonlinear superposition of solutions*). *Let $u_0(x, t)$ be some solution to the nonlinear delay equation (3.4.2.19) and let $v = V_1(x, t; b)$ be any τ-periodic solution to the linear heat equation with a source (3.4.2.9). Then the sum*

$$u = u_0(x, t) + V_1(x, t; b) \qquad (3.4.2.30)$$

is also a solution to equation (3.4.2.19). The general form of the functions $V_1(x, t; b)$ is defined by formulas (3.4.2.10)–(3.4.2.11).

This theorem can be proved by a direct substitution of formula (3.4.2.30) into the original delay equation (3.4.2.19) while using the equation for v (3.4.2.9).

Remark 3.28. In formula (3.4.2.30), one can use, for example, a spatially homogeneous solution $u_0(t)$, a stationary solution $u_0(x)$, or a traveling wave solution $u_0 = u_0(\alpha x + \beta t)$ as the particular solution $u_0(x, t)$ to the nonlinear equation (3.4.2.19).

Equation 4. Let us look at the delay PDE

$$u_t = au_{xx} + bu + f(u - kw), \qquad k > 0, \qquad (3.4.2.31)$$

which is a special case of equation (3.4.1.1) with $f(z) = b$, $g = 0$, and $z = u - kw$; for convenience, the function h has been renamed f.

1°. The functional constraint of the first kind (3.4.1.3) for equation (3.4.2.31) is

$$u - kw = p(x), \qquad w = u(x, t - \tau). \qquad (3.4.2.32)$$

The linear difference equation (3.4.2.32) can be satisfied if, for example, we take a generalized separable solution of the form

$$u = e^{ct}\varphi(x) + \psi(x), \qquad c = \frac{1}{\tau}\ln k, \qquad (3.4.2.33)$$

which gives $p(x) = (1 - k)\psi(x)$.

Substituting (3.4.2.33) into (3.4.2.31) yields ordinary differential equations for $\varphi = \varphi(x)$ and $\psi = \psi(x)$:

$$a\varphi''_{xx} + (b - c)\varphi = 0, \qquad (3.4.2.34)$$

$$a\psi''_{xx} + b\psi + f(\eta) = 0, \qquad \eta = (1 - k)\psi. \qquad (3.4.2.35)$$

Up to obvious renaming, the linear ODE (3.4.2.34) coincides with equation (3.4.2.22) at $k = 0$. Its solution is given by formulas (3.4.2.24) in which b must be replaced with $b - c$. The nonlinear ODE (3.4.2.35) is autonomous and so its general solution can be expressed in an implicit form.

2°. One can obtain more complicated exact solutions to equation (3.4.2.31), involving any number of arbitrary parameters, by employing the above solution (3.4.2.33)–(3.4.2.35) and the following theorem.

Theorem 2 (generalizes Theorem 1). *Let $u_0(x, t)$ be some solution to the nonlinear delay equation (3.4.2.31) and let $v = V_1(x, t; b)$ be any τ-periodic solution to the linear heat equation with a source (3.4.2.9). Then the sum*

$$u = u_0(x, t) + e^{ct}V_1(x, t; b - c), \qquad (3.4.2.36)$$

where $c = (\ln k)/\tau$, is also a solution to equation (3.4.2.31). The general form of the function $V_1(x, t; b)$ is given by formulas (3.4.2.10)–(3.4.2.11).

This theorem can be proved by a direct substitution of formula (3.4.2.36) into the original delay equation (3.4.2.31) while using the equation for v (3.4.2.9).

Formula (3.4.2.36) allows one to obtain a broad class of exact solutions to the nonlinear delay equation (3.4.2.31). As $u_0(x, t)$ in (3.4.2.36), one can take a constant u_0 that satisfies the algebraic (or transcendental) equation $bu_0 + f((1-k)u_0) = 0$ and so is the simplest particular solution of equation (3.4.2.31). As $u_0(x, t)$ in (3.4.2.36), one can also take simple particular solutions $u_0 = u_0(x)$ or $u_0 = u_0(t)$, or the more complex traveling wave solution $u_0 = \theta(\alpha x + \beta t)$, where α and β are arbitrary constants, and $\theta(y)$ is a function satisfying the delay ODE

$$a\alpha^2\theta''(y) - \beta\theta'(y) + b\theta(y) + f\big(\theta(y) - k\theta(y - \sigma)\big) = 0, \qquad y = \alpha x + \beta t, \quad \sigma = \beta\tau.$$

Furthermore, exact solution (3.4.2.33) is also suitable.

Remark 3.29. Let us look at a more general equation than (3.4.2.31),

$$u_t = au_{xx} + bu + f(w_1 - kw_2), \quad w_1 = u(x, t-\tau_1), \quad w_2 = u(x, t-\tau_2), \quad k > 0, \quad (3.4.2.37)$$

with two delays, τ_1 and τ_2 ($\tau_1 \neq \tau_2$). The difference equation $w_1 - kw_2 = p(x)$ can be satisfied by taking a generalized separable solution of the form (3.4.2.33) in which $c = \frac{1}{\tau} \ln k$ must be replaced with $c = \frac{1}{\tau_2 - \tau_1} \ln k$. Then a theorem similar to Theorem 2 will hold if we set $c = \frac{1}{\tau_2 - \tau_1} \ln k$ in formula (3.4.2.36).

Equation 5. Consider the delay PDE

$$u_t = au_{xx} + bu + f(u + kw), \quad k > 0, \quad (3.4.2.38)$$

which differs from equation (3.4.2.31) in the sign of k in the kinetic term. The following theorem holds.

Theorem 3. Suppose that $u_0(x, t)$ is some solution to the nonlinear delay equation (3.4.2.38) and $v = V_2(x, t; b)$ is any τ-antiperiodic solution to the linear heat equation with a source (3.4.2.14). Then the sum

$$u = u_0(x, t) + e^{ct} V_2(x, t; b - c), \quad c = \frac{1}{\tau} \ln k, \quad (3.4.2.39)$$

is also a solution to equation (3.4.2.38). The general form of $V_2(x, t; b)$ is defined by formulas (3.4.2.15)–(3.4.2.16).

Theorem 3 is proved by direct verification. It allows one to obtain a broad class of exact solutions to the nonlinear delay equation (3.4.2.38). As the particular solution $u_0(x, t)$ in (3.4.2.39), one can take, for example, a spatially homogeneous solution $u_0(t)$, a stationary solution $u_0(x)$, or a traveling wave solution $u_0 = u_0(\alpha x + \beta t)$.

Equations involving two arbitrary functions dependent on a linear combination of u and w.

Equation 6. Now we look at the more complicated nonlinear delay PDE

$$u_t = au_{xx} + uf(u - w) + wg(u - w) + h(u - w), \quad w = u(x, t - \tau), \quad (3.4.2.40)$$

where $f(z)$, $g(z)$, and $h(z)$ are arbitrary functions. (Without loss of generality, either function f or g can be set equal to zero.)

1°. The functional constraint of the first kind (3.4.1.3) for equation (3.4.2.40) has the form (3.4.2.26). The linear difference equation (3.4.2.26) can be satisfied if, as before, we take a generalized separable solution of the form (3.4.2.27). As a result, we obtain equations for the functions $\varphi(x)$ and $\psi(x)$; these are not written out, since we give a significantly more general result below.

2°. The linear difference equation (3.4.2.26) can be satisfied if we set

$$u = \sum_{n=1}^{N} [\varphi_n(x) \cos(\beta_n t) + \psi_n(x) \sin(\beta_n t)] + t\theta(x) + \xi(x), \quad \beta_n = \frac{2\pi n}{\tau}, \quad (3.4.2.41)$$

where N is any positive integer. In this case, we have $p(x) = \tau \varphi(x)$ on the right-hand side of equation (3.4.2.26).

Substituting (3.4.2.41) into equation (3.4.2.40) and rearranging, we arrive at the equation

$$\sum_{n=1}^{N}[A_n \cos(\beta_n t) + B_n \sin(\beta_n t)] + Ct + D = 0 \qquad (3.4.2.42)$$

in which the functional coefficients A_n, B_n, C, and D depend on $\varphi_n(x)$, $\psi_n(x)$, $\theta(x)$, and $\xi(x)$ and their derivatives but are independent of time t. In (3.4.2.42), equating all functional coefficients with zero, $A_n = B_n = C = D = 0$, we obtain the following ODEs for the unknown functions:

$$a\varphi_n'' + \varphi_n[f(\tau\theta) + g(\tau\theta)] - \beta_n\psi_n = 0,$$
$$a\psi_n'' + \psi_n[f(\tau\theta) + g(\tau\theta)] + \beta_n\varphi_n = 0,$$
$$a\theta'' + \theta[f(\tau\theta) + g(\tau\theta)] = 0,$$
$$a\xi'' + \xi f(\tau\theta) + (\xi - \tau\theta)g(\tau\theta) + h(\tau\theta) - \theta = 0.$$

where a prime denotes a derivative with respect to x.

It is noteworthy that the third nonlinear equation admits a trivial solution, $\theta = 0$; in this case, the other three equations become linear ODEs with constant coefficients.

Equation 7. Consider the nonlinear delay PDE

$$u_t = au_{xx} + uf(u - kw) + wg(u - kw) + h(u - kw), \quad k > 0, \qquad (3.4.2.43)$$

where $f(z)$, $g(z)$, and $h(z)$ are arbitrary functions, which is a generalization of equation (3.4.2.31).

1°. The functional constraint of the first kind (3.4.1.3) for equation (3.4.2.43) has the form (3.4.2.32). The linear difference equation (3.4.2.32) can be satisfied if, as before, we take a generalized separable solution of the form (3.4.2.33). As a result, we can obtain equations for the functions $\varphi(x)$ and $\psi(x)$; these are not written out, because we give a significantly more general result below.

2°. The linear difference equation (3.4.2.32) can be satisfied by setting

$$u = e^{ct}\left\{\theta(x) + \sum_{n=1}^{N}[\varphi_n(x)\cos(\beta_n t) + \psi_n(x)\sin(\beta_n t)]\right\} + \xi(x), \qquad (3.4.2.44)$$

$$c = \frac{1}{\tau}\ln k, \quad \beta_n = \frac{2\pi n}{\tau},$$

where N is any positive integer. The right-hand side of equation (3.4.2.32) in this case is $p(x) = (1 - k)\xi(x)$.

Substituting (3.4.2.44) into equation (3.4.2.43) and reasoning in the same fashion as for equation (3.4.2.40), we obtain the following ODEs for the functions $\theta(x)$,

$\varphi_n(x)$, $\psi_n(x)$, and $\xi(x)$:

$$a\theta'' + \theta\left[f(\eta) + \frac{1}{k}g(\eta) - c\right] = 0, \quad \eta = (1-k)\xi,$$

$$a\varphi_n'' + \varphi_n\left[f(\eta) + \frac{1}{k}g(\eta) - c\right] - \beta_n\psi_n = 0,$$

$$a\psi_n'' + \psi_n\left[f(\eta) + \frac{1}{k}g(\eta) - c\right] + \beta_n\varphi_n = 0,$$

$$a\xi'' + \xi[f(\eta) + g(\eta)] + h(\eta) = 0,$$

where a prime denotes a derivative with respect to x.

Equation 8. Consider the nonlinear delay PDE

$$u_t = au_{xx} + uf(u+kw) + wg(u+kw) + h(u+kw), \quad k > 0, \quad (3.4.2.45)$$

where $f(z)$, $g(z)$, and $h(z)$ are arbitrary functions, which is a generalization of equation (3.4.2.38).

The functional constraint of the first kind (3.4.1.3) for equation (3.4.2.45) has the form

$$u + kw = p(x), \quad w = u(x, t - \tau). \quad (3.4.2.46)$$

The linear difference equation (3.4.2.46) can be satisfied by setting

$$u = e^{ct}\sum_{n=1}^{N}[\varphi_n(x)\cos(\beta_n t) + \psi_n(x)\sin(\beta_n t)] + \xi(x), \quad (3.4.2.47)$$

$$c = \frac{1}{\tau}\ln k, \quad \beta_n = \frac{\pi(2n-1)}{\tau},$$

where N is any positive integer. The right-hand side of equation (3.4.2.46) in this case is $p(x) = (1+k)\xi(x)$.

Substituting (3.4.2.47) into equation (3.4.2.45) and reasoning in the same fashion as for equation (3.4.2.40), we obtain the following ODEs for the functions $\varphi_n(x)$, $\psi_n(x)$, and $\xi(x)$:

$$a\varphi_n'' + \varphi_n\left[f(\eta) - \frac{1}{k}g(\eta) - c\right] - \beta_n\psi_n = 0,$$

$$a\psi_n'' + \psi_n\left[f(\eta) - \frac{1}{k}g(\eta) - c\right] + \beta_n\varphi_n = 0,$$

$$a\xi'' + \xi[f(\eta) + g(\eta)] + h(\eta) = 0, \quad \eta = (1+k)\xi.$$

Notably, the last equation is isolated as it does not depend on the others.

Equations involving two arbitrary functions that depend on the sum of squares of u and w.

Equation 9. Now let us look at the nonlinear delay PDE

$$u_t = au_{xx} + uf(u^2 + w^2) + wg(u^2 + w^2), \quad w = u(x, t - \tau), \quad (3.4.2.48)$$

where $f(z)$ and $g(z)$ are arbitrary functions, dependent on the nonlinear (quadratic) argument $z = u^2 + w^2$.

The functional constraint of the first kind (3.4.1.3) for equation (3.4.2.48) has the form
$$u^2 + w^2 = p(x), \qquad w = u(x, t - \tau). \tag{3.4.2.49}$$

The nonlinear difference equation (3.4.2.49) can be satisfied if we set
$$u = \varphi_n(x) \cos(\lambda_n t) + \psi_n(x) \sin(\lambda_n t), \tag{3.4.2.50}$$
$$\lambda_n = \frac{\pi(2n+1)}{2\tau}, \quad n = 0, \pm 1, \pm 2, \ldots$$

It can be verified that the relation
$$w = (-1)^n \varphi_n(x) \sin(\lambda_n t) + (-1)^{n+1} \psi_n(x) \cos(\lambda_n t)$$
holds, and also
$$u^2 + w^2 = \varphi_n^2(x) + \psi_n^2(x) = p(x).$$

Substituting (3.4.2.50) into (3.4.2.48) and splitting the resulting expression with respect to $\cos(\lambda_n t)$ and $\sin(\lambda_n t)$, we obtain a nonlinear system of ordinary differential equations for the functions $\varphi_n = \varphi_n(x)$ and $\psi_n = \psi_n(x)$:
$$a\varphi_n'' + \varphi_n f(\varphi_n^2 + \psi_n^2) + (-1)^{n+1} \psi_n g(\varphi_n^2 + \psi_n^2) - \lambda_n \psi_n = 0,$$
$$a\psi_n'' + \psi_n f(\varphi_n^2 + \psi_n^2) + (-1)^n \varphi_n g(\varphi_n^2 + \psi_n^2) + \lambda_n \varphi_n = 0,$$
where a prime denotes a derivative with respect to x.

Some generalizations. The above results admit various generalizations in the cases where functional constraints of the second kind (3.4.1.4) with $q(t) \neq \text{const}$ were used to construct exact solutions. Let us illustrate this by specific examples.

▶ Example 3.14. The nonlinear PDE with constant delay
$$u_t = au_{xx} + uf(t, w/u), \qquad w = u(x, t - \tau), \tag{3.4.2.51}$$
which is more general than equation (3.4.2.1) as its kinetic function depends on an additional argument, t, also has a multiplicative separable solution of the form (3.4.2.3).

In turn, equation (3.4.2.51) also admits a further generalization. Specifically, the constant delay time τ can be replaced with a variable delay $\tau = \tau(t)$. The resulting more complex equation (3.4.2.51) with variable delay will also have a multiplicative separable solution (3.4.2.3). ◀

▶ Example 3.15. The nonlinear PDE with constant delay
$$u_t = au_{xx} + bu + f(t, u - w), \qquad w = u(x, t - \tau), \tag{3.4.2.52}$$
which is more general than equation (3.4.2.19) as its kinetic function involves an additional argument, t, also has a multiplicative separable solution (3.4.2.21).

Equation (3.4.2.52) can be further generalized. Specifically, equation (3.4.2.52) with a variable delay $\tau = \tau(t)$ also has an additive separable solution of the form (3.4.2.21). ◀

3.4.3. Exact Solutions to More Complicated Nonlinear Delay Reaction-Diffusion Equations

Let us now consider the following nonlinear reaction-diffusion equations with constant delay that are more complicated than (3.4.1.1) [435]:

$$u_t = [g_0(u)u_x]_x + g_1(u)f_1(z) + g_2(u)f_2(z) + g_3(u),$$
$$w = u(x, t - \tau), \quad z = z(u, w), \tag{3.4.3.1}$$

where $g_i(u)$ ($i = 0, 1, 2, 3$) and $z = z(u, w)$ are some given functions, and $f_1(z)$ and $f_2(z)$ are arbitrary functions of a single argument. We will also be looking at related equations in which the function $g_3(u)$ is replaced with $g_3(w)$.

We will be using the method of functional constraints to construct exact solutions of nonlinear delay PDEs (3.4.3.1). This method is described in Subsection 3.4.1. The exact solutions of such equations specified below were obtained in [435].

Remark 3.30. Apart from exact solutions to one-dimensional delay reaction-diffusion equations, we will describe a few solutions to more complicated related equations with several space variables.

One-dimensional equations involving a single arbitrary function.
Equation 1. Consider the nonlinear reaction-diffusion equation with constant delay that involves a single arbitrary function dependent on the ratio w/u:

$$u_t = a(u^k u_x)_x + uf(w/u), \quad w = u(x, t - \tau). \tag{3.4.3.2}$$

It is a special case of equation (3.4.3.1) with $g_0(u) = au^k$, $g_1(u) = u$, $g_2 = g_3 = 0$, $z = w/u$, and $f_1(z) = f(z)$. If $k = 0$, see equation (3.4.2.1), which admits more exact solutions than equation (3.4.3.2) with $k \neq 0$.

In this case, the functional constraint of the second kind (3.4.1.4) coincides with (3.4.2.2) and admits a multiplicative separable solution (3.4.2.3). Therefore, we look for exact solutions to the original delay reaction-diffusion equations in the form

$$u = \varphi(x)\psi(t). \tag{3.4.3.3}$$

Substituting (3.4.3.3) into (3.4.3.2) and rearranging, we obtain the following ODE and delay ODE for the functions $\varphi = \varphi(x)$ and $\psi = \psi(t)$:

$$a(\varphi^k \varphi'_x)'_x = b\varphi, \tag{3.4.3.4}$$
$$\psi'(t) = b\psi^{k+1}(t) + \psi(t)f\big(\psi(t-\tau)/\psi(t)\big), \tag{3.4.3.5}$$

where b is an arbitrary constant.

The general solution of the autonomous ODE (3.4.3.4) can be represented in implicit form. For $k \neq 0$ and $k \neq -2$, its particular solution is

$$\varphi = Ax^{2/k}, \quad A = \left[\frac{bk^2}{2a(k+2)}\right]^{1/k}.$$

Remark 3.31. *Equation (3.4.3.2) also admits an exact solution of the form*

$$u = (x+C)^{2/k}\theta(\zeta), \quad \zeta = t + \lambda\ln(x+C),$$

where C and λ are arbitrary constants, and the function $\theta = \theta(\zeta)$ is described by the delay ODE

$$\theta'(\zeta) = a\Big\{\frac{2(k+2)}{k^2}\theta^{k+1}(\zeta) + \frac{(3k+4)\lambda}{k}\theta^k(\zeta)\theta'(\zeta)$$
$$+ k\lambda^2\theta^{k-1}(\zeta)[\theta'(\zeta)]^2 + \lambda^2\theta^k(\zeta)\theta''(\zeta)\Big\} + \theta(\zeta)f(\theta(\zeta-\tau)/\theta(\zeta)).$$

Equation 2. Consider the delay reaction-diffusion equation

$$u_t = a(u^k u_x)_x + bu^{k+1} + uf(w/u), \tag{3.4.3.6}$$

which is a generalization of equation (3.4.3.2). In this case, the functional constraint of the second kind (3.4.1.4) also coincides with (3.4.2.2), and the original equation (3.4.3.6) admits separable solutions of the form (3.4.3.3), which are given below.

1°. For $b(k+1) > 0$, equation (3.4.3.6) has a multiplicative separable solution

$$u = [C_1\cos(\beta x) + C_2\sin(\beta x)]^{1/(k+1)}\psi(t), \quad \beta = \sqrt{b(k+1)/a}, \tag{3.4.3.7}$$

where C_1 and C_2 are arbitrary constants, and the function $\psi = \psi(t)$ is described by the delay ODE

$$\psi'(t) = \psi(t)f(\psi(t-\tau)/\psi(t)). \tag{3.4.3.8}$$

Equation (3.4.3.8) has an exponential particular solution

$$\psi(t) = Ae^{\lambda t}, \tag{3.4.3.9}$$

where A is an arbitrary constant, and λ is a solution of the algebraic (transcendental) equation $\lambda - f(e^{-\lambda\tau}) = 0$.

2°. For $b(k+1) < 0$, equation (3.4.3.6) has a multiplicative separable solution

$$u = [C_1\exp(-\beta x) + C_2\exp(\beta x)]^{1/(k+1)}\psi(t), \quad \beta = \sqrt{-b(k+1)/a}, \tag{3.4.3.10}$$

where C_1 and C_2 are arbitrary constants, and the function $\psi = \psi(t)$ is described by the delay ODE (3.4.3.8).

3°. Equation (3.4.3.6) with $k = -1$ admits a multiplicative separable solution

$$u = C_1\exp\Big(-\frac{b}{2a}x^2 + C_2 x\Big)\psi(t), \tag{3.4.3.11}$$

where C_1 and C_2 are arbitrary constants, and the function $\psi = \psi(t)$ is described by the delay ODE (3.4.3.8).

Equation 3. Consider the delay reaction-diffusion equation

$$u_t = a(u^k u_x)_x + b + u^{-k}f(u^{k+1} - w^{k+1}), \quad k \neq -1, \tag{3.4.3.12}$$

which involves an arbitrary function $f(z)$, where $z = u^{k+1} - w^{k+1}$.

In this case, the functional constraint of the second kind (3.4.1.4) is written as

$$u^{k+1} - w^{k+1} = q(t), \qquad w = u(x, t - \tau). \tag{3.4.3.13}$$

The difference equation (3.4.3.13) can be satisfied if we take a functional separable solution of the form

$$u = [\varphi(x) + \psi(t)]^{1/(k+1)}, \tag{3.4.3.14}$$

which gives $q(t) = \psi(t) - \psi(t - \tau)$. Substituting (3.4.3.14) into the original delay equation (3.4.3.12) and analyzing, we obtain the following results:

1°. Equation (3.4.3.12) admits a functional separable solution

$$u = \left[At - \frac{b(k+1)}{2a} x^2 + C_1 x + C_2 \right]^{1/(k+1)}, \tag{3.4.3.15}$$

where C_1 and C_2 are arbitrary constants, and A is a solution of the algebraic (transcendental) equation $A = (k+1)f(A\tau)$.

2°. Equation (3.4.3.12) admits a more complicated functional separable solution of the form

$$u = \left[\psi(t) - \frac{b(k+1)}{2a} x^2 + C_1 x + C_2 \right]^{1/(k+1)}, \tag{3.4.3.16}$$

where C_1 and C_2 are arbitrary constants, and the function $\psi(t)$ is described by the delay ODE

$$\psi'(t) = (k+1) f\big(\psi(t) - \psi(t - \tau)\big). \tag{3.4.3.17}$$

Equation 4. Consider the nonlinear delay PDE

$$u_t = a(u^{-1/2} u_x)_x + b u^{1/2} + f(u^{1/2} - w^{1/2}), \tag{3.4.3.18}$$

where $f(z)$ is an arbitrary function and $z = u^{1/2} - w^{1/2}$.

The functional constraint of the first kind (3.4.1.3) is

$$u^{1/2} - w^{1/2} = p(x), \qquad w = u(x, t - \tau). \tag{3.4.3.19}$$

The difference equation (3.4.3.19) can be satisfied by setting

$$u = [\varphi(x) t + \psi(x)]^2, \tag{3.4.3.20}$$

which gives $p(x) = \tau \varphi(x)$.

Substituting (3.4.3.20) into the delay equation (3.4.3.18) yields the following ordinary differential equations for $\varphi = \varphi(x)$ and $\psi = \psi(x)$:

$$2a\varphi''_{xx} + b\varphi - 2\varphi^2 = 0,$$
$$2a\psi''_{xx} + b\psi - 2\varphi\psi + f(\tau\varphi) = 0.$$

These equations admit the simple particular solution

$$\varphi = \frac{1}{2} b, \qquad \psi = -\frac{1}{4a} f\!\left(\frac{b\tau}{2}\right) x^2 + Ax + B,$$

where A and B are arbitrary constants.

Equation 5. Consider the delay reaction-diffusion equation

$$u_t = a(e^{\lambda u} u_x)_x + f(u - w), \qquad (3.4.3.21)$$

which involves an arbitrary function $f(z)$ with $z = u - w$.

In this case, the functional constraint of the second kind (3.4.1.4) coincides with (3.4.2.20) and admits an additive separable solution (3.4.2.21). Therefore, we seek a solution to the original delay PDE (3.4.3.21) in the form

$$u = \varphi(x) + \psi(t). \qquad (3.4.3.22)$$

Substituting (3.4.3.22) into the original delay equation (3.4.3.21) gives the exact solution

$$u = \frac{1}{\lambda} \ln(Ax^2 + Bx + C) + \psi(t), \qquad (3.4.3.23)$$

where A, B, and C are arbitrary constants, and the function $\psi = \psi(t)$ satisfies the delay ODE

$$\psi'(t) = 2a(A/\lambda)e^{\lambda \psi(t)} + f\big(\psi(t) - \psi(t - \tau)\big). \qquad (3.4.3.24)$$

Remark 3.32. *Equation (3.4.3.21) also admits a more complicated exact solution of the form*

$$u = \frac{2}{\lambda} \ln(x + C) + \theta(\zeta), \quad \zeta = t + \beta \ln(x + C), \qquad (3.4.3.25)$$

where C and β are arbitrary constants, and the function $\theta = \theta(\zeta)$ is described by the delay ODE

$$\theta'(\zeta) = ae^{\lambda \theta(\zeta)} \Big\{ \frac{2}{\lambda} + 3\beta \theta'(\zeta) + \beta^2 \lambda [\theta'(\zeta)]^2 + \beta^2 \theta''(\zeta) \Big\} + f\big(\theta(\zeta) - \theta(\zeta - \tau)\big).$$

Equation 6. Consider the equation

$$u_t = a(e^{\lambda u} u_x)_x + be^{\lambda u} + f(u - w), \qquad (3.4.3.26)$$

which is a generalization of equation (3.4.3.21).

1°. For $b\lambda > 0$, equation (3.4.3.26) admits an additive separable solution

$$u = \frac{1}{\lambda} \ln[C_1 \cos(\beta x) + C_2 \sin(\beta x)] + \psi(t), \quad \beta = \sqrt{b\lambda/a}, \qquad (3.4.3.27)$$

where C_1 and C_2 are arbitrary constants, and the function $\psi(t)$ is described by the delay ODE

$$\psi'(t) = f\big(\psi(t) - \psi(t - \tau)\big). \qquad (3.4.3.28)$$

Equation (3.4.3.28) has a simple particular solution $\psi = A + kt$, where A is an arbitrary constant, and k is a solution of the algebraic (transcendental) equation $k - f(k\tau) = 0$.

2°. For $b\lambda < 0$, equation (3.4.3.26) admits another additive separable solution

$$u = \frac{1}{\lambda} \ln[C_1 \exp(-\beta x) + C_2 \exp(\beta x)] + \psi(t), \quad \beta = \sqrt{-b\lambda/a}, \qquad (3.4.3.29)$$

where C_1 and C_2 are arbitrary constants, and the function $\psi(t)$ is described by the delay ODE (3.4.3.28).

Equation 7. Consider the nonlinear delay PDE

$$u_t = a(e^{\lambda u} u_x)_x + b + e^{-\lambda u} f(e^{\lambda u} - e^{\lambda w}). \tag{3.4.3.30}$$

In this case, the functional constraint of the second kind (3.4.1.4) is written as

$$e^{\lambda u} - e^{\lambda w} = q(t), \qquad w = u(x, t - \tau). \tag{3.4.3.31}$$

The functional equation (3.4.3.31) can be satisfied by taking a functional separable solution of the form

$$u = \frac{1}{\lambda} \ln[\varphi(x) + \psi(t)], \tag{3.4.3.32}$$

which gives $q(t) = \psi(t) - \psi(t - \tau)$. Substituting (3.4.3.32) into the original delay equation (3.4.3.30) and analyzing, we obtain the following results:

1°. Equation (3.4.3.30) admits a functional separable solution

$$u = \frac{1}{\lambda} \ln\left[At - \frac{b\lambda}{2a} x^2 + C_1 x + C_2\right], \tag{3.4.3.33}$$

where C_1 and C_2 are arbitrary constants, and A is a solution of the algebraic (transcendental) equation $A - \lambda f(A\tau) = 0$.

2°. Equation (3.4.3.30) admits a more complicated functional separable solution

$$u = \frac{1}{\lambda} \ln\left[\psi(t) - \frac{b\lambda}{2a} x^2 + C_1 x + C_2\right], \tag{3.4.3.34}$$

where C_1 and C_2 are arbitrary constants, and the function $\psi(t)$ is described by the delay ODE

$$\psi'(t) = \lambda f\big(\psi(t) - \psi(t - \tau)\big). \tag{3.4.3.35}$$

Equation 8. The nonlinear delay PDE

$$u_t = [(a \ln u + b) u_x]_x - cu \ln u + u f(w/u) \tag{3.4.3.36}$$

admits two multiplicative separable solutions

$$u = \exp(\pm \lambda x) \psi(t), \quad \lambda = \sqrt{c/a}, \tag{3.4.3.37}$$

where the function $\psi(t)$ is described by the delay ODE

$$\psi'(t) = \lambda^2 (a + b) \psi(t) + \psi(t) f\big(\psi(t - \tau)/\psi(t)\big). \tag{3.4.3.38}$$

Equation 9. Consider the nonlinear delay PDE

$$u_t = [u f'(u) u_x]_x + \frac{1}{f'(u)} [a f(u) + b f(w) + c], \tag{3.4.3.39}$$

where $f(u)$ is an arbitrary function, and the prime stands for a derivative with respect to the argument.

Equation (3.4.3.39) admits a generalized traveling wave solution (a functional separable solution of special form) that can be represented in the implicit form

$$f(u) = \varphi(t)x + \psi(t), \qquad (3.4.3.40)$$

where the functions $\varphi(t)$ and $\psi(t)$ satisfy the delay ODEs

$$\varphi'(t) = a\varphi(t) + b\varphi(t-\tau), \qquad (3.4.3.41)$$
$$\psi'(t) = a\psi(t) + b\psi(t-\tau) + c + \varphi^2(t). \qquad (3.4.3.42)$$

Equation 10. The nonlinear delay PDE

$$u_t = [uf'(u)u_x]_x + (a+b)u + \frac{2}{f'(u)}[af(u) + bf(w) + c] \qquad (3.4.3.43)$$

admits a functional separable solution that can be represented in the implicit form

$$f(u) = -\tfrac{1}{2}(a+b)x^2 + \varphi(t)x + \psi(t), \qquad (3.4.3.44)$$

where the functions $\varphi(t)$ and $\psi(t)$ are described by the delay ODEs

$$\varphi'(t) = -2b\varphi(t) + 2b\varphi(t-\tau), \qquad (3.4.3.45)$$
$$\psi'(t) = 2a\psi(t) + 2b\psi(t-\tau) + 2c + \varphi^2(t). \qquad (3.4.3.46)$$

Equation (3.4.3.45) has an exponential particular solution

$$\varphi(t) = C_1 e^{\lambda t} + C_2, \qquad (3.4.3.47)$$

where C_1 and C_2 are arbitrary constants, and λ is a root of the transcendental equation $\lambda + 2b(1 - e^{-\lambda\tau}) = 0$.

Equation 11. The nonlinear delay PDE

$$u_t = [f'(u)u_x]_x + a_1 f(u) + a_2 f(w) + a_3 + \frac{b}{f'(u)}[f(u) - f(w)] \qquad (3.4.3.48)$$

admits a functional separable solution that can be represented in the implicit form

$$f(u) = e^{\lambda t}\varphi(x) - \frac{a_3}{a_1 + a_2}, \qquad (3.4.3.49)$$

where λ is a root of the transcendental equation

$$\lambda = b(1 - e^{-\lambda\tau}), \qquad (3.4.3.50)$$

and the function $\varphi = \varphi(x)$ satisfies the linear second-order ODE with constant coefficients

$$\varphi''_{xx} + (a_1 + a_2 e^{-\lambda\tau})\varphi = 0. \qquad (3.4.3.51)$$

3.4. Method of Functional Constraints

Equation 12. The nonlinear delay PDE

$$u_t = [f'(u)u_x]_x + a[f(u) - f(w)] + \frac{1}{f'(u)}[b_1 f(u) + b_2 f(w) + b_3] \qquad (3.4.3.52)$$

admits a functional separable solution in implicit form

$$f(u) = e^{\lambda t}\varphi(x) - \frac{b_3}{b_1 + b_2}, \qquad (3.4.3.53)$$

where λ is a root of the transcendental equation

$$\lambda - b_1 - b_2 e^{-\lambda \tau} = 0, \qquad (3.4.3.54)$$

and the function $\varphi = \varphi(x)$ satisfies the linear second-order ODE with constant coefficients

$$\varphi''_{xx} + a(1 - e^{-\lambda \tau})\varphi = 0. \qquad (3.4.3.55)$$

Equation 13. Consider the nonlinear delay PDE

$$u_t = [f'(u)u_x]_x + a_1 f(u) + a_2 f(w) + a_3 + \frac{1}{f'(u)}[b_1 f(u) + b_2 f(w) + b_3], \qquad (3.4.3.56)$$

which generalizes the two preceding equations.

Let the coefficients of equation (3.4.3.56) satisfy the condition

$$(a_1 + a_2)b_3 = a_3(b_1 + b_2). \qquad (3.4.3.57)$$

Then equation (3.4.3.56) admits a functional separable solution that can be represented in the implicit form

$$f(u) = e^{\lambda t}\varphi(x) + c. \qquad (3.4.3.58)$$

Here

$$c = -\frac{a_3}{a_1 + a_2} \quad \text{if} \quad a_1 \neq -a_2 \quad \text{and} \quad c = -\frac{b_3}{b_1 + b_2} \quad \text{if} \quad b_1 \neq -b_2,$$

λ is a root of the transcendental equation

$$\lambda - b_1 - b_2 e^{-\lambda \tau} = 0, \qquad (3.4.3.59)$$

and the function $\varphi = \varphi(x)$ satisfies the linear second-order ODE with constant coefficients

$$\varphi''_{xx} + (a_1 + a_2 e^{-\lambda \tau})\varphi = 0. \qquad (3.4.3.60)$$

Equation 14. Consider the nonlinear delay PDE

$$u_t = [g(u)u_x]_x + \frac{1}{f'(u)}[c_1 f(u) + c_2 f(w) + c_3], \qquad (3.4.3.61)$$

$$g(u) = f'(u)\int [af(u) + b]\,du,$$

where $f(u)$ is an arbitrary function, and the prime stands for a derivative with respect to the argument.

Equation (3.4.3.61) admits a functional separable solution in implicit form
$$f(u) = \varphi(t)x + \psi(t), \tag{3.4.3.62}$$
where the functions $\varphi = \varphi(t)$ and $\psi = \psi(t)$ satisfy the delay ODEs
$$\varphi'(t) = a\varphi^3(t) + c_1\varphi(t) + c_2\varphi(t-\tau),$$
$$\psi'(t) = \varphi^2(t)[a\psi(t) + b] + c_1\psi(t) + c_2\psi(t-\tau) + c_3.$$

One-dimensional equations involving two arbitrary functions. Below we briefly describe several exact solutions to more general nonlinear reaction-diffusion type equations with delay that involve two arbitrary functions.

Equation 15. Consider the nonlinear delay PDE
$$u_t = a(u^k u_x)_x + uf(w/u) + u^{k+1}g(w/u), \tag{3.4.3.63}$$
where $f(z)$ and $g(z)$ are arbitrary functions.

Equation (3.4.3.63) admits a multiplicative separable solution
$$u = e^{\lambda t}\varphi(x), \tag{3.4.3.64}$$
where λ is a solution of the algebraic (transcendental) equation
$$\lambda = f(e^{-\lambda\tau}),$$
and the function $\varphi = \varphi(x)$ is described by the nonlinear second-order ODE
$$a(\varphi^k \varphi'_x)'_x + g(e^{-\lambda\tau})\varphi^{k+1} = 0.$$
For $k \neq -1$, the change of variable $\theta = \varphi^{k+1}$ reduces this equation to a linear second-order ODE with constant coefficients. If $k = -1$, one should use the change of variable $\theta = \ln\varphi$.

Equation 16. The nonlinear delay PDE
$$u_t = a(u^{-1/2}u_x)_x + f(u^{1/2} - w^{1/2}) + u^{1/2}g(u^{1/2} - w^{1/2}) \tag{3.4.3.65}$$
admits a generalized separable solution
$$u = [\varphi(x)t + \psi(x)]^2, \tag{3.4.3.66}$$
where the functions $\varphi = \varphi(x)$ and $\psi = \psi(x)$ are described by the system of ODEs
$$2a\varphi''_{xx} + \varphi g(\tau\varphi) - 2\varphi^2 = 0,$$
$$2a\psi''_{xx} + \psi g(\tau\varphi) - 2\varphi\psi + f(\tau\varphi) = 0.$$

A particular solution of this system is
$$\varphi = k, \quad \psi = -\frac{1}{4a}f(k\tau)x^2 + Ax + B,$$

where A and B are arbitrary constants, and the constant k is determined from the algebraic (transcendental) equation $g(k\tau) - 2k = 0$.

Equation 17. The nonlinear delay PDE

$$u_t = a(u^k u_x)_x + f(u^{k+1} - w^{k+1}) + u^{-k} g(u^{k+1} - w^{k+1}), \quad k \neq -1, \quad (3.4.3.67)$$

admits a functional separable solution

$$u = (At + Bx^2 + C_1 x + C_2)^{1/(k+1)}, \quad B = -\frac{(k+1)}{2a} f(A\tau), \quad (3.4.3.68)$$

where C_1 and C_2 are arbitrary constants, and the constant A is determined from the algebraic (transcendental) equation $A - (k+1)g(A\tau) = 0$.

Equation 18. The nonlinear delay PDE

$$u_t = a(e^{\lambda u} u_x)_x + f(u - w) + e^{\lambda u} g(u - w) \quad (3.4.3.69)$$

admits an additive separable solution

$$u = \beta t + \varphi(x), \quad (3.4.3.70)$$

where the constant β is determined from the algebraic (transcendental) equation

$$\beta = f(\beta\tau).$$

Function $\varphi = \varphi(x)$ appearing in solution (3.4.3.70) is described by the ODE

$$a(e^{\lambda\varphi} \varphi'_x)'_x + g(\beta\tau) e^{\lambda\varphi} = 0.$$

With the change of variable $\theta = e^{\lambda\varphi}$, this equation reduces to the linear second-order ODE with constant coefficients $a\theta''_{xx} + \lambda g(\beta\tau)\theta = 0$.

Equation 19. The nonlinear delay PDE

$$u_t = a(e^{\lambda u} u_x)_x + f(e^{\lambda u} - e^{\lambda w}) + e^{-\lambda u} g(e^{\lambda u} - e^{\lambda w}) \quad (3.4.3.71)$$

admits a functional separable solution

$$u = \frac{1}{\lambda} \ln(At + Bx^2 + C_1 x + C_2), \quad B = -\frac{\lambda}{2a} f(A\tau), \quad (3.4.3.72)$$

where C_1 and C_2 are arbitrary constants, and the constant A is determined from the algebraic (transcendental) equation $A - \lambda g(A\tau) = 0$.

Equation 20. Consider the nonlinear delay PDE

$$u_t = a[g'(u) u_x]_x + b + \frac{1}{g'(u)} f\big(g(u) - g(w)\big), \quad (3.4.3.73)$$

where $g(u)$ and $f(z)$ are arbitrary functions, and the prime stands for a derivative with respect to u.

Equation (3.4.3.73) admits a functional separable solution in implicit form

$$g(u) = \psi(t) - \frac{b}{2a}x^2 + C_1 x + C_2, \qquad (3.4.3.74)$$

where C_1 and C_2 are arbitrary constants. The function $\psi = \psi(t)$ is described by the delay ODE (3.4.3.28), which has a particular solution $\psi(t) = At$, where the constant A is determined from the algebraic (transcendental) equation $A - f(A\tau) = 0$.

Equation 22. Consider the nonlinear delay PDE

$$u_t = a[g'(u)u_x]_x + bg(u) + \frac{g(u)}{g'(u)} f(g(w)/g(u)), \qquad (3.4.3.75)$$

where $g(u)$ and $f(z)$ are arbitrary functions, and the prime stands for a derivative with respect to u.

1°. For $ab > 0$, equation (3.4.3.75) admits a functional separable solution in implicit form

$$g(u) = [C_1 \cos(\lambda x) + C_2 \sin(\lambda x)]\psi(t), \quad \lambda = \sqrt{b/a}, \qquad (3.4.3.76)$$

where C_1 and C_2 are arbitrary constants. The function $\psi = \psi(t)$ is described by the delay ODE (3.4.3.8), which has an exponential particular solution $\psi(t) = e^{\lambda t}$, where λ is a root of the algebraic (transcendental) equation $\lambda - f(e^{-\lambda \tau}) = 0$.

2°. For $ab < 0$, equation (3.4.3.75) admits a functional separable solution in implicit form

$$g(u) = [C_1 \exp(-\lambda x) + C_2 \exp(\lambda x)]\psi(t), \quad \lambda = \sqrt{-b/a}, \qquad (3.4.3.77)$$

where C_1 and C_2 are arbitrary constants, and the function $\psi = \psi(t)$ is described by the delay ODE (3.4.3.8).

3°. For $b = 0$, equation (3.4.3.75) admits a functional separable solution in implicit form

$$g(u) = (C_1 x + C_2)\psi(t),$$

where the function $\psi = \psi(t)$ is described by the delay ODE (3.4.3.8).

Equation 22. Consider the nonlinear delay PDE

$$u_t = [h(u)u_x]_x - \frac{1}{g'(u)}[c_1 g(u) + c_2 g(w)] + \frac{1}{g'(u)} f(g(u) - g(w)), \qquad (3.4.3.78)$$

$$h(u) = g'(u) \int [ag(u) + b]\, du,$$

where $g(u)$ and $f(z)$ are arbitrary functions, and the prime denotes a derivative with respect to u.

Equation (3.4.3.78) admits two generalized traveling wave solutions that can be represented in the implicit form

$$g(u) = \pm kx + \psi(t), \quad k = \sqrt{(c_1 + c_2)/a}. \qquad (3.4.3.79)$$

The function $\psi = \psi(t)$ is described by the delay ODE

$$\psi'(t) = c_2\psi(t) + bk^2 - c_2\psi(t-\tau) + f\big(\psi(t) - \psi(t-\tau)\big).$$

One-dimensional equations involving three arbitrary functions.
Equation 23. Consider the nonlinear delay PDE

$$u_t = a[f'(u)u_x]_x + g\big(f(u) - f(w)\big) + \frac{1}{f'(u)}h\big(f(u) - f(w)\big), \qquad (3.4.3.80)$$

where $f(u)$, $g(z)$, and $h(z)$ are arbitrary functions, and the prime denotes a derivative with respect to u.

Equation (3.4.3.80) admits a functional separable solution in implicit form

$$f(u) = At - \frac{g(A\tau)}{2a}x^2 + C_1 x + C_2, \qquad (3.4.3.81)$$

where C_1 and C_2 are arbitrary constants, and the constant A is a root of the algebraic (transcendental) equation $A - h(A\tau) = 0$.

Equation 24. Consider the nonlinear delay PDE

$$u_t = a[f'(u)u_x]_x + f(u)g\big(f(w)/f(u)\big) + \frac{f(u)}{f'(u)}h\big(f(w)/f(u)\big), \qquad (3.4.3.82)$$

where $f(u)$, $g(z)$, and $h(z)$ are arbitrary functions.

We assume that β is a root of the algebraic (transcendental) equation

$$\beta - h(e^{-\beta\tau}) = 0.$$

1°. For $ag(e^{-\beta\tau}) > 0$, equation (3.4.3.82) admits a functional separable solution in implicit form

$$f(u) = \big[C_1 \cos(\lambda x) + C_2 \sin(\lambda x)\big]e^{\beta t}, \quad \lambda = \sqrt{g(e^{-\beta\tau})/a}, \qquad (3.4.3.83)$$

where C_1 and C_2 are arbitrary constants.

2°. For $ag(e^{-\beta\tau}) < 0$, equation (3.4.3.82) admits another functional separable solution in implicit form

$$f(u) = \big[C_1 \exp(-\lambda x) + C_2 \exp(\lambda x)\big]e^{\beta t}, \quad \lambda = \sqrt{-g(e^{-\beta\tau})/a}, \qquad (3.4.3.84)$$

where C_1 and C_2 are arbitrary constants.

3°. For $g(e^{-\beta\tau}) = 0$, equation (3.4.3.82) admits a degenerate functional separable solution in implicit form

$$f(u) = (C_1 x + C_2)e^{\beta t}.$$

Equation 25. The nonlinear delay PDE

$$u_t = [g(u)u_x]_x - \frac{a^2}{f'(u)}\frac{d}{du}\left[\frac{g(u)}{f'(u)}\right] + \frac{1}{f'(u)}h\big(f(u)-f(w)\big) \qquad (3.4.3.85)$$

admits two functional separable solutions in implicit form

$$f(u) = \pm ax + \psi(t),$$

where the function $\psi = \psi(t)$ is described by the delay ODE

$$\psi'(t) = h\big(\psi(t) - \psi(t-\tau)\big).$$

Multi-dimensional equations involving a single arbitrary function. Below we describe multi-dimensional generalizations of some of the above one-dimensional delay reaction-diffusion equations and their exact solutions. We will use the notations: $u = u(\mathbf{x},t)$, $w = u(\mathbf{x},t-\tau)$, and $\mathbf{x} = (x_1,\ldots,x_n)$. The values $n=2$ and $n=3$ correspond to two- and three-dimensional equations.

Remark 3.33. *The exact solutions to the multi-dimensional nonlinear delay equations described below are often expressed in terms of solutions to the Laplace, Poisson, or Helmholtz equations, which are simpler than the original ones. Many solutions to these linear elliptic equations can be found in [404, 514].*

Equation 26. Consider the nonlinear delay PDE

$$u_t = a\,\mathrm{div}(u^k \nabla u) + bu^{k+1} + uf(w/u). \qquad (3.4.3.86)$$

1°. For $k \neq -1$, equation (3.4.3.86) admits a multiplicative separable solution of the form

$$u = \psi(t)\varphi^{1/(k+1)}(\mathbf{x}). \qquad (3.4.3.87)$$

The function $\psi = \psi(t)$ is described by the delay ODE (3.4.3.8) and the function $\varphi = \varphi(\mathbf{x})$ satisfies the Helmholtz equation

$$\Delta\varphi + \frac{b(k+1)}{a}\varphi = 0, \qquad (3.4.3.88)$$

where Δ is the Laplace operator.

2°. If $k=-1$, equation (3.4.3.86) admits a multiplicative separable solution

$$u = \psi(t)\ln\varphi(\mathbf{x}), \qquad (3.4.3.89)$$

where the function $\psi=\psi(t)$ is described by the delay ODE (3.4.3.8) and the function $\varphi = \varphi(\mathbf{x})$ satisfies Poisson's equation

$$\Delta\varphi + (b/a) = 0. \qquad (3.4.3.90)$$

3.4. Method of Functional Constraints

Equation 27. The nonlinear delay PDE

$$u_t = a \operatorname{div}(u^k \nabla u) + b + u^{-k} f(u^{k+1} - w^{k+1}), \quad k \neq -1, \qquad (3.4.3.91)$$

admits a functional separable solution

$$u = \left[\varphi(\mathbf{x}) + \psi(t)\right]^{1/(k+1)}, \qquad (3.4.3.92)$$

where the function $\psi = \psi(t)$ is described by the delay ODE (3.4.3.17) and the function $\varphi = \varphi(\mathbf{x})$ satisfies Poisson's equation

$$\Delta \varphi + \frac{b(k+1)}{a} = 0. \qquad (3.4.3.93)$$

Equation 28. The nonlinear delay PDE

$$u_t = a \operatorname{div}(u^{-1/2} \nabla u) + b u^{1/2} + f(u^{1/2} - w^{1/2}) \qquad (3.4.3.94)$$

admits a generalized separable solution

$$u = [\varphi(\mathbf{x}) t + \psi(\mathbf{x})]^2. \qquad (3.4.3.95)$$

The functions $\varphi = \varphi(\mathbf{x})$ and $\psi = \psi(\mathbf{x})$ are described by the stationary second-order PDEs

$$2a\Delta\varphi + b\varphi - 2\varphi^2 = 0, \qquad (3.4.3.96)$$
$$2a\Delta\psi + b\psi - 2\varphi\psi + f(\tau\varphi) = 0. \qquad (3.4.3.97)$$

Equation (3.4.3.96) has a simple particular solution $\varphi = \frac{1}{2}b = \text{const}$. In this case, equation (3.4.3.97) is Poisson's equation:

$$a\Delta\psi + \tfrac{1}{2} f\bigl(\tfrac{1}{2} b \tau\bigr) = 0.$$

Equation 29. Consider the nonlinear delay PDE

$$u_t = a \operatorname{div}(e^{\lambda u} \nabla u) + b e^{\lambda u} + f(u - w), \qquad (3.4.3.98)$$

which generalizes equation (3.4.3.26).

Equation (3.4.3.98) admits an additive separable solution

$$u = \psi(t) + \frac{1}{\lambda} \ln \varphi(\mathbf{x}), \qquad (3.4.3.99)$$

in which the function $\psi = \psi(t)$ is described by the delay ODE (3.4.3.28) and the function $\varphi = \varphi(\mathbf{x})$ satisfies the Helmholtz equation

$$\Delta \varphi + \lambda (b/a) \varphi = 0. \qquad (3.4.3.100)$$

Equation 30. The nonlinear delay PDE

$$u_t = a\,\mathrm{div}(e^{\lambda u}\nabla u) + b + e^{-\lambda u}f(e^{\lambda u} - e^{\lambda w}) \tag{3.4.3.101}$$

admits a functional separable solution

$$u = \frac{1}{\lambda}\ln\bigl[\varphi(\mathbf{x}) + \psi(t)\bigr], \tag{3.4.3.102}$$

where the function $\psi = \psi(t)$ is described by the delay ODE (3.4.3.35) and the function $\varphi = \varphi(\mathbf{x})$ satisfies Poisson's equation

$$\Delta\varphi + \lambda(b/a) = 0. \tag{3.4.3.103}$$

Equation 31. Consider the nonlinear delay PDE

$$u_t = \mathrm{div}[uf'(u)\nabla u] + \frac{1}{f'(u)}[af(u) + bf(w) + c], \tag{3.4.3.104}$$

where $f(u)$ is an arbitrary function, and the prime denotes a derivative with respect to u.

Equation (3.4.3.104) admits a functional separable solution in implicit form

$$f(u) = \sum_{k=1}^{n}\varphi_k(t)x_k + \psi(t), \tag{3.4.3.105}$$

where the functions $\varphi_k = \varphi_k(t)$ and $\psi = \psi(t)$ are described by the delay ODEs

$$\varphi'_k(t) = a\varphi_k(t) + b\varphi_k(t-\tau), \quad k = 1,\ldots,n, \tag{3.4.3.106}$$

$$\psi'(t) = a\psi(t) + b\psi(t-\tau) + c + \sum_{k=1}^{n}\varphi_k^2(t). \tag{3.4.3.107}$$

Equation 32. The nonlinear delay PDE

$$u_t = \mathrm{div}[f'(u)\nabla u] + a_1 f(u) + a_2 f(w) + a_3 + \frac{b}{f'(u)}\bigl[f(u) - f(w)\bigr] \tag{3.4.3.108}$$

admits a functional separable solution in implicit form

$$f(u) = e^{\lambda t}\varphi(\mathbf{x}) - \frac{a_3}{a_1 + a_2}, \tag{3.4.3.109}$$

where λ is a root of the transcendental equation

$$\lambda = b(1 - e^{-\lambda\tau}), \tag{3.4.3.110}$$

and the function $\varphi = \varphi(\mathbf{x})$ is described by the Helmholtz equation

$$\Delta\varphi + (a_1 + a_2 e^{-\lambda\tau})\varphi = 0. \tag{3.4.3.111}$$

Equation 33. The nonlinear delay PDE

$$u_t = \text{div}[f'(u)\nabla u] + a[f(u) - f(w)] + \frac{1}{f'(u)}\big[b_1 f(u) + b_2 f(w) + b_3\big] \quad (3.4.3.112)$$

admits a functional separable solution in implicit form

$$f(u) = e^{\lambda t}\varphi(\mathbf{x}) - \frac{b_3}{b_1 + b_2}, \quad (3.4.3.113)$$

where λ is a root of the transcendental equation

$$\lambda - b_1 - b_2 e^{-\lambda \tau} = 0, \quad (3.4.3.114)$$

and the function $\varphi = \varphi(\mathbf{x})$ is described by the Helmholtz equation

$$\Delta\varphi + a(1 - e^{-\lambda\tau})\varphi = 0. \quad (3.4.3.115)$$

Multi-dimensional equations involving two arbitrary functions.
Equation 34. The nonlinear delay PDE

$$u_t = a\,\text{div}(u^{-1/2}\nabla u) + f(u^{1/2} - w^{1/2}) + u^{1/2}g(u^{1/2} - w^{1/2}) \quad (3.4.3.116)$$

admits a generalized separable solution

$$u = [\varphi(\mathbf{x})t + \psi(\mathbf{x})]^2, \quad (3.4.3.117)$$

where the functions $\varphi = \varphi(\mathbf{x})$ and $\psi = \psi(\mathbf{x})$ are described by the stationary second-order PDEs

$$2a\Delta\varphi + \varphi g(\tau\varphi) - 2\varphi^2 = 0, \quad (3.4.3.118)$$
$$2a\Delta\psi + \psi g(\tau\varphi) - 2\varphi\psi + f(\tau\varphi) = 0. \quad (3.4.3.119)$$

Equation (3.4.3.118) has a simple particular solution $\varphi = \varphi_0 = \text{const}$, where φ_0 is a root of the algebraic (transcendental) equation $g(\tau\varphi_0) - 2\varphi_0 = 0$. In this case, equation (3.4.3.119) is Poisson's equation:

$$a\Delta\psi + \tfrac{1}{2}f(\tau\varphi_0) = 0.$$

Equation 35. The nonlinear delay PDE

$$u_t = a\,\text{div}(u^k \nabla u) + f(u^{k+1} - w^{k+1}) + u^{-k}g(u^{k+1} - w^{k+1}), \quad k \neq -1, \quad (3.4.3.120)$$

admits a functional separable solution

$$u = [At + \varphi(\mathbf{x})]^{1/(k+1)}, \quad (3.4.3.121)$$

where A is a root of the algebraic (transcendental) equation $A - (k+1)g(A\tau) = 0$, and the function $\varphi = \varphi(\mathbf{x})$ is described by Poisson's equation

$$a\,\Delta\varphi + (k+1)f(A\tau) = 0. \quad (3.4.3.122)$$

Equation 36. Consider the nonlinear delay PDE
$$u_t = a\,\text{div}(e^{\lambda u}\nabla u) + f(u-w) + e^{\lambda u}g(u-w), \tag{3.4.3.123}$$
where $f(z)$ and $g(z)$ are arbitrary functions.

Equation (3.4.3.123) admits an additive separable solution
$$u = \beta t + \frac{1}{\lambda}\ln\varphi(\mathbf{x}), \tag{3.4.3.124}$$
where β is a root of the algebraic (transcendental) equation $\beta - f(\beta\tau) = 0$, and the function $\varphi = \varphi(\mathbf{x})$ satisfies the Helmholtz equation
$$a\,\Delta\varphi + \lambda g(\beta\tau)\varphi = 0. \tag{3.4.3.125}$$

Equation 37. The nonlinear delay PDE
$$u_t = a\,\text{div}(e^{\lambda u}\nabla u) + f(e^{\lambda u} - e^{\lambda w}) + e^{-\lambda u}g(e^{\lambda u} - e^{\lambda w}) \tag{3.4.3.126}$$
admits a functional separable solution
$$u = \frac{1}{\lambda}\ln[At + \varphi(\mathbf{x})], \tag{3.4.3.127}$$
where A is a root of the algebraic (transcendental) equation $A - \lambda g(A\tau) = 0$, and $\varphi = \varphi(\mathbf{x})$ is a function described by Poisson's equation
$$a\Delta\varphi + \lambda f(A\tau) = 0. \tag{3.4.3.128}$$

Equation 38. Consider the nonlinear delay PDE
$$u_t = a\,\text{div}[g'(u)\nabla u] + b + \frac{1}{g'(u)}f\bigl(g(u)-g(w)\bigr), \tag{3.4.3.129}$$
where $g(u)$ and $f(z)$ are arbitrary functions, and the prime denotes a derivative with respect to u.

Equation (3.4.3.129) admits a functional separable solution in implicit form
$$g(u) = \varphi(\mathbf{x}) + \psi(t). \tag{3.4.3.130}$$
The function $\psi = \psi(t)$ is described by the delay ODE (3.4.3.28), and the function $\varphi = \varphi(\mathbf{x})$ satisfies Poisson's equation (3.4.3.90).

Equation 39. Consider the nonlinear delay PDE
$$u_t = a\,\text{div}[g'(u)\nabla u] + bg(u) + \frac{g(u)}{g'(u)}f\bigl(g(w)/g(u)\bigr), \tag{3.4.3.131}$$
where $g(u)$ and $f(z)$ are arbitrary functions, and the prime denotes a derivative with respect to u.

Equation (3.4.3.131) admits a functional separable solution in implicit form
$$g(u) = \varphi(\mathbf{x})\psi(t). \tag{3.4.3.132}$$
The function $\psi = \psi(t)$ is described by the delay ODE (3.4.3.8) and the function $\varphi = \varphi(\mathbf{x})$ satisfies the Helmholtz equation
$$a\Delta\varphi + b\varphi = 0. \tag{3.4.3.133}$$

Multi-dimensional equations involving three arbitrary functions.
Equation 40. Consider the nonlinear delay PDE

$$u_t = a \operatorname{div}[f'(u)\nabla u] + g\bigl(f(u) - f(w)\bigr) + \frac{1}{f'(u)} h\bigl(f(u) - f(w)\bigr), \qquad (3.4.3.134)$$

where $f(u)$, $g(z)$, and $h(z)$ are arbitrary functions.

Equation (3.4.3.134) admits a functional separable solution in implicit form

$$f(u) = At + \varphi(\mathbf{x}), \qquad (3.4.3.135)$$

where A is a root of the algebraic (transcendental) equation $A - h(A\tau) = 0$, and the function $\varphi = \varphi(\mathbf{x})$ is described by Poisson's equation

$$a\Delta\varphi + g(A\tau) = 0. \qquad (3.4.3.136)$$

Equation 41. Consider the nonlinear delay PDE

$$u_t = a \operatorname{div}[f'(u)\nabla u] + f(u) g\bigl(f(w)/f(u)\bigr) + \frac{f(u)}{f'(u)} h\bigl(f(w)/f(u)\bigr), \qquad (3.4.3.137)$$

where $f(u)$, $g(z)$, and $h(z)$ are arbitrary functions.

Equation (3.4.3.137) admits a functional separable solution in implicit form

$$f(u) = \varphi(\mathbf{x}) e^{\beta t}, \qquad (3.4.3.138)$$

where β is a root of the algebraic (transcendental) equation $\beta - h(e^{-\beta\tau}) = 0$, and $\varphi = \varphi(\mathbf{x})$ is a function described by the Helmholtz equation

$$a\Delta\varphi + g(e^{-\beta\tau})\varphi = 0. \qquad (3.4.3.139)$$

Nonlinear reaction-diffusion type equations with a variable delay of general form. The majority of the results presented above can be generalized to more complex nonlinear reaction-diffusion type equations with a variable delay $\tau = \tau(t)$, where $\tau(t)$ is an arbitrary function. Table 3.9 displays some of such equations, involving one or two arbitrary functions, and their exact solutions. In all the determining delay ODEs referred to in the last column of Table 3.9, one should set $\tau = \tau(t)$.

▶ **Example 3.16.** Consider the first equation in Table 3.9. By setting $\tau = \tau(t)$ in the determining equation (3.4.3.5) for $\psi(t)$, we obtain the delay ODE

$$\psi'(t) = b\psi^{k+1}(t) + \psi(t) f\bigl(\psi(t-\tau)/\psi(t)\bigr), \quad \tau = \tau(t).$$

◀

Some exact solutions to multi-dimensional nonlinear reaction-diffusion equations with a variable delay of general form $\tau = \tau(t)$ can be found in Table 3.10.

Table 3.9. Exact solutions to reaction-diffusion equations with a variable delay $u_t = [G(u)u_x]_x + F(u,w)$, where $w = u(x, t-\tau)$ and $\tau = \tau(t)$.

Reaction-diffusion equation	Form of exact solution	Determining equations
$u_t = a(u^k u_x)_x + uf(w/u)$	$u = \varphi(x)\psi(t)$	(3.4.3.4) (3.4.3.5)
$u_t = a(u^k u_x)_x + bu^{k+1} + uf(w/u)$	$u = \varphi(x)\psi(t)$, see (3.4.3.7), (3.4.3.10), (3.4.3.11)	(3.4.3.8)
$u_t = a(u^k u_x)_x + b + u^{-k} f(u^{k+1} - w^{k+1})$	$u = [\varphi(x) + \psi(t)]^{1/(k+1)}$, see (3.4.3.16)	(3.4.3.17)
$u_t = a(e^{\lambda u} u_x)_x + f(u-w)$	$u = \varphi(x) + \psi(t)$, see (3.4.3.23)	(3.4.3.24)
$u_t = a(e^{\lambda u} u_x)_x + be^{\lambda u} + f(u-w)$	$u = \varphi(x) + \psi(t)$, see (3.4.3.27), (3.4.3.29)	(3.4.3.28)
$u_t = a(e^{\lambda u} u_x)_x + b + e^{-\lambda u} f(e^{\lambda u} - e^{\lambda w})$	$u = \frac{1}{\lambda} \ln[\varphi(x) + \psi(t)]$, see (3.4.3.34)	(3.4.3.35)
$u_t = [(a \ln u + b)u_x]_x - cu \ln u + uf(w/u)$	$u = \exp(\pm\sqrt{c/a}\,x)\psi(t)$	(3.4.3.38)
$u_t = [uf'(u)u_x]_x + \frac{1}{f'(u)}[af(u) + bf(w) + c]$	$f(u) = \varphi(t)x + \psi(t)$	(3.4.3.41) (3.4.3.42)
$u_t = a[g'(u)u_x]_x + b + \frac{1}{g'(u)} f(g(u) - g(w))$	$g(u) = \varphi(x) + \psi(t)$, see (3.4.3.74)	(3.4.3.28)
$u_t = a[g'(u)u_x]_x + bg(u) + \frac{g(u)}{g'(u)} f(g(w)/g(w))$	$g(u) = \varphi(x)\psi(t)$, see (3.4.3.76), (3.4.3.77)	(3.4.3.8)

3.4.4. Exact Solutions to Nonlinear Delay Klein–Gordon Type Wave Equations

The method of functional constraints is also effective in constructing exact solutions to nonlinear wave equations with delay. These equations can formally be obtained from reaction-diffusion equations (3.4.1.1) and (3.4.3.1) by replacing the first derivative u_t on the left-hand side with the second derivative u_{tt} or a linear combination of these derivatives, $au_{tt} + bu_t$.

To illustrate the above, we will briefly describe some nonlinear delay Klein–Gordon type wave equations and their exact solutions obtained by the method of functional constraints; see [419, 429, 497] for other nonlinear delay hyperbolic equations and their exact solutions.

Table 3.10. Exact solutions to multi-dimensional reaction-diffusion equations with an arbitrary variable delay $u_t = \text{div}[G(u)\nabla u] + F(u,w)$, where $w = u(\mathbf{x}, t - \tau)$ and $\tau = \tau(t)$.

Reaction-diffusion equation	Form of exact solutions	Determining equations
$u_t = a\,\text{div}(u^k\nabla u) + uf(w/u) + bu^{k+1}$	$u = \psi(t)\varphi^{1/(k+1)}(\mathbf{x})$	(3.4.3.8) (3.4.3.88)
$u_t = a\,\text{div}(u^k\nabla u) + b + u^{-k}f(u^{k+1} - w^{k+1})$	$u = [\varphi(\mathbf{x}) + \psi(t)]^{1/(k+1)}$	(3.4.3.17) (3.4.3.93)
$u_t = a\,\text{div}(e^{\lambda u}\nabla u) + f(u - w) + be^{\lambda u}$	$u = \psi(t) + \frac{1}{\lambda}\ln\varphi(\mathbf{x})$	(3.4.3.28) (3.4.3.100)
$u_t = a\,\text{div}(e^{\lambda u}\nabla u) + b + e^{-\lambda u}f(e^{\lambda u} - e^{\lambda w})$	$u = \frac{1}{\lambda}\ln[\varphi(\mathbf{x}) + \psi(t)]$	(3.4.3.35) (3.4.3.103)
$u_t = a\,\text{div}[g'(u)\nabla u] + b + \frac{1}{g'(u)}f(g(u) - g(w))$	$g(u) = \varphi(\mathbf{x}) + \psi(t)$	(3.4.3.28) (3.4.3.90)
$u_t = a\,\text{div}[g'(u)\nabla u] + bg(u) + \frac{g(u)}{g'(u)}f(g(w)/g(u))$	$g(u) = \varphi(\mathbf{x})\varphi(t)$	(3.4.3.8) (3.4.3.133)

Equation 1. Consider the nonlinear Klein–Gordon type wave equation with constant delay that involves a single arbitrary function dependent on the ratio w/u:

$$u_{tt} = au_{xx} + uf(w/u), \quad w = u(x, t - \tau). \quad (3.4.4.1)$$

It differs from the reaction-diffusion equation (3.4.2.1) on the left-hand side, where u_t is replaced with u_{tt}.

1°. Equation (3.4.4.1) admits, just as equation (3.4.2.1), a multiplicative separable solution

$$u = \varphi(x)\psi(t), \quad (3.4.4.2)$$

where the functions $\varphi = \varphi(x)$ and $\psi = \psi(t)$ are described by the ODE and delay ODE

$$\varphi''_{xx} = k\varphi, \quad (3.4.4.3)$$

$$\psi''(t) = ak\psi(t) + \psi(t)f(\psi(t - \tau)/\psi(t)), \quad (3.4.4.4)$$

where k is an arbitrary constant.

The general solution of ODE (3.4.4.3) is defined by formulas (3.4.2.6). The delay ODE (3.4.4.4) admits exponential particular solutions of the form

$$\psi(t) = C_3 e^{\lambda t},$$

where C_3 is an arbitrary constant, and λ is a root of the transcendental equation

$$\lambda^2 = ak + f(e^{-\lambda \tau}).$$

2°. Equation (3.4.4.1) admits exact solutions of the form

$$u = e^{ct}v(x, t), \quad v(x, t) = v(x, t - \tau), \quad (3.4.4.5)$$

where c is an arbitrary constant and $v = v(x, t)$ is a τ-periodic function. Substituting (3.4.4.5) into equation (3.4.4.1) gives a linear problem for v:

$$v_{tt} + sv_t = av_{xx} + bv, \quad v(x, t) = v(x, t - \tau), \quad (3.4.4.6)$$

where $s = 2c$ and $b = f(e^{-c\tau}) - c^2$.

The general solution of problem (3.4.4.6), which we denote by $v = U_1(x, t; b, s)$, can be represented as [429]:

$$U_1(x, t; b, s) = \sum_{n=0}^{\infty} \exp(-\lambda_n x)\left[A_n \cos(\beta_n t - \gamma_n x) + B_n \sin(\beta_n t - \gamma_n x)\right]$$
$$+ \sum_{n=1}^{\infty} \exp(\lambda_n x)\left[C_n \cos(\beta_n t + \gamma_n x) + D_n \sin(\beta_n t + \gamma_n x)\right],$$

(3.4.4.7)

$$\beta_n = \frac{2\pi n}{\tau}, \quad \gamma_n = \left[\frac{\sqrt{(b + \beta_n^2)^2 + s^2 \beta_n^2} + b + \beta_n^2}{2a}\right]^{1/2}, \quad \lambda_n = \frac{s \beta_n}{2a \gamma_n},$$

(3.4.4.8)

where A_n, B_n, C_n, and D_n are arbitrary constants such that the series (3.4.4.7)–(3.4.4.8) and its derivatives $(U_1)_t$, $(U_1)_{tt}$, and $(U_1)_{xx}$ are convergent. In particular, the convergence occurs if $A_n = B_n = C_n = D_n = 0$ for $n > N$, where N is any positive integer.

Considering the above, we eventually arrive at the following exact solution to equation (3.4.4.1):

$$u = e^{ct} U_1(x, t; b, s), \quad b = f(e^{-c\tau}) - c^2, \quad s = 2c,$$

(3.4.4.9)

where c is an arbitrary constant and $U_1(x, t; b, s)$ is a τ-periodic function defined by formulas (3.4.4.7) and (3.4.4.8). If $c = 0$, solution (3.4.4.9) is a τ-periodic function.

3°. Equation (3.4.4.1) also admits exact solutions of the form

$$u = e^{ct} v(x, t), \quad v(x, t) = -v(x, t - \tau),$$

(3.4.4.10)

where c is an arbitrary constant and $v = v(x, t)$ is a τ-antiperiodic function. Substituting (3.4.4.10) into (3.4.4.1) yields a linear problem for determining v:

$$v_{tt} + s v_t = a v_{xx} + b v, \quad v(x, t) = -v(x, t - \tau),$$

(3.4.4.11)

where $s = 2c$ and $b = f(-e^{-c\tau}) - c^2$.

The general solution of problem (3.4.4.11), which we denote by $v = U_2(x, t; b, s)$, can be represented as the series [429]:

$$U_2(x, t; b, s) = \sum_{n=1}^{\infty} \exp(-\lambda_n x)\left[A_n \cos(\beta_n t - \gamma_n x) + B_n \sin(\beta_n t - \gamma_n x)\right]$$
$$+ \sum_{n=1}^{\infty} \exp(\lambda_n x)\left[C_n \cos(\beta_n t + \gamma_n x) + D_n \sin(\beta_n t + \gamma_n x)\right],$$

(3.4.4.12)

$$\beta_n = \frac{\pi(2n-1)}{\tau}, \quad \gamma_n = \left[\frac{\sqrt{(b + \beta_n^2)^2 + s^2 \beta_n^2} + b + \beta_n^2}{2a}\right]^{1/2}, \quad \lambda_n = \frac{s \beta_n}{2a \gamma_n},$$

(3.4.4.13)

where A_n, B_n, C_n, and D_n are arbitrary constants such that the series (3.4.4.12)–(3.4.4.13) and its derivatives $(U_1)_t$, $(U_1)_{tt}$, and $(U_1)_{xx}$ are convergent.

As a result, we arrive at the following exact solution to equation (3.4.4.1):

$$u = e^{ct}U_2(x,t;b,s), \quad b = f(-e^{-c\tau}) - c^2, \quad s = 2c, \quad (3.4.4.14)$$

where c is an arbitrary constant and $U_2(x,t;b,s)$ is a τ-antiperiodic function defined by formulas (3.4.4.12) and (3.4.4.13).

Equation 2. Consider the nonlinear Klein–Gordon type wave equation with constant delay that involves a single arbitrary function dependent on the difference $u-w$:

$$u_{tt} = au_{xx} + bu + f(u - w), \quad w = u(x, t - \tau). \quad (3.4.4.15)$$

It differs from the reaction-diffusion equation (3.4.2.19) on the left-hand side, where the first derivative is replaced with the second derivative.

Equation (3.4.4.15) admits, just as equation (3.4.2.19), an additive separable solution

$$u = \varphi(x) + \psi(t). \quad (3.4.4.16)$$

The functions $\varphi = \varphi(x)$ and $\psi = \psi(t)$ satisfy the ODE and delay ODE

$$a\varphi''_{xx} + b\varphi = k, \quad (3.4.4.17)$$

$$\psi''_{tt}(t) = b\psi(t) + k + f(\psi(t) - \psi(t - \tau)), \quad (3.4.4.18)$$

where k is an arbitrary constant.

Notably, the general solution of equation (3.4.4.17) is described by formulas (3.4.2.24), with $b \neq 0$ and $k = 0$, and (3.4.2.25), with $b = 0$ and $k \neq 0$.

Equation 3. Consider the nonlinear delay Klein–Gordon type wave equation

$$u_{tt} = au_{xx} + bu + f(u - kw), \quad k > 0. \quad (3.4.4.19)$$

1°. Table 3.11 presents a few relatively simple generalized separable solutions to equation (3.4.4.19).

2°. More complex exact solutions to nonlinear delay Klein–Gordon type wave equations (3.4.4.19) can be obtained with the following theorem.

Theorem (on nonlinear superposition of solutions) [429]. *Let $u_0(x,t)$ be a solution to the nonlinear equation (3.4.4.19) and let $v = U_1(x,t;b,s)$ be any τ-periodic solution to the linear telegraph equation (3.4.4.6), where b and s are free parameters. Then the function*

$$u = u_0(x,t) + e^{ct}U_1(x,t;b-c^2,2c), \quad c = \frac{1}{\tau}\ln k, \quad (3.4.4.20)$$

is also a solution to equation (3.4.4.19). The general form of the function $U_1(x,t;b,s)$ is defined by formulas (3.4.4.7) and (3.4.4.8).

Table 3.11. Generalized separable solutions to nonlinear delay Klein–Gordon type wave equation (3.4.4.19). Notation: A and B are arbitrary constants.

No.	Form of exact solution	Equation for determining function
1	$u = e^{ct}[A\cos(\lambda x) + B\sin(\lambda x)] + \psi(t)$, $c = \frac{1}{\tau}\ln k, \lambda = [(b-c^2)/a]^{1/2}, b > c^2$	$\psi''_{tt} = b\psi + f(\psi - k\bar\psi)$, $\bar\psi = \psi(t-\tau)$
2	$u = e^{ct}[A\exp(-\lambda x) + B\exp(\lambda x)] + \psi(t)$, $c = \frac{1}{\tau}\ln k, \lambda = [(c^2-b)/a]^{1/2}, c^2 > b$	$\psi''_{tt} = b\psi + f(\psi - k\bar\psi)$, $\bar\psi = \psi(t-\tau)$
3	$u = e^{ct}[A\cos(\lambda x) + B\sin(\lambda x)] + \varphi(x)$, $c = \frac{1}{\tau}\ln k, \lambda = [(b-c^2)/a]^{1/2}, b > c^2$	$a\varphi''_{xx} + b\varphi + f((1-k)\varphi) = 0$
4	$u = e^{ct}[A\exp(-\lambda x) + B\exp(\lambda x)] + \varphi(x)$, $c = \frac{1}{\tau}\ln k, \lambda = [(c^2-b)/a]^{1/2}, c^2 > b$	$a\varphi''_{xx} + b\varphi + f((1-k)\varphi) = 0$

Formula (3.4.4.20) allows one to obtain a broad class of exact solutions to nonlinear delay Klein–Gordon type wave equations by taking advantage of simpler particular solutions.

The simplest particular solutions of equation (3.4.4.19) are constants, $u_0 = \text{const}$, which are roots of the algebraic (transcendental) equation

$$bu_0 + f((1-k)u_0) = 0.$$

In the special case $k = 1$, there is only one constant solution: $u_0 = -f(0)/b$.

As the function $u_0(x,t)$ in (3.4.4.20), one can also take a spatially homogeneous solution, $u_0 = u_0(t)$, a stationary solution, $u_0 = u_0(x)$, or a traveling wave solution, $u_0 = \theta(\alpha x + \beta t)$, where α and β are arbitrary constants. The solutions displayed in Table 3.11 are also suitable.

Equation 4. Consider the nonlinear delay Klein–Gordon type wave equation

$$u_{tt} = a(u^k u_x)_x + uf(w/u), \quad w = u(x, t-\tau). \qquad (3.4.4.21)$$

It is easy to verify that it admits a multiplicative separable solution

$$u = \varphi(x)\psi(t).$$

The determining functions $\varphi = \varphi(x)$ and $\psi = \psi(t)$ satisfy the nonlinear ODE and delay ODE

$$a(\varphi^k \varphi'_x)'_x = b\varphi,$$
$$\psi''(t) = b\psi^{k+1}(t) + \psi(t)f(\psi(t-\tau)/\psi(t)),$$

where b is an arbitrary constant.

Equation 5. The nonlinear delay Klein–Gordon type wave equation

$$u_{tt} = a(e^{\lambda u} u_x)_x + f(u-w), \quad w = u(x, t-\tau), \qquad (3.4.4.22)$$

admits an additive separable solution of the form

$$u = \frac{1}{\lambda}\ln(Ax^2 + Bx + C) + \psi(t),$$

where A, B, and C are arbitrary constants; the function $\psi = \psi(t)$ satisfies the delay ODE

$$\psi''(t) = 2a(A/\lambda)e^{\lambda\psi(t)} + f\bigl(\psi(t) - \psi(t-\tau)\bigr).$$

Equation 6. The nonlinear delay PDE

$$u_{tt} = [(a\ln u + b)u_x]_x - cu\ln u + uf(w/u) \tag{3.4.4.23}$$

admits two multiplicative separable solutions

$$u = \exp(\pm\lambda x)\psi(t), \quad \lambda = \sqrt{c/a},$$

where the function $\psi(t)$ is described by the delay ODE

$$\psi''(t) = \lambda^2(a+b)\psi(t) + \psi(t)f\bigl(\psi(t-\tau)/\psi(t)\bigr).$$

Remark 3.34. *Many other exact solutions to nonlinear delay Klein–Gordon type wave equations and more complex nonlinear telegraph-type equations (hyperbolic reaction-diffusion equations with delay) can be found in [316, 419, 429, 497].*

4. Analytical Methods and Exact Solutions to Nonlinear Delay PDEs. Part II

4.1. Methods for Constructing Exact Solutions to Nonlinear Delay PDEs Using Solutions to Simpler Non-Delay PDEs

This section describes the methods developed in [414, 415] for constructing exact solutions of nonlinear delay PDEs, which rely on employing special solutions to simpler auxiliary PDEs without delay. We give examples of applying these methods to construct solutions of nonlinear reaction-diffusion and wave-type equations with delay dependent on arbitrary functions.

Remark 4.1. Methods for constructing solutions of complex nonlinear PDEs without delay using solutions of simpler PDEs can be found in [9]. Examples of applying these methods are also given there.

4.1.1. The First Method for Constructing Exact Solutions to Delay PDEs. General Description and Simple Examples

General description of the method. We will be dealing with nonlinear PDEs without delay in two independent variables of the form

$$\Phi(x, u, u_x, u_t, u_{xx}, u_{xt}, u_{tt}, \ldots; \beta_1, \ldots, \beta_m) = 0, \qquad (4.1.1.1)$$

where $u = u(x, t)$ is the unknown function and $\beta_1, \ldots, \beta_m$ are free parameters.

We will show that in certain cases, exact solutions to equations (4.1.1.1) can be used to construct exact solutions to more complicated nonlinear delay equations. The following statement is true.

Proposition 1. Let equation (4.1.1.1) admit a generalized traveling wave solution that can be represented in the implicit form

$$F(u) = kt + \theta(x), \qquad (4.1.1.2)$$

where k is a constant determined from an algebraic (transcendental) equation

$$P(k, \beta_1, \ldots, \beta_m) = 0, \qquad (4.1.1.3)$$

and the function $\theta = \theta(x)$ satisfies an ordinary differential equation

$$Q(x, \theta, \theta'_x, \theta''_{xx}, \ldots; \beta_1, \ldots, \beta_m) = 0. \tag{4.1.1.4}$$

Then the more complex nonlinear delay PDE obtained from (4.1.1.1) by formally replacing the free parameters $\beta_1, \ldots, \beta_m$ with functions $\varphi_1, \ldots, \varphi_m$ by the rule

$$\beta_i \implies \varphi_i\bigl(F(u) - F(w)\bigr), \quad w = u(x, t - \tau), \quad i = 1, \ldots, m, \tag{4.1.1.5}$$

where $\varphi_i(z)$ are some functions (defined quite arbitrarily), also admits exact solutions of the form (4.1.1.2). In this case, the constant k and function $\theta = \theta(x)$ are determined from equations (4.1.1.3) and (4.1.1.4) in which one should set

$$\beta_i = \varphi_i(k\tau), \quad i = 1, \ldots, m. \tag{4.1.1.6}$$

Proof. Using solutions (4.1.1.2), we get $F(w) = k(t - \tau) + \theta(x) = F(u) - k\tau$, or

$$F(u) - F(w) = k\tau = \text{const}. \tag{4.1.1.7}$$

Then any delay PDE obtained from (4.1.1.1) by replacing the parameters $\beta_1, \ldots, \beta_m$ with functions $\varphi_1, \ldots, \varphi_m$ by the rule (4.1.1.5) is equivalent, by virtue of solution (4.1.1.2) and relation (4.1.1.7), to equation (4.1.1.1) under condition (4.1.1.6).

Proposition 1 can be employed to construct exact solutions in explicit or implicit form for some delay PDEs.

Remark 4.2. In degenerate cases, equation (4.1.1.4) can be algebraic or transcendental (i.e., involving no derivatives of θ) or even define the function θ explicitly. In particular, any traveling wave solution can be written in the form (4.1.1.2) with $\theta(x) = \alpha x$, where α is an arbitrary constant. Furthermore, equation (4.1.1.3) can sometimes be absent at all, and then the constant k will serve as a free parameter.

Simple illustrative examples of the practical application of the method.

▶ **Example 4.1.** To illustrate the practical use of Proposition 1, we will look at the linear diffusion equation without delay

$$u_t = u_{xx} + a, \tag{4.1.1.8}$$

where $\beta = a$ is a free parameter.

Equation (4.1.1.8) admits a simple additive separable solution that can be written explicitly as

$$u = kt + \lambda x^2 + C_1 x + C_2, \tag{4.1.1.9}$$

where C_1, C_2, and λ are arbitrary constants, and the parameter k is expressed in terms of a and λ as follows:

$$k = 2\lambda + a. \tag{4.1.1.10}$$

Solution (4.1.1.9) is a special case of solution (4.1.1.2) with $F(u) = u$ and $\theta(x) = \lambda x^2 + C_1 x + C_2$. Substituting the function $F(u) = u$ into (4.1.1.7) gives $F(u) - F(w) = u - w = k\tau$. Applying Proposition 1 to equation (4.1.1.8), we replace the parameter a with $\varphi(u - w)$, where $\varphi(z)$ is an arbitrary function. As a result, we arrive at the nonlinear delay PDE

$$u_t = u_{xx} + \varphi(u - w).$$

It admits the exact solution (4.1.1.9) where the constant k is determined from the algebraic (transcendental) equation

$$k = 2\lambda + \varphi(k\tau),$$

which is obtained from (4.1.1.10) with $a = \varphi(k\tau)$. ◀

▶ **Example 4.2.** Consider the nonlinear reaction-diffusion equation without delay

$$u_t = (u^n u_x)_x + au^{1-n}, \qquad (4.1.1.11)$$

where $\beta = a$ is a free parameter.

Equation (4.1.1.11) admits the traveling wave solution

$$u = (kt + \lambda x + C_1)^{1/n}, \qquad (4.1.1.12)$$

where C_1 and λ are arbitrary constants, and the parameter k is expressed via a, λ, and n as follows:

$$k = an + \frac{\lambda^2}{n}. \qquad (4.1.1.13)$$

Solution (4.1.1.12) is a special case of solution (4.1.1.2) with $F(u) = u^n$. Substituting this function into (4.1.1.7) gives $F(u) - F(w) = u^n - w^n = k\tau$. Applying Proposition 1 to equation (4.1.1.11), we replace the parameter a with $\varphi(u^n - w^n)$, where $\varphi(z)$ is an arbitrary function. As a result, we arrive at the nonlinear delay equation

$$u_t = (u^n u_x)_x + u^{1-n}\varphi(u^n - w^n).$$

It admits an exact solution of the form (4.1.1.12) where the constant k is determined from the algebraic (transcendental) equation

$$k = n\varphi(k\tau) + \frac{\lambda^2}{n},$$

which derives from (4.1.1.13) with $a = \varphi(k\tau)$. ◀

4.1.2. Using the First Method for Constructing Exact Solutions to Nonlinear Delay PDEs

Equation 1. The nonlinear reaction-diffusion type equation without delay

$$u_t = [a(x)f(u)u_x]_x + \sigma + \frac{\beta}{f(u)}, \qquad (4.1.2.1)$$

which involves two arbitrary functions, $a(x)$ and $f(u)$, and two free parameters, σ and β, admits a generalized traveling wave solution in implicit form [396]:

$$\int f(u)\,du = kt - \sigma \int \frac{x\,dx}{a(x)} + C_1 \int \frac{dx}{a(x)} + C_2, \qquad (4.1.2.2)$$

where C_1 and C_2 are arbitrary constants, and the constant k is linked to the parameter β by the linear relation

$$k = \beta. \qquad (4.1.2.3)$$

Solution (4.1.2.2) is a solution of the form (4.1.1.2) with $F(u) = \int f(u)\,du$.

Applying Proposition 1 to equation (4.1.2.1), we replace the parameters σ and β with arbitrary functions $\varphi(F(u) - F(w))$ and $\psi(F(u) - F(w))$, respectively. As a result, we arrive at the new nonlinear reaction-diffusion type equation with delay

$$u_t = [a(x)f(u)u_x]_x + \varphi(F(u) - F(w)) + \frac{1}{f(u)}\psi(F(u) - F(w)),$$
$$F(u) = \int f(u)\,du. \qquad (4.1.2.4)$$

It depends on four arbitrary functions and has the exact solution

$$\int f(u)\,du = kt - \varphi(k\tau) \int \frac{x\,dx}{a(x)} + C_1 \int \frac{dx}{a(x)} + C_2, \qquad (4.1.2.5)$$

where the constant k is determined from the algebraic (transcendental) equation

$$k = \psi(k\tau), \qquad (4.1.2.6)$$

which derives from (4.1.2.3) with $\beta = \psi(k\tau)$.

▶ **Example 4.3.** By setting

$$F(u) = u^{n+1}, \quad f(u) = (n+1)u^n, \quad a(x) = a_0/(n+1) = \text{const}, \quad \psi(z) = (n+1)\bar{\psi}(z)$$

in (4.1.2.4)–(4.1.2.6), we obtain the nonlinear delay PDE

$$u_t = a_0(u^n u_x)_x + \varphi(u^{n+1} - w^{n+1}) + u^{-n}\bar{\psi}(u^{n+1} - w^{n+1}).$$

It depends on two arbitrary functions, $\varphi(z)$ and $\bar{\psi}(z)$, and has the exact solution in explicit form

$$u = \left[kt - \frac{n+1}{2a_0}\varphi(k\tau)x^2 + C_1 x + C_2\right]^{\frac{1}{n+1}}, \qquad (4.1.2.7)$$

in which the constant k is a root of the algebraic (transcendental) equation $k = (n+1)\bar{\psi}(k\tau)$. ◀

▶ **Example 4.4.** Setting

$$F(u) = e^{\lambda u}, \quad f(u) = \lambda e^{\lambda u}, \quad a(x) = a_0/\lambda = \text{const}, \quad \psi(z) = \lambda \bar{\psi}(z)$$

in (4.1.2.4)–(4.1.2.6) yields the nonlinear delay PDE

$$u_t = a_0(e^{\lambda u} u_x)_x + \varphi(e^{\lambda u} - e^{\lambda w}) + e^{-\lambda u}\bar\psi(e^{\lambda u} - e^{\lambda w}),$$

which has the exact solution

$$u = \frac{1}{\lambda}\ln\left[kt - \frac{\lambda}{2a_0}\varphi(k\tau)x^2 + C_1 x + C_2\right], \qquad (4.1.2.8)$$

where k is a root of the algebraic (transcendental) equation $k = \lambda\bar\psi(k\tau)$. ◂

Equation 2. The nonlinear delay PDE

$$u_t = [a(x)f(u)u_x]_x + b(x)\varphi\big(F(u) - F(w)\big) + \frac{1}{f(u)}\psi\big(F(u) - F(w)\big),$$

which is more general than (4.1.2.4) and dependent on five arbitrary functions $a(x)$, $b(x)$, $f(u)$, $\varphi(z)$, and $\psi(z)$, has the exact solution

$$\int f(u)\,du = kt - \varphi(k\tau)\int \frac{1}{a(x)}\left(\int b(x)\,dx\right)dx + C_1 \int \frac{dx}{a(x)} + C_2,$$

where the constant k is a root of the algebraic (transcendental) equations (4.1.2.6).

Omitting details, we will list below a few more nonlinear PDEs without delay that admit exact solutions of the form (4.1.1.2) together with more complex nonlinear delay PDEs they generate with their exact solutions.

Equation 3. Consider the nonlinear PDE without delay

$$u_t = [a(x)f(u)u_x]_x - \mu a(x)f(u)u_x + \sigma + \frac{\beta}{f(u)},$$

which admits the exact solution [400]:

$$\int f(u)\,du = kt + \frac{\sigma}{\mu}\int \frac{dx}{a(x)} + C_1 \int \frac{e^{\mu x}}{a(x)}\,dx + C_2, \qquad (4.1.2.9)$$

where the constant k is linked to the parameter β through the linear relation (4.1.2.3).

By reasoning as for equation (4.1.2.1), we obtain the nonlinear delay PDE

$$u_t = [a(x)f(u)u_x]_x - \mu a(x)f(u)u_x + \varphi\big(F(u) - F(w)\big) + \frac{1}{f(u)}\psi\big(F(u) - F(w)\big),$$

whose exact solution is given by formula (4.1.2.9) with $\sigma = \varphi(k\tau)$, and the constant k is a root of the algebraic (transcendental) equation (4.1.2.6).

Equation 4. The nonlinear Klein–Gordon type wave equation without delay

$$u_{tt} = [a(x)f(u)u_x]_x + \sigma - \beta \frac{f'_u(u)}{f^3(u)},$$

which involves two arbitrary functions, $a(x)$ and $f(u)$, and two free parameters, β and σ, admits the generalized traveling wave solution [399]:

$$\int f(u)\,du = kt - \sigma \int \frac{x\,dx}{a(x)} + C_1 \int \frac{dx}{a(x)} + C_2, \qquad (4.1.2.10)$$

where C_1 and C_2 are arbitrary constants, and the constant k is linked to the parameter β by

$$k^2 = \beta.$$

For $\beta > 0$, we have two real solutions: $k = \pm\sqrt{\beta}$.

Solution (4.1.2.10) is a solution of the form (4.1.1.2) with $F(u) = \int f(u)\,du$.

By reasoning as for equation (4.1.2.1), we arrive at the nonlinear delay Klein–Gordon type wave equation

$$u_{tt} = [a(x)f(u)u_x]_x + \varphi\bigl(F(u) - F(w)\bigr) - \frac{f'_u(u)}{f^3(u)}\psi\bigl(F(u) - F(w)\bigr), \qquad (4.1.2.11)$$

whose exact solution can be written in the implicit form (4.1.2.10) with $\sigma = \varphi(k\tau)$; the constant k is determined from the algebraic (transcendental) equation $k^2 = \psi(k\tau)$.

▶ **Example 4.5.** By setting

$$F(u) = u^{n+1}, \quad f(u) = (n+1)u^n, \quad a(x) = \frac{a_0}{n+1} = \text{const}, \quad \psi(z) = n^{-1}(n+1)^2 \bar\psi(z)$$

in (4.1.2.11), we arrive at the nonlinear delay PDE

$$u_{tt} = a_0(u^n u_x)_x + \varphi(u^{n+1} - w^{n+1}) - u^{-2n-1}\bar\psi(u^{n+1} - w^{n+1}),$$

which admits the exact solution (4.1.2.7), with the constant k determined from the algebraic (transcendental) equation $nk^2 = (n+1)^2 \bar\psi(k\tau)$. ◀

▶ **Example 4.6.** By setting

$$F(u) = e^{\lambda u}, \quad f(u) = \lambda e^{\lambda u}, \quad a(x) = a_0/\lambda = \text{const}, \quad \psi(z) = \lambda \bar\psi(z)$$

in (4.1.2.11), we get the nonlinear delay PDE

$$u_{tt} = a_0(e^{\lambda u} u_x)_x + \varphi(e^{\lambda u} - e^{\lambda w}) - e^{-2\lambda u}\bar\psi(e^{\lambda u} - e^{\lambda w}),$$

which admits the exact solution (4.1.2.8), with the constant k determined from the algebraic (transcendental) equation $k^2 = \lambda\bar\psi(k\tau)$. ◀

Equation 5. The nonlinear Klein–Gordon type wave equation without delay

$$u_{tt} = [a(x)u_x]_x + \beta \frac{a'_x(x)}{\sqrt{a(x)}} f(u), \qquad (4.1.2.12)$$

involving two arbitrary functions, $a(x)$ and $f(u)$, admits two exact solutions [399]:

$$\int \frac{du}{f(u)} = \pm 2kt - 2k \int \frac{dx}{\sqrt{a(x)}} + C, \qquad (4.1.2.13)$$

where the constant k is linked to the parameter β by the linear relation (4.1.2.3). Solution (4.1.2.13) is a solution of the form (4.1.1.2) with $F(u) = \int [du/f(u)]$.

Equation (4.1.2.12) generates the more complex delay PDE

$$u_{tt} = [a(x)u_x]_x + \frac{a'_x(x)}{\sqrt{a(x)}} f(u)\psi\big(F(u) - F(w)\big), \quad F(u) = \int \frac{du}{f(u)},$$

which has exact solutions defined by formula (4.1.2.13) with the constant k found from the algebraic (transcendental) equation $k = \psi(k\tau)$.

4.1.3. The Second Method for Constructing Exact Solutions to Delay PDEs. General Description and Simple Examples

General description of the method. The second method for constructing exact solutions to nonlinear delay PDEs relies on the following statement.

Proposition 2. Let equation (4.1.1.1) have a functional separable solution of the special form

$$F(u) = e^{kt}\theta(x), \qquad (4.1.3.1)$$

where the constant k is determined from the algebraic (transcendental) equation (4.1.1.3), and the function $\theta = \theta(x)$ satisfies ODE (4.1.1.4). Then the more complex nonlinear delay PDE obtained from (4.1.1.1) by formally replacing the free parameters $\beta_1, \ldots, \beta_m$ with functions by the rule

$$\beta_i \implies \varphi_i\big(F(w)/F(u)\big), \quad w = u(x, t - \tau), \quad i = 1, \ldots, m, \qquad (4.1.3.2)$$

where $\varphi_i(z)$ are some (quite arbitrary) given functions, also admits an exact solution of the form (4.1.3.1). The constant k and function $\theta = \theta(x)$ are determined from equations (4.1.1.3) and (4.1.1.4), in which one should set

$$\beta_i = \varphi_i(e^{-k\tau}), \quad i = 1, \ldots, m. \qquad (4.1.3.3)$$

Proof. With solutions of the form (4.1.3.1), we have $F(w) = e^{k(t-\tau)}\theta(x) = e^{-k\tau}F(u)$, or

$$F(w)/F(u) = e^{-k\tau} = \text{const}. \qquad (4.1.3.4)$$

It follows that any delay equation obtained from (4.1.1.1) by replacing the constants $\beta_1, \ldots, \beta_m$ with functions according to (4.1.3.2) is equivalent, by virtue of solutions (4.1.3.1) and relation (4.1.3.4), to equation (4.1.1.1) under condition (4.1.3.3).

Proposition 2 can be employed to construct exact solutions, in explicit or implicit form, to some delay PDEs.

Remark 4.3. Proposition 2 reduces to Proposition 1. To this end, while assuming that $F(u) > 0$, one should take the logarithm of solution (4.1.3.1) and then rename $\ln F(u) \implies F(u)$ and $\ln \theta \implies \theta$; the case $F(u) < 0$ is treated likewise. However, in practice, one often deals directly with a solution representation in the form (4.1.3.1), which is simpler and more convenient to use.

Simple illustrative examples of the practical application of the method.

▶ **Example 4.7.** Consider the linear diffusion equation

$$u_t = u_{xx} + au, \qquad (4.1.3.5)$$

where $\beta = a$ is a free parameter.

Equation (4.1.3.5) admits a separable solution

$$u = e^{kt}\theta(x), \qquad (4.1.3.6)$$

where k is an arbitrary constant, and the function $\theta = \theta(x)$ satisfies the second-order linear ODE with constant coefficients

$$\theta''_{xx} + (a - k)\theta = 0. \qquad (4.1.3.7)$$

Solution (4.1.3.6) is a special case of solution (4.1.3.1) with $F(u) = u$. Substituting this function into (4.1.3.4) gives $F(w)/F(u) = w/u = e^{-k\tau}$. Using Proposition 2, we replace in equation (4.1.3.5) the parameter a with $\varphi(w/u)$, where $\varphi(z)$ is an arbitrary function. As a result, we arrive at the nonlinear delay PDE

$$u_t = u_{xx} + u\varphi(w/u),$$

which admits the exact solution (4.1.3.6), where k is an arbitrary constant, and the function $\theta = \theta(x)$ satisfies the linear ODE

$$\theta''_{xx} + [\varphi(e^{-k\tau}) - k]\theta = 0.$$

This equation is obtained by substituting the constant $a = \varphi(e^{-k\tau})$ into (4.1.3.7) and is easy to integrate. ◀

▶ **Example 4.8.** Consider the reaction-diffusion equation with a quadratic nonlinearity

$$u_t = (uu_x)_x + au + bu^2, \qquad (4.1.3.8)$$

where a and b are free parameters.

For $b > 0$, equation (4.1.3.8) admits a separable solution in explicit form

$$u = e^{kt}\sqrt{|C_1\cos(\beta x) + C_2\sin(\beta x)|}, \quad \beta = \sqrt{2b}, \qquad (4.1.3.9)$$

which is a special case of solution (4.1.3.6), where the parameter k satisfies the linear relation

$$k = a. \qquad (4.1.3.10)$$

As in example 4.7, we have $F(u) = u$ and, hence, $F(w)/F(u) = w/u = e^{-k\tau}$. Using Proposition 2, we replace in equation (4.1.3.8) the parameters a and b with

$\varphi(w/u)$ and $\psi(w/u)$, respectively, where $\varphi(z)$ and $\psi(z)$ are arbitrary functions. As a result, we arrive at the nonlinear delay PDE

$$u_t = (uu_x)_x + u\varphi(w/u) + u^2\psi(w/u),$$

which admits an exact solution of the form (4.1.3.9), where k is determined from the algebraic (transcendental) equation

$$k = \varphi(e^{-k\tau}),$$

obtained from (4.1.3.10) by substituting $a = \varphi(e^{-k\tau})$. ◀

4.1.4. Employing the Second Method to Construct Exact Solutions to Nonlinear Delay PDEs

Equation 1. Consider the nonlinear reaction-diffusion type equation without delay

$$u_t = [f(u)u_x]_x + \left[b + \frac{c}{f(u)}\right]F(u), \quad F(u) = \int f(u)\,du, \qquad (4.1.4.1)$$

dependent on an arbitrary function, $f(u)$, and two free parameters, b and c. This equation admits an exact solution representable in implicit form [422]:

$$\int f(u)\,du = e^{kt}\theta(x), \qquad (4.1.4.2)$$

where

$$k = c, \qquad (4.1.4.3)$$

and the function $\theta = \theta(x)$ is determined from the second-order linear ODE

$$\theta''_{xx} + b\theta = 0. \qquad (4.1.4.4)$$

Solution (4.1.4.2) is a solution of the form (4.1.3.1) with $F(u) = \int f(u)\,du$.

Using Proposition 2, we replace in equation (4.1.4.1) the parameters b and c with arbitrary functions $\varphi(F(w)/F(u))$ and $\psi(F(w)/F(u))$, respectively. As a result, we arrive at the more complex nonlinear reaction-diffusion type equation with delay

$$u_t = [f(u)u_x]_x + F(u)\left[\varphi\left(\frac{F(w)}{F(u)}\right) + \frac{1}{f(u)}\psi\left(\frac{F(w)}{F(u)}\right)\right], \quad F(u) = \int f(u)\,du, \qquad (4.1.4.5)$$

which involves three arbitrary functions and has exact solutions of the form (4.1.4.2). The constant k in (4.1.4.2) is a root of the algebraic (transcendental) equation

$$k = \psi(e^{-k\tau}), \qquad (4.1.4.6)$$

obtained from (4.1.4.3) by substituting $c = \psi(e^{-k\tau})$. The function $\theta = \theta(x)$ is determined by the second-order linear ODE (4.1.4.4) with $b = \varphi(e^{-k\tau})$.

Equation 2. The nonlinear delay PDE

$$u_t = [a(x)f(u)u_x]_x + b(x)F(u)\varphi\left(\frac{F(w)}{F(u)}\right) + \frac{F(u)}{f(u)}\psi\left(\frac{F(w)}{F(u)}\right),$$

more general than (4.1.4.5) and dependent on five arbitrary functions, $a(x)$, $b(x)$, $f(u)$, $\varphi(z)$, and $\psi(z)$, admits an exact solution of the form (4.1.4.2), where the constant k is a root of the algebraic (transcendental) equation (4.1.4.6), and the function $\theta = \theta(x)$ satisfies the second-order linear ODE

$$[a(x)\theta'_x]'_x + \varphi(e^{-k\tau})b(x)\theta = 0.$$

Equation 3. The nonlinear PDE without delay

$$u_t = [f(u)u_x]_x + \mu f(u)u_x + \frac{\lambda}{f(u)}F(u), \quad F(u) = \int f(u)\,du, \qquad (4.1.4.7)$$

dependent on an arbitrary function, $f(u)$, and two free parameters, μ and λ, admits the exact solution [400]:

$$\int f(u)\,du = e^{kt}(C_1 + C_2 e^{-\mu x}), \qquad (4.1.4.8)$$

where $k = \lambda$.

Using Proposition 2, one can see, for example, that the nonlinear delay PDE

$$u_t = [f(u)u_x]_x + \mu f(u)u_x + \frac{F(u)}{f(u)}\varphi\left(\frac{F(w)}{F(u)}\right), \quad F(u) = \int f(u)\,du, \quad (4.1.4.9)$$

admits an exact solution of the form (4.1.4.8), where the constant k is determined from the algebraic (transcendental) equation $k = \varphi(e^{-k\tau})$.

▶ **Example 4.9.** By setting $\varphi(z) = \lambda z$ in (4.1.4.9), we get the equation

$$u_t = [f(u)u_x]_x + \mu f(u)u_x + \frac{\lambda}{f(u)}F(w),$$

which follows from equation (4.1.4.7) if we formally rename $F(u) \implies F(w)$. ◀

Equation 4. The nonlinear PDE without delay

$$u_t = [f(u)u_x]_x - 2\alpha f(u)u_x + \left[\alpha^2 + \frac{\beta}{f(u)}\right]F(u), \quad F(u) = \int f(u)\,du, \quad (4.1.4.10)$$

dependent on an arbitrary function, $f(u)$, and two free parameters, α and β, admits the exact solution in implicit form [400]:

$$\int f(u)\,du = e^{kt+\alpha x}(C_1 x + C_2), \qquad (4.1.4.11)$$

where C_1 and C_2 are arbitrary constants and $k = \beta$.

The nonlinear delay PDE

$$u_t = [f(u)u_x]_x - 2\alpha f(u)u_x + \alpha^2 F(u) + \frac{F(u)}{f(u)}\varphi\left(\frac{F(w)}{F(u)}\right),$$

which is more complex than equation (4.1.4.10), also admits the exact solution (4.1.4.11), where the constant k is determined from the algebraic (transcendental) equation $k = \varphi(e^{-k\tau})$.

Equation 5. Consider the nonlinear non-delay wave-type PDE with variable coefficients

$$u_{tt} = [f(x)u^n u_x]_x + g(x,a)u^{n+1} + bu, \tag{4.1.4.12}$$

where $f(x)$ and $g(x,a)$ are arbitrary functions, while a and b are free parameters.

Equation (4.1.4.12) admits a separable solution of the form (4.1.3.6), where the parameter k satisfies the quadratic relation

$$k^2 = b, \tag{4.1.4.13}$$

and the function $\theta = \theta(x)$ is described by the nonlinear second-order ODE

$$[f(x)\theta^n \theta'_x]'_x + g(x,a)\theta^{n+1} = 0. \tag{4.1.4.14}$$

In this case, we have $F(u) = u$ and $F(w)/F(u) = w/u = e^{-k\tau}$. Using Proposition 2, we replace in equation (4.1.4.12) the parameters a and b with arbitrary functions $\varphi(w/u)$ and $\psi(w/u)$, respectively. As a result, we arrive at the nonlinear delay PDE

$$u_{tt} = [f(x)u^n u_x]_x + u^{n+1}g(x,\varphi(w/u)) + u\psi(w/u),$$

which has an exact solution of the form (4.1.3.6), where k is determined from the algebraic (transcendental) equation

$$k^2 = \psi(e^{-k\tau}),$$

obtained from (4.1.4.13) by substituting $b = \psi(e^{-k\tau})$. The function $\theta = \theta(x)$ satisfies the nonlinear ODE

$$[f(x)\theta^n \theta'_x]'_x + g(x,a)\theta^{n+1} = 0, \quad a = \varphi(e^{-k\tau}).$$

With the change of variable $\xi(x) = \theta^{n+1}(x)$, this equation can be reduced to a linear second-order ODE.

Remark 4.4. *The article [415] lists several exact solutions to more complicated nonlinear PDEs and systems of delay PDEs.*

4.2. Systems of Nonlinear Delay PDEs. Generating Equations Method

4.2.1. General Description of the Method and Application Examples

Preliminary remarks. Developed in [436], the method of generating equations allows one to find solutions to nonlinear systems of delay PDEs by employing simpler exact solutions of isolated delay PDEs. The method relies on the method of functional constraints (see Section 3.4).

Description of the method. We consider two different independent (isolated) nonlinear PDEs with constant delay

$$u_t = F\big(u, \bar{u}, u_x, u_{xx}, f(z_1)\big), \quad \bar{u} = u(x, t-\tau), \quad z_1 = z_1(u, \bar{u}); \quad (4.2.1.1)$$

$$v_t = G\big(v, \bar{v}, v_x, v_{xx}, g(z_2)\big), \quad \bar{v} = v(x, t-\tau), \quad z_2 = z_2(v, \bar{v}), \quad (4.2.1.2)$$

involving arbitrary functions of a single argument, $f(z_1)$ and $g(z_2)$, with $\tau > 0$.

We will assume that equations (4.2.1.1) and (4.2.1.2) have generalized separable solutions of the form

$$u = \sum_{n=1}^{N_1} \varphi_{1n}(x)\psi_{1n}(t), \quad v = \sum_{n=1}^{N_2} \varphi_{2n}(x)\psi_{2n}(t) \quad (4.2.1.3)$$

and both these solutions satisfy any functional constraints of the same type (see Section 3.4). For example, either of the two variants is possible:

$$z_1(u,\bar{u}) = p_1(x), \quad z_2(v,\bar{v}) = p_2(x) \quad \text{(functional constraints of the first kind)};$$
$$z_1(u,\bar{u}) = q_1(t), \quad z_2(v,\bar{v}) = q_2(t) \quad \text{(functional constraints of the second kind)}.$$
$$(4.2.1.4)$$

The generating equations method relies on the following principle.

The principle of constructing delay systems and their exact solutions. Suppose that the isolated nonlinear delay PDEs (4.2.1.1) and (4.2.1.2) admit the generalized separable solutions (4.2.1.3) each satisfying two functional constraints of the same form (4.2.1.4). Then the more complicated nonlinear system of two coupled nonlinear delay PDEs

$$u_t = F\big(u, \bar{u}, u_x, u_{xx}, f(z_1, z_2)\big), \quad \bar{u} = u(x, t-\tau), \quad z_1 = z_1(u, \bar{u}); \quad (4.2.1.5)$$

$$v_t = G\big(v, \bar{v}, v_x, v_{xx}, g(z_1, z_2)\big), \quad \bar{v} = v(x, t-\tau), \quad z_2 = z_2(v, \bar{v}), \quad (4.2.1.6)$$

where $f(z_1, z_2)$ and $g(z_1, z_2)$ are arbitrary functions of two arguments, admits exact solutions of the form (4.2.1.3).

In what follows, we will call the initial independent equations (4.2.1.1) and (4.2.1.2) *generating equations*.

Notably, in simple cases, the generating equations (4.2.1.1) and (4.2.1.2) can be the same up to obvious renaming of the determining parameters and arbitrary functions.

4.2. Systems of Nonlinear Delay PDEs. Generating Equations Method

Remark 4.5. *Rather than nonlinear delay PDEs admitting generalized separable solutions (4.2.1.3), one can also take, as generating equations, nonlinear delay PDEs that have more complicated, functional separable solutions [436].*

Illustrative examples. Below we describe in more detail the procedure of employing the generating equations method with two specific examples.

▶ **Example 4.10.** To obtain both generating equations, we use one reaction-diffusion type equation with delay (3.4.2.1). After obvious renaming, we will rewrite it as two similar independent equations:

$$u_t = a_1 u_{xx} + u f(\bar{u}/u), \quad \bar{u} = u(x, t - \tau); \\ v_t = a_2 v_{xx} + v g(\bar{v}/v), \quad \bar{v} = v(x, t - \tau). \tag{4.2.1.7}$$

These involve two arbitrary functions of a single argument, $f(z_1)$ and $g(z_2)$, where $z_1 = \bar{u}/u$ and $z_2 = \bar{v}/v$.

$1°$. Equations (4.2.1.7) coincide, up to obvious renaming, with equation (3.4.2.1) and, therefore, admit, for example, a multiplicative separable solution of the form:*

$$u = \cos(\beta x)\psi_1(t), \quad v = \cos(\beta x)\psi_2(t), \tag{4.2.1.8}$$

where β is an arbitrary constant. Functions (4.2.1.8) satisfy the functional constraints of the second kind (4.2.1.4) with $z_1 = \bar{u}/u = \psi_1(t-\tau)/\psi_1(t)$ and $z_2 = \bar{v}/v = \psi_2(t-\tau)/\psi_2(t)$.

In this case, the associated nonlinear system of delay reaction-diffusion equations (4.2.1.5)–(4.2.1.6) is written as

$$u_t = a_1 u_{xx} + u f(\bar{u}/u, \bar{v}/v), \\ v_t = a_2 v_{xx} + v g(\bar{u}/u, \bar{v}/v), \tag{4.2.1.9}$$

where $f(z_1, z_2)$ and $g(z_1, z_2)$ are arbitrary functions of two arguments.

Following the generating equations method, we look for exact solutions to the system of delay PDEs (4.2.1.9) in the form (4.2.1.8). On substituting (4.2.1.8) into (4.2.1.9), we obtain the following system of delay ODEs for $\psi_1(t)$ and $\psi_2(t)$:

$$\psi_1'(t) = -a_1 \beta^2 \psi_1(t) + \psi_1(t) f\big(\psi_1(t-\tau)/\psi_1(t), \psi_2(t-\tau)/\psi_2(t)\big), \\ \psi_2'(t) = -a_2 \beta^2 \psi_2(t) + \psi_2(t) g\big(\psi_1(t-\tau)/\psi_1(t), \psi_2(t-\tau)/\psi_2(t)\big). \tag{4.2.1.10}$$

$2°$. The generating equations (4.2.1.7) also admit other multiplicative separable solutions (see solutions to equation (3.4.2.1)):

$$u = \sinh(\gamma x)\psi_1(t), \quad v = \sinh(\gamma x)\psi_2(t), \tag{4.2.1.11}$$

where γ is an arbitrary constant. Solution (4.2.1.11) satisfies the same functional constraints of the second kind as solution (4.2.1.8). Therefore, the nonlinear system of delay reaction-diffusion equations (4.2.1.9) admits exact solutions of the form

---

*Recall that additive and multiplicative separable solutions are the simplest generalized separable solutions.

(4.2.1.11). In this case, the functions $\psi_1(t)$ and $\psi_2(t)$ are described by the system of delay ODEs that derives from (4.2.1.10) if one formally replaces β^2 with $-\gamma^2$.

3°. The generating equations (4.2.1.7) admit two classes of different exact solutions, (4.2.1.8) and (4.2.1.11), with both classes satisfying functional constraints of the second kind. Therefore, the nonlinear system of delay reaction-diffusion equations (4.2.1.9) also admits exact solutions of mixed type:

$$u = \cos(\beta x)\psi_1(t), \quad v = \sinh(\gamma x)\psi_2(t). \tag{4.2.1.12}$$

The functions $\psi_1(t)$ and $\psi_2(t)$ are described by the system of delay ODEs consisting of the first equations of system (4.2.1.10) and a modified second equation of the same system obtained by formally replacing β^2 with $-\gamma^2$.

4°. The system of delay reaction-diffusion equations (4.2.1.9) admits a more general multiplicative separable solution

$$u = \varphi_1(x)\psi_1(t), \quad v = \varphi_2(x)\psi_2(t), \tag{4.2.1.13}$$

which includes solutions (4.2.1.8), (4.2.1.11), and (4.2.1.12). ◀

Remark 4.6. *The nonlinear system of delay reaction-diffusion equations*

$$u_t = a_1 u_{xx} + u f(u/v, \bar{u}/u, \bar{v}/v),$$
$$v_t = a_2 v_{xx} + v g(u/v, \bar{u}/u, \bar{v}/v),$$

which is more general than system (4.2.1.9) and in which $f(z_1, z_2, z_3)$ *and* $g(z_1, z_2, z_3)$ *are arbitrary functions of three arguments, admits exact solutions of the form (4.2.1.8) and (4.2.1.11) as well as solution (4.2.1.13) with* $\varphi_1(x) = \varphi_2(x)$.

▶ **Example 4.11.** To obtain two generating equations, we take one reaction-diffusion type equation with delay (3.4.2.19). After obvious renaming, we rewrite it as two similar independent equations

$$u_t = a_1 u_{xx} + b_1 u + f(u - \bar{u}), \quad \bar{u} = u(x, t - \tau);$$
$$v_t = a_2 v_{xx} + b_2 v + g(v - \bar{v}), \quad \bar{v} = v(x, t - \tau), \tag{4.2.1.14}$$

which involve two arbitrary functions of a single argument, $f(z_1)$ and $g(z_2)$, where $z_1 = u - \bar{u}$ and $z_2 = v - \bar{v}$.

1°. Equations (4.2.1.14) coincide, up to notation, with equation (3.4.2.19) and, therefore, have the additive separable solutions

$$u = \varphi_1(x) + \psi_1(t), \quad v = \varphi_2(x) + \psi_2(t). \tag{4.2.1.15}$$

The functions (4.2.1.15) satisfy the functional constraints of the second kind (4.2.1.4) with $z_1 = u - \bar{u} = \psi_1(t) - \psi_1(t - \tau)$ and $z_2 = v - \bar{v} = \psi_2(t) - \psi_2(t - \tau)$.

The independent equations (4.2.1.14) generate a nonlinear system of reaction-diffusion equations with delay

$$u_t = a_1 u_{xx} + b_1 u + f(u - \bar{u}, v - \bar{v}),$$
$$v_t = a_2 v_{xx} + b_2 v + g(u - \bar{u}, v - \bar{v}), \tag{4.2.1.16}$$

where $f(z_1, z_2)$ and $g(z_1, z_2)$ are arbitrary functions of two arguments.

Following the generating equations method, we seek an exact solution to system (4.2.1.16) in the form (4.2.1.15). After separating the variables, we obtain two independent systems for the determining functions:

(i) a system of independent linear second-order ODEs with constant coefficients for $\varphi_1(x)$ and $\varphi_2(x)$:
$$\begin{aligned} a_1\varphi_1'' + b_1\varphi_1 &= 0, \\ a_2\varphi_2'' + b_2\varphi_2 &= 0, \end{aligned} \qquad (4.2.1.17)$$

(ii) a system of coupled nonlinear first-order delay ODEs for $\psi_1(t)$ and $\psi_2(t)$:
$$\begin{aligned} \psi_1'(t) &= b_1\psi_1(t) + f\big(\psi_1(t) - \psi_1(t-\tau), \psi_2(t) - \psi_2(t-\tau)\big), \\ \psi_2'(t) &= b_2\psi_2(t) + g\big(\psi_1(t) - \psi_1(t-\tau), \psi_2(t) - \psi_2(t-\tau)\big). \end{aligned} \qquad (4.2.1.18)$$

Notably, the general solution of equations (4.2.1.17) with $b_1 > 0$ and $b_2 < 0$ is given by
$$\begin{aligned} \varphi_1 &= C_{11}\cos(\beta_1 x) + C_{12}\sin(\beta_1 x), & \beta_1 &= \sqrt{b_1/a_1}, \\ \varphi_2 &= C_{21}\exp(-\beta_2 x) + C_{22}\exp(\beta_2 x), & \beta_2 &= \sqrt{-b_2/a_2}, \end{aligned}$$
where C_{ij} are arbitrary constants. Solutions of equations (4.2.1.17) with other signs of the coefficients b_1 and b_2 can be obtained likewise.

2°. The generating equations (4.2.1.14) also admit generalized separable solutions (see solution (3.4.2.27) of equation (3.4.2.19)):
$$u = \xi_1(x)t + \eta_1(x), \quad v = \xi_2(x)t + \eta_2(x), \qquad (4.2.1.19)$$

where the functions (4.2.1.19) satisfy the functional constraints of the first kind (4.2.1.4) with $z_1 = u - \bar{u} = \tau\xi_1(x)$ and $z_2 = v - \bar{v} = \tau\xi_2(x)$.

On substituting (4.2.1.19) into the system of delay reaction-diffusion equations (4.2.1.16), we obtain the following system of ODEs for the determining functions:
$$\begin{aligned} a_1\xi_1'' + b_1\xi_1 &= 0, \\ a_2\xi_2'' + b_2\xi_2 &= 0, \\ a_1\eta_1'' + b_1\eta_1 &= \xi_1 - f(\tau\xi_1, \tau\xi_2), \\ a_2\eta_2'' + b_2\eta_2 &= \xi_2 - g(\tau\xi_1, \tau\xi_2). \end{aligned}$$

Up to obvious renaming, the first two equations coincide with equations (4.2.1.17) and are easy to integrate. Then one substitutes the resulting functions ξ_1 and ξ_2 into the right-hand sides of the last two equations. As a result, they become linear nonhomogeneous second-order ODEs with constant coefficients and so are also easy to integrate. ◀

4.2.2. Quasilinear Systems of Delay Reaction-Diffusion Equations and Their Exact Solutions

This subsection briefly describes a few quasilinear systems of reaction-diffusion equations with constant delay that are linear in all derivatives. These systems and their exact solutions are obtained using the method of generating equations.

System 1. To get two generating equations, we use one reaction-diffusion type equation with constant delay (3.4.2.31) that admits generalized separable solutions satisfying a functional constraint of the first kind. As a result, we arrive at the quasilinear system of delay reaction-diffusion equations

$$u_t = a_1 u_{xx} + b_1 u + f(u - k_1 \bar{u}, v - k_2 \bar{v}), \quad \bar{u} = u(x, t - \tau);$$
$$v_t = a_2 v_{xx} + b_2 v + g(u - k_1 \bar{u}, v - k_2 \bar{v}), \quad \bar{v} = v(x, t - \tau), \qquad (4.2.2.1)$$

where k_1 and k_2 are arbitrary positive constants.

System (4.2.2.1) admits the generalized separable solutions

$$u = \xi_1(x) \exp(s_1 t) + \eta_1(x), \quad s_1 = (\ln k_1)/\tau,$$
$$v = \xi_2(x) \exp(s_2 t) + \eta_2(x), \quad s_2 = (\ln k_2)/\tau. \qquad (4.2.2.2)$$

The system of ODEs for the functions $\xi_{1,2}(x)$ and $\eta_{1,2}(x)$ is omitted.

Remark 4.7. *The quasilinear system of delay reaction-diffusion equations*

$$u_t = a_1 u_{xx} + f_1(u - k_1 \bar{u}, v - k_2 \bar{v}) + u f_2(u - k_1 \bar{u}, v - k_2 \bar{v}) + v f_3(u - k_1 \bar{u}, v - k_2 \bar{v}),$$
$$v_t = a_2 v_{xx} + g_1(u - k_1 \bar{u}, v - k_2 \bar{v}) + u g_2(u - k_1 \bar{u}, v - k_2 \bar{v}) + v g_3(u - k_1 \bar{u}, v - k_2 \bar{v}),$$

which is more general than system (4.2.2.1) and involves six arbitrary functions of two arguments, $f_m(z_1, z_2)$ and $g_m(z_1, z_2)$ with $m = 1, 2, 3$, also admits exact solutions of the form (4.2.2.2).

System 2. We take the two different reaction-diffusion type equations with constant delay

$$u_t = a_1 u_{xx} + bu + f(u - \bar{u}), \quad \bar{u} = u(x, t - \tau);$$
$$v_t = a_2 v_{xx} + v g(\bar{v}/v), \quad \bar{v} = v(x, t - \tau), \qquad (4.2.2.3)$$

as the generating equations. The former coincides with the first equation of (4.2.1.14) at $b = b_1$ and admits an additive separable solution. The latter one coincides with the second equation of (4.2.1.7) and admits a multiplicative separable solution. Both solutions to equations (4.2.2.3) satisfy appropriate functional constraints of the second kind.

By applying the generating equations method to equations (4.2.2.3), we arrive at the system of coupled reaction-diffusion type equations with constant delay

$$u_t = a_1 u_{xx} + bu + f(u - \bar{u}, \bar{v}/v),$$
$$v_t = a_2 v_{xx} + v g(u - \bar{u}, \bar{v}/v), \qquad (4.2.2.4)$$

where $f(z_1, z_2)$ and $g(z_1, z_2)$ are arbitrary functions of two arguments.

By combining solutions to equations (4.2.2.3), we get an exact solution to system (4.2.2.4):

$$u = \varphi_1(x) + \psi_1(t), \quad v = \varphi_2(x)\psi_2(t). \qquad (4.2.2.5)$$

In this case, the system components u and v have different structures. We will refer to such solutions as *mixed-type solutions*.

4.2. Systems of Nonlinear Delay PDEs. Generating Equations Method

On substituting (4.2.2.5) into system (4.2.2.4) and separating the variables, we arrive at the following ODEs and delay ODEs:

$$\begin{aligned}
& a_1\varphi_1'' + b\varphi_1 = A_1, \\
& \varphi_2'' - A_2\varphi_2 = 0, \\
& \psi_1'(t) = b\psi_1(t) + A_1 + f\big(\psi_1(t) - \psi_1(t-\tau), \psi_2(t-\tau)/\psi_2(t)\big), \\
& \psi_2'(t) = A_2 a_2 \psi_2(t) + \psi_2(t) g\big(\psi_1(t) - \psi_1(t-\tau), \psi_2(t-\tau)/\psi_2(t)\big),
\end{aligned} \qquad (4.2.2.6)$$

where A_1 and A_2 are arbitrary constants. The first two equations of (4.2.2.6) are independent and easy to integrate, since both are linear second-order ODEs with constant coefficients. The last two equations of (4.2.2.6) make up a system of coupled nonlinear delay ODEs. At $b = 0$, this system admits exact solutions of the form

$$\psi_1(t) = \beta t + C_1, \quad \psi_2(t) = C_2 e^{\lambda t},$$

where C_1 and C_2 are arbitrary constants, and the coefficients β and λ are determined from the algebraic (transcendental) system

$$\beta = A_1 + f(\beta\tau, e^{-\lambda\tau}), \quad \lambda = A_2 a_2 + g(\beta\tau, e^{-\lambda\tau}).$$

System 3. As the generating equations, we use two special cases of one reaction-diffusion type equation with delay (3.4.2.48), which has generalized separable solutions satisfying a functional constraint of the first kind. As a result, we arrive at the quasilinear system of delay reaction-diffusion equations

$$\begin{aligned}
u_t &= a_1 u_{xx} + u f(u^2 + \bar{u}^2, v^2 + \bar{v}^2), \\
v_t &= a_2 v_{xx} + v g(u^2 + \bar{u}^2, v^2 + \bar{v}^2),
\end{aligned} \qquad (4.2.2.7)$$

which admits the generalized separable solutions

$$\begin{aligned}
u &= \varphi_1(x)\cos(\lambda t) + \psi_1(x)\sin(\lambda t), \quad \lambda = \frac{\pi}{2\tau}, \\
v &= \varphi_2(x)\cos(\lambda t) + \psi_2(x)\sin(\lambda t).
\end{aligned} \qquad (4.2.2.8)$$

The functions $\varphi_{1,2}(x)$ and $\psi_{1,2}(x)$ are described by the system of ODEs

$$\begin{aligned}
& a_1\varphi_1'' + \varphi_1 f(\varphi_1^2 + \psi_1^2, \varphi_2^2 + \psi_2^2) - \lambda\psi_1 = 0, \\
& a_1\psi_1'' + \psi_1 f(\varphi_1^2 + \psi_1^2, \varphi_2^2 + \psi_2^2) + \lambda\varphi_1 = 0, \\
& a_2\varphi_2'' + \varphi_2 g(\varphi_1^2 + \psi_1^2, \varphi_2^2 + \psi_2^2) - \lambda\psi_2 = 0, \\
& a_2\psi_2'' + \psi_2 g(\varphi_1^2 + \psi_1^2, \varphi_2^2 + \psi_2^2) + \lambda\varphi_2 = 0.
\end{aligned} \qquad (4.2.2.9)$$

Remark 4.8. *The more general nonlinear system of delay PDEs*

$$\begin{aligned}
u_t &= a_1 u_{xx} + u f_1(u^2 + \bar{u}^2, v^2 + \bar{v}^2) + \bar{u} f_2(u^2 + \bar{u}^2, v^2 + \bar{v}^2) \\
& \quad + v f_3(u^2 + \bar{u}^2, v^2 + \bar{v}^2) + \bar{v} f_4(u^2 + \bar{u}^2, v^2 + \bar{v}^2), \\
v_t &= a_2 v_{xx} + u g_1(u^2 + \bar{u}^2, v^2 + \bar{v}^2) + \bar{u} g_2(u^2 + \bar{u}^2, v^2 + \bar{v}^2) \\
& \quad + v g_3(u^2 + \bar{u}^2, v^2 + \bar{v}^2) + \bar{v} g_4(u^2 + \bar{u}^2, v^2 + \bar{v}^2),
\end{aligned}$$

involving eight arbitrary functions of two arguments, $f_n(z_1, z_2)$ and $g_n(z_1, z_2)$, also admits exact solutions of the form (4.2.2.8).

Remark 4.9. The constant $\lambda = \frac{\pi}{2\tau}$ in (4.2.2.8) and (4.2.2.9) can be replaced with

$$\lambda_n = \frac{\pi(2n+1)}{2\tau}, \quad n = 0, \pm 1, \pm 2, \ldots.$$

System 4. As the generating equations, we take two different reaction-diffusion type equations with constant delay: the first one coincides with the first equation of (4.2.1.14) at $b_1 = b$ and the second is a special case of equation (3.4.2.48), up to notation. Both these equations have generalized separable solutions that satisfy functional constraints of the first kind. By employing the generating equations method and by reasoning as previously, we arrive at the following system of reaction-diffusion type equations with delay:

$$\begin{aligned} u_t &= a_1 u_{xx} + bu + f(u - \bar{u}, v^2 + \bar{v}^2), \\ v_t &= a_2 v_{xx} + vg(u - \bar{u}, v^2 + \bar{v}^2). \end{aligned} \quad (4.2.2.10)$$

It admits the mixed-type exact solutions

$$\begin{aligned} u &= t\xi(x) + \eta(x), \\ v &= \varphi(x)\cos(\lambda t) + \psi(x)\sin(\lambda t), \quad \lambda = \tfrac{\pi}{2\tau}. \end{aligned}$$

The associated system of ODEs for the functions $\xi(x)$, $\eta(x)$, $\varphi(x)$, and $\psi(x)$ is omitted here.

4.2.3. Nonlinear Systems of Delay Reaction-Diffusion Equations and Their Exact Solutions

System 1. We use one reaction-diffusion type equation with constant delay (3.4.3.2) to get two generating equations. It admits a multiplicative separable solution satisfying a functional constraint of the second kind. Following the procedure described in Subsection 4.2.1, we arrive at the nonlinear system of delay reaction-diffusion equations

$$\begin{aligned} u_t &= a_1(u^k u_x)_x + uf(\bar{u}/u, \bar{v}/v), \quad \bar{u} = u(x, t-\tau), \\ v_t &= a_2(v^m v_x)_x + vg(\bar{u}/u, \bar{v}/v), \quad \bar{v} = v(x, t-\tau). \end{aligned} \quad (4.2.3.1)$$

System (4.2.3.1) inherits the solution structure of the generating equations. Consequently, it admits the multiplicative separable solutions

$$u = \varphi_1(x)\psi_1(t), \quad v = \varphi_2(x)\psi_2(t). \quad (4.2.3.2)$$

Substituting (4.2.3.2) into (4.2.3.1) and rearranging, we obtain the following system of two independent ODEs and two coupled delay ODEs for $\varphi_1 = \varphi_1(x)$, $\varphi_2 = \varphi_2(x)$, $\psi_1 = \psi_1(t)$, and $\psi_2 = \psi_2(t)$:

$$\begin{aligned} a_1(\varphi_1^k \varphi_1')' &= C_1 \varphi_1, \\ a_2(\varphi_2^m \varphi_2')' &= C_2 \varphi_2, \\ \psi_1' &= C_1 \psi_1^{k+1} + \psi_1 f(\bar{\psi}_1/\psi_1, \bar{\psi}_2/\psi_2), \quad \bar{\psi}_1 = \psi_1(t-\tau), \\ \psi_2' &= C_2 \psi_2^{m+1} + \psi_2 g(\bar{\psi}_1/\psi_1, \bar{\psi}_2/\psi_2), \quad \bar{\psi}_2 = \psi_2(t-\tau), \end{aligned} \quad (4.2.3.3)$$

where C_1 and C_2 are arbitrary constants.

The general solutions of the first two autonomous ODEs in (4.2.3.3) can be represented in implicit form. If $k, m \neq 0$ and $k, m \neq -2$, these equations have the particular solutions

$$\varphi_1 = \left[\frac{C_1 k^2 x^2}{2a_1(k+2)}\right]^{1/k}, \quad \varphi_2 = \left[\frac{C_2 m^2 x^2}{2a_2(m+2)}\right]^{1/m}.$$

System 2. If we use one PDE with constant delay (3.4.3.6), which has a multiplicative separable solution satisfying a functional constraint of the second kind, as two generating equations, then we will arrive at the nonlinear system of delay reaction-diffusion equations

$$\begin{aligned} u_t &= a_1(u^k u_x)_x + b_1 u^{k+1} + u f(\bar{u}/u, \bar{v}/v), \quad \bar{u} = u(x, t-\tau), \\ v_t &= a_2(v^m v_x)_x + b_2 v^{m+1} + v g(\bar{u}/u, \bar{v}/v), \quad \bar{v} = v(x, t-\tau), \end{aligned} \quad (4.2.3.4)$$

which is more general than system (4.2.3.1).

This system admits a multiplicative separable solution of the form (4.2.3.2), where the functions $\varphi_1 = \varphi_1(x)$ and $\varphi_2 = \varphi_2(x)$ are described by the independent autonomous ODEs:

$$\begin{aligned} a_1(\varphi_1^k \varphi_1')' + b_1 \varphi_1^{k+1} &= C_1 \varphi_1, \\ a_2(\varphi_2^m \varphi_2')' + b_2 \varphi_2^{m+1} &= C_2 \varphi_2. \end{aligned} \quad (4.2.3.5)$$

The functions $\psi_1 = \psi_1(t)$ and $\psi_2 = \psi_2(t)$ satisfy the coupled delay ODEs coinciding with the last two equations of system (4.2.3.3).

Let us focus on a few special cases where the system of delay reaction-diffusion equations (4.2.3.4) has simple exact solutions expressible in terms of elementary functions; all these cases correspond to $C_1 = C_2 = 0$ in (4.2.3.5).

$1°$. For $b_1(k+1) > 0$ and $b_2(m+1) > 0$, the system of PDEs (4.2.3.4) has the multiplicative separable solution

$$\begin{aligned} u &= [A_1 \cos(\beta_1 x) + A_2 \sin(\beta_1 x)]^{1/(k+1)} \psi_1(t), \quad \beta_1 = \sqrt{b_1(k+1)/a_1}, \\ v &= [B_1 \cos(\beta_2 x) + B_2 \sin(\beta_2 x)]^{1/(m+1)} \psi_2(t), \quad \beta_2 = \sqrt{b_2(m+1)/a_2}, \end{aligned} \quad (4.2.3.6)$$

where A_1, A_2, B_1, and B_2 are arbitrary constants, and the functions $\psi_1 = \psi_1(t)$ and $\psi_2 = \psi_2(t)$ are described by the system of the last two delay ODEs in (4.2.3.3) with $C_1 = C_2 = 0$. This system has exponential particular solutions

$$\psi_1(t) = D_1 \exp(\lambda_1 t), \quad \psi_2(t) = D_2 \exp(\lambda_2 t), \quad (4.2.3.7)$$

where D_1 and D_2 are arbitrary constants, and the constants λ_1 and λ_2 are determined by the algebraic (transcendental) system of equations

$$\lambda_1 - f(e^{-\lambda_1 \tau}, e^{-\lambda_2 \tau}) = 0, \quad \lambda_2 - g(e^{-\lambda_1 \tau}, e^{-\lambda_2 \tau}) = 0. \quad (4.2.3.8)$$

$2°$. For $b_1(k+1) < 0$ and $b_2(m+1) < 0$, the system of PDEs (4.2.3.4) has the multiplicative separable solution

$$\begin{aligned} u &= [A_1 \exp(-\beta_1 x) + A_2 \exp(\beta_1 x)]^{\frac{1}{k+1}} \psi_1(t), \quad \beta_1 = \sqrt{-b_1(k+1)/a_1}, \\ v &= [B_1 \exp(-\beta_2 x) + B_2 \exp(\beta_2 x)]^{\frac{1}{m+1}} \psi_2(t), \quad \beta_2 = \sqrt{-b_2(m+1)/a_2}, \end{aligned} \quad (4.2.3.9)$$

where A_1, A_2, B_1, and B_2 are arbitrary constants, and the functions $\psi_1 = \psi_1(t)$ and $\psi_2 = \psi_2(t)$ are described by the system of the last two delay ODEs of (4.2.3.3) with $C_1 = C_2 = 0$. This system has exponential particular solutions of the form (4.2.3.7), where the constants λ_1 and λ_2 are determined by the algebraic (transcendental) system of equations (4.2.3.8).

3°. The system of PDEs (4.2.3.4) also admits two multiplicative separable solutions of mixed type. Specifically, if $b_1(k+1) > 0$ and $b_2(m+1) < 0$, system (4.2.3.4) has an exact solution determined by the first formula of (4.2.3.6) and second formula of (4.2.3.9). The functions $\psi_1 = \psi_1(t)$ and $\psi_2 = \psi_2(t)$ appearing in this mixed solution are described, as before, by the system consisting of the last two delay ODEs of (4.2.3.3) with $C_1 = C_2 = 0$. This system has exponential particular solutions of the form (4.2.3.7), in which the constants λ_1 and λ_2 are determined from the algebraic (transcendental) system of equations (4.2.3.8).

For $b_1(k+1) < 0$ and $b_2(m+1) > 0$, system (4.2.3.4) has an exact solution defined by the second formula of (4.2.3.6) and first formula of (4.2.3.9).

System 3. To obtain the generating equations, we use one partial differential equation with constant delay (3.4.3.12), which admits a functional separable solution satisfying a functional constraint of the second kind. As a result, we arrive at the nonlinear system of delay reaction-diffusion equations

$$\begin{aligned} u_t &= a_1(u^k u_x)_x + b_1 + u^{-k} f(u^{k+1} - \bar{u}^{k+1}, v^{m+1} - \bar{v}^{m+1}), \\ v_t &= a_2(v^m v_x)_x + b_2 + v^{-m} g(u^{k+1} - \bar{u}^{k+1}, v^{m+1} - \bar{v}^{m+1}). \end{aligned} \quad (4.2.3.10)$$

System (4.2.3.10) inherits the form of solutions of the generating equations and for $k, m \neq -1$, it has functional separable solutions of the form

$$u = [\varphi_1(x) + \psi_1(t)]^{1/(k+1)}, \quad v = [\varphi_2(x) + \psi_2(t)]^{1/(m+1)}. \quad (4.2.3.11)$$

Substituting (4.2.3.11) into PDEs (4.2.3.10) gives the following system of two independent linear ODEs, for $\varphi_1 = \varphi_1(x)$ and $\varphi_2 = \varphi_2(x)$, and two coupled nonlinear delay ODEs, for $\psi_1 = \psi_1(t)$ and $\psi_2 = \psi_2(t)$:

$$\begin{aligned} & a_1 \varphi_1'' + b_1(k+1) = 0, \\ & a_2 \varphi_2'' + b_2(m+1) = 0, \\ & \psi_1' = (k+1) f(\psi_1 - \bar{\psi}_1, \psi_2 - \bar{\psi}_2), \quad \bar{\psi}_1 = \psi_1(t-\tau), \\ & \psi_2' = (m+1) g(\psi_1 - \bar{\psi}_1, \psi_2 - \bar{\psi}_2), \quad \bar{\psi}_2 = \psi_2(t-\tau). \end{aligned} \quad (4.2.3.12)$$

The general solutions of the first two ODEs (4.2.3.12) are expressed as

$$\varphi_1 = -\frac{b_1(k+1)}{2a_1} x^2 + C_1 x + C_2, \quad \varphi_2 = -\frac{b_2(m+1)}{2a_2} x^2 + C_3 x + C_4,$$

where $C_1, \ldots, C_4$ are arbitrary constants. The system of the last two coupled delay ODEs (4.2.3.12) has the simple particular solution

$$\psi_1 = A_1 t, \quad \psi_2 = A_2 t,$$

where the constants A_1 and A_2 are determined from the algebraic (transcendental) system
$$A_1 = (k+1)f(A_1\tau, A_2\tau), \quad A_2 = (m+1)g(A_1\tau, A_2\tau).$$

System 4. To get the generating equations, we use one reaction-diffusion type equation with constant delay (3.4.3.21). It admits an additive separable solution satisfying a functional constraint of the second kind. Following the procedure described in Subsection 4.2.1, we arrive at the nonlinear system of delay reaction-diffusion equations

$$\begin{aligned} u_t &= a_1(e^{\lambda u}u_x)_x + f(u - \bar{u}, v - \bar{v}), & \bar{u} &= u(x, t-\tau), \\ v_t &= a_2(e^{\beta v}v_x)_x + g(u - \bar{u}, v - \bar{v}), & \bar{v} &= v(x, t-\tau). \end{aligned} \qquad (4.2.3.13)$$

Since the system of PDEs (4.2.3.13) inherits the form of solutions of the generating equations, it admits the additive separable solutions

$$u = \varphi_1(x) + \psi_1(t), \quad v = \varphi_2(x) + \psi_2(t), \qquad (4.2.3.14)$$

where the functions $\varphi_1 = \varphi_1(x)$, $\varphi_2 = \varphi_2(x)$, $\psi_1 = \psi_1(t)$, $\psi_2 = \psi_2(t)$ are described by the system of two independent ODEs and two coupled nonlinear delay ODEs

$$\begin{aligned} (e^{\lambda\varphi_1}\varphi_1')' &= C_1, \\ (e^{\beta\varphi_2}\varphi_2')' &= C_2, \\ \psi_1' &= a_1 C_1 e^{\lambda\psi_1} + f(\psi_1 - \bar{\psi}_1, \psi_2 - \bar{\psi}_2), & \bar{\psi}_1 &= \psi_1(t-\tau), \\ \psi_2' &= a_2 C_2 e^{\beta\psi_2} + g(\psi_1 - \bar{\psi}_1, \psi_2 - \bar{\psi}_2), & \bar{\psi}_2 &= \psi_2(t-\tau), \end{aligned} \qquad (4.2.3.15)$$

where C_1 and C_2 are arbitrary constants.

The general solutions of the first two ODEs of (4.2.3.15) are given by

$$\varphi_1 = \frac{1}{\lambda}\ln\Big(\frac{1}{2}C_1\lambda x^2 + A_1 x + B_1\Big), \quad \varphi_2 = \frac{1}{\beta}\ln\Big(\frac{1}{2}C_2\beta x^2 + A_2 x + B_2\Big),$$

where A_1, B_1, A_2, B_2 are arbitrary constants.

System 5. As the generating equations, we take two different reaction-diffusion type equations with constant delay (3.4.3.2) and (3.4.3.21). The former one has a multiplicative separable solution and the latter one has an additive separable solution; both solutions satisfy functional constraints of the second kind. Employing the generating equations method and reasoning as previously, we arrive at the following system of reaction-diffusion type equations with delay:

$$\begin{aligned} u_t &= a_1(u^k u_x)_x + uf(\bar{u}/u, v - \bar{v}), & \bar{u} &= u(x, t-\tau), \\ v_t &= a_2(e^{\lambda v}v_x)_x + g(\bar{u}/u, v - \bar{v}), & \bar{v} &= v(x, t-\tau), \end{aligned} \qquad (4.2.3.16)$$

where $f(z_1, z_2)$ and $g(z_1, z_2)$ are arbitrary functions of two arguments.

Since the system of delay PDEs (4.2.3.16) inherits the form of solutions of the generating equations, it admits exact solutions of the form

$$u = \varphi_1(x)\psi_1(t), \quad v = \varphi_2(x) + \psi_2(t). \qquad (4.2.3.17)$$

The functions $\varphi_1 = \varphi_1(x)$, $\varphi_2 = \varphi_2(x)$, $\psi_1 = \psi_1(t)$, and $\psi_2 = \psi_2(t)$ are described by the system of two independent ODEs and two coupled delay ODEs

$$(\varphi_1^k \varphi_1')' = C_1 \varphi_1,$$
$$(e^{\lambda \varphi_2} \varphi_2')' = C_2,$$
$$\psi_1' = a_1 C_1 \psi_1^{k+1} + \psi_1 f(\bar{\psi}_1/\psi_1, \psi_2 - \bar{\psi}_2), \quad \bar{\psi}_1 = \psi_1(t-\tau),$$
$$\psi_2' = a_2 C_2 e^{\lambda \psi_2} + g(\bar{\psi}_1/\psi_1, \psi_2 - \bar{\psi}_2), \quad \bar{\psi}_2 = \psi_2(t-\tau),$$

where C_1 and C_2 are arbitrary constants.

System 6. To get two generating equations, we use one reaction-diffusion type equation with constant delay (3.4.3.30). It admits a functional separable solution satisfying a functional constraint of the second kind. As a result, we arrive at the nonlinear system of delay reaction-diffusion equations

$$u_t = a_1(e^{\lambda_1 u} u_x)_x + b_1 + e^{-\lambda_1 u} f(e^{\lambda_1 u} - e^{\lambda_1 \bar{u}}, e^{\lambda_2 v} - e^{\lambda_2 \bar{v}}),$$
$$v_t = a_2(e^{\lambda_2 v} v_x)_x + b_2 + e^{-\lambda_2 v} g(e^{\lambda_1 u} - e^{\lambda_1 \bar{u}}, e^{\lambda_2 v} - e^{\lambda_2 \bar{v}}).$$
(4.2.3.18)

System (4.2.3.18) inherits the form of solutions of the generating equations. Therefore, it admits functional separable solutions of the form

$$u = \frac{1}{\lambda_1} \ln[\varphi_1(x) + \psi_1(t)], \quad \varphi_1(x) = -\frac{b_1 \lambda_1}{2 a_1} x^2 + C_1 x + C_2,$$
$$v = \frac{1}{\lambda_2} \ln[\varphi_2(x) + \psi_2(t)], \quad \varphi_2(x) = -\frac{b_2 \lambda_2}{2 a_2} x^2 + C_3 x + C_4,$$

where $C_1, \ldots, C_4$ are arbitrary constants, and the functions $\psi_1(t)$ and $\psi_2(t)$ are described by the nonlinear system of delay ODEs

$$\psi_1'(t) = \lambda_1 f(\psi_1(t) - \psi_1(t-\tau), \psi_2(t) - \psi_2(t-\tau)),$$
$$\psi_2'(t) = \lambda_2 g(\psi_1(t) - \psi_1(t-\tau), \psi_2(t) - \psi_2(t-\tau)).$$

This system has simple particular solutions $\psi_1(t) = A_1 t$ and $\psi_2(t) = A_2 t$, where the coefficients A_1 and A_2 are determined from the algebraic (transcendental) system of equations

$$A_1 = \lambda_1 f(A_1 \tau, A_2 \tau), \quad A_2 = \lambda_2 g(A_1 \tau, A_2 \tau).$$

Remark 4.10. *By using the equations and their exact solutions listed in Section 3.4, one can obtain other systems of nonlinear reaction-diffusion type equations that admit exact solutions (see also [436]).*

4.2.4. Some Generalizations

Below we describe a few generalizations associated with the method of generating equations.

Systems with two constant delay times. The common constant delay time τ in a system of delay PDEs obtained by the method of generating equations can be replaced with two different delay times as follows:

$$\begin{aligned} \bar{u} = u(x, t - \tau) &\implies \bar{u} = u(x, t - \tau_1), \\ \bar{v} = v(x, t - \tau) &\implies \bar{v} = v(x, t - \tau_2), \end{aligned} \qquad (4.2.4.1)$$

where $\tau_1 > 0$ and $\tau_2 > 0$ are arbitrary constants. Then the form of the exact solution (4.2.1.3) will remain unchanged. What will change is the delay time in the delay ODEs for the determining functions $\psi_{1n}(t)$ and $\psi_{2n}(t)$:

$$\begin{aligned} \psi_{1n}(t - \tau) &\implies \psi_{1n}(t - \tau_1), \\ \psi_{2n}(t - \tau) &\implies \psi_{2n}(t - \tau_2). \end{aligned} \qquad (4.2.4.2)$$

▶ **Example 4.12.** In the system of PDEs (4.2.3.1), we will substitute the constant delay time τ with different delay times by rule (4.2.4.1). As a result, we get the more complicated system

$$\begin{aligned} u_t &= a_1(u^k u_x)_x + u f(\bar{u}/u, \bar{v}/v), \quad \bar{u} = u(x, t - \tau_1), \\ v_t &= a_2(v^m v_x)_x + v g(\bar{u}/u, \bar{v}/v), \quad \bar{v} = v(x, t - \tau_2). \end{aligned} \qquad (4.2.4.3)$$

The form of the exact solution of system (4.2.4.3), just as that of system (4.2.3.1), is given by (4.2.3.2). Eventually, we arrive at the determining system of equations consisting of the first two ODEs in (4.2.3.3), for $\varphi_1(x)$ and $\varphi_2(x)$, and the system of two ODEs for $\psi_1(t)$ and $\psi_2(t)$ with different delays

$$\begin{aligned} \psi_1' &= C_1 \psi_1^{k+1} + \psi_1 f(\bar{\psi}_1/\psi_1, \bar{\psi}_2/\psi_2), \quad \bar{\psi}_1 = \psi_1(t - \tau_1), \\ \psi_2' &= C_2 \psi_2^{m+1} + \psi_2 g(\bar{\psi}_1/\psi_1, \bar{\psi}_2/\psi_2), \quad \bar{\psi}_2 = \psi_2(t - \tau_2). \end{aligned} \qquad (4.2.4.4)$$

System (4.2.4.4) is obtained from the last two delay ODEs of (4.2.3.3) by rule (4.2.4.2). ◀

Systems with variable delay times. Suppose that both determining equations have exact solutions satisfying the functional constraints of the second kind (4.2.1.4), where $q_1(t) \neq \text{const}$ and $q_2(t) \neq \text{const}$. Then the common constant delay time τ in the resulting system of delay PDEs can be replaced with two different delay times of general form by rule (4.2.4.1), where $\tau_1 = \tau_1(t)$ and $\tau_2 = \tau_2(t)$ are arbitrary positive continuous functions. The form of the exact solution (4.2.1.3) will remain unchanged, while the delay terms in the delay ODEs for the determining function $\psi_{1n}(t)$ and $\psi_{2n}(t)$ will change by rule (4.2.4.2), where $\tau_1 = \tau_1(t)$ and $\tau_2 = \tau_2(t)$.

▶ **Example 4.13.** Exact solutions to the generating equations used to derive system (4.2.3.1) satisfy functional constraints of the second kind. Therefore, the constant delay time τ in (4.2.3.1) can be replaced with two different delay times of general form by rule (4.2.4.1). As a result, we arrive at system (4.2.4.3) in which one should set $\tau_1 = \tau_1(t)$ and $\tau_2 = \tau_2(t)$. In this case, the determining system of equations consists of the first two ODEs of (4.2.3.3) and system of two delay ODEs (4.2.4.4) where $\tau_1 = \tau_1(t)$ and $\tau_2 = \tau_2(t)$. ◀

Systems with any number of space variables. The method of generating equations can also be extended to systems of equations with any number of space variables. We will illustrate this with a specific example.

▶ **Example 4.14.** We use one multi-dimensional reaction-diffusion type equation with constant delay (3.4.3.86) with $b = k = 0$ to obtain two generating equations. It admits a multiplicative separable solution satisfying a functional constraint of the second kind. As a result, we arrive at the nonlinear system of delay reaction-diffusion equations [436]:

$$u_t = a_1 \Delta u + u f(\bar{u}/u, \bar{v}/v),$$
$$v_t = a_2 \Delta v + v g(\bar{u}/u, \bar{v}/v), \qquad (4.2.4.5)$$

which is a multi-dimensional generalization of system (4.2.1.9). We have used the following notations:

$$u = u(\mathbf{x}, t), \quad \bar{u} = u(\mathbf{x}, t - \tau), \quad v = v(\mathbf{x}, t), \quad \bar{v} = v(\mathbf{x}, t - \tau),$$
$$\mathbf{x} = (x_1, \ldots, x_n), \quad \Delta \equiv \sum_{m=1}^{n} \frac{\partial^2}{\partial x_m^2}.$$

System (4.2.4.5) admits the multiplicative separable solution

$$u = \varphi_1(\mathbf{x}) \psi_1(t), \quad v = \varphi_2(\mathbf{x}) \psi_2(t), \qquad (4.2.4.6)$$

which generalizes solution (4.2.1.13) to the one-dimensional system (4.2.1.9).

Substituting (4.2.4.6) into system (4.2.4.5) and separating the variables, we obtain two independent linear equations (Helmholtz equations) for $\varphi_1(\mathbf{x})$ and $\varphi_2(\mathbf{x})$,

$$\Delta \varphi_1 = \lambda_1 \varphi_1, \quad \Delta \varphi_2 = \lambda_2 \varphi_2,$$

where λ_1 and λ_2 are arbitrary constants, and a system of coupled nonlinear ODEs for $\psi_1(t)$ and $\psi_2(t)$:

$$\psi_1'(t) = a_1 \lambda_1 \psi_1(t) + \psi_1(t) f\big(\psi_1(t-\tau)/\psi_1(t),\ \psi_2(t-\tau)/\psi_2(t)\big),$$
$$\psi_2'(t) = a_2 \lambda_2 \psi_2(t) + \psi_2(t) g\big(\psi_1(t-\tau)/\psi_1(t),\ \psi_2(t-\tau)/\psi_2(t)\big).$$

Notably, this system admits exponential particular solutions:

$$\psi_1(t) = C_1 \exp(\beta_1 t),$$
$$\psi_2(t) = C_2 \exp(\beta_2 t),$$

where C_1 and C_2 are arbitrary constants, and the constants β_1 and β_2 are described by an algebraic (transcendental) system of equations. ◂

Systems with any number of equations and systems of higher-order equations. The method of generating equations extends to systems of hyperbolic equations, systems of higher-order equations, and systems with any number of equations. The generalizations are obvious and so are omitted. For examples of constructing systems of hyperbolic second-order and higher-order PDEs, see the article [436].

4.3. Reductions and Exact Solutions of Lotka–Volterra Type Systems and More Complex Systems of PDEs with Several Delays

4.3.1. Reaction-Diffusion Systems with Several Delays. The Lotka–Volterra System

Preliminary remarks. Subsections 4.3.1–4.3.3 deal with a nonlinear system consisting of two reaction-diffusion equations of a reasonably general form that involve three arbitrary functions and several delays. It includes, as an important special case, a multiparameter Lotka–Volterra diffusion system with several delays. These nonlinear systems are shown to reduce in different ways to simpler systems: (i) a system of stationary equations, (ii) a system of delay ODEs, (iii) a system of stationary equations with a linear Schrödinger equation, and (iv) a system of delay ODEs with a linear heat equation. A number of exact solutions to the nonlinear Lotka–Volterra system with several arbitrary delays are obtained. All of them are generalized separable or incomplete separable solutions with several free parameters. An exact solution involving infinitely many free parameters is also described.

The reaction-diffusion system of PDEs concerned. The Lotka–Volterra system. Following [417], we will look at the nonlinear reaction-diffusion system of partial differential equations with several constant delays

$$u_t = a_1 \Delta u + b_1 u + c_1 u f(k_1 \bar{u}_1 - k_2 \bar{v}_2) + g(k_1 \bar{u}_1 - k_2 \bar{v}_2),$$
$$v_t = a_2 \Delta v + b_2 v + c_2 v f(k_1 \bar{u}_3 - k_2 \bar{v}_4) + h(k_1 \bar{u}_3 - k_2 \bar{v}_4),$$
(4.3.1.1)

where $u = u(\mathbf{x}, t)$ and $v = v(\mathbf{x}, t)$ are the unknown functions, t is time, $\mathbf{x} = (x_1, \ldots, x_n)$, $\bar{u}_i = u(\mathbf{x}, t - \tau_i)$ $(i = 1, 3)$, $\bar{v}_j = v(\mathbf{x}, t - \tau_j)$ $(j = 2, 4)$, $\tau_i \geq 0$ and $\tau_j \geq 0$ are delay times, $f = f(z)$ is a monotonic arbitrary function such that $f(0) = 0$, $g = g(z_1)$ and $h = h(z_2)$ are arbitrary functions; $a_1 > 0$, $a_2 > 0$, b_1, b_2, $k_1 \neq 0$, and $k_2 \neq 0$ are free parameters; $c_1 \neq 0$ and $c_2 \neq 0$ are some constant to be defined below; and $\Delta = \sum_{k=1}^{n} \frac{\partial^2}{\partial x_k^2}$ is the n-dimensional Laplace operator.

▶ **Example 4.15.** In the case of a single space variable with $f(z_{1,2}) = z_{1,2}$ and $g(z_1) = h(z_2) = 0$, the delay reaction-diffusion system (4.3.1.1) becomes

$$u_t = a_1 u_{xx} + u[b_1 + c_1(k_1 \bar{u}_1 - k_2 \bar{v}_2)],$$
$$v_t = a_2 v_{xx} + v[b_2 + c_2(k_1 \bar{u}_3 - k_2 \bar{v}_4)].$$
(4.3.1.2)

This is a special case of the Lotka–Volterra diffusive system of delay PDEs [224], which describes the interaction between two species. The studies [95, 151] investigated the stability of solutions of a simpler system with two delays (with $\tau_1 = \tau_4 = 0$).

The articles [98, 99, 101] (see also [422]) dealt with symmetries and exact solutions to the nonlinear system (4.3.1.2) and related Lotka–Volterra type systems in the case of no delays, $\tau_1 = \tau_2 = \tau_3 = \tau_4 = 0$. ◀

The case of $a_1 = a_2$ and $b_1 = b_2 = 0$ with no delays, $\tau_1 = \tau_2 = \tau_3 = \tau_4 = 0$, some reductions and exact solutions to system (4.3.1.1) with one space variable are described in [422].

The main idea used below to find exact solutions of system (4.3.1.1) is that the form of the unknowns u and v is chosen so that the arguments of the functions $f(\dots)$, $g(\dots)$, and $h(\dots)$ only depend on $\mathbf{x}$ or t alone.

Simplest solutions (stationary points). The constants (stationary points)

$$u = u^\circ = \text{const}, \quad v = v^\circ = \text{const} \tag{4.3.1.3}$$

are the simplest solutions of system (4.3.1.1). These are determined from the nonlinear algebraic system

$$\begin{aligned} u^\circ[b_1 + c_1 f(k_1 u^\circ - k_2 v^\circ)] + g(k_1 u^\circ - k_2 v^\circ) &= 0, \\ v^\circ[b_2 + c_2 f(k_1 u^\circ - k_2 v^\circ)] + h(k_1 u^\circ - k_2 v^\circ) &= 0. \end{aligned} \tag{4.3.1.4}$$

System (4.3.1.4) normally has several roots. In particular, if $g = h = 0$, it splits into four independent subsystems whose roots possess the properties

$$\begin{aligned} &(a) \quad u^\circ = v^\circ = 0; \quad (b) \quad u^\circ \neq 0, \ v^\circ = 0; \quad (c) \quad u^\circ = 0, \ v^\circ \neq 0; \\ &(d) \quad u^\circ = (k_2/k_1)v^\circ + \text{const}, \quad v^\circ \text{ is any} \quad \text{if} \quad b_1/c_1 = b_2/c_2. \end{aligned} \tag{4.3.1.5}$$

▶ **Example 4.16.** For the Lotka–Volterra type system (4.3.1.1) with $f(z) = z$ and $g(z) = h(z) = 0$, the associated system (4.3.1.4) has the following solutions:

$$\begin{aligned} &(a) \quad u^\circ = v^\circ = 0; \quad (b) \quad u^\circ = -\frac{b_1}{c_1 k_1}, \ v^\circ = 0; \quad (c) \quad u^\circ = 0, \ v^\circ = \frac{b_2}{c_2 k_2}; \\ &(d) \quad u^\circ = \frac{k_2}{k_1}v^\circ - \frac{b_1}{c_1 k_1}, \ v^\circ \text{ is any}, \quad \text{If} \quad \frac{b_1}{c_1} = \frac{b_2}{c_2}. \end{aligned} \tag{4.3.1.6}$$

In what follows, the stationary points (4.3.1.3) will be used to construct more complex, nonstationary spatially nonhomogeneous solutions to system (4.3.1.1). ◀

4.3.2. Reductions and Exact Solutions of Systems of PDEs with Different Diffusion Coefficients ($a_1 \neq a_2$)

Reduction of the system of PDEs with three arbitrary delays to the Helmholtz equation. Suppose that the four delay times in system (4.3.1.1) are linked by a single relation

$$\tau_2 - \tau_1 = \tau_4 - \tau_3, \tag{4.3.2.1}$$

implying that any three of them can be set arbitrarily. Notably, relation (4.3.2.1) holds, for example, in the following three special cases:

$$\begin{aligned} &\tau_2 = \tau_1, \quad \tau_4 = \tau_3, \quad \tau_1, \tau_3 \text{ are arbitrary}; \\ &\tau_3 = \tau_1, \quad \tau_4 = \tau_2, \quad \tau_1, \tau_2 \text{ are arbitrary}; \\ &\tau_m = m\tau, \quad m = 1, 2, 3, 4, \quad \tau \text{ is arbitrary}. \end{aligned}$$

We seek a generalized separable solution of the nonlinear delay system (4.3.1.1) in the form

$$u = k_2 e^{\lambda(t+\tau_1)} \theta(\mathbf{x}) + u^\circ, \quad v = k_1 e^{\lambda(t+\tau_2)} \theta(\mathbf{x}) + v^\circ, \qquad (4.3.2.2)$$

where u° and v° are the stationary points (4.3.1.3) of system (4.3.1.1). We impose an additional restriction (linear differential constraint) on the function $\theta = \theta(\mathbf{x})$:

$$\Delta \theta = \mu \theta, \qquad (4.3.2.3)$$

which is the Helmholtz equation. The constants λ and μ involved in (4.3.2.2) and (4.3.2.3) are found in a subsequent analysis, and the function θ is determined from equation (4.3.2.3). For exact solutions to this equation with several space variables, see, for example, [404].

▶ **Example 4.17.** In the one-dimensional case, we have $\Delta\theta = \theta''_{xx}$, and the general solution to the linear ODE (4.3.2.3) with $\mu \neq 0$ is written as

$$\theta(x) = \begin{cases} C_1 \cos(\sqrt{|\mu|}\, x) + C_2 \sin(\sqrt{|\mu|}\, x) & \text{if } \mu < 0, \\ C_1 \exp(-\sqrt{\mu}\, x) + C_2 \exp(\sqrt{\mu}\, x) & \text{if } \mu > 0, \end{cases} \qquad (4.3.2.4)$$

where C_1 and C_2 are arbitrary constants. ◀

The functions (4.3.2.2) are chosen so that the arguments of the functions f, g, and h, appearing in system (4.3.1.1), become constant. Indeed, considering (4.3.2.1), we get

$$\begin{aligned} k_1 \bar{u}_1 - k_2 \bar{v}_2 &= k_1 u^\circ - k_2 v^\circ = \text{const}, \\ k_1 \bar{u}_3 - k_2 \bar{v}_4 &= k_1 u^\circ - k_2 v^\circ = \text{const}. \end{aligned} \qquad (4.3.2.5)$$

We substitute (4.3.2.2) into (4.3.1.1) and use relations (4.3.1.4) and (4.3.2.5) as well as equation (4.3.2.3). After simple rearrangements, we obtain the linear algebraic system

$$\begin{aligned} \lambda &= a_1 \mu + b_1 + c_1 f(k_1 u^\circ - k_2 v^\circ), \\ \lambda &= a_2 \mu + b_2 + c_2 f(k_1 u^\circ - k_2 v^\circ) \end{aligned} \qquad (4.3.2.6)$$

that serves to determine the parameters λ and μ. If $a_1 \neq a_2$, system (4.3.2.6) has the solution

$$\lambda = \frac{a_2 b_1 - a_1 b_2 + (a_2 c_1 - a_1 c_2) f^\circ}{a_2 - a_1}, \quad \mu = \frac{b_1 - b_2 + (c_1 - c_2) f^\circ}{a_2 - a_1}, \qquad (4.3.2.7)$$

where $f^\circ = f(k_1 u^\circ - k_2 v^\circ)$, and u° and v° are stationary points satisfying the algebraic system (4.3.1.4).

▶ **Example 4.18.** In the case $g = h = 0$, the coefficients (4.3.2.7) associated with the first three points (4.3.1.5) are independent of the form of the kinetic function f and can be expressed as

$$\begin{aligned}(a)\quad &\lambda = \frac{a_2 b_1 - a_1 b_2}{a_2 - a_1}, \quad &\mu &= \frac{b_1 - b_2}{a_2 - a_1}; \\ (b)\quad &\lambda = \frac{a_1(b_1 c_2 - b_2 c_1)}{c_1(a_2 - a_1)}, \quad &\mu &= \frac{\lambda}{a_1} = \frac{b_1 c_2 - b_2 c_1}{c_1(a_2 - a_1)}; \\ (c)\quad &\lambda = \frac{a_2(b_1 c_2 - b_2 c_1)}{c_2(a_2 - a_1)}, \quad &\mu &= \frac{\lambda}{a_2} = \frac{b_1 c_2 - b_2 c_1}{c_2(a_2 - a_1)}.\end{aligned} \qquad (4.3.2.8)$$

In case (d), the parameters degenerate, $\lambda = \mu = 0$; the associated solutions are of no interest.

Formulas (4.3.2.2), (4.3.2.4), and (4.3.2.8) and the first three stationary points (4.3.1.5) define six nondegenerate exact solutions (three for $\mu > 0$ and three for $\mu < 0$) to the nonlinear system (4.3.1.1) with $g = h = 0$, $a_1 \neq a_2$, and four constant delays satisfying condition (4.3.2.1). For exact solutions to the one-dimensional system (4.3.1.1) associated with the stationary point (b) in (4.3.1.5), see [417]; these are also given below.

$1°$. Solution with $\mu = \frac{b_1 c_2 - b_2 c_1}{c_1(a_2 - a_1)} < 0$:

$$\begin{aligned}u &= k_2 e^{a_1 \mu (t + \tau_1)} \left[C_1 \cos(\sqrt{|\mu|}\, x) + C_2 \sin(\sqrt{|\mu|}\, x) \right] + u°, \\ v &= k_1 e^{a_1 \mu (t + \tau_2)} \left[C_1 \cos(\sqrt{|\mu|}\, x) + C_2 \sin(\sqrt{|\mu|}\, x) \right],\end{aligned} \qquad (4.3.2.9)$$

where C_1 and C_2 are arbitrary constants.

$2°$. Solution with $\mu = \frac{b_1 c_2 - b_2 c_1}{c_1(a_2 - a_1)} > 0$:

$$\begin{aligned}u &= k_2 e^{a_1 \mu (t + \tau_1)} \left[C_1 \exp(-\sqrt{\mu}\, x) + C_2 \exp(\sqrt{\mu}\, x) \right] + u°, \\ v &= k_1 e^{a_1 \mu (t + \tau_2)} \left[C_1 \exp(-\sqrt{\mu}\, x) + C_2 \exp(\sqrt{\mu}\, x) \right].\end{aligned} \qquad (4.3.2.10)$$

Notably, for the Lotka–Volterra type system (4.3.1.2) defined by the functions $f(z) = z$ and $g = h = 0$, one should set $u° = -b_1/(c_1 k_1)$ in formulas (4.3.2.9) and (4.3.2.10). ◂

Remark 4.11. The more general, than (4.3.1.1), reaction-diffusion system of PDEs with four delays

$$\begin{aligned}u_t &= a_1 \Delta u + b_1 u + u f_1(k_1 \bar{u}_1 - k_2 \bar{v}_2) + g(k_1 \bar{u}_1 - k_2 \bar{v}_2), \\ v_t &= a_2 \Delta v + b_2 v + v f_2(k_1 \bar{u}_3 - k_2 \bar{v}_4) + h(k_1 \bar{u}_3 - k_2 \bar{v}_4),\end{aligned} \qquad (4.3.2.11)$$

where $f_1 = f_1(z_1)$ and $f_2 = f_2(z_2)$ are arbitrary functions such that $f_1(0) = f_2(0) = 0$, can be treated likewise. The other notations are the same as in system (4.3.1.1).

Suppose that the four delay times in system (4.3.2.11) are constrained by one relation (4.3.2.1). As before, we look for exact solutions to system (4.3.2.11) in the form (4.3.2.2), where $\theta = \theta(\mathbf{x})$ satisfies the Helmholtz equation (4.3.2.3). It can be shown that if $a_1 \neq a_2$, the parameters λ and μ appearing in solution (4.3.2.2) and the linear PDE (4.3.2.3) are expressed as

$$\lambda = \frac{a_2 b_1 - a_1 b_2 + a_2 f_1° - a_1 f_2°}{a_2 - a_1}, \quad \mu = \frac{b_1 - b_2 + f_1° - f_2°}{a_2 - a_1}, \qquad (4.3.2.12)$$

4.3. Reductions and Exact Solutions of Lotka–Volterra Type Systems with Several Delays 229

where $f_1^\circ = f_1(k_1 u^\circ - k_2 v^\circ)$ and $f_2^\circ = f_2(k_1 u^\circ - k_2 v^\circ)$, and u° and v° are stationary points satisfying the algebraic system

$$u^\circ[b_1 + f_1(k_1 u^\circ - k_2 v^\circ)] + g(k_1 u^\circ - k_2 v^\circ) = 0,$$
$$v^\circ[b_2 + f_2(k_1 u^\circ - k_2 v^\circ)] + h(k_1 u^\circ - k_2 v^\circ) = 0.$$

Two reductions of the system of PDEs with three arbitrary delays to a stationary system. We will assume that the four delays in system (4.3.1.1) are constrained by one relation (4.3.2.1).

1°. *Generalized separable solutions exponential in t.* We look for generalized separable solutions to system (4.3.1.1) in the form [417]:

$$u = k_2 e^{\lambda(t+\tau_1)}\theta(\mathbf{x}) + \varphi(\mathbf{x}), \quad v = k_1 e^{\lambda(t+\tau_2)}\theta(\mathbf{x}) + \psi(\mathbf{x}), \qquad (4.3.2.13)$$

where the functions $\theta = \theta(\mathbf{x})$, $\varphi = \varphi(\mathbf{x})$, and $\psi = \psi(\mathbf{x})$ and parameter λ are to be determined in the subsequent analysis.

The functions (4.3.2.13) are chosen so that the arguments of the functions f, g, and h in system (4.3.1.1) are only dependent on $\mathbf{x}$. Indeed, considering relation (4.3.2.1), we get

$$k_1\bar{u}_1 - k_2\bar{v}_2 = k_1\varphi(\mathbf{x}) - k_2\psi(\mathbf{x}),$$
$$k_1\bar{u}_3 - k_2\bar{v}_4 = k_1\varphi(\mathbf{x}) - k_2\psi(\mathbf{x}). \qquad (4.3.2.14)$$

Substituting (4.3.2.13) into (4.3.1.1) and taking into account (4.3.2.14), we obtain

$$k_2 e^{\lambda(t+\tau_1)}\big[a_1\Delta\theta + (b_1 - \lambda + c_1\hat{f})\theta\big] + a_1\Delta\varphi + (b_1 + c_1\hat{f})\varphi + \hat{g} = 0,$$
$$k_1 e^{\lambda(t+\tau_2)}\big[a_2\Delta\theta + (b_2 - \lambda + c_2\hat{f})\theta\big] + a_2\Delta\psi + (b_2 + c_2\hat{f})\psi + \hat{h} = 0, \qquad (4.3.2.15)$$

where the short notations $\hat{f} = f(k_1\varphi - k_2\psi)$, $\hat{g} = g(k_1\varphi - k_2\psi)$, and $\hat{h} = h(k_1\varphi - k_2\psi)$ have been introduced.

Relations (4.3.2.15) can be satisfied by setting

$$a_1\Delta\varphi + (b_1 + c_1\hat{f})\varphi + \hat{g} = 0,$$
$$a_2\Delta\psi + (b_2 + c_2\hat{f})\psi + \hat{h} = 0,$$
$$a_1\Delta\theta + (b_1 - \lambda + c_1\hat{f})\theta = 0,$$
$$a_2\Delta\theta + (b_2 - \lambda + c_2\hat{f})\theta = 0. \qquad (4.3.2.16)$$

The first two equations in (4.3.2.16) form a closed system for φ and ψ, and the last two equations in (4.3.2.16) form an overdetermined system for one function θ. Requiring that the last two equations of system (4.3.2.16) must coincide, we find the parameter λ and other constants:

$$\lambda = \frac{a_2 b_1 - a_1 b_2}{a_2 - a_1}, \quad c_1 = a_1, \quad c_2 = a_2 \quad (a_1 \neq a_2). \qquad (4.3.2.17)$$

In selecting c_1 and c_2, it has been taken into account that the function f is defined up to a constant multiplier.

Substituting (4.3.2.17) into (4.3.2.16) gives the system of stationary PDEs

$$a_1\Delta\varphi + (b_1 + a_1\hat{f})\varphi + \hat{g} = 0,$$
$$a_2\Delta\psi + (b_2 + a_2\hat{f})\psi + \hat{h} = 0, \qquad (4.3.2.18)$$
$$\Delta\theta + \left(\frac{b_2 - b_1}{a_2 - a_1} + \hat{f}\right)\theta = 0,$$

where $\hat{f} = f(k_1\varphi - k_2\psi)$, $\hat{g} = g(k_1\varphi - k_2\psi)$, and $\hat{h} = h(k_1\varphi - k_2\psi)$.

Remark 4.12. *The determining system of stationary PDEs (4.3.2.18) is independent of the delays. Therefore, solutions of the form (4.3.2.13) to the nonstationary systems (4.3.1.1) and (4.3.1.2) with no delays ($\tau_1 = \tau_2 = \tau_3 = \tau_4 = 0$) generate exact solutions to more complicated systems (4.3.1.1) and (4.3.1.2) with several delays satisfying condition (4.3.2.1).*

If $\hat{g}=0$ (or $\hat{h}=0$), system (4.3.2.18) significantly simplifies so that the function φ (or ψ) can be set equal to zero. In these cases, there are only two equations remaining.

▶ **Example 4.19.** For the one-dimensional delay Lotka–Volterra type system (4.3.1.2), one should set $f(z_{1,2}) = z_{1,2}$ and $g(z_1) = h(z_2) = 0$ in the respective truncated system (4.3.2.18). If $\psi = 0$, the system reduces to two ODEs

$$\varphi''_{xx} + (b + k\varphi)\varphi = 0, \qquad (4.3.2.19)$$
$$\theta''_{xx} + (\beta + k\varphi)\theta = 0, \qquad (4.3.2.20)$$

in which the notations $b = b_1/a_1$, $k = k_1$, and $\beta = (b_2 - b_1)/(a_2 - a_1)$ are employed. In what follows, we assume that $\varphi(x) \not\equiv \text{const}$ (the special case $\varphi = u^\circ = \text{const}$ was discussed in Example 4.18).

The linear transformation

$$\varphi = \varphi_1 - \frac{b}{k}, \quad \theta = \theta_1 \quad \text{(transformation } \mathfrak{G}_1\text{)}$$

converts the system of ODEs (4.3.2.19)–(4.3.2.20) to a similar system with different determining parameters

$$(\varphi_1)''_{xx} + (-b + k\varphi_1)\varphi_1 = 0,$$
$$(\theta_1)''_{xx} + (\beta - b + k\varphi_1)\theta_1 = 0.$$

The linear transformation

$$\varphi = -\varphi_2, \quad \theta = \theta_2, \quad z = ix, \quad i^2 = -1 \quad \text{(transformation } \mathfrak{G}_2\text{)},$$

also converts system (4.3.2.19)–(4.3.2.20) to a similar system with different determining parameters

$$(\varphi_2)''_{zz} + (-b + k\varphi_2)\varphi_2 = 0,$$
$$(\theta_2)''_{zz} + (-\beta + k\varphi_2)\theta_2 = 0.$$

Let (b, β) denote system (4.3.2.19)–(4.3.2.20). Then the transformations $\mathfrak{G}_1$ and $\mathfrak{G}_2$ and their composition $\mathfrak{G}_1 \circ \mathfrak{G}_2$ relate the system to three other similar systems, which can be schematically displayed as

$$(b, \beta) \xrightarrow{\mathfrak{G}_1} (-b, \beta - b); \quad (b, \beta) \xrightarrow{\mathfrak{G}_2} (-b, -\beta); \quad (b, \beta) \xrightarrow{\mathfrak{G}_1 \circ \mathfrak{G}_2} (b, b - \beta).$$

4.3. Reductions and Exact Solutions of Lotka–Volterra Type Systems with Several Delays

These transformations allow one to construct exact solutions to system (4.3.2.19)–(4.3.2.20). The following statement is true.

Proposition. Let the functions

$$\varphi = \Phi(x, b, k), \qquad \theta = \Theta(x, b, \beta, k)$$

represent an exact solution to the system of ODEs (4.3.2.19)–(4.3.2.20). Then the three pairs of formulas specified in the last three rows of Table 4.1 define exact solutions to this system with other values of the determining parameters b and β (k remains unchanged).

Table 4.1. Exact solutions to related systems of ODEs (4.3.2.19)–(4.3.2.20) that can be expressed through the solution to the original system (b, β).

System of ODEs	Function φ	Function θ
(b, β), original system	$\Phi(x, b, k)$	$\Theta(x, b, \beta, k)$
$(-b, \beta - b)$	$\Phi(x, b, k) + bk^{-1}$	$\Theta(x, b, \beta, k)$
$(-b, -\beta)$	$-\Phi(ix, b, k)$	$\Theta(ix, b, \beta, k)$
$(b, b - \beta)$	$-\Phi(ix, b, k) - bk^{-1}$	$\Theta(ix, b, \beta, k)$

The general solution of the autonomous ODE (4.3.2.19) can be represented in the implicit form

$$C_1 \pm x = \int \frac{d\varphi}{\sqrt{C_2 - b\varphi^2 - \tfrac{2}{3}k\varphi^3}}, \qquad (4.3.2.21)$$

where C_1 and C_2 are arbitrary constants. If $C_2 = 0$ and $b = 0$, $C_2 = 0$ and $b \neq 0$, or $C_2 = \tfrac{1}{3}b^3/k^2$ and $b \neq 0$, the integral in (4.3.2.21) is computable in terms of elementary functions, and the function φ can be expressed in explicit form. In general, the integral on the right-hand side of (4.3.2.21) is not expressible in terms of elementary functions.

If for some $\varphi(x)$, a nontrivial particular solution $\theta_0 = \theta_0(x)$ of the linear homogeneous equation (4.3.2.20) is known, the general solution of the equation is expressed as [423]:

$$\theta = \theta_0 \left(C_3 + C_4 \int \frac{dx}{\theta_0^2} \right), \qquad (4.3.2.22)$$

where C_3 and C_4 are arbitrary constants.

Remark 4.13. For $\beta = b$, equation (4.3.2.20) with $\theta = \varphi$ coincides with equation (4.3.2.19). It follows that in this case, the function $\theta_0 = \varphi$ is a particular solution to equation (4.3.2.20). Substituting $\theta_0 = \varphi$ into (4.3.2.22) yields the general solution of equation (4.3.2.20) with $\beta = b$.

Table 4.2 summarizes exact elementary-function solutions to the system of ODEs (4.3.2.19)–(4.3.2.20) for various values of the determining parameters b and β. These solutions were obtained in [431] using formulas (4.3.2.21) and (4.3.2.22), transformations $\mathfrak{G}_1$, $\mathfrak{G}_2$, and $\mathfrak{G}_1 \circ \mathfrak{G}_2$, and the handbooks [421, 423].

Notably, the Bessel functions and modified Bessel functions of the fractional order $\nu = 5/2$, which define the function θ in Table 4.2 (see rows 2 and 3), are expressible in terms of elementary functions by the formulas [423]:

$$J_{5/2}(\zeta) = \sqrt{\frac{2}{\pi}} \frac{3\sin\zeta - 3\zeta\cos\zeta - \zeta^2\sin\zeta}{\zeta^{5/2}},$$

$$Y_{5/2}(\zeta) = -\sqrt{\frac{2}{\pi}} \frac{3\cos\zeta + 3\zeta\sin\zeta - \zeta^2\cos\zeta}{\zeta^{5/2}},$$

$$I_{5/2}(\zeta) = \sqrt{\frac{2}{\pi}} \frac{3\sinh\zeta - 3\zeta\cosh\zeta + \zeta^2\sinh\zeta}{\zeta^{5/2}},$$

$$K_{5/2}(\zeta) = \sqrt{\frac{\pi}{2}} \frac{3 + 3\zeta + \zeta^2}{e^\zeta \zeta^{5/2}}.$$

◀

Remark 4.14. *The value $\beta = b$ in Table 4.2 corresponds to stationary solutions of the form (4.3.2.13) with $\psi = 0$ to the system of PDEs (4.3.1.2) with parameters (4.3.2.17) where $\lambda = 0$.*

$2°$. *Generalized separable solutions linear in t.* System (4.3.1.1) with four delays satisfying one relation (4.3.2.1) also admit exact generalized separable solutions of the form

$$u = k_2(t + \tau_1)\theta(\mathbf{x}) + \varphi(\mathbf{x}), \quad v = k_1(t + \tau_2)\theta(\mathbf{x}) + \psi(\mathbf{x}). \quad (4.3.2.23)$$

An analysis similar to that performed previously allows us to find the parameters of equations (4.3.1.1):

$$b_1 = \sigma a_1, \quad b_2 = \sigma a_2, \quad c_1 = a_1, \quad c_2 = a_2, \quad (4.3.2.24)$$

where σ is an arbitrary constant. In this case, the functions $\theta = \theta(\mathbf{x})$, $\varphi = \varphi(\mathbf{x})$, and $\psi = \psi(\mathbf{x})$ are described by the stationary system of PDEs

$$a_1[\Delta\varphi + (\sigma + \hat{f})\varphi] + \hat{g} - k_2\theta = 0,$$
$$a_2[\Delta\psi + (\sigma + \hat{f})\psi] + \hat{h} - k_1\theta = 0, \quad (4.3.2.25)$$
$$\Delta\theta + (\sigma + \hat{f})\theta = 0.$$

where $\hat{f} = f(k_1\varphi - k_2\psi)$, $\hat{g} = g(k_1\varphi - k_2\psi)$, $\hat{h} = h(k_1\varphi - k_2\psi)$.

▶ **Example 4.20.** For the one-dimensional delay Lotka–Volterra type system (4.3.1.2), one should set $f(z_{1,2}) = z_{1,2}$ and $g(z_1) = h(z_2) = 0$ in the respective reduced system (4.3.2.25). Then, the functions $\varphi = \varphi(x)$, $\psi = \psi(x)$, and $\theta = \theta(x)$ are described by the stationary system of ODEs

$$\varphi''_{xx} + (\sigma + \rho)\varphi - (k_2/a_1)\theta = 0,$$
$$\psi''_{xx} + (\sigma + \rho)\psi - (k_1/a_2)\theta = 0, \quad (4.3.2.26)$$
$$\theta''_{xx} + (\sigma + \rho)\theta = 0, \quad \rho = k_1\varphi - k_2\psi.$$

Let us look at the special case $a_1 = a_2 = a$. Adding up the first two equations in (4.3.2.26) multiplied by k_1 and $-k_2$, respectively, we obtain an isolated ODE

4.3. Reductions and Exact Solutions of Lotka–Volterra Type Systems with Several Delays

Table 4.2. Solutions to the system of ODEs (4.3.2.19), (4.3.2.20) at various values of the determining parameters. Notations: $J_{5/2}(\xi)$ and $Y_{5/2}(\xi)$ are the Bessel functions of the first and second kind, $I_{5/2}(\xi)$ and $K_{5/2}(\xi)$ are the modified Bessel functions of the first and second kind, and A and B are arbitrary constants.

No.	Parameters b and β	Function φ, ODE (4.3.2.19)	Function θ, ODE (4.3.2.20)	Function $\xi = \xi(x)$
1	$b=0, \beta=0$	$-\frac{6}{k}x^{-2}$	$Ax^{-2}+Bx^3$	—
2	$b=0, \beta>0$	$-\frac{6}{k}x^{-2}$	$\sqrt{x}\,[AJ_{5/2}(\sqrt{\beta}\,x)+BY_{5/2}(\sqrt{\beta}\,x)]$	—
3	$b=0, \beta<0$	$-\frac{6}{k}x^{-2}$	$\sqrt{x}\,[AI_{5/2}(\sqrt{-\beta}\,x)+BK_{5/2}(\sqrt{-\beta}\,x)]$	—
4	$b<0, \beta=0$	$-\frac{3b}{2k}+\frac{3b}{2k}\tanh^2\xi$	$A(3\tanh^2\xi-1)$ $+B[3\tanh\xi+(3\tanh^2\xi-1)\operatorname{artanh}\xi]$	$\frac{1}{2}\sqrt{-b}\,x$
5	$b<0, \beta=0$	$-\frac{3b}{2k}+\frac{3b}{2k}\coth^2\xi$	$A(3\coth^2\xi-1)$ $+B[3\coth\xi+(3\coth^2\xi-1)\operatorname{arcoth}\xi]$	$\frac{1}{2}\sqrt{-b}\,x$
6	$b>0, \beta=0$	$-\frac{3b}{2k}-\frac{3b}{2k}\tan^2\xi$	$A(3\tan^2\xi+1)$ $+B[3\tan\xi+(3\tan^2\xi+1)\arctan\xi]$	$\frac{1}{2}\sqrt{b}\,x$
7	$b>0, \beta=0$	$-\frac{3b}{2k}-\frac{3b}{2k}\cot^2\xi$	$A(3\cot^2\xi+1)$ $+B[3\cot\xi+(3\cot^2\xi+1)\operatorname{arccot}\xi]$	$\frac{1}{2}\sqrt{b}\,x$
8	$b<0, \beta=0$	$\frac{b}{2k}+\frac{3b}{2k}\tan^2\xi$	$A\cos^{-2}\xi+B[\sin(2\xi)+3\tan\xi+3\xi\cos^{-2}\xi]$	$\frac{1}{2}\sqrt{-b}\,x$
9	$b<0, \beta=0$	$\frac{b}{2k}+\frac{3b}{2k}\cot^2\xi$	$A\sin^{-2}\xi+B[\cos(2\xi)+3\cot\xi+3\xi\sin^{-2}\xi]$	$\frac{1}{2}\sqrt{-b}\,x$
10	$b>0, \beta=0$	$\frac{b}{2k}-\frac{3b}{2k}\tanh^2\xi$	$A\cosh^{-2}\xi$ $+B[\sinh(2\xi)+3\tanh\xi+3\xi\cosh^{-2}\xi]$	$\frac{1}{2}\sqrt{b}\,x$
11	$b>0, \beta=0$	$\frac{b}{2k}-\frac{3b}{2k}\coth^2\xi$	$A\sinh^{-2}\xi$ $+B[\cosh(2\xi)+3\coth\xi+3\xi\sinh^{-2}\xi]$	$\frac{1}{2}\sqrt{b}\,x$
12	$b<0, \beta=-3b$	$\frac{b}{2k}+\frac{3b}{2k}\tan^2\xi$	$A\sin\xi\cos^3\xi$ $+B(2+\cos^{-2}\xi+8\cos^2\xi-16\cos^4\xi)$	$\frac{1}{2}\sqrt{-b}\,x$
13	$b<0, \beta=-3b$	$\frac{b}{2k}+\frac{3b}{2k}\cot^2\xi$	$A\cos\xi\sin^3\xi$ $+B(2+\sin^{-2}\xi+8\sin^2\xi-16\sin^4\xi)$	$\frac{1}{2}\sqrt{-b}\,x$
14	$b>0, \beta=-3b$	$\frac{b}{2k}-\frac{3b}{2k}\tanh^2\xi$	$A\sinh\xi\cosh^3\xi$ $+B(2+\cosh^{-2}\xi+8\cosh^2\xi-16\cosh^4\xi)$	$\frac{1}{2}\sqrt{b}\,x$
15	$b>0, \beta=-3b$	$\frac{b}{2k}-\frac{3b}{2k}\coth^2\xi$	$A\cosh\xi\sinh^3\xi$ $+B(2+\sinh^{-2}\xi+8\sinh^2\xi-16\sinh^4\xi)$	$\frac{1}{2}\sqrt{b}\,x$
16	$b<0, \beta=-\frac{5}{4}b$	$\frac{b}{2k}+\frac{3b}{2k}\tan^2\xi$	$A\cos^3\xi+B\sin\xi\,(4+3\cos^{-2}\xi+8\cos^2\xi)$	$\frac{1}{2}\sqrt{-b}\,x$
17	$b<0, \beta=-\frac{5}{4}b$	$\frac{b}{2k}+\frac{3b}{2k}\cot^2\xi$	$A\sin^3\xi+B\cos\xi\,(4+3\sin^{-2}\xi+8\sin^2\xi)$	$\frac{1}{2}\sqrt{-b}\,x$
18	$b>0, \beta=-\frac{5}{4}b$	$\frac{b}{2k}-\frac{3b}{2k}\tanh^2\xi$	$A\cosh^3\xi+B\sinh\xi\,(4+3\cosh^{-2}\xi+8\cosh^2\xi)$	$\frac{1}{2}\sqrt{b}\,x$
19	$b>0, \beta=-\frac{5}{4}b$	$\frac{b}{2k}-\frac{3b}{2k}\coth^2\xi$	$A\sinh^3\xi+B\cosh\xi\,(4+3\sinh^{-2}\xi+8\sinh^2\xi)$	$\frac{1}{2}\sqrt{b}\,x$

Table 4.2. (*Continued*)

No.	Parameters b and β	Function φ, ODE (4.3.2.19)	Function θ, ODE (4.3.2.20)	Function $\xi = \xi(x)$
20	$b<0$, $\beta=\frac{1}{4}b$	$-\frac{3b}{2k}+\frac{3b}{2k}\tanh^2\xi$	$A\sinh\xi\cosh^{-2}\xi$ $+B\cosh^{-1}\xi\left(3\xi\tanh\xi-3+\cosh^2\xi\right)$	$\frac{1}{2}\sqrt{-b}\,x$
21	$b<0$, $\beta=\frac{1}{4}b$	$-\frac{3b}{2k}+\frac{3b}{2k}\coth^2\xi$	$A\cosh\xi\sinh^{-2}\xi$ $+B\sinh^{-1}\xi\left(3\xi\coth\xi-3+\sinh^2\xi\right)$	$\frac{1}{2}\sqrt{-b}\,x$
22	$b>0$, $\beta=\frac{1}{4}b$	$-\frac{3b}{2k}-\frac{3b}{2k}\tan^2\xi$	$A\sin\xi\cos^{-2}\xi$ $+B\cos^{-1}\xi\left(3\xi\tan\xi+3-\cos^2\xi\right)$	$\frac{1}{2}\sqrt{b}\,x$
23	$b>0$, $\beta=\frac{1}{4}b$	$-\frac{3b}{2k}-\frac{3b}{2k}\cot^2\xi$	$A\cos\xi\sin^{-2}\xi$ $+B\sin^{-1}\xi\left(3\xi\cot\xi+3-\sin^2\xi\right)$	$\frac{1}{2}\sqrt{b}\,x$
24	$b<0$, $\beta=\frac{3}{4}b$	$\frac{b}{2k}+\frac{3b}{2k}\tan^2\xi$	$A\sin\xi\cos^{-2}\xi$ $+B\cos^{-1}\xi\left(3\xi\tan\xi+3-\cos^2\xi\right)$	$\frac{1}{2}\sqrt{-b}\,x$
25	$b<0$, $\beta=\frac{3}{4}b$	$\frac{b}{2k}+\frac{3b}{2k}\cot^2\xi$	$A\cos\xi\sin^{-2}\xi$ $+B\sin^{-1}\xi\left(3\xi\cot\xi+3-\sin^2\xi\right)$	$\frac{1}{2}\sqrt{-b}\,x$
26	$b>0$, $\beta=\frac{3}{4}b$	$\frac{b}{2k}-\frac{3b}{2k}\tanh^2\xi$	$A\sinh\xi\cosh^{-2}\xi$ $+B\cosh^{-1}\xi\left(3\xi\tanh\xi-3+\cosh^2\xi\right)$	$\frac{1}{2}\sqrt{b}\,x$
27	$b>0$, $\beta=\frac{3}{4}b$	$\frac{b}{2k}-\frac{3b}{2k}\coth^2\xi$	$A\cosh\xi\sinh^{-2}\xi$ $+B\sinh^{-1}\xi\left(3\xi\coth\xi-3+\sinh^2\xi\right)$	$\frac{1}{2}\sqrt{b}\,x$
28	$b<0$, $\beta=b$	$\frac{b}{2k}+\frac{3b}{2k}\tan^2\xi$	$A\left(1+3\tan^2\xi\right)+B\left[\xi\left(1+3\tan^2\xi\right)+3\tan\xi\right]$	$\frac{1}{2}\sqrt{-b}\,x$
29	$b<0$, $\beta=b$	$\frac{b}{2k}+\frac{3b}{2k}\cot^2\xi$	$A\left(1+3\cot^2\xi\right)+B\left[\xi\left(1+3\cot^2\xi\right)+3\cot\xi\right]$	$\frac{1}{2}\sqrt{-b}\,x$
30	$b>0$, $\beta=b$	$\frac{b}{2k}-\frac{3b}{2k}\tanh^2\xi$	$A\left(1-3\tanh^2\xi\right)$ $+B\left[\xi\left(1-3\tanh^2\xi\right)+3\tanh\xi\right]$	$\frac{1}{2}\sqrt{b}\,x$
31	$b>0$, $\beta=b$	$\frac{b}{2k}-\frac{3b}{2k}\coth^2\xi$	$A\left(1-3\coth^2\xi\right)$ $+B\left[\xi\left(1-3\coth^2\xi\right)+3\coth\xi\right]$	$\frac{1}{2}\sqrt{b}\,x$
32	$b<0$, $\beta=b$	$-\frac{3b}{2k}+\frac{3b}{2k}\tanh^2\xi$	$A\cosh^{-2}\xi$ $+B\left[\left(3+2\cosh^2\xi\right)\tanh\xi+3\xi\cosh^{-2}\xi\right]$	$\frac{1}{2}\sqrt{-b}\,x$
33	$b<0$, $\beta=b$	$-\frac{3b}{2k}+\frac{3b}{2k}\coth^2\xi$	$A\sinh^{-2}\xi$ $+B\left[\left(3+2\sinh^2\xi\right)\coth\xi+3\xi\sinh^{-2}\xi\right]$	$\frac{1}{2}\sqrt{-b}\,x$
34	$b>0$, $\beta=b$	$-\frac{3b}{2k}-\frac{3b}{2k}\tan^2\xi$	$A\cos^{-2}\xi+B\left[\left(3+2\cos^2\xi\right)\tan\xi+3\xi\cos^{-2}\xi\right]$	$\frac{1}{2}\sqrt{b}\,x$
35	$b>0$, $\beta=b$	$-\frac{3b}{2k}-\frac{3b}{2k}\cot^2\xi$	$A\sin^{-2}\xi+B\left[\left(3+2\sin^2\xi\right)\cot\xi+3\xi\sin^{-2}\xi\right]$	$\frac{1}{2}\sqrt{b}\,x$
36	$b<0$, $\beta=\frac{9}{4}b$	$-\frac{3b}{2k}+\frac{3b}{2k}\tanh^2\xi$	$A\cosh^3\xi+B\sinh\xi\left(4+3\cosh^{-2}\xi+8\cosh^2\xi\right)$	$\frac{1}{2}\sqrt{-b}\,x$
37	$b<0$, $\beta=\frac{9}{4}b$	$-\frac{3b}{2k}+\frac{3b}{2k}\coth^2\xi$	$A\sinh^3\xi+B\cosh\xi\left(4+3\sinh^{-2}\xi+8\sinh^2\xi\right)$	$\frac{1}{2}\sqrt{-b}\,x$
38	$b>0$, $\beta=\frac{9}{4}b$	$-\frac{3b}{2k}-\frac{3b}{2k}\tan^2\xi$	$A\cos^3\xi+B\sin\xi\left(4+3\cos^{-2}\xi+8\cos^2\xi\right)$	$\frac{1}{2}\sqrt{b}\,x$
39	$b>0$, $\beta=\frac{9}{4}b$	$-\frac{3b}{2k}-\frac{3b}{2k}\cot^2\xi$	$A\sin^3\xi+B\cos\xi\left(4+3\sin^{-2}\xi+8\sin^2\xi\right)$	$\frac{1}{2}\sqrt{b}\,x$

Table 4.2. (*Continued*)

No.	Parameters b and β	Function φ, ODE (4.3.2.19)	Function θ, ODE (4.3.2.20)	Function $\xi = \xi(x)$
40	$b<0$, $\beta=4b$	$-\frac{3b}{2k}+\frac{3b}{2k}\tanh^2\xi$	$A\sinh\xi\cosh^3\xi$ $+B(2+\cosh^{-2}\xi+8\cosh^2\xi-16\cosh^4\xi)$	$\frac{1}{2}\sqrt{-b}\,x$
41	$b<0$, $\beta=4b$	$-\frac{3b}{2k}+\frac{3b}{2k}\coth^2\xi$	$A\cosh\xi\sinh^3\xi$ $+B(2+\sinh^{-2}\xi+8\sinh^2\xi-16\sinh^4\xi)$	$\frac{1}{2}\sqrt{-b}\,x$
42	$b>0$, $\beta=4b$	$-\frac{3b}{2k}-\frac{3b}{2k}\tan^2\xi$	$A\sin\xi\cos^3\xi$ $+B(2+\cos^{-2}\xi+8\cos^2\xi-16\cos^4\xi)$	$\frac{1}{2}\sqrt{b}\,x$
43	$b>0$, $\beta=4b$	$-\frac{3b}{2k}-\frac{3b}{2k}\cot^2\xi$	$A\cos\xi\sin^3\xi$ $+B(2+\sin^{-2}\xi+8\sin^2\xi-16\sin^4\xi)$	$\frac{1}{2}\sqrt{b}\,x$

for $\rho = k_1\varphi - k_2\psi$. Together with the third and first ODEs of (4.3.2.26) and the algebraic relation for ρ, we get the following mixed algebraic-differential system of equations:

$$\begin{aligned} \rho''_{xx} + (\sigma + \rho)\rho &= 0, \\ \theta''_{xx} + (\sigma + \rho)\theta &= 0, \\ \varphi''_{xx} + (\sigma + \rho)\varphi - (k_2/a)\theta &= 0, \\ \rho &= k_1\varphi - k_2\psi. \end{aligned} \quad (4.3.2.27)$$

We will give two simple classes of exact solutions to system (4.3.2.27) that correspond to $\rho = \text{const}$ and then describe a number of more complicated solutions.

1°. Exact solution of system (4.3.2.27) with $\rho = -\sigma$:

$$\rho = -\sigma, \quad \varphi = C_1 x + C_2 + \frac{k_2}{a}\left(\frac{1}{6}C_3 x^3 + \frac{1}{2}C_4 x^2\right),$$

$$\psi = \frac{1}{k_2}(k_1\varphi + \sigma), \quad \theta = C_3 x + C_4,$$

where $C_1, \ldots, C_4$ are arbitrary constants.

2°. Exact solution of system (4.3.2.27) with $\rho = 0$:

$$\rho = 0, \quad \varphi = \varphi_p(x) + \begin{cases} C_1\cos(\sqrt{\sigma}\,x) + C_2\sin(\sqrt{\sigma}\,x) & \text{if } \sigma > 0, \\ C_1\cosh(\sqrt{|\sigma|}\,x) + C_2\sinh(\sqrt{|\sigma|}\,x) & \text{if } \sigma < 0, \end{cases}$$

$$\psi = \frac{k_1}{k_2}\varphi, \quad \theta = \begin{cases} C_3\cos(\sqrt{\sigma}\,x) + C_4\sin(\sqrt{\sigma}\,x) & \text{if } \sigma > 0, \\ C_3\cosh(\sqrt{|\sigma|}\,x) + C_4\sinh(\sqrt{|\sigma|}\,x) & \text{if } \sigma < 0, \end{cases}$$

$$\varphi_p(x) = \begin{cases} -\dfrac{k_2 x}{2a\sqrt{\sigma}}\left[C_4\cos(\sqrt{\sigma}\,x) - C_3\sin(\sqrt{\sigma}\,x)\right] + \dfrac{C_3 k_2}{2a\sigma}\cos(\sqrt{\sigma}\,x) \\ \hfill \text{if } \sigma > 0, \\ \dfrac{k_2 x}{2a\sqrt{|\sigma|}}\left[C_4\cosh(\sqrt{|\sigma|}\,x) + C_3\sinh(\sqrt{|\sigma|}\,x)\right] + \dfrac{C_3 k_2}{2a\sigma}\cosh(\sqrt{|\sigma|}\,x) \\ \hfill \text{if } \sigma < 0. \end{cases}$$

3°. Up to notation, the first two ODEs of system (4.3.2.27) coincide with system (4.3.2.19)–(4.3.2.20) where $k = 1$ and $\beta = b$, whose exact solutions were described above. For given $\rho = \rho(x)$, the second equation in (4.3.2.27) for θ is a second-order linear homogeneous ODE, which admits the particular solution $\theta_0 = \rho(x)$. Hence, its general solution is expressed by formula (4.3.2.22). For given θ, the third equation in (4.3.2.27) is a second-order linear nonhomogeneous ODE for $\varphi = \varphi(x)$, with $\varphi_0 = \rho(x)$ being a particular solution of the respective homogeneous ODE. Considering the above and employing relevant formulas presented in [423], we can find the general solution of the ODE for φ. As a result, we obtain a solution to system (4.3.2.27) in the form

$$\rho = \rho(x), \quad \theta = C_1\rho(x) + C_2\omega(x), \quad \varphi = C_3\rho(x) + C_4\omega(x) + \varphi_p(x),$$
$$\psi = \frac{1}{k_2}[(C_3k_1 - 1)\rho(x) + C_4k_1\omega(x) + k_1\varphi_p(x)], \quad \omega(x) = \rho(x)\int \frac{dx}{\rho^2(x)}, \quad (4.3.2.28)$$

where $C_1, \ldots, C_4$ are arbitrary constants, and $\varphi_p(x)$ is a particular solution to the third ODE of system (4.3.2.27), which is expressed as

$$\varphi_p(x) = \frac{k_2}{a}\left[\omega(x)\int \rho(x)\theta(x)\frac{dx}{\mathcal{W}(x)} - \rho(x)\int \omega(x)\theta(x)\frac{dx}{\mathcal{W}(x)}\right]. \quad (4.3.2.29)$$

The function $\mathcal{W}(x) = \rho\omega'_x - \omega\rho'_x$ is the Wronskian determinant. Simple computations show that $\mathcal{W}(x) = 1$.

To obtain exact solutions of system (4.3.2.27), one should substitute the functions $\rho(x)$ and $\omega(x)$ displayed in Table 4.3 into formulas (4.3.2.28) and (4.3.2.29) with $\mathcal{W}(x) = 1$. ◀

Reduction of the system of PDEs with a single delay to a nonstationary system of delay ODEs. Suppose that the four delays in system (4.3.1.1) are all equal:

$$\tau_1 = \tau_2 = \tau_3 = \tau_4 = \tau. \quad (4.3.2.30)$$

We look for generalized separable solutions to system (4.3.1.1) under condition (4.3.2.30) in the form [417]:

$$u = k_2\xi(t)\theta(\mathbf{x}) + \varphi(t), \quad v = k_1\xi(t)\theta(\mathbf{x}) + \psi(t), \quad (4.3.2.31)$$

where the functions $\theta = \theta(\mathbf{x})$, $\xi = \xi(t)$, $\varphi = \varphi(t)$, and $\psi = \psi(t)$ are to be determined in the subsequent analysis.

The functions (4.3.2.31) are chosen so that the arguments of the functions f, g, and h appearing in system (4.3.1.1) become dependent on t alone. We impose an additional condition (a linear differential constraint) on the function $\theta = \theta(\mathbf{x})$:

$$\Delta\theta = \mu\theta + \varepsilon, \quad (4.3.2.32)$$

where the constant μ and ε are found in the subsequent investigation.

Notably, in the nondegenerate case of $\mu \neq 0$ in (4.3.2.32), we can set $\varepsilon = 0$ without loss of generality, since a translation of θ by a constant simply leads, by

Table 4.3. The functions $\rho(x)$ and $\omega(x)$ determining exact solutions of the form (4.3.2.28) to the system of ODEs (4.3.2.27) (according to [431]).

No.	Parameter σ	Function ρ	Function ω	Function ξ
1	$\sigma = 0$	$-\frac{6}{k}x^{-2}$	x^3	—
2	$\sigma < 0$	$\frac{\sigma}{2} + \frac{3\sigma}{2}\tan^2\xi$	$\xi(1+3\tan^2\xi) + 3\tan\xi$	$\frac{1}{2}\sqrt{-\sigma}\,x$
3	$\sigma < 0$	$\frac{\sigma}{2} + \frac{3\sigma}{2}\cot^2\xi$	$\xi(1+3\cot^2\xi) + 3\cot\xi$	$\frac{1}{2}\sqrt{-\sigma}\,x$
4	$\sigma > 0$	$\frac{\sigma}{2} - \frac{3\sigma}{2}\tanh^2\xi$	$\xi(1-3\tanh^2\xi) + 3\tanh\xi$	$\frac{1}{2}\sqrt{\sigma}\,x$
5	$\sigma > 0$	$\frac{\sigma}{2} - \frac{3\sigma}{2}\coth^2\xi$	$\xi(1-3\coth^2\xi) + 3\coth\xi$	$\frac{1}{2}\sqrt{\sigma}\,x$
6	$\sigma < 0$	$-\frac{3\sigma}{2} + \frac{3\sigma}{2}\tanh^2\xi$	$(3+2\cosh^2\xi)\tanh\xi + 3\xi\cosh^{-2}\xi$	$\frac{1}{2}\sqrt{-\sigma}\,x$
7	$\sigma < 0$	$-\frac{3\sigma}{2} + \frac{3\sigma}{2}\coth^2\xi$	$(3+2\sinh^2\xi)\coth\xi + 3\xi\sinh^{-2}\xi$	$\frac{1}{2}\sqrt{-\sigma}\,x$
8	$\sigma > 0$	$-\frac{3\sigma}{2} - \frac{3\sigma}{2}\tan^2\xi$	$(3+2\cos^2\xi)\tan\xi + 3\xi\cos^{-2}\xi$	$\frac{1}{2}\sqrt{\sigma}\,x$
9	$\sigma > 0$	$-\frac{3\sigma}{2} - \frac{3\sigma}{2}\cot^2\xi$	$(3+2\sin^2\xi)\cot\xi + 3\xi\sin^{-2}\xi$	$\frac{1}{2}\sqrt{\sigma}\,x$

virtue of (4.3.2.31), to redefinitions of the functions $\varphi(t)$ and $\psi(t)$. In the one-dimensional case, the general solution of equation (4.3.2.32) is defined by formulas (4.3.2.4).

Substituting (4.3.2.31) into (4.3.1.1) and taking into account (4.3.2.32), we obtain the relations

$$k_2[(a_1\mu + b_1 + c_1\bar{f})\xi - \xi'_t]\theta + a_1k_2\varepsilon\xi + b_1\varphi + c_1\varphi\bar{f} + \bar{g} - \varphi'_t = 0,$$
$$k_1[(a_2\mu + b_2 + c_2\bar{f})\xi - \xi'_t]\theta + a_2k_1\varepsilon\xi + b_2\psi + c_2\psi\bar{f} + \bar{h} - \psi'_t = 0, \quad (4.3.2.33)$$

where we have used the short notations $\bar{f} = f(k_1\bar{\varphi} - k_2\bar{\psi})$, $\bar{g} = g(k_1\bar{\varphi} - k_2\bar{\psi})$, $\bar{h} = h(k_1\bar{\varphi} - k_2\bar{\psi})$, $\bar{\varphi} = \varphi(t-\tau)$, and $\bar{\psi} = \psi(t-\tau)$.

Relations (4.3.2.33) can be satisfied by setting

$$\begin{aligned}\varphi'_t &= a_1k_2\varepsilon\xi + b_1\varphi + c_1\varphi\bar{f} + \bar{g}, \\ \psi'_t &= a_2k_1\varepsilon\xi + b_2\psi + c_2\psi\bar{f} + \bar{h}, \\ \xi'_t &= (a_1\mu + b_1 + c_1\bar{f})\xi, \\ \xi'_t &= (a_2\mu + b_2 + c_2\bar{f})\xi.\end{aligned} \quad (4.3.2.34)$$

The system of ODEs (4.3.2.34) is overdetermined, since it consists of four equations for three functions, ξ, φ, and ψ. Requiring that the last two equations in (4.3.2.34) must coincide, we find the parameter μ and other constants:

$$\mu = \frac{b_1 - b_2}{a_2 - a_1}, \quad c_1 = c_2 = 1 \quad (a_1 \neq a_2). \quad (4.3.2.35)$$

The values of c_1 and c_2 have been chosen considering that the function f is defined up to a constant multiplier.

If $b_1 \neq b_2$, it follows from (4.3.2.35) that $\mu \neq 0$, and hence, we can set $\varepsilon = 0$ in equations (4.3.2.32) and (4.3.2.34). For $\varepsilon = 0$, the first two equations of (4.3.2.34) are independent of $\xi = \xi(t)$ and make up a closed nonlinear system of the first-order delay ODEs for $\varphi = \varphi(t)$ and $\psi = \psi(t)$:

$$\begin{aligned} \varphi'_t &= b_1\varphi + \varphi f(k_1\bar{\varphi} - k_2\bar{\psi}) + g(k_1\bar{\varphi} - k_2\bar{\psi}), \\ \psi'_t &= b_2\psi + \psi f(k_1\bar{\varphi} - k_2\bar{\psi}) + h(k_1\bar{\varphi} - k_2\bar{\psi}), \end{aligned} \qquad (4.3.2.36)$$

where $\bar{\varphi} = \varphi(t-\tau)$ and $\bar{\psi} = \psi(t-\tau)$. Integrating the last equation of (4.3.2.34) and taking into account relations (4.3.2.35), we find that $\xi = \xi(t)$ is expressed via $\varphi = \varphi(t)$ and $\psi = \psi(t)$ as

$$\xi = A \exp\left[\frac{a_2 b_1 - a_1 b_2}{a_2 - a_1} t + \int f(k_1\bar{\varphi} - k_2\bar{\psi})\,dt\right], \qquad (4.3.2.37)$$

where A is an arbitrary constant.

Notably, if $g = 0$ or $h = 0$, the system of delay ODEs (4.3.2.36) admits one-component solutions of the form $\varphi = 0$, $\psi = \psi(t)$ or $\varphi = \varphi(t)$, $\psi = 0$, respectively. If there are no delays and the functions f and h (or g) are given, these solutions can be represented in implicit form, as they are described by first-order separable ODEs.

If $g = h = 0$, it is not difficult to show that the system of delay ODEs (4.3.2.36) with arbitrary f admits the first integral

$$\psi = C_3 e^{(b_2 - b_1)t}\varphi, \qquad (4.3.2.38)$$

where C_3 is an arbitrary constant that may depend on the delay time τ. In this case, system (4.3.2.36) reduces to a single equation

$$\varphi'_t = b_1\varphi + \varphi f\big((k_1 - k_2 C_3 e^{(b_2 - b_1)(t-\tau)})\bar{\varphi}\big), \quad \bar{\varphi} = \varphi(t-\tau). \qquad (4.3.2.39)$$

▶ Example 4.21. For the Lotka–Volterra type system of PDEs (4.3.1.1) with $\tau = 0$, $f(z) = z$, and $g = h = 0$, ODE (4.3.2.39) becomes a Bernoulli equation. Integrating this equation and taking into account (4.3.2.38), we obtain the exact solution to the system of ODEs (4.3.2.36):

$$\begin{aligned} \varphi &= \left[\frac{C_3 k_2}{b_2} e^{(b_2 - b_1)t} + C_4 e^{-b_1 t} - \frac{k_1}{b_1}\right]^{-1}, \\ \psi &= C_3 e^{(b_2 - b_1)t}\left[\frac{C_3 k_2}{b_2} e^{(b_2 - b_1)t} + C_4 e^{-b_1 t} - \frac{k_1}{b_1}\right]^{-1}, \end{aligned} \qquad (4.3.2.40)$$

where C_4 is an arbitrary constant. The associated function $\xi(t)$ is determined by formula (4.3.2.37) with $\tau = 0$ and $f(z) = z$. This function can be expressed in terms of elementary functions if, for example,

$$b_1 = 0; \quad b_2 = 0; \quad b_1 = b_2; \quad C_3 = 0; \quad C_4 = 0.$$

In particular, by setting $\tau = 0$ and $C_3 = 0$ in (4.3.2.37) and (4.3.2.40), we find that

$$\varphi = \frac{b_1 e^{b_1 t}}{b_1 C_4 - k_1 e^{b_1 t}}, \quad \psi = 0, \quad \xi = \frac{A}{b_1 C_4 - k_1 e^{b_1 t}} \exp\left(\frac{a_2 b_1 - a_1 b_2}{a_2 - a_1} t\right). \quad (4.3.2.41)$$

Formulas (4.3.2.4), (4.3.2.31), (4.3.2.35), and (4.3.2.41) describe an exact solution to the one-dimensional Lotka–Volterra type system of PDEs (4.3.1.1) without delay.

Likewise, at $b_1 = b_2 = b$ and $\tau = 0$, we have

$$\varphi = \frac{be^{bt}}{C_4 b + (C_3 k_2 - k_1)e^{bt}}, \quad \psi = \frac{bC_3 e^{bt}}{C_4 b + (C_3 k_2 - k_1)e^{bt}}, \quad \xi = \frac{A}{b}\varphi. \quad (4.3.2.42)$$

◂

In the general case, the function φ in the delay ODEs (4.3.2.39) must be set on the interval $[-\tau, 0]$:

$$\varphi = \varphi_0(t) \quad \text{at} \quad -\tau \leq t \leq 0. \quad (4.3.2.43)$$

A Cauchy-type problem for equation (4.3.2.39) with the initial data (4.3.2.43) can be solved by the method of steps (see Subsection 1.1.5 and [37, 144]). To this end, we subdivide time t into intervals of length τ and denote

$$\varphi(t) = \varphi_m(t) \quad \text{for} \quad t_{m-1} \leq t \leq t_m, \quad (4.3.2.44)$$

where $t_m = m\tau$, $m = 0, 1, 2, \ldots$. On integrating equation (4.3.2.39) from t_{m-1} to t, we obtain

$$\varphi_m(t) = \varphi_m^\circ \exp\left[b_1(t - t_{m-1}) + \int_{t_{m-1}}^{t} f\big((k_1 - k_2 C_3 e^{(b_2 - b_1)(t - \tau)})\varphi_{m-1}(t)\big)\, dt\right],$$

$$\varphi_m^\circ = \varphi_m(t_{m-1}) = \varphi_{m-1}(t_m),$$

(4.3.2.45)

on the interval $[t_{m-1}, t_m]$. The left-hand side of formula (4.3.2.45) is the function $\varphi_m(t)$, which is sought on the interval $[t_{m-1}, t_m]$, while the right-hand side includes $\varphi_{m-1}(t)$, which is defined on the preceding interval $[t_{m-2}, t_{m-1}]$. The computation is carried out sequentially starting from $m = 1$, when the known function defined on the initial interval (4.3.2.43) is used on the right-hand side. This results in $\varphi_1(t)$. Then one proceeds to $m = 2$ to determine $\varphi_2(t)$, which is expressed in terms of the already known function $\varphi_1(t)$ using (4.3.2.45). The procedure is then repeated likewise.

If $\varepsilon \neq 0$, the functions $\xi = \xi(t)$, $\varphi = \varphi(t)$, and $\psi = \psi(t)$ are determined through the solution of the nonlinear system (4.3.2.34) with $b_1 = b_2 = b$, $c_1 = c_2 = 1$, and $\mu = 0$ (in this case, the last two equations coincide).

4.3.3. Reductions and Exact Solutions of Systems of PDEs with Equal Diffusion Coefficients ($a_1 = a_2$)

Reduction of the system of PDEs with three arbitrary delays to a stationary system of PDEs and a linear Schrödinger equation. We will assume that the four delay times in system (4.3.1.1) are constrained by one relation (4.3.2.1).

We look for exact solutions to system (4.3.1.1) with $a_1 = a_2 = a$, $b_1 = b_2 = b$, and $c_1 = c_2 = 1$ in the form

$$u = k_2 e^{\lambda(t+\tau_1)}\theta(\mathbf{x},t) + \varphi(\mathbf{x}), \quad v = k_1 e^{\lambda(t+\tau_2)}\theta(\mathbf{x},t) + \psi(\mathbf{x}), \quad (4.3.3.1)$$

where λ is a free parameter, and the functions $\theta = \theta(\mathbf{x},t)$, $\varphi = \varphi(\mathbf{x})$, and $\psi = \psi(\mathbf{x})$ are to be determined in the subsequent analysis.

Solutions of the form (4.3.3.1), in which the function θ depends on all its arguments and the parameter λ is arbitrary, substantially generalize solutions of the form (4.3.2.13), where θ is independent of time t and λ is expressed in terms of the system constants.

We impose a periodicity condition on the function $\theta = \theta(\mathbf{x},t)$:

$$\theta(\mathbf{x}, t + \tau_2 - \tau_1) = \theta(\mathbf{x}, t). \quad (4.3.3.2)$$

It is not difficult to verify that if conditions (4.3.2.1) and (4.3.3.2) hold, relations (4.3.2.14) remain valid. Considering the above and substituting (4.3.3.1) into (4.3.1.1), we obtain the following closed system of two stationary equations for φ and ψ:

$$\begin{aligned} a\,\Delta\varphi + [b + f(k_1\varphi - k_2\psi)]\varphi + g(k_1\varphi - k_2\psi) &= 0, \\ a\,\Delta\psi + [b + f(k_1\varphi - k_2\psi)]\psi + h(k_1\varphi - k_2\psi) &= 0. \end{aligned} \quad (4.3.3.3)$$

The function $\theta = \theta(\mathbf{x},t)$ is described by the linear Schrödinger equation

$$\theta_t = a\,\Delta\theta + (b - \lambda + \hat{f})\theta, \quad \hat{f} = f(k_1\varphi - k_2\psi), \quad (4.3.3.4)$$

and the periodic condition (4.3.3.2). Notably, the function $\hat{f}$ only depends on the space coordinates.

Below we note two important cases where the periodicity condition (4.3.3.2) holds automatically.

1°. Condition (4.3.3.2) can be satisfied by seeking a stationary solution $\theta = \theta(\mathbf{x})$ of equation (4.3.3.4).

2°. Condition (4.3.3.2) also holds if $\tau_2 = \tau_1$ and $\tau_3 = \tau_4$, in which case relation (4.3.2.1) is satisfied automatically.

▶ **Example 4.22.** We set $f(z_{1,2}) = z_{1,2}$ and $g(z_1) = h(z_2) = 0$ in the reduced system (4.3.3.3)–(4.3.3.4) of the one-dimensional Lotka–Volterra type system of delay PDEs (4.3.1.2). Then the functions $\varphi = \varphi(x)$, $\psi = \psi(x)$, and $\theta = \theta(x,t)$ are described by the system of two stationary ODEs and one nonstationary PDE

$$a\varphi''_{xx} + (b + k_1\varphi - k_2\psi)\varphi = 0, \quad (4.3.3.5)$$

$$a\psi''_{xx} + (b + k_1\varphi - k_2\psi)\psi = 0, \quad (4.3.3.6)$$

$$\theta_t = a\theta''_{xx} + (b - \lambda + k_1\varphi - k_2\psi)\theta. \quad (4.3.3.7)$$

Notably, the coefficients of equation (4.3.3.7) are only dependent on the space variable x, and the θ must satisfy the one-dimensional periodic condition (4.3.3.2).

4.3. Reductions and Exact Solutions of Lotka–Volterra Type Systems with Several Delays 241

Let us multiply ODE (4.3.3.5) by k_1 and ODE (4.3.3.6) by $-k_2$ and add together. This results in an ODE for $\rho = k_1\varphi - k_2\psi$. We supplement this equation with ODE (4.3.3.5) and PDE (4.3.3.7) to obtain the mixed algebraic-differential system

$$a\rho''_{xx} + (b + \rho)\rho = 0, \tag{4.3.3.8}$$
$$a\varphi''_{xx} + (b + \rho)\varphi = 0, \tag{4.3.3.9}$$
$$\rho = k_1\varphi - k_2\psi, \tag{4.3.3.10}$$
$$\theta_t = a\theta''_{xx} + (b - \lambda + \rho)\theta. \tag{4.3.3.11}$$

System (4.3.3.8)–(4.3.3.11) can be solved step by step from one equation to another. We start with the isolated stationary ODE (4.3.3.8), which, when divided by a, coincides with equation (4.3.2.19) up to obvious renaming. PDE (4.3.3.11) should be equipped with the additional conditions (4.3.2.1) and (4.3.3.2), which hold, for example, in the stationary case of $\theta = \theta(x)$ and the nonstationary case if $\tau_2 = \tau_1$ and $\tau_4 = \tau_3$.

Below we specify two simple classes of stationary exact solutions to system (4.3.3.8)–(4.3.3.11) that correspond to $\rho = $ const and then describe a number of more complicated solutions.

$1°$. Stationary exact solution to system (4.3.3.8)–(4.3.3.11) with $\rho = 0$:

$$\rho = 0, \quad \varphi = \begin{cases} C_1 \cos(\sqrt{b}\,x) + C_2 \sin(\sqrt{b}\,x) & \text{if } b > 0, \\ C_1 \cosh(\sqrt{-b}\,x) + C_2 \sinh(\sqrt{-b}\,x) & \text{if } b < 0; \end{cases} \tag{4.3.3.12}$$

$$\psi = \begin{cases} \dfrac{k_1}{k_2}[C_1 \cos(\sqrt{b}\,x) + C_2 \sin(\sqrt{b}\,x)] & \text{if } b > 0, \\ \dfrac{k_1}{k_2}[C_1 \cosh(\sqrt{-b}\,x) + C_2 \sinh(\sqrt{-b}\,x)] & \text{if } b < 0; \end{cases} \tag{4.3.3.13}$$

$$\theta = \begin{cases} C_3 \cos[\sqrt{(b-\lambda)/a}\,x] + C_4 \sin[\sqrt{(b-\lambda)/a}\,x] & \text{if } \lambda < b, \\ C_3 \cosh[\sqrt{(\lambda-b)/a}\,x] + C_4 \sinh[\sqrt{(\lambda-b)/a}\,x] & \text{if } \lambda > b, \\ C_3 x + C_4 & \text{if } \lambda = b, \end{cases} \tag{4.3.3.14}$$

where $C_1, \ldots, C_4$ are arbitrary constants.

$2°$. Stationary exact solution to system (4.3.3.8)–(4.3.3.11) with $\rho = -b$:

$$\rho = -b, \quad \varphi = C_1 x + C_2, \quad \psi = \frac{k_1}{k_2}(C_1 x + C_2) + \frac{b}{k_2}, \tag{4.3.3.15}$$

$$\theta = \begin{cases} C_3 \cosh(\sqrt{\lambda/a}\,x) + C_4 \sinh(\sqrt{\lambda/a}\,x) & \text{if } \lambda > 0, \\ C_3 \cos(\sqrt{-\lambda/a}\,x) + C_4 \sin(\sqrt{-\lambda/a}\,x) & \text{if } \lambda < 0. \end{cases} \tag{4.3.3.16}$$

$3°$. More complex stationary solutions of system (4.3.3.8)–(4.3.3.11) with $\rho = \rho(x) \neq $ const and $\theta = \theta(x)$ can be constructed in several steps as described below.

(i) The isolated subsystem of two ODEs (4.3.3.8) and (4.3.3.9) for ρ and φ coincides, up to notation, with system (4.3.2.19)–(4.3.2.20) where $\beta = b$. It follows that one can obtain exact solutions to equations (4.3.3.8) and (4.3.3.9) using the formulas from Table 4.2 with $\beta = b$ by replacing the functions and determining parameters as follows: $\varphi \Rightarrow \rho$, $\theta \Rightarrow \varphi$, $b \Rightarrow b/a$, $\beta \Rightarrow b/a$, and $k \Rightarrow 1/a$.

(ii) The function ψ is determined by substituting the functions ρ and φ obtained in step (i) into (4.3.3.10), which results in

$$\psi = (k_1 \varphi - \rho)/k_2.$$

(iii) The isolated subsystem of two equations (4.3.3.8) and (4.3.3.11) for ρ and θ, with $\theta_t = 0$, coincides, up to notation, with system (4.3.2.19)–(4.3.2.20). Therefore, one can obtain exact solutions to equations (4.3.3.8) and (4.3.3.11) using formulas from Table 4.2 by replacing the functions and determining parameters as follows: $\varphi \Rightarrow \rho$, $\psi \Rightarrow \theta$, $b \Rightarrow b/a$, $\beta \Rightarrow (b-\lambda)/a$, and $k \Rightarrow 1/a$.

Table 4.4 summarizes stationary solutions of system (4.3.3.8)–(4.3.3.11) represented in terms of elementary functions for various values of the determining parameters a, b, and λ.

4°. For $\tau_2 = \tau_1$ and $\tau_2 = \tau_4$, some nonstationary exact solutions of system (4.3.3.8)–(4.3.3.11) with $\rho = 0$ are defined by formulas (4.3.3.12) and (4.3.3.13) for φ and ψ and any of the following expressions of $\theta = \theta(x, t)$:

$$\begin{aligned}
\theta &= \left[(x^2 + 2at) + C\right] e^{(b-\lambda)t}, \\
\theta &= \exp\left[(a\mu^2 + b - \lambda)t \pm \mu x\right], \\
\theta &= \frac{1}{\sqrt{t}} \exp\left[-\frac{x^2}{4at} + (b-\lambda)t\right], \\
\theta &= \exp\left[(b - \lambda - a\mu^2)t\right] \cos(\mu x), \\
\theta &= \exp\left[(b - \lambda - a\mu^2)t\right] \sin(\mu x), \\
\theta &= \exp[-\mu x + (b-\lambda)t] \cos(\mu x - 2a\mu^2 t), \\
\theta &= \exp[-\mu x + (b-\lambda)t] \sin(\mu x - 2a\mu^2 t), \\
\theta &= \exp(-\mu x) \cos(\beta x - 2a\beta\mu t), \quad \beta = \sqrt{\mu^2 + (b-\lambda)/a}, \\
\theta &= \exp(-\mu x) \sin(\beta x - 2a\beta\mu t), \quad \beta = \sqrt{\mu^2 + (b-\lambda)/a},
\end{aligned} \quad (4.3.3.17)$$

where C and μ are arbitrary constants. Some other nonstationary exact solutions of system (4.3.3.8)–(4.3.3.11) with $\rho = 0$ can be obtained using formulas (4.3.3.12) and (4.3.3.13) and the expression $\theta = e^{(b-\lambda)t}\xi$, where $\xi = \xi(x, t)$ is any solution of the standard linear heat equation $\xi_t = a\xi_{xx}$.

5°. For $\tau_2 = \tau_1$ and $\tau_3 = \tau_4$, some nonstationary exact solutions of system (4.3.3.8)–(4.3.3.11) with $\rho = -b$ are defined by formulas (4.3.3.15) for φ and ψ and any of the expressions from (4.3.3.17) with $b = 0$. ◂

▶ **Example 4.23.** If $\tau_2 = \tau_1$ and $\tau_3 = \tau_4$, one can take for φ and ψ any simple solution of the form (4.3.1.3) that satisfies the algebraic system (4.3.1.4) with $b_1 = b_2 = b$ and $c_1 = c_2 = 1$. Equation (4.3.3.4) can then be reduced with the substitution

$$\theta = \exp[(b - \lambda + f°)t]\zeta, \quad f° = f(k_1 \varphi° - k_2 \psi°), \qquad (4.3.3.18)$$

4.3. Reductions and Exact Solutions of Lotka–Volterra Type Systems with Several Delays 243

Table 4.4. Stationary solutions of system (4.3.3.8)–(4.3.3.11) for various values of the determining parameters. Notations: $\xi = \frac{1}{2}\sqrt{|b/a|}\,x$, $\cosh^{-1}\xi = 1/\cosh\xi$, $\sinh^{-1}\xi = 1/\sinh\xi$, $\cos^{-1}\xi = 1/\cos\xi$, $\sin^{-1}\xi = 1/\sin\xi$, and $C_1,\ldots,C_4$ are arbitrary constants. The function ψ is expressed as $\psi = (k_1\varphi - \rho)/k_2$ (according to [431]).

No.	Parameters b and λ	Function ρ, ODE (4.3.3.8)	Function φ, ODE (4.3.3.9)	Function θ, equation (4.3.3.11)
1	$\lambda<0, b=0$	$-6ax^{-2}$	$C_1 x^{-2} + C_2 x^3$	$\sqrt{x}\,[C_3 J_{5/2}(\sqrt{-\lambda/a}\,x) + C_4 Y_{5/2}(\sqrt{-\lambda/a}\,x)]$
2	$\lambda>0, b=0$	$-6ax^{-2}$	$C_1 x^{-2} + C_2 x^3$	$\sqrt{x}\,[C_3 I_{5/2}(\sqrt{\lambda/a}\,x) + C_4 K_{5/2}(\sqrt{\lambda/a}\,x)]$
3	$\lambda=-3b,\ b<0$	$\frac{3b}{2}(\tanh^2\xi - 1)$	$C_1\cosh^{-2}\xi + C_2 \times [(3+2\cosh^2\xi)\tanh\xi + 3\xi\cosh^{-2}\xi]$	$C_3\sinh\xi\cosh^3\xi + C_4(2+\cosh^{-2}\xi + 8\cosh^2\xi - 16\cosh^4\xi)$
4	$\lambda=-3b,\ b<0$	$\frac{3b}{2}(\coth^2\xi - 1)$	$C_1\sinh^{-2}\xi + C_2 \times [(3+2\sinh^2\xi)\coth\xi + 3\xi\sinh^{-2}\xi]$	$C_3\cosh\xi\sinh^3\xi + C_4(2+\sinh^{-2}\xi + 8\sinh^2\xi - 16\sinh^4\xi)$
5	$\lambda=-3b,\ b>0$	$-\frac{3b}{2}(\tanh^2\xi + 1)$	$C_1\cos^{-2}\xi + C_2 \times [(3+2\cos^2\xi)\tan\xi + 3\xi\cos^{-2}\xi]$	$C_3\sin\xi\cos^3\xi + C_4(2+\cos^{-2}\xi + 8\cos^2\xi - 16\cos^4\xi)$
6	$\lambda=-3b,\ b>0$	$-\frac{3b}{2}(\coth^2\xi + 1)$	$C_1\sin^{-2}\xi + C_2 \times [(3+2\sin^2\xi)\cot\xi + 3\xi\sin^{-2}\xi]$	$C_3\cos\xi\sin^3\xi + C_4(2+\sin^{-2}\xi + 8\sin^2\xi - 16\sin^4\xi)$
7	$\lambda=-\frac{5}{4}b,\ b<0$	$\frac{3b}{2}(\tanh^2\xi - 1)$	$C_1\cosh^{-2}\xi + C_2 \times [(3+2\cosh^2\xi)\tanh\xi + 3\xi\cosh^{-2}\xi]$	$C_3\cosh^3\xi + C_4\sinh\xi \times (4+3\cosh^{-2}\xi + 8\cosh^2\xi)$
8	$\lambda=-\frac{5}{4}b,\ b<0$	$\frac{3b}{2}(\coth^2\xi - 1)$	$C_1\sinh^{-2}\xi + C_2 \times [(3+2\sinh^2\xi)\coth\xi + 3\xi\sinh^{-2}\xi]$	$C_3\sinh^3\xi + C_4\cosh\xi \times (4+3\sinh^{-2}\xi + 8\sinh^2\xi)$
9	$\lambda=-\frac{5}{4}b,\ b>0$	$-\frac{3b}{2}(\tan^2\xi + 1)$	$C_1\cos^{-2}\xi + C_2 \times [(3+2\cos^2\xi)\tan\xi + 3\xi\cos^{-2}\xi]$	$C_3\cos^3\xi + C_4\sin\xi \times (4+3\cos^{-2}\xi + 8\cos^2\xi)$
10	$\lambda=-\frac{5}{4}b,\ b>0$	$-\frac{3b}{2}(\cot^2\xi + 1)$	$C_1\sin^{-2}\xi + C_2 \times [(3+2\sin^2\xi)\cot\xi + 3\xi\sin^{-2}\xi]$	$C_3\sin^3\xi + C_4\cos\xi \times (4+3\sin^{-2}\xi + 8\sin^2\xi)$
11	$\lambda=\frac{1}{4}b,\ b<0$	$\frac{b}{2}(1+3\tan^2\xi)$	$C_1(1+3\tan^2\xi) + C_2[\xi(1+3\tan^2\xi) + 3\tan\xi]$	$C_3\sin\xi\cos^{-2}\xi + C_4\cos^{-1}\xi \times (3\xi\tan\xi + 3 - \cos^2\xi)$
12	$\lambda=\frac{1}{4}b,\ b<0$	$\frac{b}{2}(1+3\cot^2\xi)$	$C_1(1+3\cot^2\xi) + C_2[\xi(1+3\cot^2\xi) + 3\cot\xi]$	$C_3\cos\xi\sin^{-2}\xi + C_4\sin^{-1}\xi \times (3\xi\tan\xi + 3 - \sin^2\xi)$
13	$\lambda=\frac{1}{4}b,\ b>0$	$\frac{b}{2}(1-3\tanh^2\xi)$	$C_1(1-3\tanh^2\xi) + C_2[\xi(1-3\tanh^2\xi) + 3\tanh\xi]$	$C_3\sinh\xi\cosh^{-2}\xi + C_4\cosh^{-1}\xi \times (3\xi\tanh\xi - 3 + \cosh^2\xi)$
14	$\lambda=\frac{1}{4}b,\ b>0$	$\frac{b}{2}(1-3\coth^2\xi)$	$C_1(1-3\coth^2\xi) + C_2[\xi(1-3\coth^2\xi) + 3\coth\xi]$	$C_3\cosh\xi\sinh^{-2}\xi + C_4\sinh^{-1}\xi \times (3\xi\coth\xi - 3 + \sinh^2\xi)$

Table 4.4. (*Continued*)

No.	Parameters b and λ	Function ρ, ODE (4.3.3.8)	Function φ, ODE (4.3.3.9)	Function θ, equation (4.3.3.11)
15	$\lambda=\tfrac{3}{4}b$, $b<0$	$\tfrac{3b}{2}(\tanh^2\xi-1)$	$C_1\cosh^{-2}\xi+C_2$ $\times[(3+2\cosh^2\xi)\tanh\xi$ $+3\xi\cosh^{-2}\xi]$	$C_3\sinh\xi\cosh^{-2}\xi+C_4\cosh^{-1}\xi$ $\times(3\xi\tanh\xi-3+\cosh^2\xi)$
16	$\lambda=\tfrac{3}{4}b$, $b<0$	$\tfrac{3b}{2}(\coth^2\xi-1)$	$C_1\sinh^{-2}\xi+C_2$ $\times[(3+2\sinh^2\xi)\coth\xi$ $+3\xi\sinh^{-2}\xi]$	$C_3\cosh\xi\sinh^{-2}\xi+C_4\sinh^{-1}\xi$ $\times(3\xi\coth\xi-3+\sinh^2\xi)$
17	$\lambda=\tfrac{3}{4}b$, $b>0$	$-\tfrac{3b}{2}(\tan^2\xi+1)$	$C_1\cos^{-2}\xi+C_2$ $\times[(3+2\cos^2\xi)\tan\xi$ $+3\xi\cos^{-2}\xi]$	$C_3\sin\xi\cos^{-2}\xi+C_4\cos^{-1}\xi$ $\times(3\xi\tan\xi+3-\cos^2\xi)$
18	$\lambda=\tfrac{3}{4}b$, $b>0$	$-\tfrac{3b}{2}(\cot^2\xi+1)$	$C_1\sin^{-2}\xi+C_2$ $\times[(3+2\sin^2\xi)\cot\xi$ $+3\xi\sin^{-2}\xi]$	$C_3\cos\xi\sin^{-2}\xi+C_4\sin^{-1}\xi$ $\times(3\xi\cot\xi+3-\sin^2\xi)$
19	$\lambda=b$, $b<0$	$\tfrac{3b}{2}(\tanh^2\xi-1)$	$C_1\cosh^{-2}\xi+C_2$ $\times[(3+2\cosh^2\xi)\tanh\xi$ $+3\xi\cosh^{-2}\xi]$	$C_3(3\tanh^2\xi-1)+C_4[3\tanh\xi$ $+(3\tanh^2\xi-1)\operatorname{artanh}\xi]$
20	$\lambda=b$, $b<0$	$\tfrac{3b}{2}(\coth^2\xi-1)$	$C_1\sinh^{-2}\xi+C_2$ $\times[(3+2\sinh^2\xi)\coth\xi$ $+3\xi\sinh^{-2}\xi]$	$C_3(3\coth^2\xi-1)+C_4[3\coth\xi$ $+(3\coth^2\xi-1)\operatorname{arcoth}\xi]$
21	$\lambda=b$, $b<0$	$\tfrac{b}{2}(1+3\tan^2\xi)$	$C_1(1+3\tan^2\xi)+C_2$ $\times[\xi(1+3\tan^2\xi)$ $+3\tan\xi]$	$C_3\cos^{-2}\xi+C_4[\sin(2\xi)$ $+3\tan\xi+3\xi\cos^{-2}\xi]$
22	$\lambda=b$, $b<0$	$\tfrac{b}{2}(1+3\cot^2\xi)$	$C_1(1+3\cot^2\xi)+C_2$ $\times[\xi(1+3\cot^2\xi)$ $+3\cot\xi]$	$C_3\sin^{-2}\xi+C_4[\cos(2\xi)$ $+3\tan\xi+3\xi\sin^{-2}\xi]$
23	$\lambda=b$, $b>0$	$-\tfrac{3b}{2}(\tan^2\xi+1)$	$C_1\cos^{-2}\xi+C_2$ $\times[(3+2\cos^2\xi)\tan\xi$ $+3\xi\cos^{-2}\xi]$	$C_3(3\tan^2\xi+1)+C_4[3\tan\xi$ $+(3\tan^2\xi+1)\arctan\xi]$
24	$\lambda=b$, $b>0$	$-\tfrac{3b}{2}(\cot^2\xi+1)$	$C_1\sin^{-2}\xi+C_2$ $\times[(3+2\sin^2\xi)\cot\xi$ $+3\xi\sin^{-2}\xi]$	$C_3(3\cot^2\xi+1)+C_4[3\cot\xi$ $+(3\cot^2\xi+1)\operatorname{arccot}\xi]$
25	$\lambda=b$, $b>0$	$\tfrac{b}{2}(1-3\tanh^2\xi)$	$C_1(1-3\tanh^2\xi)$ $+C_2[\xi(1-3\tanh^2\xi)$ $+3\tanh\xi]$	$C_3\cosh^{-2}\xi+C_4[\sinh(2\xi)$ $+3\tanh\xi+3\xi\cosh^{-2}\xi]$
26	$\lambda=b$, $b>0$	$\tfrac{b}{2}(1-3\coth^2\xi)$	$C_1(1-3\coth^2\xi)$ $+C_2[\xi(1-3\coth^2\xi)$ $+3\coth\xi]$	$C_3\sinh^{-2}\xi+C_4[\cosh(2\xi)$ $+3\coth\xi+3\xi\sinh^{-2}\xi]$
27	$\lambda=\tfrac{9}{4}b$, $b<0$	$\tfrac{b}{2}(1+3\tan^2\xi)$	$C_1(1+3\tan^2\xi)$ $+C_2[\xi(1+3\tan^2\xi)$ $+3\tan\xi]$	$C_3\cos^3\xi+C_4\sin\xi\,(4+3\cos^{-2}\xi$ $+8\cos^2\xi)$

Table 4.4. (*Continued*)

No.	Parameters b and λ	Function ρ, ODE (4.3.3.8)	Function φ, ODE (4.3.3.9)	Function θ, equation (4.3.3.11)
28	$\lambda = \tfrac{9}{4}b$, $b<0$	$\tfrac{b}{2}(1+3\cot^2\xi)$	$C_1(1+3\cot^2\xi)$ $+C_2[\xi(1+3\cot^2\xi)$ $+3\cot\xi]$	$C_3\sin^3\xi + C_4\cos\xi\,(4+3\sin^{-2}\xi$ $+8\sin^2\xi)$
29	$\lambda = \tfrac{9}{4}b$, $b>0$	$\tfrac{b}{2}(1-3\tanh^2\xi)$	$C_1(1-3\tanh^2\xi)$ $+C_2[\xi(1-3\tanh^2\xi)$ $+3\tanh\xi]$	$C_3\cosh^3\xi + C_4\sinh\xi\,(4+3\cosh^{-2}\xi$ $+8\cosh^2\xi)$
30	$\lambda = \tfrac{9}{4}b$, $b>0$	$\tfrac{b}{2}(1-3\coth^2\xi)$	$C_1(1-3\coth^2\xi)$ $+C_2[\xi(1-3\coth^2\xi)$ $+3\coth\xi]$	$C_3\sinh^3\xi + C_4\cosh\xi\,(4+3\sinh^{-2}\xi$ $+8\sinh^2\xi)$
31	$\lambda = 4b$, $b<0$	$\tfrac{b}{2}(1+3\tan^2\xi)$	$C_1(1+3\tan^2\xi)$ $+C_2[\xi(1+3\tan^2\xi)$ $+3\tan\xi]$	$C_3\sin\xi\cos^3\xi + C_4(2+\cos^{-2}\xi$ $+8\cos^2\xi - 16\cos^4\xi)$
32	$\lambda = 4b$, $b<0$	$\tfrac{b}{2}(1+3\cot^2\xi)$	$C_1(1+3\cot^2\xi)$ $+C_2[\xi(1+3\cot^2\xi)$ $+3\cot\xi]$	$C_3\cos\xi\sin^3\xi + C_4(2+\sin^{-2}\xi$ $+8\sin^2\xi - 16\sin^4\xi)$
33	$\lambda = 4b$, $b>0$	$\tfrac{b}{2}(1-3\tanh^2\xi)$	$C_1(1-3\tanh^2\xi)$ $+C_2[\xi(1-3\tanh^2\xi)$ $+3\tanh\xi]$	$C_3\sinh\xi\cosh^3\xi + C_4(2+\cosh^{-2}\xi$ $+8\cosh^2\xi - 16\cosh^4\xi)$
34	$\lambda = 4b$, $b>0$	$\tfrac{b}{2}(1-3\coth^2\xi)$	$C_1(1-3\coth^2\xi)$ $+C_2[\xi(1-3\coth^2\xi)$ $+3\coth\xi]$	$C_3\cosh\xi\sinh^3\xi + C_4(2+\sinh^{-2}\xi$ $+8\sinh^2\xi - 16\sinh^4\xi)$

to the standard linear heat equation

$$\zeta_t = a\,\Delta\zeta, \tag{4.3.3.19}$$

whose exact solutions are described, for example, in [404]. ◀

For $\tau_1 \neq \tau_2$, exact solutions to equation (4.3.3.4) that satisfy the periodicity condition (4.3.3.2) can be sought in the form

$$\theta_m(\mathbf{x}, t) = \xi_m(\mathbf{x})\cos(\beta_m t) + \eta_m(\mathbf{x})\sin(\beta_m t), \tag{4.3.3.20}$$

$$\beta_m = \frac{2\pi m}{\tau_2 - \tau_1}, \quad m = 0, 1, 2, \ldots,$$

where $m = 0$ corresponds to a stationary solution. Substituting (4.3.3.20) into (4.3.3.4) gives the following linear stationary system of PDEs for $\xi_m = \xi_m(\mathbf{x})$ and $\eta_m = \eta_m(\mathbf{x})$:

$$\begin{aligned} a\,\Delta\xi_m + [b - \lambda + f(k_1\varphi - k_2\psi)]\xi_m - \beta_m\eta_m &= 0, \\ a\,\Delta\eta_m + [b - \lambda + f(k_1\varphi - k_2\psi)]\eta_m + \beta_m\xi_m &= 0. \end{aligned} \tag{4.3.3.21}$$

Equation (4.3.3.4) is linear in θ; therefore, an arbitrary linear combination of exact solutions of the form (4.3.3.20),

$$\theta = \sum_{m=1}^{\infty} \alpha_m \theta_m(\mathbf{x},t) = \sum_{m=1}^{\infty} \alpha_m [\xi_m(\mathbf{x}) \cos(\beta_m t) + \eta_m(\mathbf{x}) \sin(\beta_m t)],$$

where α_m are arbitrary constants, is also an exact solution to this equation.

If we take, as φ and ψ, any simple solution (4.3.1.3) that satisfies the algebraic system (4.3.1.4) with $b_1 = b_2 = b$ and $c_1 = c_2 = 1$, then system (4.3.3.21) will become a system with constant coefficients, whose exact solutions are sought as a linear combination of exponentials.

▶ **Example 4.24.** In the one-dimensional case, the general solution of the linear PDE (4.3.3.4) corresponding to any solution (4.3.1.3) and satisfying the periodic condition (4.3.3.2) can be represented as

$$\theta(x,t) = \sum_{m=0}^{\infty} \exp(-\mu_m x)\big[A_m \cos(\beta_m t - \gamma_m x) + B_m \sin(\beta_m t - \gamma_m x)\big]$$
$$+ \sum_{m=1}^{\infty} \exp(\mu_m x)\big[C_m \cos(\beta_m t + \gamma_m x) + D_m \sin(\beta_m t + \gamma_m x)\big], \quad (4.3.3.22)$$

where

$$\mu_m = \left(\frac{\sqrt{d^2 + \beta_m^2} - d}{2a}\right)^{1/2}, \quad \gamma_m = \left(\frac{\sqrt{d^2 + \beta_m^2} + d}{2a}\right)^{1/2},$$

$$d = b - \lambda + f(k_1 \varphi^\circ - k_2 \psi^\circ), \quad \beta_m = \frac{2\pi m}{\tau_2 - \tau_1},$$

and A_m, B_m, C_m, and D_m are arbitrary constants such that the series (4.3.3.22) and the respective derivatives θ_t and θ_{xx} are convergent; the convergence can, for example, be ensured by choosing $A_m = B_m = C_m = D_m = 0$ for $m > M$, where M is an arbitrary positive integer.

Below we note two special cases:
  (i) formula (4.3.3.22) with $A_0 = B_0 = 0$ and $C_m = D_m = 0$, $m = 1, 2, \ldots$ defines time periodic solutions decaying as $x \to \infty$;
  (ii) formula (4.3.3.22) with $C_m = D_m = 0$, $m = 1, 2, \ldots$ defines time periodic solutions bounded as $x \to \infty$. ◀

Reduction of the system of PDEs with one delay to a nonstationary system of ODEs and a linear heat equation. We look for incomplete separable solutions to system (4.3.1.1) with $a_1 = a_2 = a$, $b_1 = b_2 = b$, $c_1 = c_2 = 1$, and one common delay time (4.3.2.30) in the form

$$u = k_2 \theta(\mathbf{x},t) + \varphi(t), \quad v = k_1 \theta(\mathbf{x},t) + \psi(t), \qquad (4.3.3.23)$$

where the functions $\theta = \theta(\mathbf{x},t)$, $\varphi = \varphi(t)$, and $\psi = \psi(t)$ are to be determined in the subsequent analysis. The functions (4.3.3.23) are chosen so that the arguments of the functions f, g, and h depend on t alone.

Substituting (4.3.3.23) into the system of PDEs (4.3.1.1) yields a nonlinear system of first-order delay ODEs for φ and ψ,

$$\begin{aligned}\varphi'_t &= b\varphi + \varphi f(k_1\bar\varphi - k_2\bar\psi) + g(k_1\bar\varphi - k_2\bar\psi), & \bar\varphi &= \varphi(t-\tau), \\ \psi'_t &= b\psi + \psi f(k_1\bar\varphi - k_2\bar\psi) + h(k_1\bar\varphi - k_2\bar\psi), & \bar\psi &= \psi(t-\tau),\end{aligned} \qquad (4.3.3.24)$$

and a linear parabolic PDE with variable coefficients for θ:

$$\theta_t = a\,\Delta\theta + \bigl[b + f(k_1\bar\varphi - k_2\bar\psi)\bigr]\theta. \qquad (4.3.3.25)$$

The system of delay ODEs (4.3.3.24) coincides with system (4.3.2.36) at $b_1 = b_2 = b$. For a procedure to integrate this system and for some of its exact solutions, see Subsection 4.3.2. With the substitution

$$\theta = \exp\!\left[bt + \int f(k_1\bar\varphi_1 - k_2\bar\psi_1)\,dt\right]\xi(\mathbf{x},t), \qquad (4.3.3.26)$$

equation (4.3.3.25) can be reduced to the standard linear heat equation

$$\xi_t = a\,\Delta\xi, \qquad (4.3.3.27)$$

whose exact solutions can be found, for example, in [404].

4.3.4. Systems of Delay PDEs Homogeneous in the Unknown Functions

This subsection describes a few exact solutions of two nonlinear homogeneous systems of PDEs with two delays that remain unchanged under transformations of the form $u = C_1 U$, $v = C_1 V$, $x = X + C_2$, $t = T + C_3$, where C_1, C_2, and C_3 are arbitrary constants.

System 1. Consider the reaction-diffusion system of equations with two delays

$$\begin{aligned}u_t &= au_{xx} + uf(u/v,\bar u_1/u,\bar v_2/v), \\ v_t &= bv_{xx} + vg(u/v,\bar u_1/u,\bar v_2/v),\end{aligned} \qquad (4.3.4.1)$$

where $f(z,z_1,z_2)$ and $g(z,z_1,z_2)$ are arbitrary functions; $\bar u_1 = u(x,t-\tau_1)$ and $\bar v_2 = v(x,t-\tau_2)$.

In the special case of $f(z,1,1) = k_1 - k_2 z^{-1}$ and $g(z,1,1) = k_2 - k_1 z$ with no delays, $\tau_1 = \tau_2 = 0$, system (4.3.4.1) describes a two-component diffusion complicated by a first-order reversible chemical reaction [122]. The Eigen–Schuster model describes the competition of populations for a nutrient substrate at constant breeding rates, variable population size and no delay. This model leads to the above system with $f(z,1,1) = \frac{kz}{z+1}$ and $g(z,1,1) = -\frac{k}{z+1}$, where k is the difference between the breeding coefficients [142] (see also [457], pp. 31 and 32).

Below we describe several exact solutions to the nonlinear system of PDEs (4.3.4.1).

1°. A multiplicative separable solution periodic in x:
$$u = [C_1 \sin(kx) + C_2 \cos(kx)]\varphi(t),$$
$$v = [C_1 \sin(kx) + C_2 \cos(kx)]\psi(t), \qquad (4.3.4.2)$$

where C_1, C_2, and k are arbitrary constants, and the functions $\varphi = \varphi(t)$ and $\psi = \psi(t)$ are described by the system of first-order delay ODEs

$$\varphi'_t = -ak^2\varphi + \varphi f(\varphi/\psi, \bar{\varphi}_1/\varphi, \bar{\psi}_2/\psi),$$
$$\psi'_t = -bk^2\psi + \psi g(\varphi/\psi, \bar{\varphi}_1/\varphi, \bar{\psi}_2/\psi). \qquad (4.3.4.3)$$

Notations: $\bar{\varphi}_1 = \varphi(t-\tau_1)$ and $\bar{\psi}_2 = \psi(t-\tau_2)$.

2°. A multiplicative separable solution:
$$u = [C_1 \exp(kx) + C_2 \exp(-kx)]\varphi(t),$$
$$v = [C_1 \exp(kx) + C_2 \exp(-kx)]\psi(t), \qquad (4.3.4.4)$$

where C_1, C_2, and k are arbitrary constants, and the functions $\varphi = \varphi(t)$ and $\psi = \psi(t)$ are described by the system of first-order delay ODEs

$$\varphi'_t = ak^2\varphi + \varphi f(\varphi/\psi, \bar{\varphi}_1/\varphi, \bar{\psi}_2/\psi),$$
$$\psi'_t = bk^2\psi + \psi g(\varphi/\psi, \bar{\varphi}_1/\varphi, \bar{\psi}_2/\psi). \qquad (4.3.4.5)$$

Remark 4.15. *The systems of delay ODEs (4.3.4.3) and (4.3.4.5) admit exponential exact solutions of the form*
$$\varphi(t) = Ae^{-\lambda t}, \quad \psi(t) = Be^{-\lambda t}, \qquad (4.3.4.6)$$
where λ is an arbitrary constant. The constants A and B are determined from the algebraic (transcendental) equations
$$\pm ak^2 + \lambda + f(A/B, e^{\lambda\tau_1}, e^{\lambda\tau_2}) = 0,$$
$$\pm bk^2 + \lambda + g(A/B, e^{\lambda\tau_1}, e^{\lambda\tau_2}) = 0,$$
where the lower sign corresponds to system (4.3.4.3), while the upper sign corresponds to system (4.3.4.5).

3°. Also, there is a degenerate solution of the form
$$u = (C_1 x + C_2)\varphi(t), \quad v = (C_1 x + C_2)\psi(t).$$

4°. A multiplicative separable solution:
$$u = e^{-\lambda t}y(x), \quad v = e^{-\lambda t}z(x), \qquad (4.3.4.7)$$

where λ is an arbitrary constant, and the functions $y = y(x)$ and $z = z(x)$ are described by the system of second-order ODEs

$$ay''_{xx} + \lambda y + yf(y/z, e^{\lambda\tau_1}, e^{\lambda\tau_2}) = 0,$$
$$bz''_{xx} + \lambda z + zg(y/z, e^{\lambda\tau_1}, e^{\lambda\tau_2}) = 0.$$

5°. Also, there are solutions of the form
$$u = e^{-\lambda t}y(\xi), \quad v = e^{-\lambda t}z(\xi), \quad \xi = x + kt,$$

where k is an arbitrary constant, which generalize the solution of Item 4°.

4.3. Reductions and Exact Solutions of Lotka–Volterra Type Systems with Several Delays

System 2. Consider the reaction-diffusion system of equations with two delays

$$u_t = a u_{xx} + u f(u/v, \bar{u}_1/\bar{v}_1, \bar{u}_2/\bar{v}_2),$$
$$v_t = a v_{xx} + v g(u/v, \bar{u}_1/\bar{v}_1, \bar{u}_2/\bar{v}_2),$$
(4.3.4.8)

where $\bar{u}_1 = u(x, t-\tau_1)$, $\bar{u}_2 = u(x, t-\tau_2)$, $\bar{v}_1 = v(x, t-\tau_1)$, and $\bar{v}_2 = v(x, t-\tau_2)$.

1°. The system of delay PDEs (4.3.4.8) admits three sets of multiplicative separable solutions (4.3.4.2), (4.3.4.4), and (4.3.4.7) (the details are omitted).

2°. An incomplete separable solution:

$$u = \varphi(t)\theta(x,t), \quad v = \psi(t)\theta(x,t).$$

The functions $\varphi = \varphi(t)$ and $\psi = \psi(t)$ are described by the system of first-order delay ODEs

$$\varphi'_t = \varphi f(\varphi/\psi, \bar{\varphi}_1/\bar{\psi}_1, \bar{\varphi}_2/\bar{\psi}_2), \quad \psi'_t = \psi g(\varphi/\psi, \bar{\varphi}_1/\bar{\psi}_1, \bar{\varphi}_2/\bar{\psi}_2),$$
$$\bar{\varphi}_j = \varphi(t-\tau_j), \quad \bar{\psi}_j = \psi(t-\tau_j), \quad j=1,2.$$
(4.3.4.9)

The function $\theta = \theta(x,t)$ satisfies the linear heat equation

$$\theta_t = a\theta_{xx}.$$
(4.3.4.10)

Notably, the system of delay ODEs (4.3.4.9) admits an exponential exact solution of the form (4.3.4.6).

In the general case, system (4.3.4.9) can be reduced with the substitution $\varphi = \omega\psi$ to a single equation for the function $\omega = \omega(t)$:

$$\omega'_t = \omega[f(\omega, \bar{\omega}_1, \bar{\omega}_2) - g(\omega, \bar{\omega}_1, \bar{\omega}_2)],$$
(4.3.4.11)

where $\bar{\omega}_1 = \omega(t-\tau_1)$ and $\bar{\omega}_2 = \omega(t-\tau_2)$. Once ω is determined, the functions φ and ψ are found as

$$\varphi = \omega\psi, \quad \psi = C\exp\left[\int g(\omega, \bar{\omega}_1, \bar{\omega}_2)\, dt\right],$$

where C is an arbitrary constant. Equation (4.3.4.11) is integrable in terms of elementary functions if, for example,

$$f(\omega, \bar{\omega}_1, \bar{\omega}_2) = b_1\omega^k + c_1 + h(\bar{\omega}_1, \bar{\omega}_2), \quad g(\omega, \bar{\omega}_1, \bar{\omega}_2) = b_2\omega^k + c_2 + h(\bar{\omega}_1, \bar{\omega}_2),$$

where $h(\bar{\omega}_1, \bar{\omega}_2)$ is an arbitrary function, and b_1, b_2, c_1, c_2, and k are arbitrary constants.

3°. An incomplete separable solution with $g(z, z_1, z_2) = -z^2 f(z, z_1, z_2)$:

$$u = \theta(x,t)\sin\varphi(t), \quad v = \theta(x,t)\cos\varphi(t).$$

The function $\varphi = \varphi(t)$ satisfies the first-order delay ODE

$$\varphi'_t = f(\tan\varphi, \tan\bar{\varphi}_1, \tan\bar{\varphi}_2)\tan\varphi, \quad \bar{\varphi}_1 = \varphi(t-\tau_1), \quad \bar{\varphi}_2 = \varphi(t-\tau_2).$$

The function $\theta = \theta(x,t)$ is described by the linear heat equation (4.3.4.10).

4°. An incomplete separable solution with $g(z, z_1, z_2) = z^2 f(z, z_1, z_2)$:

$$u = \theta(x,t) \sinh \varphi(t), \quad v = \theta(x,t) \cosh \varphi(t).$$

The function $\varphi = \varphi(t)$ satisfies the first-order delay ODE

$$\varphi'_t = f(\tanh \varphi, \tanh \bar{\varphi}_1, \tanh \bar{\varphi}_2) \tanh \varphi, \quad \bar{\varphi}_1 = \varphi(t - \tau_1), \quad \bar{\varphi}_2 = \varphi(t - \tau_2).$$

The function $\theta = \theta(x,t)$ is described by the linear heat equation (4.3.4.10).

5°. Also, there are solutions of the form

$$u = \theta(x,t) \cosh \varphi(t), \quad v = \theta(x,t) \sinh \varphi(t).$$

4.4. Nonlinear PDEs with Proportional Arguments. Principle of Analogy of Solutions

This section deals with nonlinear partial differential equations with proportional arguments that involve, besides the unknown function $u = u(x,t)$, a few functions with one or more scaled independent variables such as $u(px, t)$, $u(x, qt)$, or $u(px, qt)$, where p and q are the scaling parameters ($0 < p < 1$, $0 < q < 1$).

4.4.1. Principle of Analogy of Solutions

Below we describe a reasonably general method for constructing exact solutions to nonlinear PDEs with proportional arguments. It relies on the following principle [9, 416].

The principle of analogy of solutions. The structure of exact solutions to PDEs with proportional arguments

$$\begin{aligned} F(x, t, u, u_x, u_t, u_{xx}, u_{xt}, u_{tt}, \ldots, w, w_x, w_t, w_{xx}, w_{xt}, w_{tt}, \ldots) &= 0, \\ w &= u(px, qt), \end{aligned} \quad (4.4.1.1)$$

is frequently (but not always) determined by the solution structure of the simpler PDEs with regular arguments

$$F(x, t, u, u_x, u_t, u_{xx}, u_{xt}, u_{tt}, \ldots, u, u_x, u_t, u_{xx}, u_{xt}, u_{tt}, \ldots) = 0. \quad (4.4.1.2)$$

Equation (4.4.1.2) does not involve the unknown function with proportional arguments; it is formally obtained from (4.4.1.1) by replacing w with u.

Figure 4.1 displays a schematic of applying the analogy principle to a second-order proportional argument PDE solved for u_t.

Below we illustrate the application of the analogy principle with three examples of proportional argument PDEs that have different types of solution.

4.4. Nonlinear PDEs with Proportional Arguments. Principle of Analogy of Solutions

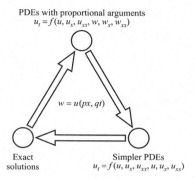

Figure 4.1. A schematic of using the analogy principle to construct exact solutions for nonlinear PDEs with proportional arguments.

▶ **Example 4.25.** Consider the reaction-diffusion equation with proportional arguments

$$u_t = au_{xx} + bu^m w^k, \quad w = u(px, qt), \tag{4.4.1.3}$$

that involve a power-law nonlinearity.

Following the analogy principle, we set $w = u$ in equation (4.4.1.3) to obtain a simpler nonlinear heat (diffusion) equation with a power-law nonlinearity and without proportional arguments:

$$u_t = au_{xx} + bu^{m+k}. \tag{4.4.1.4}$$

The equation admits a self-similar solution [134]:

$$u(x,t) = t^{\frac{1}{1-m-k}} U(z), \quad z = xt^{-1/2}, \quad k \neq 1 - m. \tag{4.4.1.5}$$

While using the analogy principle, we look for a solution to the nonlinear equation with proportional arguments (4.4.1.3) in the same form (4.4.1.5). As a result, we obtain the following nonlinear ODE with proportional argument for the function $U = U(z)$:

$$aU''_{zz} + \frac{1}{2}zU'_z - \frac{1}{1-m-k}U + bq^{\frac{k}{1-m-k}}U^m W^k = 0,$$
$$W = U(sz), \quad s = pq^{-1/2}. \tag{4.4.1.6}$$

◀

Remark 4.16. Interestingly, in the special case $p = q^{1/2}$, the PDE with proportional arguments (4.4.1.3) with $0 < p < 1$ and $0 < q < 1$ has the exact solution (4.4.1.5), which is expressed through a solution to the ODE without delay (4.4.1.6), at $s = 1$. If $p < q^{1/2}$, equation (4.4.1.3) reduces to a delay ODE with $s < 1$ and if $p > q^{1/2}$, it reduces to an advanced ODE, with $s > 1$. Furthermore, if $p > 1$ and $q > 1$, solutions to equation (4.4.1.3) can also be expressed, for suitable values of p and q, through solutions of a delay ODE ($s < 1$), ODE without delay ($s = 1$), or advanced ODE ($s > 1$).

▶ **Example 4.26.** Now consider the reaction-diffusion equation with proportional arguments and exponential nonlinearities

$$u_t = au_{xx} + be^{\mu u + \lambda w}, \quad w = u(px, qt). \tag{4.4.1.7}$$

Following the analogy principle, we set $w = u$ in equation (4.4.1.7) to get the simpler nonlinear heat (diffusion) equation with ordinary arguments and an exponential source

$$u_t = au_{xx} + be^{(\mu+\lambda)u}. \tag{4.4.1.8}$$

This equation admits an invariant solution of the form [134]:

$$u(x,t) = U(z) - \frac{1}{\mu+\lambda}\ln t, \quad z = xt^{-1/2}, \quad \mu \neq -\lambda. \tag{4.4.1.9}$$

According to the analogy principle, we seek a solution to the nonlinear PDE with proportional arguments (4.4.1.7) in the form (4.4.1.9). As a result, we obtain the following nonlinear ODE with proportional argument for $U = U(z)$:

$$aU''_{zz} + \frac{1}{2}zU'_z + \frac{1}{\mu+\lambda} + bq^{-\frac{\lambda}{\mu+\lambda}}e^{\mu U + \lambda W} = 0,$$
$$W = U(sz), \quad s = pq^{-1/2}.$$

◀

▶ **Example 4.27.** Consider the PDE with proportional arguments and a logarithmic nonlinearity

$$u_t = au_{xx} + u(b\ln u + c\ln w + d), \quad w = u(px, qt). \tag{4.4.1.10}$$

By setting $w = u$ in equation (4.4.1.10), we get the simpler nonlinear equation without delay

$$u_t = au_{xx} + u[(b+c)\ln u + d],$$

which admits a multiplicative separable solution [422]:

$$u(x,t) = \varphi(x)\psi(t). \tag{4.4.1.11}$$

According to the analogy principle, we look for a solution to the nonlinear PDE with proportional arguments (4.4.1.10) in the form (4.4.1.11). On separating the variables, we obtain the following nonlinear ODEs with proportional arguments for $\varphi = \varphi(x)$ and $\psi = \psi(t)$:

$$a\varphi''_{xx} + \varphi(b\ln\varphi + c\ln\bar{\varphi}) = K\varphi, \quad \bar{\varphi} = \varphi(px);$$
$$\psi'_t = \psi(b\ln\psi + c\ln\bar{\psi}) + (d+K)\psi, \quad \bar{\psi} = \psi(qt),$$

where K is an arbitrary constant. ◀

In the next subsection, we describe some nonlinear reaction-diffusion and wave type equations with proportional delay and their exact solutions. Many of them (but not all) were obtained in [411, 416]. Most of these solutions were constructed using the analogy principle.

4.4.2. Exact Solutions to Quasilinear Diffusion Equations with Proportional Delay

This subsection deals with quasilinear reaction-diffusion equations with proportional delay that are linear in both derivatives.

Equations involving free parameters.

Most of the solutions given below were obtained in [416].

Equation 1. The PDE with proportional delay and a logarithmic nonlinearity

$$u_t = au_{xx} + u(b\ln u + c\ln w + d), \quad w = u(x, qt), \qquad (4.4.2.1)$$

admits functional separable solutions of the form

$$u(x,t) = \exp[\psi_2(t)x^2 + \psi_1(t)x + \psi_0(t)],$$

where the functions $\psi_n = \psi_n(t)$ are described by the nonlinear system of ODEs with proportional delay

$$\psi_2' = 4a\psi_2^2 + b\psi_2 + c\bar{\psi}_2, \quad \bar{\psi}_2 = \psi_2(qt),$$
$$\psi_1' = 4a\psi_1\psi_2 + b\psi_1 + c\bar{\psi}_1, \quad \bar{\psi}_1 = \psi_1(qt),$$
$$\psi_0' = a[\psi_1^2 + 2\psi_2] + b\psi_0 + c\bar{\psi}_0 + d, \quad \bar{\psi}_0 = \psi_0(qt).$$

Equation 2. The PDE with proportional delay and a logarithmic nonlinearity

$$u_t = au_{xx} + u(b\ln^2 u + c\ln u + d\ln w + s), \quad w = u(x, qt), \qquad (4.4.2.2)$$

admits two exact solutions depending on the sign of ab. These are given below.

1°. A functional separable solution for $ab > 0$:

$$u(x,t) = \exp[\psi_1(t)\varphi(x) + \psi_2(t)],$$
$$\varphi(x) = A\cos(\lambda x) + B\sin(\lambda x), \quad \lambda = \sqrt{b/a},$$

where A and B are arbitrary constants, and the functions $\psi_n = \psi_n(t)$ are described by the nonlinear system of ODEs with proportional delay

$$\psi_1' = 2b\psi_1\psi_2 + (c-b)\psi_1 + d\bar{\psi}_1, \quad \bar{\psi}_1 = \psi_1(qt),$$
$$\psi_2' = b(A^2 + B^2)\psi_1^2 + b\psi_2^2 + c\psi_2 + d\bar{\psi}_2 + s, \quad \bar{\psi}_2 = \psi_2(qt).$$

2°. A functional separable solution for $ab < 0$:

$$u(x,t) = \exp[\psi_1(t)\varphi(x) + \psi_2(t)],$$
$$\varphi(x) = A\cosh(\lambda x) + B\sinh(\lambda x), \quad \lambda = \sqrt{-b/a},$$

where A and B are arbitrary constants, and the functions $\psi_n = \psi_n(t)$ are described by the nonlinear system of ODEs with proportional delay

$$\psi_1' = 2b\psi_1\psi_2 + (c-b)\psi_1 + d\bar{\psi}_1, \quad \bar{\psi}_1 = \psi_1(qt),$$
$$\psi_2' = b(A^2 - B^2)\psi_1^2 + b\psi_2^2 + c\psi_2 + d\bar{\psi}_2 + s, \quad \bar{\psi}_2 = \psi_2(qt).$$

If $A = \pm B$, we get $\varphi(x) = Ae^{\pm\lambda x}$. In this case, the second equation of the system becomes independent, while the first one is linear in ψ_1.

Remark 4.17. Equations (4.4.2.1) and (4.4.2.2) and their solutions admit generalizations to the case of a variable delay of general form, so that $w = u(x, t - \tau(t))$, where $\tau(t)$ is an arbitrary function.

Equations involving arbitrary functions of the form $f(u - w)$.

Equation 3. The nonlinear PDE with proportional delay

$$u_t = a u_{xx} + f(u - w), \quad w = u(x, qt), \qquad (4.4.2.3)$$

admits an additive separable solution

$$u(x, t) = C_1 x^2 + C_2 x + \psi(t),$$

where C_1 and C_2 are arbitrary constants, and the function $\psi = \psi(t)$ is described by the nonlinear first-order ODE with proportional delay

$$\psi'_t = 2aC_1 + f(\psi - \bar{\psi}), \quad \bar{\psi} = \psi(qt).$$

Equation 4. The nonlinear PDE with proportional argument

$$u_t = a u_{xx} + f(u - w), \quad w = u(px, t), \qquad (4.4.2.4)$$

admits an additive separable solution

$$u(x, t) = Ct + \varphi(x),$$

where C is an arbitrary constant, and the function $\varphi = \varphi(x)$ is described by the nonlinear second-order ODE with proportional argument

$$a\varphi''_{xx} - C + f(\varphi - \bar{\varphi}) = 0, \quad \bar{\varphi} = \varphi(px).$$

Equation 5. The nonlinear PDE with proportional delay

$$u_t = a u_{xx} + bu + f(u - w), \quad w = u(x, qt), \qquad (4.4.2.5)$$

admits two exact solutions depending on the sign of ab as shown below.

1°. An additive separable solution with $ab < 0$:

$$u(x, t) = A \cosh(\lambda x) + B \sinh(\lambda x) + \psi(t), \quad \lambda = \sqrt{-b/a},$$

where A and B are arbitrary constants, and the function $\psi = \psi(t)$ is described by the nonlinear first-order ODE with proportional delay

$$\psi'_t = b\psi + f(\psi - \bar{\psi}), \quad \bar{\psi} = \psi(qt). \qquad (4.4.2.6)$$

2°. An additive separable solution with $ab > 0$:

$$u(x, t) = A \cos(\lambda x) + B \sin(\lambda x) + \psi(t), \quad \lambda = \sqrt{b/a},$$

where A and B are arbitrary constants, and the function $\psi = \psi(t)$ is described by the nonlinear first-order ODE with proportional delay (4.4.2.6).

It is noteworthy that equation (4.4.2.5) and its solutions admit generalizations to the case of a variable delay of general form, so that $w = u(x, t - \tau(t))$, where $\tau(t)$ is an arbitrary function.

Equation 6. The nonlinear PDE with proportional argument

$$u_t = au_{xx} + bu + f(u - w), \quad w = u(px, t), \tag{4.4.2.7}$$

admits an additive separable solution

$$u(x, t) = Ce^{bt} + \varphi(x),$$

where C is an arbitrary constant and $\varphi = \varphi(x)$ is a function satisfying the nonlinear second-order ODE with proportional argument

$$a\varphi''_{xx} + b\varphi + f(\varphi - \bar{\varphi}) = 0, \quad \bar{\varphi} = \varphi(px).$$

Equation 7. The nonlinear PDE with proportional arguments

$$u_t = au_{xx} + e^{\lambda u} f(u - w), \quad w = u(px, qt), \tag{4.4.2.8}$$

admits exact solutions of the form

$$u(x, t) = U(z) - \frac{1}{\lambda} \ln t, \quad z = xt^{-1/2},$$

where $U = U(z)$ is a function satisfying the nonlinear ODE with proportional argument

$$aU''_{zz} + \frac{1}{2}zU'_z + \frac{1}{\lambda} + e^{\lambda U} f\left(U - W + \frac{1}{\lambda}\ln q\right) = 0,$$
$$W = U(sz), \quad s = pq^{-1/2}.$$

Equations involving arbitrary functions of the form $f(w/u)$.
Equation 8. The nonlinear PDE with proportional delay

$$u_t = au_{xx} + uf(w/u), \quad w = u(x, qt), \tag{4.4.2.9}$$

admits a few multiplicative separable solutions listed below.

1°. A solution with hyperbolic functions:

$$u(x, t) = [A \cosh(\lambda x) + B \sinh(\lambda x)]\psi(t),$$

where A, B, and λ are arbitrary constants, and the function $\psi = \psi(t)$ is described by the nonlinear first-order ODE with proportional delay

$$\psi'_t = a\lambda^2 \psi + \psi f(\bar{\psi}/\psi), \quad \bar{\psi} = \psi(qt).$$

2°. A solution with trigonometric functions periodic in x:

$$u(x, t) = [A \cos(\lambda x) + B \sin(\lambda x)]\psi(t),$$

where A, B, and λ are arbitrary constants, and the function $\psi = \psi(t)$ is described by the nonlinear first-order ODE with proportional delay

$$\psi'_t = -a\lambda^2 \psi + \psi f(\bar{\psi}/\psi), \quad \bar{\psi} = \psi(qt).$$

3°. A degenerate solution:

$$u(x,t) = (Ax+B)\psi(t),$$

where A, B, and λ are arbitrary constants, and the function $\psi = \psi(t)$ is described by the nonlinear first-order ODE with proportional delay

$$\psi'_t = \psi f(\bar\psi/\psi), \quad \bar\psi = \psi(qt).$$

Equation 9. The nonlinear PDE with proportional argument

$$u_t = au_{xx} + uf(w/u), \quad w = u(px,t), \tag{4.4.2.10}$$

admits a multiplicative separable solution

$$u(x,t) = e^{\lambda t}\varphi(x),$$

where λ is an arbitrary constant and $\varphi = \varphi(x)$ is a function satisfying the nonlinear second-order ODE with proportional argument

$$a\varphi''_{xx} + \varphi[f(\bar\varphi/\varphi) - \lambda] = 0, \quad \bar\varphi = \varphi(px).$$

Equation 10. The nonlinear PDE with proportional delay

$$u_t = au_{xx} + bu\ln u + uf(w/u), \quad w = u(x,qt), \tag{4.4.2.11}$$

admits a multiplicative separable solution

$$u(x,t) = \varphi(x)\psi(t),$$

where $\varphi = \varphi(x)$ and $\psi = \psi(t)$ are functions satisfying the nonlinear second-order ODE and first-order ODE with proportional delay

$$\begin{aligned} a\varphi''_{xx} &= C_1\varphi - b\varphi\ln\varphi, \\ \psi'_t &= C_1\psi + \psi f(\bar\psi/\psi) + b\psi\ln\psi, \quad \bar\psi = \psi(qt), \end{aligned} \tag{4.4.2.12}$$

with C_1 being an arbitrary constant.

The first equation in (4.4.2.12) is autonomous. Its general solution can be obtained in implicit form. A particular one-parameter solution to the equation is

$$\varphi = \exp\left[-\frac{b}{4a}(x+C_2)^2 + \frac{C_1}{b} + \frac{1}{2}\right],$$

where C_2 is an arbitrary constant.

Notably, equation (4.4.2.11) and its solutions admit a generalization to the case of a variable delay of general form, so that $w = u(x, t - \tau(t))$, where $\tau(t)$ is an arbitrary function.

4.4. Nonlinear PDEs with Proportional Arguments. Principle of Analogy of Solutions 257

Equation 11. The nonlinear PDE with proportional argument

$$u_t = au_{xx} + bu\ln u + uf(w/u), \quad w = u(px, t) \tag{4.4.2.13}$$

admits a multiplicative separable solution

$$u(x,t) = \exp(Ce^{bt})\varphi(x),$$

where C is an arbitrary constant and $\varphi = \varphi(x)$ is a function satisfying the nonlinear ODE with proportional argument

$$a\varphi''_{xx} + b\varphi\ln\varphi + \varphi f(\bar{\varphi}/\varphi) = 0, \quad \bar{\varphi} = \varphi(px).$$

4.4.3. Exact Solutions to More Complicated Nonlinear Diffusion Equations with Proportional Delay

Equations with a variable power-law transfer coefficient.
Equation 1. The nonlinear PDE with proportional delay

$$u_t = a(u^k u_x)_x + uf(w/u), \quad w = u(x, qt), \tag{4.4.3.1}$$

admits a multiplicative separable solution

$$u(x,t) = \varphi(x)\psi(t).$$

The functions $\varphi = \varphi(x)$ and $\psi = \psi(t)$ are determined from the ODE without delay and ODE with proportional delay

$$a(\varphi^k \varphi'_x)'_x = b\varphi,$$
$$\psi'_t = b\psi^{k+1} + \psi f(\bar{\psi}/\psi), \quad \bar{\psi} = \psi(qt),$$

where b is an arbitrary constant.

Equation 2. The nonlinear PDE with proportional argument

$$u_t = a(u^k u_x)_x + uf(w/u), \quad w = u(px, t), \tag{4.4.3.2}$$

admits exact solutions of the form

$$u(x,t) = e^{2\lambda t}U(z), \quad z = e^{-k\lambda t}x,$$

where λ is an arbitrary constant and $U = U(z)$ is a function satisfying the ODE with proportional argument

$$2\lambda U - k\lambda z U'_z = a(U^k U'_z)'_z + Uf(W/U), \quad W = U(pz).$$

Equation 3. The nonlinear PDE with proportional delay

$$u_t = a(u^k u_x)_x + bu^{k+1} + uf(w/u), \quad w = u(x, qt), \tag{4.4.3.3}$$

admits three multiplicative separable solutions listed below.

1°. Solution with $b(k+1) > 0$:
$$u(x,t) = [C_1 \cos(\beta x) + C_2 \sin(\beta x)]^{1/(k+1)} \psi(t), \quad \beta = \sqrt{b(k+1)/a},$$

where C_1 and C_2 are arbitrary constants, and the function $\psi = \psi(t)$ is described by the ODE with proportional delay
$$\psi'_t = \psi f(\bar{\psi}/\psi), \quad \bar{\psi} = \psi(qt). \tag{4.4.3.4}$$

2°. Solution with $b(k+1) < 0$:
$$u(x,t) = [C_1 \exp(-\beta x) + C_2 \exp(\beta x)]^{1/(k+1)} \psi(t), \quad \beta = \sqrt{-b(k+1)/a},$$

where C_1 and C_2 are arbitrary constants, and the function $\psi = \psi(t)$ is described by ODE with proportional delay (4.4.3.4).

3°. Solution with $k = -1$:
$$u(x,t) = C_1 \exp\left(-\frac{b}{2a}x^2 + C_2 x\right)\psi(t),$$

where C_1 and C_2 are arbitrary constants, and the function $\psi = \psi(t)$ is described by ODE proportional delay (4.4.3.4).

Equation 4. The nonlinear PDE with proportional delay
$$u_t = a(u^k u_x)_x + u^{k+1} f(w/u), \quad w = u(x, qt), \tag{4.4.3.5}$$

admits the exact solution
$$u(x,t) = t^{-1/k} \varphi(z), \quad z = x + \lambda \ln t,$$

where λ is an arbitrary constant, and the function $\varphi = \varphi(z)$ satisfies the second-order ODE with constant delay
$$a(\varphi^k \varphi'_z)'_z - \lambda \varphi'_z + \frac{1}{k}\varphi + \varphi^{k+1} f(q^{-1/k} \bar{\varphi}/\varphi) = 0, \quad \bar{\varphi} = \varphi(z + \lambda \ln q).$$

Equation 5. The nonlinear PDE with proportional arguments
$$u_t = a(u^k u_x)_x + u^n f(w/u), \quad w = u(px, qt), \tag{4.4.3.6}$$

admits two exact solutions given below.

1°. A self-similar solution:
$$u(x,t) = t^{\frac{1}{1-n}} U(z), \quad z = x t^{\frac{n-k-1}{2(1-n)}},$$

where $U = U(z)$ is a function satisfying the second-order ODE with proportional argument
$$\frac{1}{1-n}U + \frac{n-k-1}{2(1-n)}zU'_z = a(U^k U'_z)'_z + U^n f(W/U), \quad W = U(sz), \quad s = pq^{\frac{n-k-1}{2(1-n)}}.$$

2°. A traveling wave solution with $q = p$:

$$u(x,t) = U(z), \quad z = kx - \lambda t,$$

where k and λ are arbitrary constants, and the function $U = U(z)$ satisfies the second-order ODE with proportional argument

$$ak^2(U^k U'_z)'_z + \lambda U'_z + U^n f(W/U) = 0, \quad W = U(pz).$$

Equation 6. The nonlinear PDE with proportional delay

$$u_t = a(u^k u_x)_x + b + u^{-k} f(u^{k+1} - w^{k+1}), \quad w = u(x, qt), \qquad (4.4.3.7)$$

admits a functional separable solution

$$u(x,t) = \left[\psi(t) - \frac{b(k+1)}{2a}x^2 + C_1 x + C_2\right]^{1/(k+1)},$$

where C_1 and C_2 are arbitrary constants, and the function $\psi = \psi(t)$ is described by the ODE with proportional delay

$$\psi'_t = (k+1)f(\psi - \bar\psi), \quad \bar\psi = \psi(qt).$$

Equation 7. The nonlinear PDE with proportional argument

$$u_t = a(u^k u_x)_x + bu^{-k} + f(u^{k+1} - w^{k+1}), \quad w = u(px, t), \qquad (4.4.3.8)$$

admits a functional separable solution

$$u = \left[b(k+1)t + \varphi(x)\right]^{\frac{1}{k+1}},$$

where the function $\varphi = \varphi(x)$ is described by the ODE with proportional argument

$$a\varphi''_{xx} + (k+1)f(\varphi - \bar\varphi) = 0, \quad \bar\varphi = \varphi(px).$$

Equations with an exponential transfer coefficient.
Equation 8. The nonlinear PDE with proportional delay

$$u_t = a(e^{\lambda u} u_x)_x + f(u - w), \quad w = u(x, qt), \qquad (4.4.3.9)$$

admits an additive separable solution

$$u = \frac{1}{\lambda} \ln(Ax^2 + Bx + C) + \psi(t),$$

where A, B, and C are arbitrary constants, and the function $\psi(t)$ is described by the ODE with proportional delay

$$\psi' = 2a(A/\lambda)e^{\lambda\psi} + f(\psi - \bar\psi), \quad \bar\psi = \psi(qt).$$

Equation 9. The nonlinear PDE with proportional delay

$$u_t = a(e^{\lambda u} u_x)_x + b e^{\lambda u} + f(u - w), \quad w = u(x, qt), \qquad (4.4.3.10)$$

admits two additive separable solutions.

1°. Solution with $b\lambda > 0$:

$$u(x, t) = \frac{1}{\lambda} \ln[C_1 \cos(\beta x) + C_2 \sin(\beta x)] + \psi(t), \quad \beta = \sqrt{b\lambda/a},$$

where C_1 and C_2 are arbitrary constants, and the function $\psi = \psi(t)$ is described by the ODE with proportional delay

$$\psi'_t = f(\psi - \bar{\psi}), \quad \bar{\psi} = \psi(qt). \qquad (4.4.3.11)$$

2°. Solution with $b\lambda < 0$:

$$u(x, t) = \frac{1}{\lambda} \ln[C_1 \exp(-\beta x) + C_2 \exp(\beta x)] + \psi(t), \quad \beta = \sqrt{-b\lambda/a},$$

where C_1 and C_2 are arbitrary constants, and the function $\psi = \psi(t)$ is described by the ODE with proportional delay (4.4.3.11).

Equation 10. The nonlinear PDE with proportional argument

$$u_t = a(e^{\lambda u} u_x)_x + e^{\lambda u} f(u - w), \quad w = u(px, t), \qquad (4.4.3.12)$$

admits an additive separable solution

$$u(x, t) = -\frac{1}{\lambda} \ln t + \varphi(x),$$

where the function $\varphi = \varphi(x)$ is described by the ODE with proportional argument

$$a(e^{\lambda \varphi} \varphi'_x)'_x + \frac{1}{\lambda} + e^{\lambda \varphi} f(\varphi - \bar{\varphi}) = 0, \quad \bar{\varphi} = \varphi(px).$$

Equation 11. The nonlinear PDE with proportional arguments

$$u_t = a(e^{\lambda u} u_x)_x + e^{\mu u} f(u - w), \quad w = u(px, qt), \qquad (4.4.3.13)$$

admits the exact solution

$$u(x, t) = U(z) - \frac{1}{\mu} \ln t, \quad z = x t^{\frac{\lambda - \mu}{2\mu}},$$

where the function $U = U(z)$ is described by the nonlinear ODE with proportional argument

$$\frac{\lambda - \mu}{2\mu} z U'_z - \frac{1}{\mu} = a(e^{\lambda U} U'_z)'_z + e^{\mu U} f\left(U - W + \frac{1}{\mu} \ln q\right),$$

$$W = U(sz), \quad s = pq^{\frac{\lambda - \mu}{2\mu}}.$$

4.4. Nonlinear PDEs with Proportional Arguments. Principle of Analogy of Solutions

Equation 12. The nonlinear PDE with proportional delay

$$u_t = a(e^{\lambda u} u_x)_x + b + e^{-\lambda u} f(e^{\lambda u} - e^{\lambda w}), \quad w = u(x, qt), \qquad (4.4.3.14)$$

admits a functional separable solution

$$u(x,t) = \frac{1}{\lambda} \ln\left[\psi(t) - \frac{b\lambda}{2a} x^2 + C_1 x + C_2\right],$$

where C_1 and C_2 are arbitrary constants, and the function $\psi = \psi(t)$ is described by the ODE with proportional delay

$$\psi'_t = \lambda f(\psi - \bar{\psi}), \quad \bar{\psi} = \psi(qt).$$

Other equations with a variable transfer coefficient.

Equation 13. The PDE with proportional delay and a variable logarithmic transfer coefficient

$$u_t = [(a \ln u + b) u_x]_x - cu \ln u + u f(w/u), \quad w = u(x, qt), \qquad (4.4.3.15)$$

admits two multiplicative separable solutions

$$u(x,t) = \exp(\pm\sqrt{c/a}\, x)\psi(t),$$

where the function $\psi = \psi(t)$ is described by the ODE with proportional delay

$$\psi'_t = c(1 + b/a)\psi + \psi f(\bar{\psi}/\psi), \quad \bar{\psi} = \psi(qt).$$

Equation 14. The PDE with proportional delay and a transfer coefficient of general form

$$u_t = [u f'_u(u) u_x]_x + \frac{1}{f'_u(u)} [af(u) + bf(w) + c], \quad w = u(x, qt), \qquad (4.4.3.16)$$

admits a solution in implicit form

$$f(u) = \varphi(t) x + \psi(t),$$

where the functions $\varphi = \varphi(t)$ and $\psi = \psi(t)$ satisfy the ODEs with proportional delay

$$\varphi'_t = a\varphi + b\bar{\varphi}, \quad \bar{\varphi} = \varphi(qt),$$
$$\psi'_t = a\psi + b\bar{\psi} + c + \varphi^2, \quad \bar{\psi} = \psi(qt).$$

Equation 15. The PDE with proportional delay and a transfer coefficient of general form

$$u_t = a[f'_u(u) u_x]_x + b + \frac{1}{f'_u(u)} g(f(u) - f(w)), \quad w = u(x, qt), \qquad (4.4.3.17)$$

admits a functional separable solution in implicit form

$$f(u) = \psi(t) - \frac{b}{2a}x^2 + C_1 x + C_2,$$

where the function $\psi = \psi(t)$ is described by the ODE with proportional delay

$$\psi'_t = g(\psi - \bar{\psi}), \quad \bar{\psi} = \psi(qt).$$

Equation 16. The PDE with proportional delay and a transfer coefficient of general form

$$u_t = a[f'_u(u)u_x]_x + bf(u) + \frac{f(u)}{f'_u(u)}g(f(w)/f(u)), \quad w = u(x, qt), \quad (4.4.3.18)$$

admits two functional separable solutions in implicit form depending on the sign of ab.

1°. Solution with $ab > 0$:

$$f(u) = [C_1 \cos(\lambda x) + C_2 \sin(\lambda x)]\psi(t), \quad \lambda = \sqrt{b/a},$$

where C_1 and C_2 are arbitrary constants, and the function $\psi = \psi(t)$ is described by the ODE with proportional delay

$$\psi'_t = \psi g(\bar{\psi}/\psi), \quad \bar{\psi} = \psi(qt). \quad (4.4.3.19)$$

2°. Solution with $ab < 0$:

$$f(u) = [C_1 \exp(-\lambda x) + C_2 \exp(\lambda x)]\psi(t), \quad \lambda = \sqrt{-b/a},$$

where C_1 and C_2 are arbitrary constants, and the function $\psi = \psi(t)$ is described by the ODE with proportional delay (4.4.3.19).

Remark 4.18. *Equations (4.4.3.1), (4.4.3.3), (4.4.3.7), (4.4.3.9), (4.4.3.10), and (4.4.3.14)–(4.4.3.18) and their solutions admit generalizations to the case of a variable delay of general form, so that $w = u(x, t - \tau(t))$, where $\tau(t)$ is an arbitrary function.*

Equation 17. The PDE with a proportional argument and a transfer coefficient of general form

$$u_t = [f'_u(u)u_x]_x + \frac{a}{f'_u(u)} + g(f(u) - f(w)), \quad w = u(px, t), \quad (4.4.3.20)$$

admits a functional separable solution in implicit form

$$f(u) = at + \varphi(x),$$

where the function $\varphi = \varphi(x)$ is described by the ODE with proportional argument

$$\varphi''_{xx} + g(\varphi - \bar{\varphi}) = 0, \quad \bar{\varphi} = \varphi(px).$$

Equation 18. The PDE with proportional arguments and a transfer coefficient of general form
$$u_t = [f(u,w)u_x]_x, \quad w = u(px, qt), \tag{4.4.3.21}$$
admits a self-similar solution
$$u(x,t) = U(z), \quad z = xt^{-1/2},$$
where the function $U = U(z)$ is described by the nonlinear ODE with proportional argument
$$[f(U,W)U'_z]'_z + \tfrac{1}{2}zU'_z = 0, \quad W = U(sz), \quad s = pq^{-1/2}.$$

If $f(u,w) = aw$, equation (4.4.3.21) has a simple solution that can be expressed in terms of elementary functions:
$$u(x,t) = -\frac{qx^2}{6ap^2t}.$$

Equation 19. The second-order evolution PDE of general form with proportional arguments
$$u_t = F(u, w, u_x, u_{xx}), \quad w = u(px, pt),$$
admits the traveling wave solution
$$u(x,t) = U(z), \quad z = kx - \lambda t,$$
where the function $U = U(z)$ is described by the nonlinear ODE with proportional argument
$$F(U, W, kU'_z, k^2 U''_{zz}) + \lambda U'_z = 0, \quad W = U(pz).$$

Remark 4.19. For more nonlinear reaction-diffusion type equations with proportional delay that admit exact solutions, see [416].

4.4.4. Exact Solutions to Nonlinear Wave-Type Equations with Proportional Delay

Quasilinear equations with constant speed linear in the derivatives.
Most of the solutions given below were obtained in [418].
Equation 1. The nonlinear Klein–Gordon type wave equation with proportional arguments and a power-law nonlinearity
$$u_{tt} = au_{xx} + bw^k, \quad w = u(px, qt), \tag{4.4.4.1}$$
admits, for $k \neq 1$, a self-similar solution
$$u(x,t) = t^{\frac{2}{1-k}} U(z), \quad z = x/t,$$

where the function $U = U(z)$ is described by the nonlinear ODE with proportional argument

$$\frac{2(1+k)}{(1-k)^2}U - \frac{2(1+k)}{1-k}zU'_z + z^2 U''_{zz} = aU''_{zz} + bq^{\frac{2k}{1-k}}W^k,$$
$$W = U(sz), \quad s = p/q.$$

Equation 2. The nonlinear Klein–Gordon type wave equation with proportional arguments and a power-law nonlinearity

$$u_{tt} = au_{xx} + bu^m w^k, \quad w = u(px, qt), \tag{4.4.4.2}$$

admits, for $k + m \neq 1$, a self-similar solution

$$u(x,t) = t^{\frac{2}{1-k-m}} U(z), \quad z = x/t,$$

where the function $U = U(z)$ is described by the nonlinear ODE with proportional argument

$$\frac{2(1+k+m)}{(1-k-m)^2}U - \frac{2(1+k+m)}{1-k-m}zU'_z + z^2 U''_{zz} = aU''_{zz} + bq^{\frac{2k}{1-k-m}}U^m W^k,$$
$$W = U(sz), \quad s = p/q.$$

Equation 3. The nonlinear Klein–Gordon type wave equation with proportional arguments and an exponential nonlinearity

$$u_{tt} = au_{xx} + be^{\mu u + \lambda w}, \quad w = u(px, qt), \tag{4.4.4.3}$$

admits, for $\mu + \lambda \neq 0$, an exact solution of the form

$$u(x,t) = U(z) - \frac{2}{\mu + \lambda} \ln t, \quad z = \frac{x}{t},$$

where the function $U = U(z)$ is described by the nonlinear ODE with proportional argument

$$(z^2 U'_z)'_z + \frac{2}{\mu + \lambda} = aU''_{zz} + bq^{-\frac{2\lambda}{\mu+\lambda}} e^{\mu U + \lambda W},$$
$$W = U(sz), \quad s = p/q.$$

Equation 4. The nonlinear Klein–Gordon type wave equation with proportional arguments and a logarithmic nonlinearity

$$u_{tt} = au_{xx} + u(b \ln u + c \ln w), \quad w = u(px, qt), \tag{4.4.4.4}$$

admits a multiplicative separable solution

$$u(x,t) = \varphi(x)\psi(t),$$

4.4. Nonlinear PDEs with Proportional Arguments. Principle of Analogy of Solutions

where the functions $\varphi = \varphi(x)$ and $\psi = \psi(t)$ are described by the nonlinear second-order ODEs with proportional arguments

$$a\varphi''_{xx} + \varphi(b\ln\varphi + c\ln\bar\varphi) = 0, \quad \bar\varphi = \varphi(px);$$
$$\psi''_{tt} = \psi(b\ln\psi + c\ln\bar\psi), \quad \bar\psi = \psi(qt).$$

Equation 5. The nonlinear Klein–Gordon type wave equation with proportional delay

$$u_{tt} = au_{xx} + f(u-w), \quad w = u(x, qt), \tag{4.4.4.5}$$

which involves an arbitrary function $f(z)$, admits an additive separable solution

$$u(x,t) = C_1 x^2 + C_2 x + \psi(t),$$

where C_1 and C_2 are arbitrary constants, and the function $\psi = \psi(t)$ is described by the nonlinear second-order ODE with proportional delay

$$\psi''_{tt} = 2aC_1 + f(\psi - \bar\psi), \quad \bar\psi = \psi(qt).$$

Equation 6. The nonlinear wave-type equation with proportional argument

$$u_{tt} = au_{xx} + f(u-w), \quad w = u(px, t), \tag{4.4.4.6}$$

involving an arbitrary function $f(z)$, admits an additive separable solution

$$u(x,t) = C_1 t^2 + C_2 t + \varphi(x),$$

where C_1 and C_2 are arbitrary constants, and the function $\varphi = \varphi(x)$ is described by the nonlinear second-order ODE with proportional argument

$$a\varphi''_{xx} - 2C_1 + f(\varphi - \bar\varphi) = 0, \quad \bar\varphi = \varphi(px).$$

Equation 7. The nonlinear wave-type equation with proportional delay

$$u_{tt} = au_{xx} + bu + f(u-w), \quad w = u(x, qt), \tag{4.4.4.7}$$

involving an arbitrary function $f(z)$, admits two additive separable solutions depending on the sign of ab as shown below.

1°. Solution with $ab < 0$:

$$u(x,t) = A\cosh(\lambda x) + B\sinh(\lambda x) + \psi(t), \quad \lambda = \sqrt{-b/a},$$

where A and B are arbitrary constants, and the function $\psi = \psi(t)$ is described by the nonlinear second-order ODE with proportional delay

$$\psi''_{tt} = b\psi + f(\psi - \bar\psi), \quad \bar\psi = \psi(qt). \tag{4.4.4.8}$$

2°. Solution with $ab > 0$:

$$u(x,t) = A\cos(\lambda x) + B\sin(\lambda x) + \psi(t), \quad \lambda = \sqrt{b/a},$$

where A and B are arbitrary constants, and the function $\psi = \psi(t)$ is described by the nonlinear second-order ODE with proportional delay (4.4.4.8).

Notably, equation (4.4.4.7) and its solution admit a generalization to the case of a variable delay of general form, so that $w = u(x, t - \tau(t))$, where $\tau(t)$ is an arbitrary function.

Equation 8. The nonlinear wave-type equation with proportional arguments

$$u_{tt} = au_{xx} + e^{\lambda u} f(u - w), \quad w = u(px, qt), \qquad (4.4.4.9)$$

involving an arbitrary function $f(z)$, admits exact solutions of the form

$$u(x,t) = U(z) - \frac{2}{\lambda} \ln t, \quad z = \frac{x}{t},$$

where the function $U = U(z)$ is described by the nonlinear ODE with proportional argument

$$(z^2 U_z')_z' + \frac{2}{\lambda} = aU_{zz}'' + e^{\lambda U} f\left(U - W + \frac{2}{\lambda} \ln q\right) = 0,$$
$$W = U(sz), \quad s = p/q.$$

Equation 9. The nonlinear wave-type equation with proportional argument

$$u_{tt} = au_{xx} + uf(w/u), \quad w = u(px, t), \qquad (4.4.4.10)$$

admits two nondegenerate multiplicative separable solutions given below.

1°. Exact solution:

$$u(x,t) = (Ae^{-\lambda t} + Be^{\lambda t})\varphi(x),$$

where A, B, and λ are arbitrary constants, and the function $\varphi = \varphi(x)$ is described by the nonlinear second-order ODE with proportional argument

$$a\varphi_{xx}'' + \varphi[f(\bar\varphi/\varphi) - \lambda^2] = 0, \quad \bar\varphi = \varphi(px).$$

2°. Exact solution:

$$u(x,t) = [A\cos(\lambda t) + B\sin(\lambda t)]\varphi(x),$$

where A, B, and λ are arbitrary constants, and the function $\varphi = \varphi(x)$ is described by the nonlinear second-order ODE with proportional argument

$$a\varphi_{xx}'' + \varphi[f(\bar\varphi/\varphi) + \lambda^2] = 0, \quad \bar\varphi = \varphi(px).$$

Remark 4.20. *Equations (4.4.4.6) and (4.4.4.10) and their exact solutions admit generalizations to the case of a variable delay of general form, so that $w = u(x - \tau(x), t)$, where $\tau(x)$ is an arbitrary function.*

4.4. Nonlinear PDEs with Proportional Arguments. Principle of Analogy of Solutions   267

More complicated nonlinear equations.
Equation 10. The nonlinear wave-type equation with proportional delay

$$u_{tt} = a(u^k u_x)_x + u f(w/u), \quad w = u(x, qt), \tag{4.4.4.11}$$

admits a multiplicative separable solution

$$u(x,t) = \varphi(x)\psi(t),$$

where the functions $\varphi(x)$ and $\psi(t)$ satisfy the ODE without delay and ODE with proportional delay

$$a(\varphi^k \varphi'_x)'_x = b\varphi,$$
$$\psi''_{tt} = b\psi^{k+1} + \psi f(\bar\psi/\psi), \quad \bar\psi = \psi(qt),$$

with b being an arbitrary constant.

Equation 11. The nonlinear wave-type equation with proportional argument

$$u_{tt} = a(u^k u_x)_x + u f(w/u), \quad w = u(px, t), \tag{4.4.4.12}$$

admits exact solutions of the form

$$u(x,t) = e^{2\lambda t} U(z), \quad z = e^{-k\lambda t} x,$$

where λ is an arbitrary constant and $U = U(z)$ is a function satisfying the ODE with proportional argument

$$4\lambda^2 U - 4k\lambda^2 z U'_z + k^2\lambda^2 z(zU'_z)'_z = a(U^k U'_z)'_z + U f(W/U), \quad W = U(pz).$$

Equation 12. The nonlinear wave-type equation with proportional delay

$$u_{tt} = a(u^k u_x)_x + u f(w/u) + b u^{k+1}, \quad w = u(x, qt), \tag{4.4.4.13}$$

can have three different multiplicative separable solutions depending on the values of the coefficients b and k as shown below.

1°. Solution with $b(k+1) > 0$:

$$u(x,t) = [C_1 \cos(\beta x) + C_2 \sin(\beta x)]^{1/(k+1)} \psi(t), \quad \beta = \sqrt{b(k+1)/a},$$

where C_1 and C_2 are arbitrary constants, and the function $\psi = \psi(t)$ is described by the ODE with proportional delay

$$\psi''_{tt} = \psi f(\bar\psi/\psi), \quad \bar\psi = \psi(qt). \tag{4.4.4.14}$$

2°. Solution with $b(k+1) < 0$:

$$u(x,t) = [C_1 \exp(-\beta x) + C_2 \exp(\beta x)]^{1/(k+1)} \psi(t), \quad \beta = \sqrt{-b(k+1)/a},$$

where C_1 and C_2 are arbitrary constants, and the function $\psi = \psi(t)$ is described by the ODE with proportional delay (4.4.4.14).

3°. Solution with $k = -1$:

$$u(x,t) = C_1 \exp\left(-\frac{b}{2a}x^2 + C_2 x\right)\psi(t),$$

where C_1 and C_2 are arbitrary constants, and the function $\psi = \psi(t)$ is described by the ODE with proportional delay (4.4.4.14).

Remark 4.21. Equations (4.4.4.11) and (4.4.4.13) and their solutions admit generalizations to the case of a variable delay of general form, so that $w = u(x, t - \tau(t))$, where $\tau(t)$ is an arbitrary function.

Equation 13. The nonlinear wave-type equation with proportional delay

$$u_{tt} = a(u^k u_x)_x + u^{k+1} f(w/u), \quad w = u(x, qt), \tag{4.4.4.15}$$

admits an exact solution of the form

$$u(x,t) = t^{-2/k} \varphi(z), \quad z = x + \lambda \ln t,$$

where λ is an arbitrary constant, and the function $\varphi = \varphi(z)$ satisfies the second-order ODE with constant delay

$$\frac{2(k+2)}{k^2}\varphi - \lambda \frac{k+4}{k}\varphi'_z + \lambda^2 \varphi''_{zz} = a(\varphi^k \varphi'_z)'_z + \varphi^{k+1} f(q^{-2/k} \bar{\varphi}/\varphi),$$
$$\bar{\varphi} = \varphi(z + \lambda \ln q), \quad \lambda \ln q < 0.$$

Equation 14. The nonlinear wave-type equation with proportional arguments

$$u_{tt} = a(u^k u_x)_x + u^n f(w/u), \quad w = u(px, qt), \tag{4.4.4.16}$$

admits a self-similar solution

$$u(x,t) = t^{\frac{2}{1-n}} U(z), \quad z = xt^{\frac{n-k-1}{1-n}},$$

where the function $U = U(z)$ satisfies the second-order ODE with proportional argument

$$\frac{2(1+n)}{(1-n)^2}U + \frac{(n-k-1)(2n-k+2)}{(1-n)^2}zU'_z + \frac{(n-k-1)^2}{(1-n)^2}z^2 U''_{zz}$$
$$= a(U^k U'_z)'_z + U^n f(q^{\frac{2}{1-n}} W/U), \quad W = U(sz), \quad s = pq^{\frac{n-k-1}{1-n}}.$$

Equation 15. The nonlinear wave-type equation with proportional delay

$$u_{tt} = a(e^{\lambda u} u_x)_x + f(u - w), \quad w = u(x, qt), \tag{4.4.4.17}$$

admits an additive separable solution

$$u(x,t) = \frac{1}{\lambda}\ln(Ax^2 + Bx + C) + \psi(t),$$

where A, B, and C are arbitrary constants, and the function $\psi = \psi(t)$ is described by the ODE with proportional delay

$$\psi''_{tt} = 2a(A/\lambda)e^{\lambda\psi} + f(\psi - \bar{\psi}), \quad \bar{\psi} = \psi(qt).$$

Equation 16. The nonlinear wave-type equation with proportional arguments

$$u_{tt} = a(e^{\lambda u}u_x)_x + e^{\mu u}f(u - w), \quad w = u(px, qt), \qquad (4.4.4.18)$$

admits an exact solution of the form

$$u(x,t) = U(z) - \frac{2}{\mu}\ln t, \quad z = xt^{\frac{\lambda-\mu}{\mu}},$$

where the function $U = U(z)$ is described by the nonlinear ODE with proportional argument

$$\frac{2}{\mu} + \frac{\mu - \lambda}{\mu}zU'_z + \frac{(\lambda - \mu)^2}{\mu^2}z(zU'_z)'_z = a(e^{\lambda U}U'_z)'_z + e^{\mu U}f\left(U - W + \frac{2}{\mu}\ln q\right),$$

$$W = U(sz), \quad s = pq^{\frac{\lambda-\mu}{\mu}}.$$

Equation 17. The nonlinear wave-type equation with proportional arguments

$$u_{tt} = [f(w)u_x]_x, \quad w = u(px, qt), \qquad (4.4.4.19)$$

admits a self-similar solution

$$u(x,t) = U(z), \quad z = x/t,$$

where the function $U = U(z)$ is described by the nonlinear ODE with proportional argument

$$(z^2 U'_z)'_z = [f(W)U'_z]'_z, \quad W = U(sz), \quad s = p/q.$$

This equation admits the first integral

$$z^2 U'_z = f(W)U'_z + C, \qquad (4.4.4.20)$$

where C is an arbitrary constant. In the special case $C = 0$, equation (4.4.4.20) degenerates into a transcendental equation leading to the solution $z^2 = f(W)$, which generates an exact solution in implicit form to the original equation (4.4.4.19):

$$u(x,t) = U(z), \quad z^2 = f(U(sz)), \quad z = x/t, \quad s = p/q.$$

This solution can be represented as $u = f^{-1}(x^2/(st)^2)$, where f^{-1} is the inverse of f.

Equation 18. The nonlinear PDE with proportional arguments

$$u_{tt} = F(u, w, u_x, u_{xx}), \quad w = u(px, pt),$$

admits the traveling wave solution

$$u(x,t) = U(z), \quad z = kx - \lambda t,$$

where the function $U = U(z)$ is described by the nonlinear ODE with proportional argument

$$F(U, W, kU'_z, k^2 U''_{zz}) - \lambda^2 U''_{zz} = 0, \quad W = U(pz).$$

4.5. Unstable Solutions and Hadamard Ill-Posedness of Some Delay Problems

4.5.1. Solution Instability for One Class of Nonlinear PDEs with Constant Delay

Global instability of solutions. Let us consider the class of nonlinear reaction-diffusion equations with constant delay

$$u_t = au_{xx} - \frac{bk(u-w)}{1-k} + F\left(\frac{u-kw}{1-k}\right), \quad w = u(x, t-\tau), \tag{4.5.1.1}$$

where $F(u)$ is an arbitrary function (other than a constant), $a > 0$, and $k > 0$ ($k \neq 1$). The following theorem holds true [434].

Theorem. *Let $u_0 = u_0(x, t)$ be a solution to the nonlinear equation (4.5.1.1). Then the function*

$$\begin{gathered} u = u_0(x, t) + \delta e^{ct} \sin(\gamma x + \nu), \\ c = (\ln k)/\tau, \quad \gamma = \sqrt{(b-c)/a}, \quad b - c > 0, \end{gathered} \tag{4.5.1.2}$$

where δ and ν are arbitrary constants, is also a solution to this equation.

The theorem is proved by a direct check. It is a special case of Theorem 2 for equation (3.4.2.31) in which one should set

$$f(z) \equiv -b\frac{z}{1-k} + F\left(\frac{z}{1-k}\right), \quad z = u - kw.$$

Assuming that $\delta \ll 1$ and the points x satisfy the inequality $\sin(\gamma x + \nu) \neq 0$, we find from formula (4.5.1.2) that two solutions, u_0 and u, indefinitely close in the initial stage, will exponentially diverge from each other over time under the conditions

$$k > 1, \quad b > 0, \quad \tau > (\ln k)/b. \tag{4.5.1.3}$$

The divergence conditions (4.5.1.3) are purely geometric, and they do not depend on the sign or form of the kinetic function $F(u)$, implying that we face a *global solution instability* here. Moreover, the results are exact, as they were obtained without any approximations or simplifications. Hence, they hold for any solutions to the class of equations concerned.

Some remarks. Assuming that $\tau = 0$ (or $k = 0$) in (4.5.1.1), meaning that there is no delay, we get the standard nonlinear diffusion equation with a volumetric reaction

$$u_t = au_{xx} + F(u). \tag{4.5.1.4}$$

Notably, the delay PDE (4.5.1.1) and PDE (4.5.1.4) have identical stationary solutions, $u_0 = u_0(x)$, including the simplest solution $u_0 = \text{const}$, where u_0 is a root of the algebraic (transcendental) equation $F(u_0) = 0$.

We choose the kinetic function $F(z)$ so as to ensure that the stationary solution $u_0 = u_0(x)$ to the diffusion equation without delay (4.5.1.4) is stable. This stationary solution will also solve the delay equation (4.5.1.1). In the original delay PDE (4.5.1.1), we fix the parameters $k > 1$ and $b > 0$ and will gradually increase the delay time τ starting from $\tau = 0$. For a sufficiently large τ that satisfies the last inequality in (4.5.1.3), the stationary solution will become unstable. In other words, introducing a delay into a mathematical model can make a stable solution unstable.

4.5.2. Hadamard Ill-Posedness of Some Delay Problems

Hadamard ill-posedness of some delay problems with initial data. Suppose that $u_0 = u_0(x,t)$ is a solution to the Cauchy-type problem for equation (4.5.1.1) in the domain $-\infty < x < \infty$ and with initial data of general form

$$u_0(x,t) = \varphi(x,t) \quad \text{at} \quad -\tau \leq t \leq 0. \tag{4.5.2.1}$$

We assume that the function u_0 is bounded as $x \to \pm\infty$ for any fixed $t > 0$.

Equation (4.5.1.1) also has the solution u defined by formulas (4.5.1.2). Comparing u and u_0 on the initial time interval, we find that

$$|u - u_0| \leq \delta \quad \text{at} \quad -\tau \leq t \leq 0. \tag{4.5.2.2}$$

Therefore, for fixed τ and k (with $k > 1$ and so $c > 0$), the difference between the solutions u and u_0 can be made arbitrarily small through choosing a suitable δ. This means that the initial data for these solutions will differ little on $-\tau \leq t \leq 0$. On the other hand, under conditions (4.5.1.3) and assuming that $\nu = 0$, we obtain

$$|u - u_0| = \delta e^{ct} \to \infty \quad \text{as} \quad t \to \infty$$

at the point $x = \pi/(2\gamma)$. Hence, under the global instability conditions (4.5.1.3), solutions of two Cauchy problems for equation (4.5.1.1) that are close in the initial data will indefinitely diverge over time.

The solution instability with respect to the initial data make the Cauchy problem for the delay equations (4.5.1.1) ill-posed in the sense of Hadamard (under conditions (4.5.1.3)). Notably, this instability is general (global instability), and it does not depend on the form of the kinetic function $F(u)$.

Remark 4.22. The statement of the Cauchy-type problem for equation (4.5.1.1) in the domain $-\infty < x < \infty$ and with initial conditions of general form (4.5.2.1) included the additional condition for u_0 that the solution must be bounded as $x \to \pm\infty$ for any fixed $t > 0$. Such conditions are typically used for linear heat equations without delay. It would be interesting to see what happens if we replace the boundedness condition in this problem with the stronger condition $u_0 \to 0$ as $x \to \pm\infty$ (if $f(z) \to 0$ as $z \to 0$) and assume that inequalities (4.5.1.3) hold.

Hadamard ill-posedness of some initial-boundary value problems. If conditions (4.5.1.3) hold, solutions to nonlinear initial-boundary value problems for the delay equation (4.5.1.1) can be globally unstable when boundary conditions of the first, second, or third kind are used in the region $0 \leq x \leq h$ (for some h).

Let $u_0 = u_0(x, t)$ be a solution of the initial-boundary value problem for the delay equation (4.5.1.1) subjected to the initial condition (4.5.2.1) and general boundary conditions of the first kind:

$$u_0 = \psi_1(x, t) \text{ at } x = 0, \quad u_0 = \psi_2(x, t) \text{ at } x = h \quad (t > 0), \qquad (4.5.2.3)$$

where $h = \pi/\gamma$, and the coefficient γ is defined in (4.5.1.2).

Formula (4.5.1.2) for u with $\nu = 0$ gives a solution to equation (4.5.1.1) that satisfies the boundary conditions (4.5.2.3) exactly. Through selecting an appropriate δ, this solution can be made indefinitely close to the solution $u_0 = u_0(x, t)$ in the initial data interval $-\tau \leq t \leq 0$ (see inequality (4.5.2.2)). However, if the global instability conditions (4.5.1.3) hold, the initially close solutions u_0 and u of the initial-boundary value problems will exponentially diverge as $t \to \infty$ at $x = h/2$. This instability of solutions in equation (4.5.1.1) with respect to the initial data makes the initial-boundary value problem in question *ill-posed in the sense of Hadamard*.

For other initial-boundary value problems, the solution u_0 should be compared with the solution u obtained by formula (4.5.1.2). The constant ν and the length of the interval h must be chosen so that u_0 and u satisfy the same boundary conditions. In particular, as u, one should choose solution (4.5.1.2) with $\nu = \pi/2$ in the case of boundary conditions of the second kind on the interval of length $h = \pi/\gamma$, when the derivatives u_x are set on the boundaries of the region.

Remark 4.23. *The global instability and Hadamard ill-posedness of some initial-boundary value problems also occur for a more complex nonlinear reaction-diffusion equation with constant delay. The equation is formally obtained by replacing the first time derivative u_t on the left-hand side of equation (4.5.1.1) with the linear combination of derivatives $\varepsilon u_{tt} + \sigma u_t$ [410].*

5. Numerical Methods for Solving Delay Differential Equations

5.1. Numerical Integration of Delay ODEs

5.1.1. Main Concepts and Definitions

Consider the following Cauchy problem for an ODE with constant delay:

$$u'(t) = f(t, u(t), u(t - \tau)), \quad t_0 < t \leq T, \qquad (5.1.1.1)$$
$$u(t) = \varphi(t), \quad t_0 - \tau \leq t \leq t_0. \qquad (5.1.1.2)$$

Here and henceforth, it is assumed that the range of the independent variable t is limited by a quantity T, which is set by the researcher based on the main purposes and capabilities of the computer and software used.

The numerical methods for solving problem (5.1.1.1)–(5.1.1.2) rely on replacing the equation for the continuous function $u(t)$ with an approximate equation (or system of equations) for functions of a discrete argument defined on a discrete set of points from the interval $[t_0, T]$. A set of points $G = \{t_0, t_1, \ldots, t_K = T\}$ is called a *grid*, the points t_k are called *grid nodes* (also known as *grid points*), and a discrete function of the discrete argument $u_h = \{u_k = u_h(t_k), k = 0, 1, \ldots, K\}$ is called a *grid function*. The spacing between two neighboring nodes of a grid is called a *grid step size* (or simply *step size*) and denoted $h_{k+1} = t_{k+1} - t_k$. If $h_k = $ const for any k, the grid step size is said to be *constant*; otherwise, it is *variable*. A continuous approximation (obtained with interpolation) of the function $u_h(t)$ will be denoted by $\tilde{u}_h(t)$.

Solving the Cauchy problem (5.1.1.1)–(5.1.1.2) numerically suggests the following. Let an initial function $u(t) = \varphi(t)$ be given on the interval $[-\tau, t_0]$. It is required to select a suitable step size h_k and find approximate values u_k of the unknown function $u(t)$ at the nodes t_k, where $k = 1, \ldots, K$.

For a constant delay, one should select a step size (constant or variable) such that $h_k \leq \tau$, or $t_k - \tau \leq t_{k-1}$. The value of $u(t - \tau)$ is then known at each step and equal, depending on $t - \tau$, to either the value of the initial function $\varphi(t - \tau)$ or that of the continuous approximation $\tilde{u}_h(t - \tau)$. In other words, at step $k + 1$, it is required to solve the following subproblem for the ODE without delay:

$$u'(t) = f(t, u(t), u(t - \tau)), \quad t_k < t \leq t_{k+1},$$
$$u(t_k) = u_k, \qquad (5.1.1.3)$$

where $u_k = u_h(t_k)$ and

$$u(t - \tau) = \begin{cases} \varphi(t - \tau) & \text{if } t \leq t_0 + \tau, \\ \tilde{u}_h(t - \tau) & \text{if } t_0 + \tau < t \leq t_{k+1}. \end{cases}$$

The numerical integration of problem (5.1.1.3) results in the value u_{k+1} of the grid function u_h and prolongation of the continuous approximation $\tilde{u}_h(t)$ to the interval $[t_k, t_{k+1}]$, with $\tilde{u}_h(t_{k+1}) = u_{k+1}$.

Remark 5.1. *If the delay is variable, $\tau = \tau(t)$, it may happen that $t - \tau(t) > t_k$ for $t \in [t_k, t_{k+1}]$, meaning that the value of the delayed argument lies within the interval concerned. Then one fails to reduce the problem to an ODE without delay, in which case interpolation is required (see [42, Section 3.3]).*

To characterize the properties of numerical methods, it is conventional to use the concepts and definitions specified below (see also [247, 248, 464]).

In the space of grid functions, the norm is introduced analogously to that in the space of continuous functions:

$$\|u_h\| = \max_{0 \leq k \leq K} |u_k|, \quad u_k = u_h(t_k). \tag{5.1.1.4}$$

For grid functions with three arguments, the norm is defined as

$$\|f_h\| = \max_{0 \leq k \leq K} |f_k|, \quad f_k = f(t_k, u_k, w_k).$$

A numerical method is said to be *convergent* if

$$\|u_h - u\| \to 0 \quad \text{as} \quad h \to 0.$$

A method is said to be *convergent with order $p > 0$* if the estimate

$$\|u_h - u\| \leq Ch^p$$

holds, where C is some positive constant independent of h.

Remark 5.2. *Here and henceforth, for a variable step size h_k, the expression $h \to 0$ is understood in the sense that $\|h\| \to 0$, where $\|h\| = \max_{1 \leq k \leq K} h_k$.*

It is convenient to rewrite the delay ODE (5.1.1.1) in the operator notation as

$$\mathscr{L}[u] = f.$$

Likewise, the problem of numerical integration can be represented as

$$\mathscr{L}_h[u_h] = f_h, \tag{5.1.1.5}$$

where $\mathscr{L}_h[u_h]$ is a finite-difference differentiation operator. For the first-order delay ODE (5.1.1.1), the operator reads $\mathscr{L}_h[u_h] = [u_h(t_{k+1}) - u_h(t_k)]/h_{k+1}$, where $f_h = f(t_k, u_h(t_k), \tilde{u}_h(t_k - \tau))$.

We shall call the grid function $\psi_h = \mathscr{L}_h[u] - f_h$ a *residual grid function* or *approximation error of a numerical method*. It follows that the unknown function u satisfies the approximate equation (5.1.1.5) up to the approximation error. A numerical method is said to *approximate the original equation* if

$$\|\psi_h\| \to 0 \quad \text{as} \quad h \to 0$$

or *approximates with order* $p > 0$ if the estimate

$$\|\psi_h\| \leq Ch^p$$

holds, where C is some positive constant independent of h.

Scheme (5.1.1.5) with the initial data (5.1.1.2) is called *stable* if the solution u_h depends continuously on the input data determined by the functions f and φ, and this dependence is uniform with respect to the grid step size. In other words, for any $\varepsilon > 0$ there exists a $\delta(\varepsilon)$, independent of the step h, at least for sufficiently small h, such that if two pairs of functions $f^{\mathrm{I}}, \varphi^{\mathrm{I}}$ and $f^{\mathrm{II}}, \varphi^{\mathrm{II}}$ satisfy the conditions

$$\|f^{\mathrm{I}} - f^{\mathrm{II}}\| \leq \delta \quad \text{and} \quad \|\varphi^{\mathrm{I}} - \varphi^{\mathrm{II}}\| \leq \delta,$$

then the respective grid functions satisfy the inequality

$$\|u_h^{\mathrm{I}} - u_h^{\mathrm{II}}\| \leq \varepsilon.$$

The fact that the solution is continuously dependent on f is referred to as *stability with respect to the right-hand side* and if the solution is continuously dependent on φ, it is said to be *stable with respect to the initial data*.

Numerical methods suitable for solving ODEs and delay ODEs must be stable, capable of approximating the problems well, and convergent to the exact solution.

5.1.2. Qualitative Features of the Numerical Integration of Delay ODEs

Working with delay ODEs is complicated by the presence of the unknown function with a delayed argument $t - \tau$. Its values may lie outside the grid G; this is especially true for ODEs with a variable delay $\tau = \tau(t)$. Therefore, one has to compute continuous approximations $\tilde{u}_h(t)$ of the grid function u_h. The function $\tilde{u}_h(t)$ can be constructed using a posterior interpolation of the values of u_h obtained by a discrete method or using the so-called *continuous* methods, which calculate $\tilde{u}_h(t)$ at each step.

We assume that for the problem in question, a grid G can be built such that the following condition holds: for any $t_k \in G$, we have either $t_k - \tau(t_k) < t_0$ or $t_k - \tau(t_k) \in G$. We can then employ methods that only rely on the nodes of the grid G. One of them is the explicit Euler method, which uses the formula

$$u_{k+1} = u_k + h_{k+1} f(t_k, u_k, u_q), \quad q < k.$$

For a constant delay, $\tau = \text{const}$, the condition $t_k - \tau \in G$ will hold if we choose a constant step size h such that $\tau = Nh$, where N is a positive integer.

This approach is inapplicable for ODEs with proportional delay. As an illustration, we will look at the model problem

$$u'(t) = u(t/2), \quad 0 < t \leq 1,$$
$$u(0) = 1.$$

For any $t_k \in G$, the value $t_k/2$ must also lie on the grid, which implies that no initial step is possible. Furthermore, as the grid nodes satisfy the condition $t_{k+1} = 2t_k$ for $k \geq 1$, we get $h_{k+1} = t_k$. Hence, the last step size always equals $1/2$, which means that the method convergence condition cannot be met.

In the case of a variable delay of general form, with $\tau(t) > 0$, the grid G can be built on any bounded interval $[t_0, T]$ with an arbitrarily small step size (see [42, pp. 37, 38]). To this end, starting from the node t_0, one should first identify all points of discontinuity of the derivative and include them in G. Then, one should build the grid in the reverse direction starting from the last node $t_K = T$ and using a desired maximum step size. Each node t_k generates a preceding node, $t_{k-1} = t_k - \tau(t_k)$, which must be included in the grid. For some delays, this approach may result in an irregular or redundant distribution of points (e.g., if the finite argument $\alpha(t) = t - \tau(t)$ has a horizontal asymptote).

Another essential feature to be considered when developing numerical methods for integrating delay ODEs is the propagation of discontinuities in the derivatives. As shown in Subsection 1.1.2, the solution of the Cauchy problem with delay may have a discontinuity in the derivative at the initial time t_0. This discontinuity further extends to higher-order derivatives. Therefore, for a numerical method to have the required order of accuracy, the solution of the Cauchy problem must be sufficiently smooth on each integration interval $[t_k, t_{k+1}]$: a method can have the order of accuracy p if the solution has continuous derivatives up to order $p + 1$ inclusive. To meet this requirement, all points of discontinuity of the unknown function and its derivatives up to order $p + 1$ inclusive must be added to the grid.

In the case of a constant delay, a point of discontinuity of order m is determined from the simple relation $t_m^* = t_0 + m\tau$. For a variable delay, the discontinuities are identified in two main ways. The first one is known as *tracking of discontinuities* (see [29, 154, 380, 381, 555] and the references in [42, p. 49]) and relies on seeking discontinuities $t_{m,j}^* > t_0$ that satisfy the system of equations

$$t_{m,j}^* - \tau(t_{m,j}^*) = t_{m-1,i}^* \quad \text{for some } i, \tag{5.1.2.1}$$

where j is the number of a discontinuity of order m induced by the ith discontinuity of order $m - 1$; the discontinuities are assumed to be in ascending order and such that $j > i$.

▶ **Example 5.1.** Consider the following problem for a delay ODE with a discontinuity at $t_0 = 0$:

$$u'(t) = u(t - 2t^{1/2}), \quad t > 0;$$
$$u(t) = 1, \quad -1 \leq t \leq 0.$$

The derivative has a discontinuity of order m at the point t_m^*, which is found from the equation $t_m^* - 2(t_m^*)^{1/2} = t_{m-1}^*$, where $t_m^* > t_{m-1}^*$ and $t_0^* = 0$. It follows that discontinuities occur at $t_m^* = \left(1 + \sqrt{t_{m-1}^* + 1}\right)^2$ for any positive integer m. ◀

The other approach relies on controlling the step size in the discontinuity region by estimating the local error. Although easier to program, such algorithms involve a large number of 'rejected' step sizes and can lead to numerous tiny steps in the vicinity of low-order discontinuities (see [29, Section 3.4], [380] and references therein).

Remark 5.3. *Some other approaches to integrating problems with discontinuities in the derivatives are outlined, with literature references, in [42, p. 39].*

Furthermore, difficulties also occur in problems for ODEs with proportional delay

$$\begin{aligned} u'(t) &= f(t, u(t), u(pt)), \quad t > 0; \\ u(0) &= u_0, \end{aligned} \tag{5.1.2.2}$$

where $0 < p < 1$. On the first integration interval, $0 < t \leq h_1$, an 'overlapping' arises for any step size h_1: the argument pt of the delay function $u(pt)$ lies inside the interval, which makes it impossible to use the method of steps. However, the method of steps can already be employed starting from the second step, for $t \geq h_1$, provided that the condition $h_{k+1} < h_k/p$ holds, which ensures that the argument pt lies on the preceding integration interval (see [42, Subsection 6.4.1]).

Remark 5.4. *At the first step $0 < t \leq h_1$, one can use an approximate analytical solution obtained as a truncated power series in the independent variable (see Subsection 1.4.2 and [471, 473, 474]).*

Remark 5.5. *Notably, in Cauchy problems for ODEs with proportional delay where the initial condition is set at $t = 0$, discontinuities of the first derivative do not extend to higher-order derivatives, and hence, no associated additional restrictions on the selection of grid nodes are required.*

Differential equations with proportional delay can be classified as equations with an infinite delay: as time passes, the $t - pt$ gap between the current and past moment increases indefinitely. As shown in [309], this circumstance leads to a significant shortage of RAM in numerical computations on uniform grids. To resolve this issue for the Cauchy problem with proportional delay (5.1.2.2), one uses the transformation $x = \ln t$, $v = u$, which results in a problem for an ODE with constant delay:

$$\begin{aligned} v'(x) &= e^x f(e^x, v(x), v(x - \tau)), \quad x > -\infty, \\ v(-\infty) &= u_0, \end{aligned}$$

where $\tau = -\ln p > 0$. This problem is complicated by the fact that the initial point becomes negative infinity. Therefore, the numerical solution of problem (5.1.2.2) is reasonable to carry out in two stages. First, one solves the original problem (5.1.2.2) on the interval $0 < t \leq t_0$. Then, the transformed problem is solved on the semi-infinite interval $x > x_0 = \ln t_0$:

$$v'(x) = e^x f(e^x, v(x), v(x - \tau)), \quad x > x_0,$$
$$v(x) = u(e^x), \quad x \leq x_0.$$

In view of the above, we can conclude that the presence of delay affects the accuracy and stability of numerical algorithms (see examples in [42, pp. 9–19]). Therefore, the formal application of numerical methods designed for ODEs without delay to problems with delay is not optimal. Consequently, numerical methods for integrating delay ODEs should be developed considering the equations' properties and their solutions' behavior.

5.1.3. Modified Method of Steps

Subsection 1.1.5 described the method of steps, which is a simple natural method for solving delay ODEs. The resulting ODEs without delay can be treated using appropriate numerical methods. The study [43] suggested, as a more convenient alternative for numerical integration, a modification of the methods of steps for constant delays. It was further extended in [44] to the case of a monotonically decreasing and non-vanishing variable delay. We will outline this modification below (see also [42, Section 3.4] and [29, 43, 44]).

Consider the Cauchy problem for an ODE with constant delay

$$\begin{aligned} u'(t) &= f(t, u(t), u(t - \tau)), \quad t_0 < t \leq T; \\ u(t) &= \varphi(t), \quad -\tau \leq t \leq t_0. \end{aligned} \quad (5.1.3.1)$$

Discontinuities of the derivative are found as $t_m^* = t_0 + m\tau$, $m = 1, 2, \ldots$ We will be integrating problem (5.1.3.1) on the intervals from one discontinuity to the next one until the condition $t_0 + m^*\tau \geq T$ holds at some step m^*. On the first interval $[t_0, t_0 + \tau]$, we have

$$\begin{aligned} u'(t) &= f(t, u(t), \varphi(t - \tau)), \quad t_0 < t \leq t_0 + \tau, \\ u(t_0) &= \varphi(t_0). \end{aligned} \quad (5.1.3.2)$$

By integrating problem (5.1.3.2) with a suitable numerical method, we find approximate values $u_h(t)$ of the desired function $u(t)$ on the interval $[t_0, t_0 + \tau]$.

We will now consider the second interval: $[t_0 + \tau, t_0 + 2\tau]$. By determining the functions $u_1(t) = u(t - \tau)$ and $u_2(t) = u(t)$, we represent problem (5.1.3.1) as a system of two equations:

$$\begin{aligned} u_1'(t) &= f(t - \tau, u_1(t), \varphi(t - 2\tau)), \quad t_0 + \tau < t \leq t_0 + 2\tau, \\ u_2'(t) &= f(t, u_2(t), u_1(t)), \quad t_0 + \tau < t \leq t_0 + 2\tau, \\ u_1(t_0 + \tau) &= \varphi(t_0), \\ u_2(t_0 + \tau) &= u_h(t_0 + \tau). \end{aligned} \quad (5.1.3.3)$$

The first equation of system (5.1.3.3) is independent. Integrating system (5.1.3.3) using a suitable numerical method, we find approximate values $u_h(t)$ of the desired function $u_2(t) = u(t)$ on the interval $[t_0 + \tau, t_0 + 2\tau]$.

By reasoning in the same way, we arrive at the general formula representing the original problem (5.1.3.1) on the interval $[t_0 + (m-1)\tau, t_0 + m\tau]$, $m = 1, 2, \ldots$, as a system of m equations:

$$\begin{aligned} u_i'(t) &= f(t - (m-i)\tau, u_i(t), u_{i-1}(t)), & i &= 1, \ldots, m, \\ u_i(t_0 + (m-1)\tau) &= u_h(t_0 + (i-1)\tau), & i &= 1, \ldots, m, \end{aligned} \qquad (5.1.3.4)$$

where $u_0(t) = \varphi(t - m\tau)$ and $u_i(t) = u(t - (m-i)\tau)$.

Formula (5.1.3.4) allows one to integrate the original problem (5.1.3.1) step by step with a numerical method. The set of values u_k of the function $u_h(t)$ computed at each step k represents approximate values of the desired function $u(t)$. Notably, one has to solve system (5.1.3.4) of an increasingly high order m at each step.

The method's main advantage is that system (5.1.3.4) does not include delay and so can be integrated by 'conventional' numerical methods. The main disadvantage is that previously computed values must be recalculated at each step. However, it is compensated by the fact that no interpolation is required or lack-of-memory problems arise. The method of steps is applicable to ODEs with a variable delay $\tau = \tau(t)$ (see [42, Section 3.4]).

5.1.4. Numerical Methods for ODEs with Constant Delay

Preliminary remarks. The previous section discussed a modified method of steps suitable for the numerical solution of delay ODEs. In what follows, we outline more effective numerical methods for integrating ODEs with constant delay. For simplicity, these methods will be exemplified by the following Cauchy problem for a nonlinear first-order ODE:

$$\begin{aligned} u'(t) &= f(t, u(t), u(t-\tau)), & t_0 &< t \leq T, \\ u(t) &= \varphi(t), & -\tau &\leq t \leq t_0. \end{aligned} \qquad (5.1.4.1)$$

Importantly, the methods outlined below admit natural generalizations to several constant delays, a variable delay, and higher-order ODEs and systems of ODEs.

First-order Euler methods. Integrating the delay ODE (5.1.4.1) on the grid interval $[t_k, t_{k+1}]$, we write

$$u(t_{k+1}) = u(t_k) + \int_{t_k}^{t_{k+1}} f(t, u(t), u(t-\tau))\, dt. \qquad (5.1.4.2)$$

Approximating the integral on the right-hand side by the rectangle method, we obtain [115, 144]:

$$\begin{aligned} u_{k+1} &= u_k + h f(t_k, u_k, u_{k-N}), & k &= 0, 1, \ldots, K-1, \\ u_k &= \varphi(t_k), & k &= -N, -N+1, \ldots, 0. \end{aligned} \qquad (5.1.4.3)$$

These formulas constitute the explicit Euler method. The constant step size of the grid, h, should be chosen as $h = \tau/N$, where N is a positive integer, to ensure that the points t_{k-N} always belong to the grid. Then, if the last N values of the grid function u_h are stored in RAM, the value u_{k-N} will also be known. The Euler method is the simplest explicit first-order method.

Now consider the case of a variable step, with step sizes h_k. In general, we have $t_k - \tau \neq t_{k-N}$, meaning that $t_k - \tau$ is not a grid node. The Euler method formulas will then become

$$u_{k+1} = u_k + h_{k+1} f\big(t_k, u_k, \tilde{u}_h(t_k - \tau)\big), \quad k = 0, 1, \ldots, K-1,$$
$$\tilde{u}_h(t) = \varphi(t), \quad -\tau \leq t \leq t_0, \tag{5.1.4.4}$$

where $\tilde{u}_h(t)$ is a continuous approximation of the grid function u_h. The value $\tilde{u}_h(t_k - \tau)$ is computed by interpolation on the interval $[t_q, t_{q+1}]$, where t_q is a grid node such that $t_q \leq t_k - \tau \leq t_{q+1}$. For example, the simplest piecewise linear interpolation can be used:

$$\tilde{u}_h(t) = \frac{t_{q+1} - t}{h_{q+1}} u_q + \frac{t - t_q}{h_{q+1}} u_{q+1}, \quad t_q \leq t \leq t_{q+1}. \tag{5.1.4.5}$$

There are so-called *continuous* methods that compute values of the function $\tilde{u}_h(t)$ by special algorithms. In this case, the function $\tilde{u}_h(t)$ is called an *interpolant of a numerical method*. Continuous methods are especially effective when a variable delay, $\tau = \tau(t)$, is used. The information given below on continuous methods for integrating delay ODEs relies on [42].

The continuous Euler method is defined by the formulas [40, 42, 116]:

$$u_{k+1} = \tilde{u}_h(t_k + h_{k+1}), \quad k = 0, 1, \ldots, K-1,$$
$$\tilde{u}_h(t_k + \theta h_{k+1}) = u_k + \theta h_{k+1} f\big(t_k, u_k, \tilde{u}_h(t_k - \tau)\big), \quad 0 \leq \theta \leq 1,$$
$$\tilde{u}_h(t) = \varphi(t), \quad -\tau \leq t \leq t_0.$$

The implicit Euler method with a constant step size is defined by

$$u_{k+1} = u_k + h f(t_{k+1}, u_{k+1}, u_{k+1-N}), \quad k = 0, 1, \ldots, K-1,$$
$$u_k = \varphi(t_k), \quad k = -N, -N+1, \ldots, 0. \tag{5.1.4.6}$$

Implicit methods are characterized by an extended stability region but require solving a system of algebraic equations at each step to compute u_{k+1}. A continuous analogue with variable step size h_k has the interpolant

$$\tilde{u}_h(t_k + \theta h_{k+1}) = u_k + \theta h_{k+1} f\big(t_{k+1}, u_{k+1}, \tilde{u}_h(t_{k+1} - \tau)\big), \quad 0 \leq \theta \leq 1.$$

Second-order methods. There are more accurate second-order modifications of Euler's method. For example, the *midpoint method* first computes the intermediate values

$$t_{k+\frac{1}{2}} = t_k + \tfrac{1}{2} h, \quad u_{k+\frac{1}{2}} = u_k + \tfrac{1}{2} h f(t_k, u_k, u_{k-N})$$

5.1. Numerical Integration of Delay ODEs

and then determines u_{k+1} as

$$u_{k+1} = u_k + hf\left(t_{k+\frac{1}{2}}, u_{k+\frac{1}{2}}, u_{k-N+\frac{1}{2}}\right)$$
$$\equiv u_k + hf\left(t_k + \tfrac{1}{2}h, u_k + \tfrac{1}{2}hf_k, u_{k-N} + \tfrac{1}{2}hf_{k-N}\right), \quad (5.1.4.7)$$

where $f_k = f(t_k, u_k, u_{k-N})$, $k = 0, 1, \ldots, K-1$, and N is a positive integer such that $h = \tau/N$.

The continuous midpoint method with a variable step size h_k has the interpolant

$$\tilde{u}_h(t_k + \theta h_{k+1}) = u_k + \theta h_{k+1} f\left(t_k + \tfrac{1}{2}h_{k+1}, u_k + \tfrac{1}{2}h_{k+1}\tilde{f}_k, \tilde{u}_h(t_k - \tau)\right),$$

where $0 \le \theta \le 1$ and $\tilde{f}_k = f(t_k, u_k, \tilde{u}_h(t_k - \tau))$.

Another second-order method is based on the formula

$$u_{k+1} = u_k + \tfrac{1}{2}h\bigl[f(t_k, u_k, u_{k-N}) + f(t_{k+1}, u_k + hf_k, u_{k-N} + hf_{k-N})\bigr] \quad (5.1.4.8)$$

and called *Heun's method*. Its continuous analogue is discussed in [40, 116] and has the interpolant

$$\tilde{u}_h(t_k + \theta h_{k+1}) = u_k + (\theta - \tfrac{1}{2}\theta^2) h_{k+1} \tilde{f}_k$$
$$+ \tfrac{1}{2}\theta^2 h_{k+1} f\bigl(t_{k+1}, u_k + h_{k+1}\tilde{f}_k, \tilde{u}_h(t_{k+1} - \tau)\bigr], \quad 0 \le \theta \le 1.$$

Besides (5.1.4.7) and (5.1.4.8), there is an implicit second-order method called the *trapezoidal method* and defined by

$$u_{k+1} = u_k + \tfrac{1}{2}h\bigl[f(t_k, u_k, u_{k-N}) + f(t_{k+1}, u_{k+1}, u_{k+1-N})\bigr]. \quad (5.1.4.9)$$

The interpolant of the continuous trapezoidal method with a variable step size is expressed as

$$\tilde{u}_h(t_k + \theta h_{k+1}) = u_k + (\theta - \tfrac{1}{2}\theta^2) h_{k+1} f\bigl(t_k, u_k, \tilde{u}_h(t_k - \tau)\bigr)$$
$$+ \tfrac{1}{2}\theta^2 h_{k+1} f(t_{k+1}, u_{k+1}, \tilde{u}_h(t_{k+1} - \tau)).$$

Remark 5.6. *The above first- and second-order methods are special cases of Runge–Kutta methods. For similar discrete methods for ODEs without delay, see, for example, [247, pp. 243–247], [464, pp. 214–220], and [423, pp. 64, 65]. Continuous methods for ODEs without delay are described in [42, Section 5].*

Fourth-order Runge–Kutta methods. The values u_k of the grid function u_h are calculated as

$$u_{k+1} = u_k + \tfrac{1}{6}h\bigl(r^{(1)}_{k+1} + 2r^{(2)}_{k+1} + 2r^{(3)}_{k+1} + r^{(4)}_{k+1}\bigr),$$
$$r^{(1)}_{k+1} = f(t_k, u_k, u_{k-N}),$$
$$r^{(2)}_{k+1} = f\bigl(t_k + \tfrac{1}{2}h, u_k + \tfrac{1}{2}hr^{(1)}_{k+1}, u_{k-N} + \tfrac{1}{2}hr^{(1)}_{k+1-N}\bigr),$$
$$r^{(3)}_{k+1} = f\bigl(t_k + \tfrac{1}{2}h, u_k + \tfrac{1}{2}hr^{(2)}_{k+1}, u_{k-N} + \tfrac{1}{2}hr^{(2)}_{k+1-N}\bigr),$$
$$r^{(4)}_{k+1} = f\bigl(t_{k+1}, u_k + hr^{(3)}_{k+1}, u_{k-N} + hr^{(3)}_{k+1-N}\bigr),$$

where h is a constant step size, N is a positive integer such that $h = \tau/N$. The interpolant for the continuous method with a variable step size h_k can approximately be represented as

$$\tilde{u}_h(t_k+\theta h_{k+1}) = u_k + \tfrac{1}{6}h_{k+1}\left[(4\theta-3\theta^2)r_{k+1}^{(1)}+2\theta r_{k+1}^{(2)}+2\theta r_{k+1}^{(3)}+(3\theta^2-2\theta)r_{k+1}^{(4)}\right],$$

where $0 \leq \theta \leq 1$. Alternatively, it can be defined by

$$\begin{aligned}\tilde{u}_h(t_k + \theta h_{k+1}) &= u_k + \tfrac{1}{6}h_{k+1}\big[(4\theta^3 - 9\theta^2 + 6\theta)r_{k+1}^{(1)} \\ &+ (6\theta^2 - 4\theta^3)r_{k+1}^{(2)} + (6\theta^2 - 4\theta^3)r_{k+1}^{(3)} + (4\theta^3 - 3\theta^2)r_{k+1}^{(4)}\big].\end{aligned}$$

The Runge–Kutta methods. General scheme. Below we outline the principle of constructing the family of Runge–Kutta methods. Integrating equation (5.1.4.1) on the interval $[t_k, t_{k+1}]$, we obtain equation (5.1.4.2). Let us introduce auxiliary nodes

$$t_{k+1}^{(m)} = t_k + \alpha_m h_{k+1}, \quad m = 1, 2, \ldots, M,$$

where $0 = \alpha_1 \leq \alpha_2 \leq \cdots \leq \alpha_M \leq 1$. Note that $t_{k+1}^{(1)} = t_k$ and $t_{k+1}^{(M)} \leq t_{k+1}$. Replacing the integral on the right-hand side of (5.1.4.2) by the quadrature formula involving the nodes $t_{k+1}^{(m)}$, we obtain

$$u(t_{k+1}) \approx u(t_k) + h_{k+1}\sum_{m=1}^{M} c_m f\big(t_{k+1}^{(m)}, u(t_{k+1}^{(m)}), u(t_{k+1}^{(m)} - \tau)\big), \qquad (5.1.4.10)$$

where c_m are weights of the quadrature formula ($0 \leq c_m \leq 1$). To take advantage of formula (5.1.4.10), we need to know the values $u(t_{k+1}^{(m)})$, $m = 2, 3, \ldots, M$. These can be found likewise by integrating equation (5.1.4.1):

$$u(t_{k+1}^{(m)}) = u(t_k) + \int_{t_k}^{t_{k+1}^{(m)}} f(t, u(t), u(t - \tau))\,dt, \quad m = 2, 3, \ldots, M. \quad (5.1.4.11)$$

Replacing the integral on the right-hand side of (5.1.4.11) by the quadrature formula with nodes $t_{k+1}^{(1)}, t_{k+1}^{(2)}, \ldots, t_{k+1}^{(m-1)}$, we arrive at the approximate relations

$$\begin{aligned}u(t_{k+1}^{(2)}) &\approx u(t_k) + h_{k+1}\beta_{21}f\big(t_{k+1}^{(1)}, u(t_{k+1}^{(1)}), u(t_{k+1}^{(1)} - \tau)\big), \\ u(t_{k+1}^{(3)}) &\approx u(t_k) + h_{k+1}\beta_{31}f\big(t_{k+1}^{(1)}, u(t_{k+1}^{(1)}), u(t_{k+1}^{(1)} - \tau)\big) \\ &\quad + h_{k+1}\beta_{32}f\big(t_{k+1}^{(2)}, u(t_{k+1}^{(2)}), u(t_{k+1}^{(2)} - \tau)\big), \qquad (5.1.4.12)\end{aligned}$$

$\ldots$

$$u(t_{k+1}^{(m)}) \approx u(t_k) + h_{k+1}\sum_{j=1}^{m-1}\beta_{mj}f\big(t_{k+1}^{(j)}, u(t_{k+1}^{(j)}), u(t_{k+1}^{(j)} - \tau)\big),$$

where β_{mj} are weights of the quadrature formulas.

5.1. Numerical Integration of Delay ODEs

To summarize, considering (5.1.4.11) and (5.1.4.12) and relying on formula (5.1.4.10), we can write the standard scheme of the M-staged explicit Runge–Kutta method as

$$u_{k+1} = u_k + h_{k+1} \sum_{m=1}^{M} c_m r_{k+1}^{(m)}, \qquad (5.1.4.13)$$

$$r_{k+1}^{(m)} = f\left(t_k + \alpha_m h_{k+1},\ u_k + h_{k+1} \sum_{j=1}^{m-1} \beta_{mj} r_{k+1}^{(j)},\ u(t_k + \alpha_m h_{k+1} - \tau)\right).$$

The method computes approximate values $u_k = u_h(t_k)$ of the unknown function $u(t)$ at the points of the grid G. It depends on the set of parameters c_m, α_m, and β_{mj}, whose values are selected to ensure the required order of accuracy. Examples of specific values of these parameters can be found, for example, in [203, 464] (see also the second- and fourth-order Runge–Kutta methods described above).

▶ **Example 5.2.** Heun's scheme (5.1.4.8) corresponds to $M = 2$, $c_1 = c_2 = \frac{1}{2}$, $\alpha_1 = 0$, $\alpha_2 = 1$, and $\beta_{21} = 1$ and a constant step size $h_{k+1} = h$ in (5.1.4.13), provided that $\tau = Nh$, where N is a positive integer, and $u(t_k - \tau) = u_{k-N}$. ◀

Remark 5.7. *The schemes of implicit Runge–Kutta methods can be obtained by replacing the summation limit $m - 1$ in the second formula of (5.1.4.13) with M_*, where $m \leq M_* \leq M$. Implicit methods are more stable and suitable for solving stiff problems. For details, see Subsection 5.1.7.*

The values of the function with delay $u(t_k + \alpha_m h_{k+1} - \tau)$ are generally not known and are usually computed using interpolation on the interval $[t_q, t_{q+1}]$, where q is a positive integer such that $t_q \leq t_k + \alpha_m h_{k+1} - \tau \leq t_{q+1}$. *Continuous Runge–Kutta methods* (see [595, 596] and [42, Sections 5, 6]) allow one to compute at each step an approximate continuous solution (interpolant) $\tilde{u}_h(t)$ by the special formula

$$\tilde{u}_h(t_k + \theta h_{k+1}) = u_k + h_{k+1} \sum_{m=1}^{M} c_m(\theta) r_{k+1}^{(m)}, \quad 0 \leq \theta \leq 1, \qquad (5.1.4.14)$$

where $c_m(\theta)$ are polynomials satisfying the conditions

$$c_m(0) = 0, \quad c_m(1) = c_m, \quad m = 1, \ldots, M,$$

and some additional constraints associated with the order of accuracy (e.g., see [42, pp. 118, 119]).

Different variants of continuous Runge–Kutta methods for delay ODEs that employ interpolation can be found, for example, in [42, 203, 214, 230, 366, 373, 476] (see also the references in [29, 380]).

5.1.5. Numerical Methods for ODEs with Proportional Delay. Cauchy Problem

Preliminary remarks. Equations in question. Numerical methods for integrating differential equations with proportional delay and related equations, as well as examples of their application, are discussed in many studies (e.g., see [75, 133, 149, 196, 294, 312, 471, 588, 590]). Some qualitative features of the numerical integration of ODEs with proportional delay were discussed above in Subsection 5.1.2.

For simplicity and clarity, we will exemplify relevant numerical methods by looking at the Cauchy problem for a nonlinear first-order ODE with proportional delay

$$u'(t) = f(t, u(t), u(pt)), \quad 0 < t \leq T;$$
$$u(0) = u_0, \qquad (5.1.5.1)$$

where $0 < p < 1$.

Notably, the methods outlined below admit a natural extension to linear and nonlinear higher-order ODEs with several delays as well as to systems of higher-order delay ODEs.

Quasi-geometric grid. The study [39] suggested a *quasi-geometric* grid that is useful in dealing with proportional delay ODEs. Suppose a solution is known up to a certain time $T_0 = t_0 > 0$. We build a *primary* grid by the formula

$$T_n = \frac{T_{n-1}}{p}, \quad n = 1, 2, \ldots$$

We will be using so-called *primary* intervals

$$H_n = T_n - T_{n-1} = T_0 \frac{1-p}{p^n}, \quad n = 1, 2, \ldots$$

Note that the lengths of the intervals H_n increase exponentially. Now we introduce a *global* grid by dividing each primary interval into m equal subintervals:

$$h_{k+1} = \frac{H_{[k/m]+1}}{m} = \frac{T_0}{m} \frac{1-p}{p^{[k/m]+1}}, \quad k = 0, 1, \ldots,$$

where $[A]$ stands for the integer part of the number A. Then the grid nodes are defined as

$$t_k = T_{[k/m]} + r_{k/m} h_k, \quad k = 0, 1, \ldots, \qquad (5.1.5.2)$$

where $r_{k/m} = k - m[k/m] \equiv k \bmod m$ is the integer remainder of dividing k by m.

For $k > m$, we obtain the following recurrence relation from (5.1.5.2):

$$t_k = p^{-1} t_{k-m}.$$

The global grid (5.1.5.2) depends on t_0, m, and p. It has the advantage that the delayed arguments involving pt become known at each step, as there is no 'overlapping', meaning that $pt < t_k$ for any $t \in [t_k, t_{k+1}]$.

These kinds of grids are used in Runge–Kutta methods [294, 312, 576] and weighting methods [39, 196, 309, 310], which are described below.

Remark 5.8. *A more general similar grid can be built by dividing the primary intervals into m subintervals of arbitrary length (see [39]).*

Weighting method. The simplest method for integrating ODEs with proportional delay relies on the formula

$$u_{k+1}=u_k+h_{k+1}[(1-\sigma)f(t_k,u_k,\tilde{u}_h(pt_k))+\sigma f(t_{k+1},u_{k+1},\tilde{u}_h(pt_{k+1}))], \quad 0\leq\sigma\leq 1.$$

If $\sigma = 0$, we get the explicit first-order Euler method, while $\sigma = 1$ corresponds to the implicit first-order Euler method. If $\sigma = \frac{1}{2}$, we get the second-order trapezoidal rule. A continuous approximation of the solution $\tilde{u}_h(t)$ is built using interpolation. For example, the study [309] employs piecewise linear interpolation (5.1.4.5). Also, there are continuous methods analogous to those described above for ODEs with constant delay.

For more details and the investigation of weighting methods for linear ODEs with proportional delay, see [39, 196, 309]. Weighting methods for related nonlinear ODEs are discussed in [310].

Runge–Kutta methods. The principles of constructing Runge–Kutta methods for ODEs with proportional delay (5.1.5.1) are generally the same as those described above for ODEs with constant delay. The only difference is that the proportional delay is zero at $t = 0$, and hence, the integration must be performed in two stages [42, Subsection 6.4.1]. The first stage consists of one step, with step size h_1, where the following value is found:

$$\tilde{u}_h(\theta h_1) = u_0 + h_1 \sum_{m=1}^{M} c_m(\theta) r_1^{(m)}, \quad 0 \leq \theta \leq 1,$$

$$r_1^{(m)} = f\left(t_1^{(m)},\ u_0 + h_1 \sum_{j=1}^{m-1} \beta_{mj} r_1^{(j)},\ u_0 + h_1 \sum_{j=1}^{m-1} c_j(p\alpha_m) r_1^{(j)}\right).$$

After that, the computation is carried out by the formulas

$$\tilde{u}_h(t_k + \theta h_{k+1}) = u_k + h_{k+1} \sum_{m=1}^{M} c_m(\theta) r_{k+1}^{(m)}, \quad 0 \leq \theta \leq 1,$$

$$r_{k+1}^{(m)} = f\left(t_{k+1}^{(m)},\ u_k + h_{k+1} \sum_{j=1}^{m-1} \beta_{mj} r_{k+1}^{(j)},\ \bar{U}_{k+1}^{(m)}\right),$$

where

$$\bar{U}_{k+1}^{(m)} = \begin{cases} u_k + h_{k+1} \sum_{j=1}^{m-1} c_j\left(\frac{pt_{k+1}^{(m)}-t_k}{h_{k+1}}\right) r_{k+1}^{(j)} & \text{if } pt_{k+1}^{(m)} > t_k; \\ \tilde{u}_h\left(pt_k^{(m)}\right) & \text{if } pt_{k+1}^{(m)} \leq t_k. \end{cases}$$

For questions of stability of Runge–Kutta methods for equations with proportional delay, see [279, 294, 312, 576].

Spectral collocation methods. The main idea of the collocation method was outlined in Subsection 1.4.6. Depending on the basis functions selected, the method

Table 5.1. Basis functions in spectral collocation methods for ODEs with proportional delay.

Name	Basis functions $\varphi_n(t)$	Literature
Power-law functions	t^n	[200, 471, 474]
Exponentials	e^{-nt}	[590]
Shifted Chebyshev polynomials	$T_n(t) = \cos[n \arccos(2tL^{-1} - 1)]$	[470, 582]
Hermite polynomials	$H_n(t) = (-1)^n e^{t^2} \frac{d^n}{dt^n}(e^{-t^2})$	[578]
Bessel polynomials of the first kind	$J_n(t) = \sum_{k=0}^{[\frac{N-n}{2}]} \frac{(-1)^k}{k!(k+n)!} \left(\frac{t}{2}\right)^{2k+n}$	[589]
Jacobi rational functions	$R_n^{(\alpha,\beta)}(t) = \frac{t-1}{t+1} P_n^{(\alpha,\beta)}(t)$, $P_n^{(\alpha,\beta)}(t)$ are Jacobi polynomials (see below)	[133]
Bernoulli polynomials	$B_n(t)$ (see below)	[518]

generates a number of special techniques known as spectral collocation methods. Table 5.1 gives several examples of basis functions used in the spectral collocation methods for ODEs with proportional delay.

The Jacobi polynomials appearing in Table 5.1 (see the penultimate row) are defined as [402, 553]:

$$P_n^{(\alpha,\beta)}(t) = \frac{(-1)^n}{2^n n!}(1-t)^{-\alpha}(1+t)^{-\beta}\frac{d^n}{dt^n}[(1-t)^{\alpha+n}(1+t)^{\beta+n}]$$
$$= 2^{-n}\sum_{m=0}^{n} C_{n+\alpha}^m C_{n+\beta}^{n-m}(t-1)^{n-m}(t+1)^m,$$

where $C_a^0 = 1$ and $C_a^k = \frac{a(a-1)\ldots(a-k+1)}{k!}$ for $k = 1, 2, \ldots$

The Bernoulli polynomials mentioned in the last row of Table 5.1 are defined as [369, 402, 553]:

$$B_n(t) = \sum_{k=0}^{n} B_k C_n^k x^{n-k} \quad (n = 0, 1, 2, \ldots),$$

where $C_n^k = \frac{n!}{k!(n-k)!}$ are binomial coefficients and B_k are Bernoulli numbers, which are found using the double sum

$$B_n = \sum_{k=0}^{n} \frac{1}{k+1} \sum_{m=0}^{k} (-1)^m C_k^m m^n$$

or the recurrence relations

$$B_0 = 1, \quad \sum_{k=0}^{n-1} C_n^k B_k = 0.$$

Notably, the Bernoulli numbers arise when the following generating function is expanded in a Taylor series:

$$\frac{t}{e^t - 1} = \sum_{n=0}^{\infty} B_n \frac{t^n}{n!}, \quad |t| < 2\pi.$$

Sometimes, this expansion is used as the definition of the Bernoulli numbers.

All B_n with odd n, expect for B_1, are zero; the Bernoulli numbers with even n have alternating signs. Below are several first Bernoulli numbers:

$$B_0 = 1, \quad B_1 = -\tfrac{1}{2}, \quad B_2 = \tfrac{1}{6}, \quad B_4 = -\tfrac{1}{30}, \quad B_6 = \tfrac{1}{42},$$
$$B_8 = -\tfrac{1}{30}, \quad B_{10} = -\tfrac{5}{66}, \quad B_{12} = -\tfrac{691}{2730}, \quad B_{14} = \tfrac{7}{6}.$$

5.1.6. Shooting Method (Boundary Value Problems)

Preliminary remarks. The main idea of the shooting method is to reduce solving a boundary problem for a given ODE with proportional delay to solving a series of simpler Cauchy problems of the same type for the same equation. For conditions where the shooting method applies to ODEs with variable delay, see [124]. For clarity, we will denote the independent variable by x rather than t and restrict ourselves to second-order ODEs with proportional delay, which, in addition to the unknown function $u = u(x)$, also involve $w = u(px)$, where $0 < p \leq 1$.

Boundary value problems with boundary conditions of the first, second, or third kind or with mixed boundary conditions. Suppose we deal with a boundary value problem, in the domain $x_1 \leq x \leq x_2$ (either $x_1 = 0$, $x_2 = L$ or $x_1 = L$, $x_2 = L$ allowed), for a second-order ODE with proportional delay:

$$u''_{xx} = f(x, u, u'_x, w, w'_x), \quad w = u(px). \tag{5.1.6.1}$$

The equation is subjected to the boundary conditions of the first kind

$$u(x_1) = a, \quad u(x_2) = b, \tag{5.1.6.2}$$

where a and b are given numbers.

Consider the auxiliary Cauchy problem for equation (5.1.6.1) with the initial conditions

$$u(x_1) = a, \quad u'_x(x_1) = \lambda. \tag{5.1.6.3}$$

For any λ, a solution to this problem, obtained with a Runge–Kutta method or any other suitable numerical method, will satisfy the first boundary condition (5.1.6.2) at the point $x = x_1$. The original problem will be solved if one finds a $\lambda = \lambda_*$ such that the solution $u = u(x, \lambda_*)$ coincides at $x = x_2$ with the value prescribed by the second boundary condition (5.1.6.2):

$$u(x_2, \lambda_*) = b.$$

Let us first set an arbitrary value $\lambda = \lambda_1$ (e.g., $\lambda_1 = 0$) and solve the Cauchy problem (5.1.6.1), (5.1.6.3) numerically to obtain the number

$$\Delta_1 = u(x_2, \lambda_1) - b. \tag{5.1.6.4}$$

Then we choose a different value $\lambda = \lambda_2$ and solve the problem to obtain

$$\Delta_2 = u(x_2, \lambda_2) - b. \tag{5.1.6.5}$$

Suppose that λ_2 was chosen such that Δ_1 and Δ_2 have unlike signs (perhaps, a few attempts will be required to get a suitable λ_2). By virtue of solution continuity in λ, the desired value λ_* will lie between λ_1 and λ_2. Then, for example, we set $\lambda_3 = \frac{1}{2}(\lambda_1 + \lambda_2)$ and solve the Cauchy problem again to obtain Δ_3. Out of the two preceding values λ_j ($j = 1, 2$), we only keep the one for which Δ_j and Δ_3 have unlike signs. The desired λ_* will lie between λ_j and λ_3. On setting further $\lambda_4 = \frac{1}{2}(\lambda_j + \lambda_3)$, we find Δ_4 and so on. We will repeat this procedure until we obtain λ_* with the required accuracy.

Remark 5.9. *The above algorithm can be improved by using, instead of bisection, the formulas*

$$\lambda_3 = \frac{|\Delta_2|\lambda_1 + |\Delta_1|\lambda_2}{|\Delta_2| + |\Delta_1|}, \quad \lambda_4 = \frac{|\Delta_3|\lambda_j + |\Delta_j|\lambda_3}{|\Delta_3| + |\Delta_j|}, \quad \ldots$$

Statements of initial conditions for the auxiliary Cauchy problem. Table 5.2 lists the initial conditions to be used in the auxiliary Cauchy problem for the numerical solution of boundary value problems for second-order ODEs with proportional delay (5.1.6.1) and different linear and nonlinear boundary conditions at the left endpoint. The parameter λ in the Cauchy problem is selected to satisfy the boundary condition at the right endpoint.

Table 5.2. Initial conditions in the auxiliary Cauchy problem used to solve boundary value problems by the shooting method ($x_1 \leq x \leq x_2$).

No.	Boundary value problem	Boundary condition at left endpoint	Initial conditions for the Cauchy problem
1	First boundary value problem	$u(x_1) = a$	$u(x_1) = a,\ u'_x(x_1) = \lambda$
2	Second boundary value problem	$u'_x(x_1) = a$	$u(x_1) = \lambda,\ u'_x(x_1) = a$
3	Third boundary value problem	$u'_x(x_1) - ku(x_1) = a$	$u(x_1) = \lambda,\ u'_x(x_1) = a + k\lambda$
4	Problem with a nonlinear boundary condition	$u'_x(x_1) = \varphi(u(x_1))$, $\varphi(z)$ is a given function	$u(x_1) = \lambda,\ u'_x(x_1) = \varphi(\lambda)$
5	Problem with a nonlinear boundary condition	$u(x_1) = \varphi(u'_x(x_1))$, $\varphi(z)$ is a given function	$u(x_1) = \varphi(\lambda),\ u'_x(x_1) = \lambda$

Importantly, nonlinear boundary value problems can have one, several, or no solutions. See Examples 3.14 and 3.17 in the book [423, pp. 142, 148], illustrating the three scenarios by constructing exact solutions to combustion-theory one-parameter problems without delay. Therefore, one should take special care when solving nonlinear problems: once a suitable value $\lambda = \lambda_1$ is found, one should look for other allowed values in a wide range of variation of λ. If no suitable λ_1 is found, one should consider the possibility that the problem may have no solutions. The studies [192, 252, 468] dealt with existence and uniqueness conditions for solutions to boundary value problems for nonlinear delay equations.

Linear boundary value problems. Modified shooting method. Consider the linear boundary value problem for a second-order ODE with proportional delay

$$u''_{xx} + f_1(x)u'_x + f_2(x)w'_x + f_3(x)u + f_4(x)w = g(x), \quad w = u(px), \quad (5.1.6.6)$$

subjected to general homogeneous boundary conditions of the third kind

$$a_1 u'_x + b_1 u = 0 \quad \text{at} \quad x = 0, \quad (5.1.6.7)$$

$$a_2 u'_x + b_2 u = 0 \quad \text{at} \quad x = l. \quad (5.1.6.8)$$

We assume that a solution to problem (5.1.6.6)–(5.1.6.8) exists and is unique.

The linear boundary value problem (5.1.6.6)–(5.1.6.8) is the easiest to solve using the modified shooting method outlined below.

First, let us find the auxiliary function $u_1 = u_1(x)$ that solves the auxiliary Cauchy problem for the linear nonhomogeneous equation (5.1.6.6) with the initial conditions

$$u = a_1 \quad \text{at} \quad x = 0; \quad u'_x = -b_1 \quad \text{at} \quad x = 0. \quad (5.1.6.9)$$

If follows from (5.1.6.9) that $u_1 = u_1(x)$ satisfies the left boundary condition (5.1.6.7). Then we find the auxiliary function $u_0 = u_0(x)$ that solves another auxiliary Cauchy problem, for the linear homogeneous equation (5.1.6.6) with $g(x) = 0$ and the boundary conditions (5.1.6.9). By virtue of the linearity of the problem and homogeneity of the boundary conditions, the function $Cu_0(x)$ will also solve equation (5.1.6.6) and satisfy the left boundary condition (5.1.6.7). Therefore, we seek a solution to original boundary value problem (5.1.6.6)–(5.1.6.8) as the sum

$$u(x) = u_1(x) + C u_0(x). \quad (5.1.6.10)$$

Since the function (5.1.6.10) must satisfy the right boundary condition (5.1.6.8), we get the following linear algebraic equation for the constant C:

$$a_2 u'_1(l) + b_2 u_1(l) + C[a_2 u'_0(l) + b_2 u_0(l)] = 0. \quad (5.1.6.11)$$

To sum up, solving the original boundary value problem (5.1.6.6)–(5.1.6.8) has reduced to solving two auxiliary Cauchy problems, which can be integrated using any numerical method described in this chapter. When dealing with a boundary value problem subjected to nonhomogeneous boundary conditions, one should use a transformation reducing it to a problem with homogeneous boundary conditions. This can always be done with the change of variable $u = v + A_2 x^2 + A_1 x + A_0$ by appropriately selecting the constants A_m.

5.1.7. Integration of Stiff Systems of Delay ODEs Using the Mathematica Software

Preliminary remarks. The previous sections discussed several numerical methods for integrating delay ODEs. These methods can be extended to systems of ODEs. However, not all of them work well with stiff systems. In particular, such systems arise from the spatial discretization of delay PDEs in the method of lines (see Subsection 5.2.2 below). A system is called *stiff* if it describes processes occurring on extremely different time scales [63, 464]. When solving such problems numerically, restrictions on the step size are imposed not to improve the accuracy but to ensure the algorithm stability [63], and only tiny step sizes are usually suitable. One uses implicit methods from the families of Runge–Kutta or multi-step Gear methods to solve stiff systems.

For the numerical solution of stiff systems of ODEs with constant delay, it is advisable to employ the widely used software packages Mathematica and Maple, which allow computations and programming in an analytical (symbolic) form. In what follows, we restrict ourselves to describing the practical application of Mathematica for solving such equations.

Mathematica uses the NDSolve function [558–560] to solve stiff systems of ODEs with constant delay (including those with several delays). If no additional options are specified, NDSolve employs a sophisticated approach in which the method and its parameters are automatically selected. With the Method option, the user can manually specify one of the built-in methods for solving stiff systems: the implicit Runge–Kutta method [561, 562] or the implicit multi-step Gear method. The latter is based on the backward differentiation formula (BDF) [563].

Below we will exemplify the use of the method by looking at the Cauchy problem for a system of ODEs with constant delay written in vector form:

$$\boldsymbol{u}'_t = \boldsymbol{f}(t, \boldsymbol{u}, \boldsymbol{u}(t-\tau)), \quad 0 < t \leq T;$$
$$\boldsymbol{u}(t) = \boldsymbol{\varphi}(t), \quad -\tau \leq t \leq 0, \tag{5.1.7.1}$$

where $\boldsymbol{u} = (u_1, \ldots, u_N)^T$, $\boldsymbol{\varphi} = (\varphi_1, \ldots, \varphi_N)^T$, and $\boldsymbol{f} = (f_1, \ldots, f_N)^T$ are column vectors.

Implicit Runge–Kutta method. The principles of constructing Runge–Kutta schemes for delay ODEs are described in Subsection 5.1.4. Implicit Runge–Kutta schemes for stiff systems of ODEs (5.1.7.1) are constructed likewise and defined by the formulas [42, 561, 562]:

$$\boldsymbol{u}_{k+1} = \boldsymbol{u}_k + h_{k+1} \sum_{m=1}^{M} c_m \boldsymbol{r}_{k+1}^{(m)}, \quad k = 0, \ldots, K-1, \tag{5.1.7.2}$$

$$\boldsymbol{r}_{k+1}^{(m)} = \boldsymbol{f}\left(t_k + \alpha_m h_{k+1},\ \boldsymbol{u}_k + h_{k+1} \sum_{j=1}^{M_*} \beta_{mj} \boldsymbol{r}_{k+1}^{(j)},\ \boldsymbol{u}(t_k + \alpha_m h_{k+1} - \tau)\right),$$

$$\tag{5.1.7.3}$$

where $\boldsymbol{r}_k^{(m)} = \left(r_{1,k}^{(m)}, r_{2,k}^{(m)}, \ldots, r_{N,k}^{(m)}\right)^T$ is a column vector of auxiliary functions $r_{n,k}^{(m)}$ each associated with the unknown function u_n, time layer t_k, and stage m, $n = 1, \ldots, N$, $k = 1, \ldots, K$, $m = 1, \ldots, M$; $h_{k+1} = t_{k+1} - t_k$ is the grid step size, c_m are weights of the quadrature formula ($0 \leq c_m \leq 1$), α_m are coefficients determining the nodes of the quadrature formula, and β_{mj} are weights of the intermediate quadrature formulas. The values of the delayed functions $\boldsymbol{u}(t_k + \alpha_m h_{k+1} - \tau)$ are computed using interpolation on the interval $[t_q, t_{q+1}]$, where q is a positive integer such that $t_q \leq t_k + \alpha_m h_{k+1} - \tau \leq t_{q+1}$ if $t_k + \alpha_m h_{k+1} - \tau$ lies outside the grid nodes, and coincide with the values $\boldsymbol{u}_q$ computed previously on layer t_q if $t_k + \alpha_m h_{k+1} - \tau = t_q$. The different Runge–Kutta methods are generated by different quadrature formulas defined by the sets of coefficients β_{mj}, c_m, and α_m.

The method based on formulas (5.1.7.2)–(5.1.7.3) is an implicit M-staged method. Besides being designed for systems, it differs from the explicit method (5.1.4.13) in that the sum in (5.1.7.3) is computed up to M_* rather than $m - 1$. If $M_* = m - 1$, we get the explicit method (5.1.4.13) written for systems of delay ODEs. If $M_* = m$, the values $\boldsymbol{r}_k^{(m)}$ are found consecutively from individual nonlinear equations. If $M_* = M$, the values $\boldsymbol{r}_k^{(m)}$ must be sought for all stages at once from a system of $N \times M$ equations; by default, Mathematica solves it by Newton's method (see [248, 464] for its description).

Selecting appropriate coefficients in the quadrature formulas is essential to handle stiff problems successfully. By default, Mathematica determines the values of the coefficients automatically. However, the type of the coefficients can be set manually using the property **Coefficients** of the option **Method** of NDSolve [562]. For example, these can be Lobatto IIIC coefficients, which appear in the Lobatto quadrature formula [204, 311]. The first and last nodes in the Lobatto formula coincide with the beginning and end of the integration interval. Therefore, $\alpha_1 = 0$ and $\alpha_M = 1$. The other coefficients α_m are zeros of the derivatives of shifted Legendre polynomials:

$$\frac{d^{M-2}}{d\alpha_m^{M-2}}\left(\alpha_m^{M-1}(\alpha_m - 1)^{M-1}\right) = 0. \qquad (5.1.7.4)$$

As a result, one obtains quadrature formulas of order $2M - 2$. The weights $c_1, \ldots, c_M$ and coefficients β_{mj} in the Lobatto quadrature formulas are determined from the conditions

$$\sum_{m=1}^{M} c_m \alpha_m^{\gamma-1} = \frac{1}{\gamma}, \quad \gamma = 1, \ldots, 2M - 2;$$

$$\sum_{j=1}^{M} \beta_{mj} \alpha_j^{\gamma-1} = \frac{\alpha_m^\gamma}{\gamma}, \quad m = 1, \ldots, M, \ \gamma = 1, \ldots, M - 1; \qquad (5.1.7.5)$$

$$\beta_{m1} = c_1, \quad m = 1, \ldots, M.$$

Mathematica determines the step size h_k in the method (5.1.7.2)–(5.1.7.3) automatically based on estimating the local error of the solution [561]. To this end, Mathematica compares the solutions obtained by the main method of order p with weights c_m to those obtained by the auxiliary method of order $\hat{p}$ with weights $\hat{c}_m$ (by default, $\hat{p} = p - 1$). The coefficients α_m and β_{mj} in both methods coincide, and hence, the values $s_k^{(m)}$ also coincide, which eliminates the need to solve the nonlinear system (5.1.7.3) again.

Remark 5.10. *In problems with solutions attaining huge absolute values (for example, see Test problem 2 from Section 5.2, where the solution grows exponentially and reaches absolute values of order 10^8 or higher), the NDSolve function with the Runge–Kutta method selected can take a significant time, several minutes to an hour, to construct a solution. Increasing the allowed absolute and relative errors with the options AccuracyGoal → q and PrecisionGoal → p can seriously reduce the running time to only a few seconds. For given q and p, the algorithm will attempt to ensure that the error of the numerical solution does not exceed $10^{-q} + 10^{-p}|x|$.*

Gear method. The Gear method, also known as the BDF (Backward Differentiation Formula) method, is implemented in Mathematica as part of the package IDA included in the library of methods SUNDIALS, developed by Lawrence Livermore National Laboratory, USA. IDA stands for Implicit Differential-Algebraic implicit solver, and SUNDIALS stands for SUite of Nonlinear and DIfferential/ALgebraic equation Solvers, a collection of nonlinear and differential-algebraic solvers [563]. The codes of the IDA methods (see the user manual [215]) rely on DASPK [73, 74], a collection of Fortran subroutines for solving differential-algebraic systems of high dimension.

The M-step Gear method for system (5.1.7.1) is based on the formula [63, 215, 464]:

$$\alpha_0 \boldsymbol{u}_k - h_k \boldsymbol{f}(t_k, \boldsymbol{u}_k, \boldsymbol{u}(t_k - \tau)) = -\sum_{m=1}^{M} \alpha_m \boldsymbol{u}_{k-m}. \qquad (5.1.7.6)$$

The values $\boldsymbol{u}(t_k - \tau)$ are calculated using interpolation on the interval $[t_q, t_{q+1}]$, where q is a positive integer such that $t_q \leq t_k - \tau \leq t_{q+1}$ if $t_k - \tau$ lies outside the grid nodes, and coincide with $\boldsymbol{u}_q$, computed previously on layer t_q, if $t_k - \tau = t_q$. The system of nonlinear algebraic equations (5.1.7.6) can be solved using one or another iterative method (e.g., Newton's method).

For the Gear method to have the pth order of accuracy, one should set [464, p. 255]:

$$\alpha_0 = -\sum_{m=1}^{M} \alpha_m, \quad \sum_{m=1}^{M} m\alpha_m = -1, \quad \sum_{m=1}^{M} m^j \alpha_m = 0, \quad j = 2, 3, \ldots, p. \qquad (5.1.7.7)$$

The highest achievable order of accuracy of the M-step Gear method is M.

Setting $M = 1$ in (5.1.7.6)–(5.1.7.7), we get the formula of the implicit Euler method (5.1.4.6). For $M = 2$, $M = 3$, and $M = 4$ we get the relations [464, p. 256]:

$$3\boldsymbol{u}_k - 4\boldsymbol{u}_{k-1} + \boldsymbol{u}_{k-2} = 2h_k \boldsymbol{f}(t_k, \boldsymbol{u}_k, \boldsymbol{u}(t_k - \tau)),$$
$$11\boldsymbol{u}_k - 18\boldsymbol{u}_{k-1} + 9\boldsymbol{u}_{k-2} - 2\boldsymbol{u}_{k-3} = 6h_k \boldsymbol{f}(t_k, \boldsymbol{u}_k, \boldsymbol{u}(t_k - \tau)),$$
$$25\boldsymbol{u}_k - 48\boldsymbol{u}_{k-1} + 36\boldsymbol{u}_{k-2} - 16\boldsymbol{u}_{k-3} + 3\boldsymbol{u}_{k-4} = 12h_k \boldsymbol{f}(t_k, \boldsymbol{u}_k, \boldsymbol{u}(t_k - \tau)),$$

which determine the second-, third-, and fourth-order Gear method, respectively.

In Mathematica, the Gear method computes, on each time layer, a local error estimate, E_k, and automatically selects a step size h_k and order M to ensure that the relation $E_k/\|\boldsymbol{\omega_k}\|$ holds, where the nth component $\omega_{n,k}$ of the vector $\boldsymbol{\omega_k}$ is evaluated as

$$\omega_{n,k} = \frac{1}{10^{-p}|u_{n,k}| + 10^{-q}}.$$

The values of p and q are determined with the options PrecisionGoal $\to$ p and AccuracyGoal $\to$ q of NDSolve. By default, NDSolve selects the norm $\|\cdot\|$ automatically depending on the solution method; however, the norm can be set manually. The Gear method uses the norm $\|x\|_2 = \sqrt{\sum |x_i|^2}$ [564].

NDSolve selects the step sizes h_k automatically. By default, the maximum number of steps in which the algorithm must build a solution is estimated by the initial step size [565]. This approach may fail if, for example, the solution increases indefinitely exponentially (e.g., see Test problem 2 from Section 5.2). This restriction can be eliminated by using the option MaxSteps $\to \infty$ in the function NDSolve.

5.1.8. Test Problems for Delay ODEs. Comparison of Numerical and Exact Solutions

Some exact solutions to delay ODEs can be found in [135, 136, 412, 495]. We will take advantage of the results of these studies to construct a few model problems and test Mathematica's NDSolve function.

In the test problems for delay ODEs presented below, we mostly carried out the solution on the time interval $T = 50\,\tau$ for three delay times: $\tau = 0.05$, $\tau = 0.1$, and $\tau = 0.5$. We used the following numerical methods implemented in Mathematica (Version 11.2.0):

(i) the second-order Runge–Kutta method,
(ii) the fourth-order Runge–Kutta method,
(iii) the Gear method,
(iv) an automatic method (when Mathematica automatically selects the best method for a particular computation).

The relative error, σ, of a numerical solution $u_k = u_h(t_k)$ to a test problem for a delay ODE will be understood as

$$\sigma = \max_{1 \leq k \leq K} |(u_e - u_k)/u_e|,$$

where $u_e = u_e(t_k)$ is the value obtained from the exact solution to the problem at $t = t_k$, and K is the number of time steps selected by NDSolve automatically.

Test problem 1. The Cauchy problem for a nonlinear ODE with constant delay

$$\begin{aligned} u'_t &= a(1 - ab\tau^2) + b(u-w)^2, \quad w = u(t-\tau), \quad t > 0; \\ u(t) &= at + c, \quad -\tau \leq t \leq 0, \end{aligned} \qquad (5.1.8.1)$$

has the exact solution

$$u(t) = at + c, \quad t > 0, \qquad (5.1.8.2)$$

where a, b, and c are free parameters.

The Cauchy problem (5.1.8.1) was solved on the entire time interval for all three delay times by the four numerical methods with a relative error of the order of 10^{-15}. The Runge–Kutta methods run into some problems in the initial times: the algorithm chooses the step size automatically and makes it so small that the number of iterations becomes extremely large, which can significantly increase the running time of the method. Figure 5.1 shows by open circles the numerical solutions obtained by the second-order Runge–Kutta method and Gear method for the delay time $\tau = 0.5$ and parameters $a = 0.5$, $b = 1$, and $c = 1$; the solid lines indicate the exact solutions of the form (5.1.8.2).

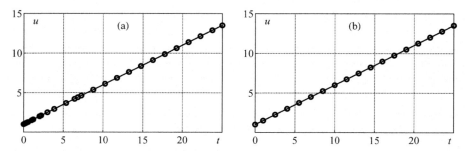

Figure 5.1. Exact solutions (solid line) and numerical solutions (open circles) of the test problem (5.1.8.1) with $a = 0.5$, $b = 1$, and $c = 1$ and $\tau = 0.5$ obtained by (a) the second-order Runge–Kutta and (b) the Gear method.

Test problem 2. The Cauchy problem for a nonlinear ODE with constant delay

$$u'_t = w^2/u, \quad w = u(t - \tau), \quad t > 0;$$
$$u(t) = e^{\beta t}, \quad -\tau \le t \le 0, \qquad (5.1.8.3)$$

has the exact solution

$$u(t) = e^{\beta t}, \quad t > 0. \qquad (5.1.8.4)$$

For given τ, the value of the parameter β is determined numerically from the transcendental equation

$$\beta - e^{-2\beta\tau} = 0$$

using the FindRoot function [566]. Note that solution (5.1.8.4) decreases as the parameter τ increases (we have $\beta = 1$ for $\tau = 0$ and $\beta = 0$ for $\tau \to \infty$).

We solved the Cauchy problem (5.1.8.3) on the entire time interval for all three delay times by the four numerical methods. The following relative errors were detected: 10^{-7} for the automatic and Gear methods, 10^{-8} for the second-order Runge–Kutta method, and 10^{-13} for the fourth-order Runge–Kutta method. Figure 5.2 plots the exact solution (5.1.8.4) and the numerical solution obtained using the fourth-order Runge–Kutta method for problem (5.1.8.3) with delay times $\tau = 0.05$ ($\beta = 0.912765$) and $\tau = 0.5$ ($\beta = 0.567143$).

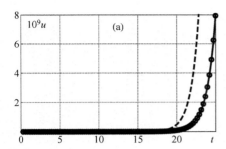 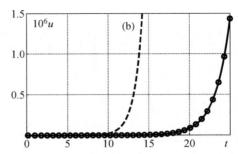

Figure 5.2. Exact solutions (solid lines) and numerical solutions (open circles), obtained by the fourth-order Runge–Kutta method, for problem (5.1.8.3) at two delay times: (a) $\tau = 0.05$ and (b) $\tau = 0.5$; the dashed line indicates e^t ($\tau = 0$).

Test problem 3. The Cauchy problem for a nonlinear ODE with constant delay

$$u'_t = w \exp(u^2 + w^2), \quad w = u(t - \tau), \quad t > 0;$$
$$u(t) = \sqrt{\ln \beta} \cos(\beta t), \quad \beta = \frac{3\pi}{2\tau}, \quad -\tau \leq t \leq 0, \qquad (5.1.8.5)$$

has a periodic exact solution

$$u(t) = \sqrt{\ln \beta} \cos(\beta t), \quad t > 0. \qquad (5.1.8.6)$$

When integrating problem (5.1.8.5), each numerical method describes several periods well, then fails with an error. The time intervals of adequate computation with the Runge–Kutta methods are broader than those of the Gear and automatic methods. As the delay time τ increases, these intervals expand. Figure 5.3 displays in open circles the numerical solutions to problem (5.1.8.5) obtained by the fourth-order Runge–Kutta method for delay times $\tau = 0.05$ and $\tau = 0.5$. The numerical solutions obtained by the other methods are qualitatively similar to those in Figure 5.3 and therefore are omitted here. The solid lines indicate the exact solutions of the form (5.1.8.6).

Notably, solution (5.1.8.6) rapidly oscillates for small τ and is singular with respect to the delay parameter because this solution does not have a limit as $\tau \to 0$. This circumstance restricts the usage of the employed numerical methods for small τ. The failure of the numerical solution at moderate τ is likely due to the instability of the periodic solution. The instability (in linear approximation) of the only stationary solution $u = 0$ can indirectly confirm the stated assumption.

Remark 5.11. *We also employed the numerical methods implemented in Mathematica and described in Subsection 5.1.7 to solve Cauchy problems for some nonlinear delay ODEs. Subsection 6.1.1 discusses a numerical solution obtained by the second-order Runge–Kutta to the Cauchy problem for Hutchinson's equation. Subsection 6.1.2 describes a solution obtained by the Gear method to the problem for Nicholson's equation. Subsection 6.1.3 solves the Cauchy problem for a Mackey–Glass type system using the fourth-order Runge–Kutta method.*

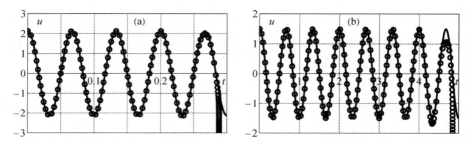

Figure 5.3. Exact solutions (solid lines) and numerical solutions (open circles), obtained by the fourth-order Runge–Kutta method, for problem (5.1.8.5) at two delay times: (a) $\tau = 0.05$ and (b) $\tau = 0.5$.

5.2. Numerical Integration of Delay PDEs

5.2.1. Preliminary Remarks. Method of Time-Domain Decomposition

Preliminary remarks. A problem for delay reaction-diffusion equations. The key features and issues in the numerical integration of delay PDEs are similar to those in delay ODEs (see Subsection 5.1.2). These complicate the numerical solution of delay PDEs as compared to respective PDEs without delay.

Notably, the statements of the initial data and boundary conditions for nonlinear delay PDEs (see Chapter 3 for examples of such equations) coincide with those for linear delay PDEs, which are specified in Section 2.2.

Consider the initial-boundary value problem for the quasilinear reaction-diffusion type equation with constant delay

$$u_t = au_{xx} + f(u, w), \quad 0 < x < L, \quad t > 0, \tag{5.2.1.1}$$

where $u = u(x, t)$, $w = u(x, t - \tau)$, and $a > 0$, subjected to the initial condition

$$u(x, t) = g(x, t), \quad 0 < x < L, \quad -\tau \leq t \leq 0, \tag{5.2.1.2}$$

and homogeneous boundary conditions of the first kind

$$u(0, t) = u(L, t) = 0, \quad t > 0. \tag{5.2.1.3}$$

Remark 5.12. *The problem described by the delay PDE (5.2.1.1) with initial data (5.2.1.2) and nonhomogeneous boundary conditions of the first kind*

$$u(0, t) = h_1(t), \quad u(L, t) = h_2(t), \quad t > 0,$$

where $h_1(t)$ and $h_2(t)$ are given functions, can be converted, with the substitution

$$u = U + h_1(t) + \frac{x}{L}[h_2(t) - h_1(t)],$$

to a delay initial-boundary value problem for $U = U(x, t)$ with homogeneous boundary conditions. The resulting equation will then explicitly depend on x and t, which is a minor complication to the problem.

The time-domain decomposition method for delay PDEs. For a constant delay, the time interval $0 \leq t \leq T$ can be broken down into several subintervals of equal length $[0, \tau]$, $[\tau, 2\tau]$, ... (e.g., see [212]). In this case, problem (5.2.1.1)–(5.2.1.3) on the first interval $0 < t \leq \tau$ becomes

$$u_t = au_{xx} + f(u, g), \quad 0 < x < L, \quad 0 < t \leq \tau,$$
$$u(x, 0) = g(x, 0), \quad 0 < x < L, \quad (5.2.1.4)$$
$$u(0, t) = u(L, t) = 0, \quad 0 \leq t \leq \tau.$$

It has been taken into account that $w(x, t) \equiv g(x, t)$ on $0 < t \leq \tau$. Suppose we have found a solution $u_1(x, t)$ to problem (5.2.1.4), which is a problem without delay. Then we can proceed to solve the problem on the next interval, $\tau < t \leq 2\tau$:

$$u_t = au_{xx} + f(u, u_1), \quad 0 < x < L, \quad \tau < t \leq 2\tau,$$
$$u(x, \tau) = u_1(x, \tau), \quad 0 < x < L, \quad (5.2.1.5)$$
$$u(0, t) = u(L, t) = 0, \quad \tau < t \leq 2\tau.$$

Here we have $w(x, t) \equiv u_1(x, t)$ on $\tau < t \leq 2\tau$. By reasoning likewise, we can eventually solve the original problem (5.2.1.1)–(5.2.1.3) on the time interval in question. Each subproblem is one without delay, which can be solved by any suitable analytical or numerical method for partial differential equations without delay (e.g., a finite difference method or the finite element method).

Remark 5.13. *The time-domain decomposition method is a natural generalization of the method of steps, which is used to solve Cauchy problems for delay ODEs (see Subsection 1.1.5). If $f(u, w) = f_1(w)u + f_0(w)$ in equation (5.2.1.1), the subproblems on all intervals $[0, \tau]$, $[\tau, 2\tau]$, ... will be linear.*

5.2.2. Method of Lines—Reduction of a Delay PDE to a System of Delay ODEs

Preliminary remarks. Presently, the theory of solving delay ODEs is fairly well developed as compared to the theory of solving delay PDEs. This applies to both analytical methods (e.g., see Chapter 1 and [37, 138, 144, 146, 205, 275, 276, 283, 482]) and numerical methods (e.g., see Section 5.1 and [42, 284, 479]). In addition, the widely used software packages such as Maple™, Mathematica®, and MATLAB® have the ability to solve first-order delay ODEs [334, 336, 560]. Therefore, it is useful to first reduce the partial differential equation with delay to a system of ordinary differential equations with delay and then solve this system rather than the original equation. This approach is often implemented with the method of lines [217, 385]. The book [469] presents a large number of codes to analyze delay PDE models using the method of lines.

Reaction-diffusion type PDEs with delay. We use a space grid $x_n = nh$, $n = 0, 1, \ldots, N$, where $h = L/N$ is the step size of the grid and N is the number of space intervals. Let us reduce problem (5.2.1.1)–(5.2.1.3) to a system of ODEs by approximating the space derivative with a finite difference analogue and writing the equation for a node x_n:

$$(u_n)'_t = a\delta_{xx}u_n + f(u_n, w_n), \quad n = 1, \ldots, N-1, \quad 0 < t \leq T;$$
$$u_0(t) = u_N(t) = 0, \quad 0 \leq t \leq T; \tag{5.2.2.1}$$
$$u_n(t) = g_n(t), \quad n = 1, \ldots, N-1, \quad -\tau \leq t \leq 0,$$

where $\delta_{xx}u_n = h^{-2}(u_{n+1} - 2u_n + u_{n-1})$. System (5.2.2.1) involves $N-1$ unknown functions $u_n(t)$ and the same number of equations as well as two known functions, $u_0(t)$ and $u_N(t)$.

The main drawback of this approach is that the resulting system of delay ODEs is often stiff, when the step size has to be reduced for stability rather than to increase the algorithm's accuracy. Such systems have to be solved using specially developed methods with increased stability [29, 479]. These are usually algorithms from the class of implicit Runge–Kutta methods [42, 230, 366]. Their use requires computations at several points $t_m + \alpha_j s_m$, where s_m is the temporal step size, and so entails calculating the values of delayed functions at the points $t_m + \alpha_j s_m - \tau$, which may not coincide with grid nodes. In this case, interpolation algorithms have to be employed. See Subsections 5.1.4 and 5.1.7 for a detailed description of Runge–Kutta methods.

Higher accuracy can be achieved via space discretization by using Chebyshev–Gauss–Lobatto nodes [58, 235, 340, 382]:

$$x_n = \frac{L}{2} + \frac{L}{2}\cos\left(\frac{\pi n}{N}\right), \quad n = 0, 1, \ldots, N.$$

Then the space derivative is approximated as

$$u_{xx}(x_n, t) \approx \sum_{i=0}^{N} c_{ni} u(x_i, t),$$

where c_{ni} are coefficients of the differential matrix; see [84, 164, 554] for details. As a result, considering the homogeneous boundary conditions, we obtain the nonlinear system of delay ODEs

$$(u_n)'_t = a\sum_{i=1}^{N-1} c_{ni} u_i + f(u_n, w_n), \quad n = 1, \ldots, N-1, \quad 0 < t \leq T;$$
$$u_n(t) = g_n(t), \quad n = 0, 1, \ldots, N, \quad -\tau \leq t \leq 0. \tag{5.2.2.2}$$

If $f(u, w) = ur(w)$, system (5.2.2.2) can be conveniently written in operator notation [235]:

$$\mathbf{u}'_t = a\mathbb{C}\mathbf{u} + \mathbb{R}\mathbf{u}, \quad 0 < t \leq T;$$
$$\mathbf{u} = \mathbf{g}, \quad -\tau \leq t \leq 0,$$

where **u** and **g** are column vectors, $\mathbb{C} = [c_{ni}]_{n,i=1}^{N-1}$ is the differential matrix, and $\mathbb{R} = \text{diag}\{r(w_1), \ldots, r(w_{N-1})\}$ is a diagonal matrix. We apply the Gauss–Jacobi decomposition:

$$\mathbb{C} = \mathbb{A} + \mathbb{B}, \quad \mathbb{A} = \text{diag}\{\mathbb{C}\}, \quad \mathbb{B} = \mathbb{C} - \mathbb{A}.$$

Approximating the delayed function **w** by the values from the preceding iteration, $\mathbf{w}^{(k)}$, we obtain the system of ODEs for the values at the next iteration, $\mathbf{u}^{(k+1)}$:

$$(\mathbf{u}^{(k+1)})'_t = (a\mathbb{A} + \mathbb{R}^{(k)})\mathbf{u}^{(k+1)} + a\mathbb{B}\mathbf{u}^{(k)}, \quad 0 < t \leq T;$$
$$\mathbf{u}^{(k+1)} = \mathbf{g}, \quad -\tau \leq t \leq 0,$$

where $k = 0, 1, \ldots$ and $\mathbf{u}^{(0)}$ is an arbitrary initial value. One usually takes

$$\mathbf{u}^{(0)} = \begin{cases} \mathbf{g}(t), & -\tau \leq t \leq 0, \\ \mathbf{g}(0), & t > 0. \end{cases}$$

The resulting linear system of ODEs is suitable for parallel computing [235] and can be solved by the methods described in Subsections 5.1.4 and 5.1.7.

Wave-type delay PDEs. Now consider the initial-boundary value problem for a wave-type nonlinear delay equation of a general form:

$$\varepsilon u_{tt} + \sigma u_t = [p(x,u)u_x]_x + q(x,u,w)u_x + f(x,u,w), \quad t > 0, \ 0 \leq x \leq L;$$
$$u(x,t) = \varphi_0(x,t), \quad u_t(x,t) = \varphi_1(x,t), \quad -\tau \leq t \leq 0; \qquad (5.2.2.3)$$
$$u(0,t) = \psi_0(t), \quad u(1,t) = \psi_1(t), \quad t > 0,$$

where $w = u(x, t - \tau)$. The functions p, q, and f can additionally depend on t. Special cases of the equation include delay reaction-diffusion equations ($\varepsilon = 0$ and $\sigma = 1$), delay Klein–Gordon equations ($\varepsilon = 1$ and $\sigma = 0$), and nonlinear delay telegraph equations ($\varepsilon = 1$ and $\sigma \neq 0$).

For the method of lines to be applicable to hyperbolic equations, we have to introduce the second unknown, $v = u_t$. As a result, we obtain

$$u_t = v, \quad t > 0, \ 0 \leq x \leq L;$$
$$\varepsilon v_t + \sigma v = [p(x,u)u_x]_x + q(x,u,w)u_x + f(x,u,w), \quad t > 0, \ 0 \leq x \leq L;$$
$$u(x,t) = \varphi_0(x,t), \quad v(x,t) = \varphi_1(x,t), \quad -\tau \leq t \leq 0; \qquad (5.2.2.4)$$
$$u(0,t) = \psi_0(t), \quad u(L,t) = \psi_1(t), \quad t > 0,$$
$$v(0,t) = (\psi_0)'_t, \quad v(L,t) = (\psi_1)'_t, \quad t > 0.$$

We use the space grid $x_n = nh$, where $n = 0, 1, \ldots, N$, $h = L/N$ is the step size, and N is the number of spatial intervals. Approximating the derivatives with respect to x by finite difference analogues and writing the equation for the node x_n, we reduce problem (5.2.2.4) to the system of ODEs

$$(u_n)'_t = v_n, \quad n=1,\ldots N-1, \quad 0<t\leq T;$$
$$\varepsilon(v_n)'_t + \sigma v_n = \delta_x[p_n\delta_x u_n] + q_n\delta_x u_n + f_n, \quad n=1,\ldots,N-1, \quad 0<t\leq T;$$
$$u_n(t) = \varphi_0(x_n,t), \quad v_n(t) = \varphi_1(x_n,t), \quad n=0,1,\ldots,N, \quad -\tau\leq t\leq 0; \quad (5.2.2.5)$$
$$u_0(t) = \psi_0(t), \quad u_N(t) = \psi_1(t), \quad 0<t\leq T;$$
$$v_0(t) = (\psi_0)'_t, \quad v_N(t) = (\psi_1)'_t, \quad 0<t\leq T,$$

where $u_n = u_n(t) = u(x_n,t)$, $v_n = v_n(t) = v(x_n,t)$, $w_n = u(x_n, t-\tau)$, $p_n = p(x_n, u_n)$, $q_n = q(x_n, u_n, w_n)$, $f_n = f(x_n, u_n, w_n)$, T is the endpoint of the computational temporal interval, and δ_x is the difference operator expressed as

$$\delta_x u_n = \frac{1}{h}(u_{n+1} - u_n),$$
$$\delta_x[p_n\delta_x u_n] = \frac{1}{h^2}[p_n(u_{n+1} - u_n) - p_{n-1}(u_n - u_{n-1})].$$

System (5.2.2.5) involves $N-1$ unknown functions $u_n(t)$, $N-1$ unknown functions $v_n(t)$, and $2N-2$ equations as well as four known functions $u_0(t)$, $u_N(t)$, $v_0(t)$, and $v_N(t)$.

A procedure for numerically solving delay problems by the method of lines using Mathematica. Figure 5.4 schematically displays a procedure for the numerical integration of the initial-boundary value problem (5.2.2.3) using the Mathematica software. The procedure involves the following sequence of actions.

1°. State the problem by specifying the equation, initial data, and boundary conditions.

2°. Select the number of spatial intervals N.

3°. If the equation is hyperbolic, introduce the new variable $v = u_t$.

4°. Apply the method of lines to obtain a system of delay ODEs consisting of (a) $N-1$ equations and $N-1$ initial conditions (plus two algebraic relations at the boundary of the domain) if the equation is parabolic or (b) $2N-2$ equations and $2N-2$ initial conditions (plus four algebraic relations at the boundary of the domain) if the equation is hyperbolic.

5°. Select the temporal interval $0 < t \leq T$ at which the system is integrated.

6°. Solve the system of ODEs from step 4° using one of the methods of the NDSolve function.

7°. In case of a computation error, reduce the temporal interval from step 5° and try to obtain a satisfactory solution on the shorter interval.

8°. Eventually obtain the values of the unknown at all desired (or possible) temporal layers, the absolute and relative error estimates of the exact solution (if known), and plots and animations of the numerical solution (as well as the exact solution).

It is noteworthy that the constant coefficient σ in equation (5.2.2.3) can be replaced with a function $\sigma = \sigma(x, u, w)$.

5.2. Numerical Integration of Delay PDEs

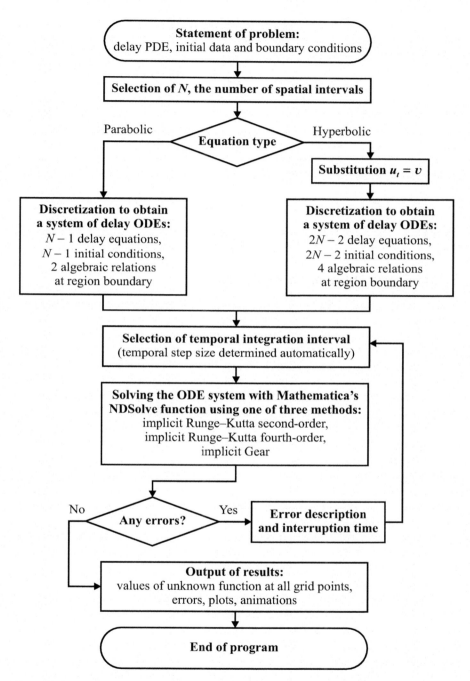

Figure 5.4. A flowchart for the numerical integration of reaction-diffusion and wave-type delay problems by the method of lines.

Remark 5.14. Apart from uniform grids in x, one can use grids with a variable step size [281]. For nonuniform grids with step size $h_n = x_n - x_{n-1}$, the second derivative u_{xx} is approximated as

$$u_{xx} \approx \frac{2}{h_n + h_{n+1}} \left(\frac{u_{n+1} - u_n}{h_{n+1}} - \frac{u_n - u_{n-1}}{h_n} \right),$$

where $\sum h_n = L$ (with $0 \leq x \leq L$).

Remark 5.15. The method of lines can also be employed to integrate 2D and 3D reaction-diffusion type delay equations. The terms $[p(x,u)u_x]_x$ and $q(x,u,w)u_x$ should then be replaced with $\nabla \cdot [p(\mathbf{x}, u)\nabla u]$ and $\mathbf{q}(\mathbf{x}, u, w) \cdot \nabla u$. In particular, in the two-dimensional case on a uniform grid, the Laplace operator, $\Delta u = u_{xx} + u_{yy}$, is approximated as follows:

$$\Delta u \approx \frac{1}{h_1^2}(u_{n+1,m} - 2u_{n,m} + u_{n-1,m}) + \frac{1}{h_2^2}(u_{n,m+1} - 2u_{n,m} + u_{n,m-1}),$$

where $u_{n,m} = u(x_n, y_m, t)$, $x_n = nh_1$, $y_m = mh_2$ ($n = 0, 1, \ldots, N$; $m = 0, 1, \ldots, M$), and $h_1 = L_1/N$ and $h_2 = L_2/M$ are the step sizes in the space variables x and y.

5.2.3. Finite Difference Methods

Basic concepts and definitions of the theory of finite difference methods. *Finite difference methods* (or *difference methods* for short) are the most common numerical methods for solving problems for partial differential equations without delay. With these, independent variables are treated on a *discrete set of points*, while suitable difference analogues are employed to approximate all continuous derivatives of the unknown function. Appropriate modifications of these methods can be used to solve problems for delay PDEs like (5.2.1.1)–(5.2.1.3). Following [248], we present below the basic concepts and definitions from the theory of difference methods.

We will use a space grid $G_x = \{x_0 = 0, x_1, \ldots, x_N = L\}$ and a temporal grid $G_t = \{t_0 = 0, t_1, \ldots, t_K = T\}$. Through these discrete points, we draw straight lines parallel to the x and t axes. The points of intersection of these lines define *nodes* of the spatio-temporal grid $G = \{(x_n, t_k)\}$. The collection of all grid nodes lying on a straight line $t = t_k$ is called a *temporal layer*. The line $t = 0$ is called the *initial layer*. The spatial points $x_0 = 0$ and $x_N = L$ are the *boundary* nodes, while $x_1, \ldots, x_{N-1}$ are *inner* nodes.

A spatial step size of the grid is denoted $h_n = x_n - x_{n-1}$ and a temporal step size is denoted $s_k = t_{k+1} - t_k$. A rectangle $[x_{n-1}, x_n; t_k, t_{k+1}]$ is called a *grid cell*. If $h_n = h = \text{const}$ and $s_k = s = \text{const}$ for any n and k, then the grid is called *uniform*, otherwise it is *nonuniform*.

All derivatives involved in the equation and boundary conditions are replaced with differences (or suitable linear combinations) of the values of the unknown $u(x,t)$ at the grid nodes. The resulting algebraic equations are called a *finite difference scheme* or *difference scheme* for short. Solving the algebraic system, one finds the values of the discrete grid functions $u_h = \{u_{n,k} = u_h(x_n, t_k), n = 0, 1, \ldots, N, k = 0, 1, \ldots, K\}$, which provide an approximate (difference) solution at the nodes.

Within the region under consideration, we only replace the partial differential equation with the difference scheme, using the same configuration of nodes called a *stencil*. A scheme with a stencil containing only one node of the new layer is called *explicit*. One computes the value at this node via the values from the original layer (or a few previous layers) in finitely many steps. A scheme is called *implicit* if it has a stencil with several new-layer nodes. The values at the new layer are found from a system of many algebraic equations. Solving such systems is often non-trivial and requires the development of special algorithms.

Suppose the equation involves only the first time derivative u_t. Then the difference scheme is usually two-layer, meaning that it only has two layers: the current one t_k and the new one t_{k+1}. However, at least three layers are required if the equation contains the second time derivative u_{tt}, in which case the previous layer t_{k-1} has to be used in addition to the current and new ones.

The numerical integration of problem (5.2.1.1)–(5.2.1.3) suggests that one has to select appropriate step sizes h_n and s_k and find approximate values $u_{n,k}$ of the unknown function $u(x,t)$ at the discrete nodes (x_n, t_k), where $n = 1, \ldots, N$ and $k = 1, \ldots, K$.

Characterizing the properties of numerical methods requires using certain concepts and definitions (e.g., see [248]) that we describe below.

For convenience, we rewrite equation (5.2.1.1) briefly in the operator form

$$\mathscr{L}[u] = f, \quad (x,t) \in Q, \qquad (5.2.3.1)$$

where $\mathscr{L}[u] = u_t - a u_{xx}$, $Q = \{0 < x < L,\ 0 < t \leq T\}$. We introduce a grid G in the domain Q. The numerical finite-difference scheme (*difference scheme* or just *scheme* for short) for problem (5.2.1.1)–(5.2.1.3) can be written as

$$\mathscr{L}_h[u_h] = f_h, \quad (x,t) \in G, \qquad (5.2.3.2)$$

where $\mathscr{L}_h[u_h]$ is a finite-difference differentiation operator, and the right-hand side is defined by $f_h = f(u_h(x_n, t_k), \tilde{w}_h(x_n, t_k))$; the tilde denotes a continuous approximation of a grid function.

The delay reaction-diffusion equations (5.2.1.1) most frequently come with the operator

$$\mathscr{L}_h[u_h] = s^{-1}[u_h(x_n, t_k + s) - u_h(x_n, t_k)]$$
$$- h^{-2}[u_h(x_{n+1}, t_k) - 2u_h(x_n, t_k) + u_h(x_{n-1}, t_k)].$$

The norm used in the space of grid functions is similar to that of continuous functions:

$$\|u_h\| = \max_{(x_n, t_k) \in G} |u_h(x_n, t_k)|.$$

The numerical solution u_h is only defined on the grid G, and hence, the continuous operator $\mathscr{L}$, defined for all $(x,t) \in Q$, is inapplicable to it. The exact solution $u(x,t)$ is defined for all (x,t), including the grid G; therefore, the difference operator $\mathscr{L}_h$ can be applied to $u(x,t)$.

The proximity of a difference scheme to the original equation is determined by the magnitude of the *residual*

$$\psi_h = \mathscr{L}_h[u] - f_h, \quad (x,t) \in G.$$

A difference scheme is said to approximate a PDE if

$$\|\psi_h\| \to 0 \quad \text{as} \quad h \to 0 \text{ and } s \to 0 \qquad (5.2.3.3)$$

and has approximation orders $p > 0$ in x and $q > 0$ in t if

$$\|\psi_h\| = O(h^p + s^q).$$

If $\|\psi_h\| \to 0$ regardless of how s and h tend to zero, the approximation is called *unconditional*. For some schemes, the norm of the residual is defined as $\|\psi_h\| = O(h^p + s^q + s^r/h^m)$. In this case, the additional condition $s^r/h^m \to 0$ must hold for $\|\psi_h\| \to 0$. This kind of approximation is called *conditional*.

If the numerical result does not tend to the exact solution as the grids condense and, on the contrary, the initially small errors indefinitely increase, the scheme is called unstable.

Scheme (5.2.3.2) with the initial data (5.2.1.2) and boundary conditions (5.2.1.3) is said to be *stable* if the solution u_h is continuously dependent on the input data, defined by the functions f and φ, and this dependence is uniform with respect to the grid step size [247]. In other words, for any $\varepsilon > 0$ there exists a $\delta(\varepsilon)$, independent of the step size h (at least for sufficiently small h), such that if two pairs of functions $f^{\mathrm{I}}, \varphi^{\mathrm{I}}$ and $f^{\mathrm{II}}, \varphi^{\mathrm{II}}$ satisfy the conditions

$$\|f^{\mathrm{I}} - f^{\mathrm{II}}\| \leq \delta \quad \text{and} \quad \|\varphi^{\mathrm{I}} - \varphi^{\mathrm{II}}\| \leq \delta,$$

then the respective grid functions satisfy the inequality

$$\|u_h^{\mathrm{I}} - u_h^{\mathrm{II}}\| \leq \varepsilon.$$

A solution continuously dependent on f is called *stable with respect to the right-hand side* and that continuously dependent on φ is called *stable with respect to the initial data*.

Numerical methods for integrating PDEs can also be conditionally or unconditionally stable. A method is called *unconditionally stable* if the above relations hold for any step sizes in time and space as long as they are small enough. If additional relations are imposed on the step sizes, then the method is called *conditionally stable*.

The grid solution of the initial layer is affected by the errors of the initial data. On each successive layer, the grid solutions are also affected by the approximation errors of the differential equation and boundary conditions. These errors are transferred to the subsequent layers and can accumulate during the computation. To obtain good accuracy, one has to ensure that all these errors remain small and do not increase much.

A difference solution is said to converge to the exact solution if

$$\|u_h - u\| \to 0 \quad \text{as} \quad h \to 0 \text{ and } s \to 0.$$

The method is *convergent with order $p > 0$ in x and order $q > 0$ in t* if

$$\|u_h - u\| = O(h^p + s^q) \quad \text{as} \quad h \to 0.$$

The difference scheme (5.2.3.2) is called *well-defined* if it is stable and its solution exists and is unique for any allowed f.

Below we discuss a few difference methods (schemes) for problems of the form (5.2.1.1)–(5.2.1.3) with constant delay.

Explicit finite-difference scheme. Data storage optimization in RAM. Consider a uniform spatial grid $x_n = nh$ ($n = 0, 1, \ldots, N$), where $h = L/N$ is the spatial step size. Let s denote a temporal step size such that $Ms = \tau$, where M is a positive integer. We approximate the time derivative with a finite difference analogue as

$$u_t \approx s^{-1}(u_{n,k+1} - u_{n,k})$$

and the space derivative with the second-order finite-difference derivative

$$u_{xx} \approx \delta_{xx} u_{n,k} = h^{-2}(u_{n+1,k} - 2u_{n,k} + u_{n-1,k}).$$

We represent problem (5.2.1.1)–(5.2.1.3) as the difference scheme

$$\begin{aligned}
&u_{n,k+1} = u_{n,k} + as\delta_{xx} u_{n,k} + sf(u_{n,k}, u_{n,k-M}), \\
&n = 1, 2, \ldots, N-1, \quad k = 0, \ldots, K-1; \\
&u_{0,k} = u_{N,k} = 0, \quad k = 0, 1, \ldots, K; \\
&u_{n,k} = g_{n,k}, \quad n = 0, 1, \ldots, N, \quad k = -M, \ldots, 0.
\end{aligned} \quad (5.2.3.4)$$

It is apparent from (5.2.3.4) that computing values at layer $k+1$ requires data from not only the preceding layer n but also layer $n - M$. Hence, one has to store the data from all temporal layers within the delay time range, since they are used for the solution. Furthermore, the situation is complicated by the need to ensure that the Currant criterion

$$s < \frac{h^2}{2\|u(x,t)\|}$$

holds for the explicit scheme (5.2.3.4) to be stable.

When solving problems with delay, one has to store a sufficiently large amount of data to which it is necessary to have constant access. Most often, the amount of data exceeds the CPU's fast cache memory, meaning that RAM must be used. Consequently, the computation speed depends on the speed of data exchange with RAM. Notably, storing data on external media is unacceptable since the reading and writing time would be prohibitively too high.

The studies [48, 70, 594] discuss an algorithm that can significantly reduce RAM-related costs. The authors propose storing data from only a few reference temporal layers rather than all layers and restoring intermediate values using interpolation. The number of reference layers within the delay range can be varied during the computation to maintain the balance between the algorithm's accuracy and the amount of data stored; this will depend on the smoothness of the functions. The type of interpolation has to be selected based on the specific parameters and properties of the model.

The smoothness of the functions will be determined by the formula [70, 594]:

$$\Gamma(t+s) = \max_{1 \leq n \leq N} \left| \frac{u(x_n, t+s) - u(x_n, t)}{u(x_n, t)} \right|.$$

We introduce a parameter p such that each pth layer inside the delay range is a reference layer. To minimize the interpolation errors while preserving a reasonably high computational speed, one should choose the value of p from the range 1 to 20. If the values of u change too quickly, then layers should be preserved to retain the required accuracy of the algorithm by choosing $p = 1$. If u varies slowly ($\Gamma < 0.01$), one should set $p = 20$. In this case, the proposed algorithm shows the highest efficiency. The studies [70, 594] suggest an empirical rule to optimize the number of reference layers for reaction-diffusion problems:

$$p = \tau s^{-1}[1 + \exp(2 - 50\,\Gamma(t))]^{-1}.$$

The total number of stored layers q in the delay range depends on p and the temporal step size s. In the simplest case of fixed p and s, one can take advantage of the formula $q = \tau p^{-1} s^{-1}$.

Suppose $\hat{t}_j$ is the time of the jth reference layer ($j = 1, \ldots, q$) and $\hat{u}_{n,j}$ is the value of u at layer j. To restore the values of u at the intermediate layers from q reference layers, or to find $u(x_n, t)$ for $t_{k-M} \leq t \leq t_{k-1}$, one can use Newton's interpolation polynomial [70, 594]:

$$u(x_n, t) = R(\hat{u}_{n,1}) + (t - \hat{t}_1) R(\hat{u}_{n,1}, \hat{u}_{n,2}) + \cdots$$
$$+ (t - \hat{t}_1)(t - \hat{t}_2) \ldots (t - \hat{t}_{q-1}) R(\hat{u}_{n,1}, \ldots, \hat{u}_{n,q}),$$

where the divided differences $R(\ldots)$ are defined as

$$R(\hat{u}_{n,1}) = \hat{u}_{n,1},$$
$$R(\hat{u}_{n,1}, \hat{u}_{n,2}) = \frac{\hat{u}_{n,2} - \hat{u}_{n,1}}{\hat{t}_2 - \hat{t}_1},$$
$$R(\hat{u}_{n,1}, \ldots, \hat{u}_{n,q}) = \frac{R(\hat{u}_{n,2}, \ldots, \hat{u}_{n,q}) - R(\hat{u}_{n,1}, \ldots, \hat{u}_{n,q-1})}{t_q - t_1}.$$

Importantly, the above algorithm is independent of the numerical scheme but only establishes the data storage order during the system's evolution (or the delay interval). Therefore, it can be easily generalized to use with other difference schemes.

Implicit finite-difference scheme. We are going to solve problem (5.2.1.1)–(5.2.1.3) consecutively on the intervals $[0, \tau]$, $[\tau, 2\tau]$, $\ldots$ using temporal decomposition. We construct a uniform grid such that $x_n = nh$ ($n = 0, 1, \ldots, N$) and $t_k = ks$ ($k = 0, 1, \ldots, K$), where $h = L/N$ is the spatial step size, $s = \tau/M$ is the temporal step size, and M is a positive integer. We approximate the time derivative with the finite difference analogue

$$u_t \approx s^{-1}(u_{n,k+1} - u_{n,k})$$

and the spatial derivative with the second-order difference derivative

$$u_{xx} \approx \delta_{xx}u_{n,k+1} = h^{-2}(u_{n+1,k+1} - 2u_{n,k+1} + u_{n-1,k+1}).$$

Considering that $w(x,t) = g(x,t)$ for $0 < t \leq \tau$, we write the implicit difference scheme of problem (5.2.1.1)–(5.2.1.3) on this interval:

$$\begin{aligned}
(1 - as\delta_{xx})u_{n,k+1} &= u_{n,k} + sf(u_{n,k+1}, g_{n,k+1-M}), \\
n &= 1, 2, \ldots, N-1, \quad k = 0, \ldots, K-1; \\
u_{0,k+1} &= u_{N,k+1} = 0, \quad k = 0, 1, \ldots, K; \\
u_{n,0} &= g_{n,0}, \quad n = 0, 1, \ldots, N.
\end{aligned} \qquad (5.2.3.5)$$

The values of $w(x,t)$ will be known on the next interval $\tau < t \leq 2\tau$ and equal to the appropriate values of $u(x,t)$ calculated on the current interval $0 < t \leq \tau$. Continuing the computation of the subsequent intervals, we will obtain a solution on the entire time interval of interest. The study [212] established uniqueness conditions for the solution obtained using scheme (5.2.3.5). The study [156] proved convergence of this scheme with order $h^2 + s$ and analyzed its stability.

Remark 5.16. *Scheme (5.2.3.5) represents a nonlinear system of difference equations, which can be solved using an iterative method such as, for example, the Picard–Schwarz method [212] or the method of upper and lower solutions [321, 322, 378, 379].*

Weighted finite-difference scheme. Let us consider problem (5.2.1.1)–(5.2.1.3) in the domain $Q = \{0 \leq x \leq L,\ 0 \leq t \leq T\}$. We use a uniform grid such that $x_n = nh$ ($n = 0, 1, \ldots, N$) and $t_k = ks$ ($k = 0, 1, \ldots, K$), where $h = L/N$ is the spatial step size, $s = \tau/M$ is the temporal step size, and M is a positive integer.

We denote $f_{n,k} = f(u_{n,k}, w_{n,k}(t))$ and write a weighted scheme (e.g., see [293, 384, 386]):

$$\begin{aligned}
(1 - \sigma as\delta_{xx})u_{n,k+1} &= [1 + (1-\sigma)as\delta_{xx}]u_{n,k} + sf_{n,k}, \\
n &= 1, 2, \ldots, N-1, \quad k = 0, 1, \ldots, K-1; \\
u_{0,k} &= u_{N,k} = 0, \quad k = 0, 1, \ldots, K; \\
u_{n,k} &= g_{n,k}, \quad n = 0, 1, \ldots, N, \quad k = -M, \ldots, -1, 0,
\end{aligned} \qquad (5.2.3.6)$$

where $0 \leq \sigma \leq 1$. The weight $\sigma = 0$ corresponds to the implicit scheme. For $0 < \sigma \leq 1$, we get a linear tridiagonal system of linear algebraic equations, which is solved by the sweep method. For $\sigma \geq 1/2$, the scheme is unconditionally stable; see [293] for details on stability, including the case $\sigma < 1/2$.

Remark 5.17. *The studies [386, 387] discuss a related weighted scheme for a hyperbolic delay equation. See [573] for a weighted scheme for a parabolic equation with a delayed diffusion term.*

Higher-order finite-difference schemes. For sufficiently smooth solutions to equation (5.2.1.1), one can use the following stable multistep difference scheme of the higher order of approximation $h^4 + s^2$ [506]:

$$(\mathcal{A} - \tfrac{1}{2}as\delta_{xx})u_{n,k+1} = (\mathcal{A} + \tfrac{1}{2}as\delta_{xx})u_{n,k}$$
$$+ s\mathcal{A}f(\tfrac{3}{2}u_{n,k} - \tfrac{1}{2}u_{n,k-1}, \tfrac{1}{2}u_{n,k+1-M} + \tfrac{1}{2}u_{n,k-M}),$$
$$n = 1, \ldots, N-1, \quad k = 0, 1, \ldots, N-1,$$

where $\mathcal{A}$ is a difference operator defined as

$$\mathcal{A}u_{n,k} = \tfrac{1}{12}(u_{n-1,k} + 10u_{n,k} + u_{n+1,k}).$$

Remark 5.18. *The studies [601, 602] deal with similar multistep higher-order difference schemes for related more complicated reaction-diffusion type equations.*

Two special finite-difference schemes for a linear problem. The paper [219] discusses two special finite-difference schemes for problem (5.2.1.1)–(5.2.1.3) with a linear kinetic function $f(u, w) = bw$. The schemes have the same region of stability as the original problem. This is ensured through selecting suitable spatial and temporal step sizes and using a special approximation technique.

Both schemes use the same grid. The spatial step size is computed as $\tilde{h} = 2\sin(\tfrac{1}{2}h)$, where h is the classical step size of a rectangular grid. The temporal step size is selected as $s = \tau/(M - \varepsilon)$, where M is a positive integer and $0 \leq \varepsilon < 1$. Notably, unlike the schemes discussed above, the intervals $[0, \tau]$, $[\tau, 2\tau]$, $\ldots$ generally involve a non-integer number of temporal steps.

The first scheme is constructed using the trapezoid rule, where the space derivative and delayed term are calculated as the mean of the two adjacent temporal layers:

$$u_{xx} \approx \frac{1}{2}\tilde{\delta}_{xx}(u_{n,k+1} + u_{n,k}), \quad bw \approx \frac{b}{2}(w_{n,k+1} + w_{n,k}),$$

where $\tilde{\delta}_{xx}u_{n,k} = \tilde{h}^{-2}(u_{n+1,k} - 2u_{n,k} + u_{n-1,k})$. The grid function $w_{n,k}$ approximates $w(x, t)$ at the node (x_n, t_k) using linear interpolation:

$$w_{n,k} = \varepsilon u_{n,k-M+1} + (1 - \varepsilon)u_{n,k-M}. \tag{5.2.3.7}$$

Then equation (5.2.1.1) is approximated as follows:

$$(2 - as\tilde{\delta}_{xx})u_{n,k+1} = (2 + as\tilde{\delta}_{xx})u_{n,k} + bs[\varepsilon u_{n,k-M+2} + u_{n,k-M+1} + (1-\varepsilon)u_{n,k-M}].$$

The other difference scheme is based on a second-order backward differentiation formula for approximating the temporal derivative. It can be written as

$$\frac{1}{2s}(3u_{n,k+2} - 4u_{n,k+1} + u_{n,k}) = a\tilde{\delta}_{xx}u_{n,k+2} + bw_{n,k+2}.$$

Using the linear interpolation (5.2.3.7) for $w_{n,k+2}$ and rearranging the terms, we obtain

$$(3 - 2as\tilde{\delta}_{xx})u_{n,k+2} = 4u_{n,k+1} - u_{n,k} + 2bs[\varepsilon u_{n,k-M+3} + (1 - \varepsilon)u_{n,k-M+2}].$$

The values $u_{n,1}$ ($n = 1, \ldots, N-1$) are required here; these can be obtained with the first difference scheme.

5.3. Construction, Selection, and Usage of Test Problems for Delay PDEs

5.3.1. Preliminary Remarks

The qualitative features of delay PDEs significantly complicate obtaining adequate numerical solutions. The point is that even in the absence of delay, theoretical estimates of the accuracy of numerical solutions to nonlinear PDEs involve constants that depend on solution smoothness and cannot usually be calculated in advance. This is especially true for non-smooth solutions, which are typical for delay equations. Moreover, the practical convergence of numerical methods based on refining the grid cannot guarantee the reliability of the employed schemes or computational accuracy. This issue mainly arises near the values of the problem parameters corresponding to unstable solutions or near the values of variables corresponding to equation singularities or large gradients of solutions.

In many cases, the most effective and apparent way to evaluate the scope and accuracy of numerical methods is a direct comparison of numerical and exact solutions to test problems. Chapters 3 and 4 discussed many classes of delay partial differential equations that admit exact solutions in terms of elementary functions. These equations and their exact solutions involve several free parameters (which can be varied). They can serve as test problems to evaluate the accuracy of numerical methods (see Sections 5.3.4 and 5.3.5).

5.3.2. Main Principles for Selecting Test Problems

For nonlinear partial differential equations with delay (or without delay), when selecting test problems intended for checking the adequacy and estimating the accuracy of the corresponding numerical and approximate analytical methods, it is helpful to be guided by the following principles [409, 499].

$1°$. The most reliable test problems are those obtained using the exact solutions to delay partial differential equations.

$2°$. Choosing simple test problems with solutions expressed in terms of elementary functions is preferable.

$3°$. It is preferable to choose test problems involving free parameters (which can vary in a wide range) or arbitrary functions.

$4°$. Selecting test problems from a broader class of equations of a similar type is allowed. (There is no need to use exact solutions of the equation in question, which are not always possible to obtain).

$5°$. Using several different test problems is advisable.

$6°$. Numerical methods should first be tested using simple problems with monotonic solutions that have small gradients of the unknowns.

$7°$. Numerical methods should be tested using problems with large gradients of the unknowns in the initial data or boundary conditions (e.g., with rapidly oscillating initial data).

8°. Testing numerical methods should be done using rapidly growing solutions for sufficiently large times.

9°. If possible, the accuracy of the numerical methods should be tested near critical values of the parameters and independent variables that define singular points of the equation, unstable solutions, or solutions with large gradients.

Importantly, well-chosen test problems allow one to compare and improve 'workable' numerical methods and weed out those of little use.

Remark 5.19. *For nonlinear delay PDEs, one should not be limited to test problems derived from exact solutions to simpler nonlinear PDEs without delay.*

5.3.3. Constructing Test Problems

Examples of exact solutions to delay PDEs that can be employed to formulate test problems. Test problems should be constructed using available exact solutions to nonlinear delay PDEs. An extensive list of such solutions can be found in Chapters 3 and 4 of this book.

Consider the nonlinear reaction-diffusion equation with delay

$$u_t = au_{xx} + bu[1 - s(u - kw)], \quad w = u(x, t - \tau), \tag{5.3.3.1}$$

which involves five parameters $a > 0$, b, k, s, and $\tau > 0$ and is a special case of equation (3.4.2.43) with $g(z) = h(z) \equiv 0$ and $f(z) = b(1 - sz)$. We choose the values of the parameters such that the equation has the stationary solutions $u_0 = 0$ and $u_0 = 1$. The trivial stationary solution $u_0 = 0$ is already present. To obtain the second stationary solution, we substitute $u = 1$ into (5.3.3.1) and rearrange the terms to find that $s = 1/(1 - k)$. As a result, we arrive at the equation

$$u_t = au_{xx} + bu\left(1 - \frac{u - kw}{1 - k}\right), \quad w = u(x, t - \tau), \tag{5.3.3.2}$$

which can be rewritten in the alternative form

$$u_t = au_{xx} + bu[1 - (\sigma_1 u + \sigma_2 w)], \quad \sigma_1 + \sigma_2 = 1,$$

where $\sigma_1 = 1/(1 - k)$.

For $k \to 0$, equation (5.3.3.2) becomes the Fisher equation and for $k \to \pm\infty$, it becomes the diffusive logistic equation with delay. Below are two groups of the simplest exact solutions to equation (5.3.3.2).

(i) Solutions for $k > 0$ ($k \neq 1$):

$$\begin{aligned}
u &= e^{ct}[A\cos(\gamma x) + B\sin(\gamma x)], & \gamma &= \sqrt{(b-c)/a} & &\text{if } b > c; \\
u &= e^{ct}(Ae^{-\gamma x} + Be^{\gamma x}), & \gamma &= \sqrt{(c-b)/a} & &\text{if } b < c; \\
u &= e^{ct}(Ax + B), & & & &\text{if } b = c; \qquad (5.3.3.3)\\
u &= 1 + e^{ct}[A\cos(\gamma x) + B\sin(\gamma x)], & \gamma &= \sqrt{-c/a} & &\text{if } c < 0; \\
u &= 1 + e^{ct}(Ae^{-\gamma x} + Be^{\gamma x}), & \gamma &= \sqrt{c/a} & &\text{if } c > 0,
\end{aligned}$$

where $c = (\ln k)/\tau$; A and B are arbitrary constants.

5.3. Construction, Selection, and Usage of Test Problems for Delay PDEs 311

(ii) Solutions for $k < 0$:

$$u = A_n e^{ct \mp \lambda_n x} \cos(\beta_n t \mp \gamma_n x + C_n), \quad \beta_n = \frac{\pi(2n-1)}{\tau},$$

$$\gamma_n = \frac{\beta_n}{2a\lambda_n}, \quad \lambda_n = \left(\frac{\sqrt{(b-c)^2 + \beta_n^2} - b + c}{2a}\right)^{1/2};$$

$$u = 1 + A_n e^{ct \mp \lambda_n x} \cos(\beta_n t \mp \gamma_n x + C_n), \quad \beta_n = \frac{\pi(2n-1)}{\tau}, \quad (5.3.3.4)$$

$$\gamma_n = \frac{\beta_n}{2a\lambda_n}, \quad \lambda_n = \left(\frac{\sqrt{c^2 + \beta_n^2} + c}{2a}\right)^{1/2},$$

where $c = (\ln|k|)/\tau$; A_n and C_n are arbitrary constants; $n = 1, 2, \dots$

To test methods for numerical integration of nonlinear delay reaction-diffusion equations, one can choose any of the above exact solutions to equation (5.3.3.2). Below we note some qualitative features of these solutions to facilitate the comparison with numerical results.

The first and fourth solutions from group (i) are periodic functions in the space variable x. They are suitable as test solutions for initial-boundary value problems with boundary conditions of the first or second kind on the interval $0 \le x \le m\pi/\gamma$ ($m = 1, 2, \dots$). An appropriate choice of the free constants A and B can make the unknown function equal to zero or one at the boundary (for boundary conditions of the first kind) or make the derivative with respect to x vanish at the boundary (for boundary conditions of the second kind). For problems with mixed boundary conditions, it is convenient to consider these solutions on the intervals $0 \le x \le \frac{1}{2}m\pi/\gamma$ ($m = 1, 2, \dots$). The initial data at $-\tau \le t \le 0$ (or $0 \le t \le \tau$) derive from the solutions used as test problems. It is reasonable to compare numerical and exact solutions of the test problem for k close to 1 (when the solutions change little over time) and for sufficiently large k (when the solutions change rapidly).

Solutions from group (ii) for $k = -1$ are periodic in time and rapidly oscillating in both variables for $\tau \to 0$. Such solutions are useful for assessing the accuracy of numerical methods in problems with large gradients.

One formulates test problems as follows: the equation and its exact solution are supplemented with the initial data at $-\tau \le t \le 0$ and boundary conditions at $x = 0$ and $x = L$ obtained from the exact solution. Below we give a few test problems formulated in this way.

Test problems for delay reaction-diffusion equations. Below we use known exact solutions to formulate several reaction-diffusion type model problems with delay. These problems can serve to determine the scope and assess the accuracy of numerical methods. All test problems involve a few free parameters.

Test problem 1. In the first formula of (5.3.3.3), we set

$$A = 1, \quad B = 2, \quad b = (\ln k)/\tau + a\pi^2/4, \quad c = (\ln k)/\tau, \quad k > 0. \quad (5.3.3.5)$$

This results in the following exact solution to equation (5.3.3.2):

$$u = U_1(x, t) \equiv e^{ct}[\cos(\pi x/2) + 2\sin(\pi x/2)], \quad c = (\ln k)/\tau. \quad (5.3.3.6)$$

Substituting $-\tau \leq t \leq 0$ and then $x = 0$ and $x = 1$, we get the initial condition

$$u(x,t) = e^{ct}[\cos(\pi x/2) + 2\sin(\pi x/2)], \quad -\tau \leq t \leq 0, \tag{5.3.3.7}$$

and boundary conditions

$$u(0,t) = e^{ct}, \quad t > 0; \quad u(1,t) = 2e^{ct}, \quad t > 0. \tag{5.3.3.8}$$

As a result, we obtain a test problem described by equation (5.3.3.2) with parameter b from (5.3.3.5), initial condition (5.3.3.7) and boundary conditions (5.3.3.8). The exact solution to this problem is defined by formula (5.3.3.6), where $0 \leq x \leq 1, t > 0$.

Remark 5.20. *Other test problems can be obtained likewise by using exact solutions (5.3.3.3) and (5.3.3.4) to the reaction-diffusion equation (5.3.3.2).*

Test problem 2. A direct verification shows that equation (5.3.3.1) with

$$b = -4a, \quad k = e^{5a\tau}, \quad s = \frac{3}{2(1-k)} \tag{5.3.3.9}$$

admits the exact solution

$$u = U_2(x,t) \equiv \cosh^{-2}(x) + e^{ct}\cosh^3(x), \quad c = (\ln k)/\tau. \tag{5.3.3.10}$$

By setting $-\tau \leq t \leq 0$ and then $x = 0$ and $x = 1$ in (5.3.3.10), we obtain the initial condition

$$u(x,t) = \cosh^{-2}(x) + e^{ct}\cosh^3(x), \quad -\tau \leq t \leq 0, \tag{5.3.3.11}$$

and boundary conditions

$$u(0,t) = 1 + e^{ct}, \quad t > 0; \quad u(1,t) = \cosh^{-2}(1) + e^{ct}\cosh^3(1), \quad t > 0. \tag{5.3.3.12}$$

This results in a test problem described by equation (5.3.3.1) with parameters (5.3.3.9), initial condition (5.3.3.11), and boundary conditions (5.3.3.12). The exact solution to this problem is given by formula (5.3.3.10), where $0 \leq x \leq 1$ and $t > 0$.

Below we consider the nonlinear five-parameter reaction-diffusion equation with delay

$$u_t = au_{xx} + bu - s(u - kw)^2, \quad w = u(x, t - \tau), \tag{5.3.3.13}$$

which is a special case of equation (3.4.2.31) with $f(z) = -sz^2$. In the degenerate cases $k = 0$ and $\tau = 0$, it becomes the non-normalized Fisher equation.

Omitting details, we will only present the statements of two test problems and their exact solutions.

Test problem 3. We set

$$k > 0, \quad k \neq 1, \quad b = (\ln k)/\tau - a, \quad s = b/(1-k)^2. \tag{5.3.3.14}$$

Then the test problem described by equation (5.3.3.13)–(5.3.3.14), the initial condition

$$u(x,t) = U_3(x,t) \equiv 1 + \frac{e^{ct+1}}{e^2 - 1}(e^x - e^{-x}), \quad c = \frac{\ln k}{\tau}, \quad -\tau \leq t \leq 0, \tag{5.3.3.15}$$

and boundary conditions
$$u(0,t) = 1, \quad t > 0; \quad u(1,t) = 1 + e^{ct}, \quad t > 0, \qquad (5.3.3.16)$$

has the exact solution $u = U_3(x,t)$ in the region $0 \le x \le 1$, $t > 0$.

The solution $u = U_3(x,t)$ has been obtained using formula (3.4.2.33), where $\psi \equiv 1$ and φ is the corresponding solution to the linear ODE (3.4.2.34).

Test problem 4. Suppose
$$k > 0, \quad k \ne 1, \quad b = 4a\pi^2 + (\ln k)/\tau - 1/(a\tau^2), \quad s = b/(1-k)^2. \quad (5.3.3.17)$$

Then the test problem described by equation (5.3.3.13) with (5.3.3.17), the initial condition
$$\begin{gathered} u(x,t) = U_4(x,t) \equiv 1 + e^{ct-\lambda x}\cos(\beta t - 2\pi x), \quad -\tau \le t \le 0, \\ c = (\ln k)/\tau, \quad \lambda = 1/(a\tau), \quad \beta = 4\pi/\tau, \end{gathered} \qquad (5.3.3.18)$$

and boundary conditions
$$u(0,t) = 1 + e^{ct}\cos(\beta t), \quad t > 0; \quad u(1,t) = 1 + e^{ct-\lambda}\cos(\beta t), \quad t > 0, \quad (5.3.3.19)$$

has the exact solution $u = U_4(x,t)$ in the region $0 \le x \le 1$, $t > 0$

The solution $u = U_4(x,t)$ has been obtained using formula (3.4.2.36), where $u_0(x,t) \equiv 1$. The function $V_1(x,t;b-c)$ is defined by formulas (3.4.2.10)–(3.4.2.11) in which $A_2 = 1$ and the other constants A_n, B_n, C_n, and D_n are all equal to zero.

Test problems for delay Klein–Gordon type wave equations. We will take advantage of the exact solutions from [429] and state a few model Klein–Gordon type wave problems with delay that can be used to estimate the accuracy of numerical methods. All test problems involve several free parameters.

Test problem 5. It is not difficult to verify that the nonlinear delay Klein–Gordon type wave equation
$$u_{tt} = au_{xx} + u(u - kw), \quad w = u(x, t - \tau), \qquad (5.3.3.20)$$

with $k > 0$ admits a simple exponential exact solution [429]:
$$u = U_5(x,t) \equiv \exp(ct + cx/\sqrt{a}), \quad c = (\ln k)/\tau. \qquad (5.3.3.21)$$

This solution satisfies the initial data
$$u(x,t) = \exp(ct + cx/\sqrt{a}), \quad u_t(x,t) = c\exp(ct + cx/\sqrt{a}), \quad -\tau \le t \le 0, \quad (5.3.3.22)$$

and boundary conditions
$$u(0,t) = \exp(ct), \quad u(1,t) = \exp(ct + c/\sqrt{a}), \quad t > 0. \qquad (5.3.3.23)$$

Thus, we have a test problem described by equation (5.3.3.20) with $0 \le x \le 1$ and $t > 0$, the initial conditions (5.3.3.22), and boundary conditions (5.3.3.23). The exact solution of this problem is given by formula (5.3.3.21).

Test problem 6. The nonlinear delay Klein–Gordon type wave equation (5.3.3.20) with $k = 1$ also admits a trigonometric periodic solution of the form

$$u(x,t) = U_6(x,t) \equiv \sin(\beta x/\sqrt{a})\cos(\beta t), \quad \beta = 2\pi/\tau, \qquad (5.3.3.24)$$

that satisfies the initial data

$$\begin{aligned} u(x,t) &= \sin(\beta x/\sqrt{a})\cos(\beta t), \\ u_t(x,t) &= -\beta \sin(\beta x/\sqrt{a})\sin(\beta t), \quad -\tau \leq t \leq 0, \end{aligned} \qquad (5.3.3.25)$$

and boundary conditions

$$u(0,t) = 0, \quad u(1,t) = \sin(\beta/\sqrt{a})\cos(\beta t), \quad t > 0. \qquad (5.3.3.26)$$

So we have a test problem described by equation (5.3.3.20) with $k = 1$ in the region $0 \leq x \leq 1$, $t > 0$ and subjected to the initial conditions (5.3.3.25) and boundary conditions (5.3.3.26). The exact solution of this problem is defined by formula (5.3.3.24).

Test problem 7. Consider another nonlinear Klein–Gordon type wave equation with delay

$$u_{tt} = au_{xx} + bu - s(u - kw)^2, \quad w = u(x, t - \tau), \qquad (5.3.3.27)$$

which depends on five parameters $a > 0$, b, k, s, and $\tau > 0$ and is a special case of equations 5 from Table 2 of the article [429] with $f(z) = -sz^2$.

We assume that

$$k > 0, \quad k \neq 1, \quad b = (\ln k)^2/\tau^2 - a, \quad s = b/(1-k)^2. \qquad (5.3.3.28)$$

Then the test problem described by equation (5.3.3.27)–(5.3.3.28), the initial conditions

$$\begin{aligned} u(x,t) &= U_7(x,t) \equiv 1 + \frac{e^{ct+1}}{e^2 - 1}(e^x - e^{-x}), \quad u_t(x,t) = \frac{\partial}{\partial t}U_7(x,t), \\ c &= \frac{\ln k}{\tau}, \quad -\tau \leq t \leq 0, \end{aligned} \qquad (5.3.3.29)$$

and boundary conditions

$$u(0,t) = 1, \quad t > 0; \quad u(1,t) = 1 + e^{ct}, \quad t > 0, \qquad (5.3.3.30)$$

has the exact solution $u = U_7(x,t)$ in the region $0 \leq x \leq 1, t > 0$.

The solution $u = U_7(x,t)$ has been obtained from the respective formula specified in the right column for equation 5 in Table 2 of the article [429].

Test problem 8. Now we will look at the nonlinear five-parameter delay Klein–Gordon type wave equation

$$u_{tt} = au_{xx} + bu[1 - s(u - kw)], \quad w = u(x, t - \tau), \qquad (5.3.3.31)$$

which is a special case of equations 2 from Table 3 of the article [429] with $f(z) = b(1 - sz)$ and $g(z) = h(z) \equiv 0$.

We assume that
$$k > 0, \quad b = (\ln k)^2/\tau^2 + a\pi^2/4. \tag{5.3.3.32}$$

Using the results of [429], we find that the test problem described by equation (5.3.3.31)–(5.3.3.32) and subjected to the initial data

$$u(x,t) = U_8(x,t) \equiv e^{ct}[\cos(\pi x/2) + 2\sin(\pi x/2)], \quad u_t(x,t) = \frac{\partial}{\partial t} U_8(x,t),$$
$$c = (\ln k)/\tau, \quad -\tau \leq t \leq 0, \tag{5.3.3.33}$$

and boundary conditions

$$u(0,t) = e^{ct}, \quad t > 0; \quad u(1,t) = 2e^{ct}, \quad t > 0, \tag{5.3.3.34}$$

has the exact solution $u = U_8(x,t)$ in the region $0 \leq x \leq 1, t > 0$.

A direct method for constructing test problems using related classes of delay PDEs. To obtain test problems for a given class of nonlinear PDEs with delay (or without delay), one can employ exact solutions from a wider class of related equations. We will illustrate this with a concrete example.

We will take the class of delay reaction-diffusion equations

$$u_t = au_{xx} + F(u,w), \quad w = u(x, t - \tau), \tag{5.3.3.35}$$

as the original class of equations. Instead of (5.3.3.35), we will look at the wider class of equations

$$u_t = au_{xx} + F(u,w) + G(x,t), \quad w = u(x, t - \tau), \tag{5.3.3.36}$$

which becomes (5.3.3.35) in the special case $G \equiv 0$. We choose a sufficiently arbitrary function $\eta = \eta(x,t)$ that satisfies prescribed boundary conditions (the initial condition is determined based on the given function η). This function will be an exact solution to equation (5.3.3.36) if

$$G(x,t) = \eta_t - a\eta_{xx} - F(\eta, \bar{\eta}), \quad \bar{\eta} = \eta(x, t - \tau). \tag{5.3.3.37}$$

Equation (5.3.3.36)–(5.3.3.37) in conjunction with appropriate initial and boundary conditions is a test problem with solution $u = \eta(x,t)$. This solution should be compared with a numerical solution to the test problem. Different functions $\eta = \eta(x,t)$ generate different equations (5.3.3.36) and different test problems.

Outlined in a broad form, the suggested simple method for constructing test problems can generally have a serious flaw. The function $\eta = \eta(x,t)$ is defined a priori and is not related to the equation in question. Therefore, it does not have the specific features inherent in exact solutions to delay PDEs. The appropriate choice of these functions depends entirely on the researcher's luck, intuition, and experience.

Remark 5.21. *With the direct method, it is perfectly acceptable to use the exact solutions to nonlinear delay reaction-diffusion and wave equations described in Chapters 3 and 4, as the function η.*

5.3.4. Comparison of Numerical and Exact Solutions to Nonlinear Delay Reaction-Diffusion Equations

The numerical solutions to all test problems discussed in the present subsection were obtained with the Mathematica software package using the method of lines in combination with the second-order Runge–Kutta method or the Gear method. The computations were carried out on the interval $0 \leq t \leq T = 50\,\tau$ for three delay times $\tau = 0.05$, $\tau = 0.1$, and $\tau = 0.5$ and, sometimes, $\tau = 1$ and $\tau = 5$. For some test problems, the integration interval was too large to construct a numerical solution: the computations were interrupted with an error and an indication of the interruption time. However, in most cases, an adequate numerical solution could be obtained by appropriately reducing the time interval.

We will understand the absolute and relative errors of a numerical solution $u_{n,k} = u_h(x_n, t_k)$ to a test problem for a delay PDE in the sense of the following quantities:

$$\sigma_{\mathrm{a}} = \max_{n,k} |u_{\mathrm{e}} - u_{n,k}|, \qquad \sigma_{\mathrm{r}} = \max_{n,k} |(u_{\mathrm{e}} - u_{n,k})/u_{\mathrm{e}}|,$$

where $u_{\mathrm{e}} = u_{\mathrm{e}}(x_n, t_k)$ is the value of the exact solution to the test problem on the temporal layer t_k at the node x_n.

Comparison of exact and numerical solutions to test problems. The preceding subsection formulated four test problems for nonlinear delay reaction-diffusion equations (Fisher-type delay equations). The current subsection presents the results of numerical integration of the test problems and compares the numerical solutions with exact solutions to these problems. The numbers and statements of the test problems discussed below coincide with those of Subsection 5.3.3.

Test problem 1. The exact solution $u = U_1(x, t)$ of test problem 1 from Subsection 5.3.3 with $a = 1$, $k = 0.5$, and $s = 0.2$ monotonically decays in time. The relative error of the numerical solution becomes noticeable at only sufficiently large times when the solution practically vanishes. For the numerical methods employed, the time interval at which the methods show a small relative error increases as the delay time τ increases (from 0.05 to 5). For $N = 100$, the numerical solution obtained with the second-order Runge–Kutta method starts to diverge from the exact solution upon reaching absolute values of the order of 10^{-5} at $\tau = 0.05$ and 10^{-20} at $\tau = 5$. The numerical solution obtained using the Gear method begins to deviate from the exact solution upon reaching values of the order of 10^{-6} at $\tau = 0.05$ and 10^{-10} at $\tau = 5$.

Figure 5.5 displays three numerical solutions (with a logarithmic scale in the vertical axis) obtained using the second-order Runge–Kutta method for test problem 1 where an approximating system of delay ODEs with $N = 100$ was solved. Also included are the graphs of the respective exact solutions. The values of the parameters were $a = 1$, $k = 0.5$, and $s = 0.2$ and the delay times were $\tau = 0.05$ and $\tau = 0.5$. The respective graphs for the results obtained with the Gear method look very similar and are not shown here.

Table 5.3 shows absolute errors of the numerical solutions obtained using the second-order Runge–Kutta method and Gear method for five different delay times.

Test problem 2. The exact solution $u = U_2(x, t)$ of test problem 2 from Subsection 5.3.3 with $a = 1$ increases exponentially in time. When the solution reaches

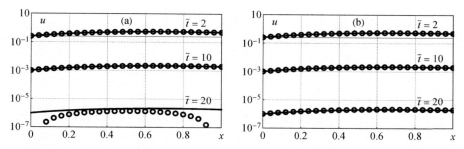

Figure 5.5. Exact solutions (solid lines) and respective numerical solutions (open circles) obtained using a combination of the method of lines and the second-order Runge–Kutta method for test problem 1 with $a = 1$, $k = 0.5$, $s = 0.2$, and $N = 100$ at different times $\bar{t} = t/\tau$. Two delay times were used: (a) $\tau = 0.05$ and (b) $\tau = 0.5$.

Table 5.3. Absolute errors of the numerical solutions to test problem 1 with $a = 1$, $k = 0.5$, and $s = 0.2$ for five different delay times τ on the interval $0 \leq t \leq T = 50\,\tau$.

Method	N	$\tau = 0.05$	$\tau = 0.1$	$\tau = 0.5$	$\tau = 1$	$\tau = 5$
Second-order Runge–Kutta	10	2.8×10^{-4}	4.6×10^{-4}	9.9×10^{-4}	1.2×10^{-3}	1.4×10^{-3}
	50	1.3×10^{-5}	2.0×10^{-5}	4.1×10^{-5}	4.8×10^{-5}	5.6×10^{-5}
	100	4.5×10^{-6}	6.1×10^{-6}	1.2×10^{-5}	1.3×10^{-5}	1.4×10^{-5}
Gear	10	2.8×10^{-4}	4.6×10^{-4}	9.9×10^{-4}	1.2×10^{-3}	1.4×10^{-3}
	50	1.3×10^{-5}	1.9×10^{-5}	4.0×10^{-5}	4.7×10^{-5}	5.6×10^{-5}
	100	2.8×10^{-6}	4.7×10^{-6}	9.9×10^{-6}	1.2×10^{-5}	1.4×10^{-5}

a large value, the program is interrupted by an error. For the second-order Runge–Kutta method, this value is of the order of 10^8. For $\tau = 0.05$, we managed to obtain a solution on the entire interval up to $T = 50\,\tau$. Also, we obtained a solution up to $T = 35\,\tau$ for $\tau = 0.1$ and up to $T = 7.3\,\tau$ for $\tau = 0.5$. The Gear method works better, up to values of the order of 10^{13}. We obtained a solution on the entire interval up to $T = 50\,\tau$ for $\tau = 0.05$ and $\tau = 0.1$ and up to $T = 11.5\,\tau$ for $\tau = 0.5$. It is noteworthy that the Gear method builds solutions in only a few seconds, while the second-order Runge–Kutta method takes a few minutes to tens of minutes to obtain a solution. See Remark 5.10 for a possible running time reduction.

Figure 5.6 depicts a few numerical solutions (with a logarithmic scale in the vertical axis) obtained using the Gear method for test problem 2 with $a = 1$ where a system of ODEs with $N = 100$ was solved. Also included are the graphs of the respective exact solutions. Two delay times were used: $\tau = 0.05$ and $\tau = 0.5$. The respective graphs of the solutions obtained with the Runge–Kutta method are qualitatively very similar and are not shown here.

Table 5.4 shows the relative errors of the numerical solutions obtained using the second-order Runge–Kutta method and Gear method.

Test problem 3. Figure 5.7 displays the exact solution and respective numerical solution obtained for three different values of X using the second-order Runge–Kutta

318   5. NUMERICAL METHODS FOR SOLVING DELAY DIFFERENTIAL EQUATIONS

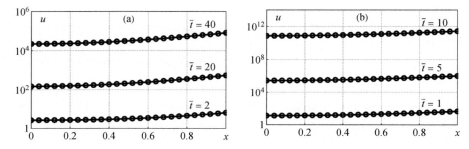

Figure 5.6. Exact solutions (solid lines) and respective numerical solutions (open circles) obtained with a combination of the method of lines and Gear method for test problem 2 with $a = 1$ and $N = 100$ at different times $\bar{t} = t/\tau$. Two delay times were used: (a) $\tau = 0.05$ and (b) $\tau = 0.5$.

Table 5.4. Relative errors of the numerical solutions for problem 2 with $a = 1$ and three different delay times τ.

Method	N	$\tau = 0.05, T = 50\tau$	$\tau = 0.1, T = 35\tau$	$\tau = 0.5, T = 7.3\tau$
Second-order Runge–Kutta	10 50 100	1.8×10^{-3} 7.5×10^{-5} 2.0×10^{-5}	1.9×10^{-3} 7.8×10^{-5} 2.0×10^{-5}	2.5×10^{-3} 1.0×10^{-4} 2.6×10^{-5}
Gear	10 50 100	1.8×10^{-3} 7.3×10^{-5} 1.8×10^{-5}	1.9×10^{-3} 7.6×10^{-5} 1.9×10^{-5}	2.5×10^{-3} 1.0×10^{-4} 2.5×10^{-5}

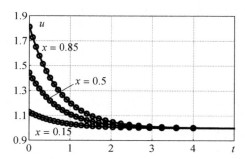

Figure 5.7. Exact solutions (solid lines) and respective numerical solutions (open circles) obtained using a combination of the method of lines and second-order Runge–Kutta method for test problem 3 with $a = 1$, $k = 0.5$, $\tau = 0.5$, and $N = 100$ at different values of x.

method for test problem 3 from Subsection 5.3.3. A system of $N = 100$ ODEs was solved with $a = 1$ and $k = 0.5$ and a moderate delay time $\tau = 0.5$. The graphs of the numerical solutions obtained using the Gear method look very similar and are not included here. Both methods demonstrate good approximation of the exact solution on the entire interval $0 \le t \le T = 50\tau$.

5.3. Construction, Selection, and Usage of Test Problems for Delay PDEs

Table 5.5. Relative errors of the numerical solutions to problem 3 with $a = 1$ and $k = 0.5$ at three different delay times τ on the interval $0 \leq t \leq T = 50\,\tau$.

Method	N	$\tau = 0.5$	$\tau = 1$	$\tau = 5$
Second-order Runge–Kutta	10	7.1×10^{-5}	5.4×10^{-5}	4.8×10^{-5}
	50	2.0×10^{-6}	1.9×10^{-6}	1.9×10^{-6}
	100	2.5×10^{-7}	3.7×10^{-7}	5.1×10^{-7}
Gear	10	7.2×10^{-5}	5.5×10^{-5}	4.8×10^{-5}
	50	2.8×10^{-6}	2.2×10^{-6}	1.9×10^{-6}
	100	6.3×10^{-7}	5.1×10^{-7}	5.1×10^{-7}

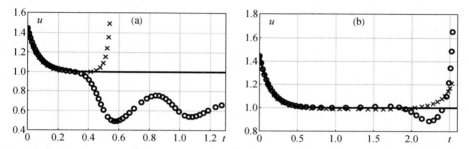

Figure 5.8. Exact solutions (solid lines) and respective numerical solutions (open circles, second-order Runge–Kutta method, and crosses, Gear method) of test problem 3 with $a = 1$ and $k = 0.5$ at the point $x = 0.5$ for $N = 100$ and two delay times: (a) $\tau = 0.05$ and (b) $\tau = 0.1$.

Table 5.5 displays the relative errors of the numerical solutions to problem 3 obtained by the second-order Runge–Kutta and Gear methods at moderate and large delay times on the interval $0 \leq t \leq T = 50\,\tau$.

For $a = 1$ and $k = 0.5$ and sufficiently small delay times $\tau = 0.05$ and $\tau = 0.1$, the numerical solution to test problem 3 starts to deviate significantly from the exact solution in the region where the solution enters the stationary mode. After that, the program is interrupted by an error. This is because the parameter b appearing in the test problem equation and defined in (5.3.3.14) increases unboundedly as $O(\tau^{-1})$ for $\tau \to 0$. As a result, the stationary solution $u = 1$, approached by the problem's solution as $t \to \infty$, becomes unstable for small τ. This fact is proved below in the linear approximation.

To illustrate the above situation, we consider the graphs of the numerical and exact solutions versus time for fixed $x = 0.5$ (Figure 5.8) with $a = 1$ and $k = 0.5$ and small delay times $\tau = 0.05$ and $\tau = 0.1$. The choice of the midpoint, $x = 0.5$, from the range of the spatial variable is because it is the point of maximum deviation between the numerical and exact solutions. For $\tau = 0.05$, both numerical methods only work adequately on a short initial interval (before their solutions start to turn away from the asymptote). Then the second-order Runge–Kutta method produces a nonmonotonic descending oscillating curve that has nothing to do with the exact solution. The Gear method produces a curve that rises sharply and deviates strongly from the exact solution. For $\tau = 0.1$, the Runge–Kutta method and Gear methods provide a fairly accurate approximation of the desired solution over a substantial time interval (staying on the asymptote for quite a while). The Gear method has a slightly larger range of applicability in t. The solution error of the Gear method is monotonic over time, while that of the Runge–Kutta method is nonmonotonic. In both cases, the errors increase sharply soon after the steady state establishes. Below we will show that the values $\tau = 0.05$ and $\tau = 0.1$ lie inside the instability region of the stationary solution.

For test problem 3, Equation (5.3.3.13) becomes

$$u_t = a u_{xx} + bu - \frac{b}{(1-k)^2}(u - kw)^2, \quad b = \frac{\ln k}{\tau} - a. \qquad (5.3.4.1)$$

It admits the stationary solution $u_0 = 1$; the test problem solution asymptotically approaches it as $t \to \infty$. To investigate the linear stability and instability of the solution $u_0 = 1$, we will look at perturbed solutions of the form [410, 498]:

$$u = 1 + \delta e^{-\lambda t} \sin(\pi n x), \quad n = 1, 2, \ldots, \qquad (5.3.4.2)$$

where δ is a small parameter and λ is a spectral parameter to be determined. The perturbed solution (5.3.4.2) equals 1 for any t at the boundaries $x = 0$ and $x = 1$. Substituting (5.3.4.2) into equation (5.3.4.1), dropping the terms of the order of δ^2 and higher, and dividing by $\sin(\pi n x)$, we arrive at the dispersion equation for λ:

$$\lambda - a(\pi n)^2 - \frac{b(1+k)}{1-k} + \frac{2bk}{1-k} e^{\lambda \tau} = 0, \quad b = \frac{\ln k}{\tau} - a. \qquad (5.3.4.3)$$

For $a = n = 1$, $k = 0.5$, and $\tau = 0.05$, the dispersion equation (5.3.4.3) has a negative root $\lambda \approx -27.0213$. It follows that the second term in formula (5.3.4.2) increases exponentially as $t \to \infty$, and hence, the stationary solution of the delay reaction-diffusion equation (5.3.3.13) with $\tau = 0.05$ is unstable in the linear approximation. Up to $\tau_* \approx 0.09153$ (the other parameters remain unchanged), the transcendental equation (5.3.4.3) has one or two real negative roots, while for $\tau > \tau_*$, the equation does not have real negative roots.

For $a = n = 1$, $k = 0.5$, and $\tau = 0.1$, the dispersion equation (5.3.4.3) has a complex root with a negative real part $\operatorname{Re}\lambda = -4.38498$. Consequently, the stationary solution of the delay reaction-diffusion equation (5.3.3.13) with $\tau = 0.1$ is also unstable in the linear approximation.

Remark 5.22. *The value $\tau = 0.5$ lies outside the instability region of the stationary solution to test problem 3. In this case, the dispersion equation (5.3.4.3) has a root with the minimum real part $\operatorname{Re}\lambda = 0.0895394$, and, as noted previously, the methods of numerical integration work well.*

Remark 5.23. *It follows from the above considerations that the methods concerned provide accurate solutions of nonlinear delay reaction-diffusion equations on a certain initial time interval even in the instability region. The length of the interval derives from comparing the numerical solutions obtained by different methods. If the solutions practically coincide over some time interval, so that the error of one numerical solution is acceptably small compared to the others, then the solutions will likely be sufficiently accurate on this time interval.*

5.3. Construction, Selection, and Usage of Test Problems for Delay PDEs

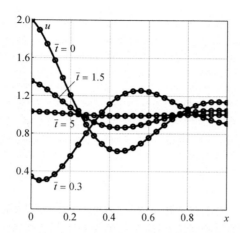

Figure 5.9. Exact solutions (solid lines) and respective numerical solutions (open circles) obtained using a combination of the method of lines and second-order Runge–Kutta method for test problem 4 with $a = 1$ and $k = \tau = 0.5$ at different times $\bar{t} = t/\tau$ for $N = 100$.

Table 5.6. Relative errors of numerical solutions for test problem 4 with $a = 1$ and $k = 0.5$ for three different delay times τ on the interval $0 \leq t \leq T = 50\,\tau$.

N	$\tau = 0.5$	$\tau = 1$	$\tau = 5$
10	7.3×10^{-3}	1.2×10^{-2}	7.4×10^{-2}
50	2.9×10^{-4}	5.1×10^{-4}	2.9×10^{-3}
100	7.3×10^{-5}	1.3×10^{-4}	7.3×10^{-4}

Test problem 4. Figure 5.9 displays the exact solution and respective numerical solution at four different times $\bar{t} = t/\tau$ for test problem 4 with $a = 1$, $k = 0.5$, and $\tau = 0.5$. The numerical solutions were obtained using the second-order Runge–Kutta method by integrating an ODE system with $N = 100$. The graphs of the solutions obtained using the Gear method look very similar and are omitted here.

Table 5.6 displays the relative errors of a few numerical solutions obtained using the second-order Runge–Kutta and Gear methods for moderate and large delay times on the interval $0 \leq t \leq T = 50\,\tau$. Both methods show the same errors coinciding up to two decimal places.

Notably, for delay times of the order of one (e.g., $\tau = 0.5$), the solution oscillations are not high frequency and do not cause problems in the numerical integration of the test problem, unlike the case of small delay times, which is discussed below.

At small delay times $\tau = 0.05$ and $\tau = 0.1$ and with $a = 1$ and $k = 0.5$, both the Runge–Kutta and Gear methods fail to provide a numerical solution to the ODE system with $N = 100$ for test problem 4. This circumstance is primarily due to the fact that the parameter b, which occurs in the test problem equation and is defined in (5.3.3.17), grows indefinitely as $O(\tau^{-2})$ for $\tau \to 0$. Here, the coefficient b grows much faster than in test problem 3. Another complicating factor is the rapid oscillations of the solution in a small neighborhood of the left boundary $x = 0$; however, we have $u \approx 1$ in the rest of the region. Obtaining adequate numerical results in such cases is possible only with small spatial step sizes in the boundary-layer type region, where the solution changes quickly. Using a variable spatial step size, that is, a different number of equations for regions with and without rapid oscillations, in the method of lines is problematic since it is difficult to anticipate in which region high-frequency oscillations arise. However, using small steps in the entire computational region is associated with an excessive increase in the running time of the method.

Remark 5.24. *The method of lines in combination with the Gear method implemented in Mathematica is employed in Subsection 6.3.4 to solve an initial-boundary value problem for a diffusive Lotka–Volterra type system.*

5.3.5. Comparison of Numerical and Exact Solutions to Nonlinear Delay Klein–Gordon Type Wave Equations

Subsection 5.3.3 stated four test problems for nonlinear delay Klein–Gordon type wave equations. The current subsection discusses the numerical results of integrating these problems using a combination of the method of lines and three solution methods for delay ODE systems implemented in the Mathematica package. These are the second-order Runge–Kutta method, fourth-order Runge–Kutta method, and Gear method. We compare the numerical solutions with the exact solutions to the test problems. The numbering and statements of the test problems discussed below coincide with those from Subsection 5.3.3.

Test problem 5. The solution $u = U_5$ of test problem 5 with $a = 1$ and $k = 0.5$ is an exponentially decaying function. Table 5.7 specifies the absolute errors of the numerical solutions obtained using a combination of the method of lines with three methods for solving a delay ODE system with different N and τ on the interval $0 \leq t \leq 50\tau$. It is apparent from Table 5.7 that all methods did well to solve the problem, with the fourth-order Runge–Kutta method having provided a better approximation of the exact solution. As N increases, the absolute errors decrease, and all methods give second-order approximation in space. It is also noteworthy that the absolute errors decrease as the delay time increases.

Figure 5.10 depicts a few exact (solid lines) and numerical (open circles) solutions for test problem 5 with $a = 1$ and $k = 0.5$. The numerical solutions are obtained using the second-order Runge–Kutta method for $\tau = 0.05$, $\tau = 0.5$, and $N = 100$ at three times $\bar{t} \approx 0.1$, $\bar{t} \approx 1$, and $\bar{t} \approx 3$, where $\bar{t} = t/\tau$. The graphs obtained with the other methods look similar and are omitted here.

Test problem 6. The solution $u = U_6$ of test problem 6 from Subsection 5.3.3 with $a = k = 1$ represents a non-decaying oscillatory process with period τ in both variables. Importantly, the solution rapidly oscillates at small τ and is singular with respect to the delay parameter (since $u = U_6$ does not have a limit as $\tau \to 0$). This circumstance restricts the capabilities of the employed numerical methods at small τ since it requires a large number of grid nodes in x, and hence, a large number of equations in the approximating system of delay ODEs. For example, at $\tau = 0.05$, one needs over 1000 nodes to achieve acceptable accuracy, while for

5.3. Construction, Selection, and Usage of Test Problems for Delay PDEs

Table 5.7. Absolute errors of numerical solutions for test problem 5 with $a = 1$ and $k = 0.5$ on the interval $0 \le t \le T = 50\,\tau$.

Method	N	$\tau = 0.05$	$\tau = 0.1$	$\tau = 0.5$	$\tau = 1$
Second-order Runge–Kutta	10	4.0×10^{-2}	1.2×10^{-2}	2.5×10^{-4}	2.7×10^{-5}
	50	2.0×10^{-3}	5.0×10^{-4}	1.0×10^{-5}	1.9×10^{-6}
	100	5.0×10^{-4}	1.2×10^{-4}	2.6×10^{-6}	9.5×10^{-7}
	200	1.2×10^{-4}	3.0×10^{-5}	7.2×10^{-7}	6.0×10^{-7}
Fourth-order Runge–Kutta	10	4.0×10^{-2}	1.2×10^{-2}	2.5×10^{-4}	2.7×10^{-5}
	50	2.0×10^{-3}	5.0×10^{-4}	9.8×10^{-6}	1.1×10^{-6}
	100	5.0×10^{-4}	1.2×10^{-4}	2.5×10^{-6}	2.7×10^{-7}
	200	1.2×10^{-4}	3.0×10^{-5}	6.1×10^{-7}	6.7×10^{-8}
Gear	10	4.0×10^{-2}	1.2×10^{-2}	2.5×10^{-4}	2.7×10^{-5}
	50	2.0×10^{-3}	5.0×10^{-4}	9.8×10^{-6}	1.3×10^{-6}
	100	5.0×10^{-4}	1.2×10^{-4}	2.5×10^{-6}	3.0×10^{-7}
	200	1.2×10^{-4}	3.1×10^{-5}	6.5×10^{-7}	1.3×10^{-7}

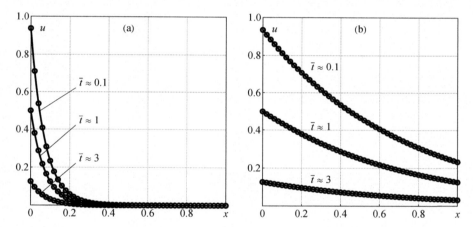

Figure 5.10. Exact solutions (solid lines) and respective numerical solutions (open circles) obtained using a combination of the method of lines and the second-order Runge–Kutta method for test problem 5 with $a = 1$ and $k = 0.5$ at three different times $\bar{t} = t/\tau$, $N = 100$, and two delay times: (a) $\tau = 0.05$ and (b) $\tau = 0.5$.

$\tau = 0.1$ and $N = 1000$, the absolute computational error of the second-order Runge–Kutta method is quite large and equal to 4.1×10^{-2}.

Table 5.8 specifies the absolute errors of numerical solutions to test problem 6 obtained using a combination of the method of lines with the Gear method and second-order Runge–Kutta method on the interval $0 \le t \le 50\,\tau$ for three moderate delay times $\tau = 0.5$, $\tau = 1$, and $\tau = 2$ and different numbers of grid nodes in the space variable ($N = 50$, $N = 100$, and $N = 200$). It is apparent that as τ and the number of equations N increase, the error of the numerical solution decreases. The error also decreases if the temporal interval length T decreases. For example, at $\tau = 0.5$, both methods show an acceptable approximation to the exact solution for $N = 100$ with an absolute error of 0.08 on the interval $0 \le t \le T = 20\,\tau$; in contrast,

Table 5.8. Absolute errors of a few numerical solutions to test problem 6 with $a = k = 1$ and moderate delay times τ on the interval $0 \le t \le T = 50\,\tau$.

Method	N	$\tau = 0.5$	$\tau = 1$	$\tau = 2$
Second-order Runge–Kutta	50	0.79	0.2	4.6×10^{-2}
	100	0.2	4.4×10^{-2}	7.7×10^{-3}
	200	4.3×10^{-2}	5.7×10^{-3}	1.9×10^{-3}
Gear	50	0.8	0.2	5.1×10^{-2}
	100	0.2	5.1×10^{-2}	1.3×10^{-2}
	200	5.1×10^{-2}	1.3×10^{-2}	3.2×10^{-3}

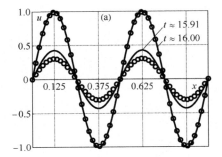 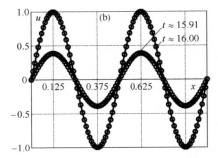

Figure 5.11. Exact solutions (solid lines) and respective numerical solutions (open circles) obtained using a combination of the method of lines and the second-order Runge–Kutta method for test problem 6 with $a = k = 1$ and $\tau = 0.5$ at times $t = 15.91$ and $t = 16.00$ for systems of (a) $N = 100$ and (b) $N = 200$ equations.

for $N = 200$, the absolute error is four times less on the same interval. The errors of the fourth-order Runge–Kutta method coincide with those of the Gear method and, therefore, are omitted from Table 5.8. The oscillations in x have period τ; that is, the oscillation frequency decreases as the delay time increases, and hence, fewer grid nodes are needed to achieve acceptable accuracy. We did not test the methods at moderate or large delay times for large N because this would require a large amount of RAM and so, given the aforesaid, it was unnecessary. Notably, the second-order Runge–Kutta method provides a slightly better approximation to the exact solution.

Figure 5.11 displays two exact solutions (solid lines) and two respective numerical solutions (open circles) at $a = 1$ and $\tau = 0.5$ for $N = 100$ and $N = 200$. The numerical solutions were obtained with the second-order Runge–Kutta method at an intermediate time $t = 15.91$ (chosen to highlight the error of the numerical solution) and the time of maximum amplitude $t = 16.00$. One can see that the error decreases as the number of equations N increases. The graphs of the respective solutions obtained with the Gear method look alike and are omitted here.

Test problem 7. The exact solution $u = U_7$ of test problem 7 from Subsection 5.3.3 with $a = 1$ and $k = 0.5$ is monotonically decaying in both variables. All three methods (second-order Runge–Kutta, fourth-order Runge–Kutta, and Gear) work adequately on the entire computational interval $0 \le t \le T = 50\,\tau$ at all delay times considered. Figure 5.12 displays the graphs of the numerical solutions obtained using

5.3. Construction, Selection, and Usage of Test Problems for Delay PDEs 325

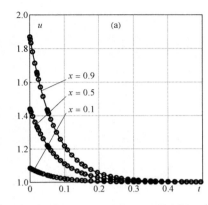
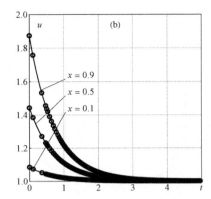

Figure 5.12. Exact solutions (solid lines) and respective numerical solutions (open circles) obtained by a combination of the method of lines and Gear method for test problem 7 with $a = 1$ and $k = 0.5$ at the points $x = 0.1$, $x = 0.5$, and $x = 0.9$ for $N = 100$ and two delay times: (a) $\tau = 0.05$ and (b) $\tau = 0.5$.

Table 5.9. Absolute errors of some numerical solutions to test problem 7 with $a = 1$ and $k = 0.5$ on the interval $0 \leq t \leq T = 50\,\tau$.

Method	N	$\tau = 0.05$	$\tau = 0.1$	$\tau = 0.5$	$\tau = 1$
Second-order Runge–Kutta	10	2.1×10^{-6}	5.3×10^{-6}	4.9×10^{-5}	2.2×10^{-3}
	50	9.4×10^{-7}	1.8×10^{-6}	2.5×10^{-6}	6.6×10^{-5}
	100	1.7×10^{-6}	1.2×10^{-6}	1.0×10^{-6}	1.1×10^{-5}
	200	1.3×10^{-6}	1.2×10^{-6}	6.5×10^{-7}	1.4×10^{-6}
Fourth-order Runge–Kutta	10	1.2×10^{-6}	4.5×10^{-6}	4.8×10^{-5}	2.3×10^{-3}
	50	5.5×10^{-8}	2.8×10^{-7}	1.9×10^{-6}	7.9×10^{-5}
	100	1.5×10^{-8}	5.6×10^{-8}	4.8×10^{-7}	2.0×10^{-5}
	200	3.4×10^{-9}	1.2×10^{-8}	1.2×10^{-7}	4.9×10^{-6}
Gear	10	1.2×10^{-6}	4.5×10^{-6}	4.8×10^{-5}	2.3×10^{-3}
	50	8.4×10^{-8}	3.2×10^{-7}	1.9×10^{-6}	8.0×10^{-5}
	100	5.0×10^{-8}	9.4×10^{-8}	4.7×10^{-7}	2.1×10^{-5}
	200	6.3×10^{-8}	3.6×10^{-8}	1.3×10^{-7}	5.3×10^{-6}

the Gear method for $N = 100$ and delay times $\tau = 0.05$ and $\tau = 0.5$. The graphs of the solutions obtained with the other methods look very similar and are omitted here. Table 5.9 specifies the absolute errors of some numerical solutions.

Test problem 8. The exact solution $u = U_8$ of test problem 8 from Subsection 5.3.3 with $a = 1$, $k = 0.5$, and $s = 0.2$ is monotonically decreasing in time. At moderate delay times, $\tau = 0.5$ and $\tau = 1$, all three methods (second-order Runge–Kutta, fourth-order Runge–Kutta, and Gear) work adequately on the entire computational interval $0 \leq t \leq T = 50\,\tau$.

At small delay times, $\tau = 0.05$ and $\tau = 0.1$, all methods only work adequately on the initial segment $0 \leq t \leq 10\,\tau$ and then, after reaching the asymptote $u = 0$, start to deviate strongly from the exact solution. This circumstance is due to the instability of the stationary solution $u = 0$ at small τ.

Figure 5.13 illustrates the behavior of the methods at the midpoint $x = 0.5$ for $N = 200$. The graphs of the numerical solutions obtained using the fourth-order

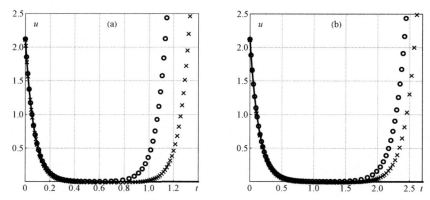

Figure 5.13. Exact solutions (solid lines) and respective numerical solutions obtained with the second-order Runge–Kutta method (open circles) and Gear method (crosses) for test problem 8 with $a = 1$, $k = 0.5$, and $s = 0.2$ at the point $x = 0.5$ for $N = 200$ and two delay times: (a) $\tau = 0.05$ and (b) $\tau = 0.1$.

Table 5.10. Absolute errors of some numerical solutions to test problem 8 with $a = 1$, $k = 0.5$, and $s = 0.2$.

Method	N	$\tau = 0.05, T = 10\tau$	$\tau = 0.1, T = 10\tau$	$\tau = 0.5, T = 50\tau$	$\tau = 1, T = 50\tau$
Second-order Runge–Kutta	10	1.3×10^{-2}	4.1×10^{-2}	2.2×10^{-3}	2.4×10^{-3}
	50	9.4×10^{-4}	1.8×10^{-3}	8.8×10^{-5}	9.6×10^{-5}
	100	5.6×10^{-4}	6.0×10^{-4}	2.3×10^{-5}	2.5×10^{-5}
	200	4.8×10^{-4}	2.9×10^{-4}	6.8×10^{-6}	7.2×10^{-6}
Fourth-order Runge–Kutta	10	1.3×10^{-2}	4.1×10^{-2}	2.2×10^{-3}	2.4×10^{-3}
	50	5.4×10^{-4}	1.6×10^{-3}	8.6×10^{-5}	9.4×10^{-5}
	100	1.4×10^{-4}	4.1×10^{-4}	2.2×10^{-5}	2.4×10^{-5}
	200	3.4×10^{-5}	1.0×10^{-4}	5.4×10^{-6}	5.9×10^{-6}
Gear	10	1.3×10^{-2}	4.1×10^{-2}	2.2×10^{-3}	2.4×10^{-3}
	50	5.4×10^{-4}	1.6×10^{-3}	8.6×10^{-5}	9.4×10^{-5}
	100	1.4×10^{-4}	4.1×10^{-4}	2.2×10^{-5}	2.4×10^{-5}
	200	3.5×10^{-5}	1.0×10^{-4}	5.4×10^{-6}	6.1×10^{-6}

Runge–Kutta method are qualitatively similar to those of the respective solutions obtained with the Gear method and omitted here. Clearly, the Gear method and the fourth-order Runge–Kutta method have a slightly wider range of applicability in t.

Remark 5.25. *Notably, the methods employed provide accurate solutions to nonlinear delay Klein–Gordon type wave equations on some initial time interval, even in the instability region. The length of the interval derives from comparing the numerical solutions obtained by different methods. For example, suppose the solutions practically coincide on a time interval so that the error of one numerical solution is acceptably small compared to the others. In that case, the solutions will likely be accurate on this interval.*

Note that the solution $u = U_8$ to test problem 8 decays very rapidly at small τ. The failure of the numerical solutions at small delay times $\tau = 0.05$ and $\tau = 0.1$ is due to linear instability of the limiting stationary state of the solution ($u \to 0$ as $t \to \infty$).

Table 5.10 shows the absolute errors of some numerical solutions at different delay times and on the relevant integration intervals.

6. Models and Delay Differential Equations Used in Applications

6.1. Models Described by Nonlinear Delay ODEs

6.1.1. Hutchinson's Equation—a Delay Logistic Equation

Preliminary remarks. In the literature, one often encounters nonlinear delay ODEs (and systems of delay equations) that describe a wide variety of processes. Models with delay arise, as a rule, from generalizing simpler models without delay. Below we will illustrate such generalizations by looking at a chain of population dynamics models described by ODEs without delay, from simple to more complex, eventually leading to a complicated model with delay.

Malthusian equation. In 1798, Thomas Malthus was the first to propose a mathematical model to characterize the dynamics of species growth. According to this model, under favorable conditions, any species increases its population by the exponential law $u(t) = u_0 e^{bt}$, and consequently satisfies the linear ODE

$$u'_t = bu. \qquad (6.1.1.1)$$

It was later called the *Malthusian exponential model*. The parameter b is the population growth rate, equal to the difference between the birth and death rates; it is also known as the Malthusian parameter of population growth and Malthusian coefficient of linear growth. Under unfavorable conditions, the constant b can be negative.

Numerous experimental data well support the Malthusian model only if the population size is small, that is, when its size is not limited by anything.

Logistic differential equation. The Malthusian equation implies unrestricted population growth. However, in reality, there are some limitations to the population size (e.g., limited food or territory, unfavorable weather, competition with other species, and so on). To take this into account, in 1835, Adolphe Quetelet and Pierre-François Verhulst proposed a more complicated mathematical model than (6.1.1.1), which is described by the nonlinear *logistic differential equation*

$$u'_t = bu(1 - u/k), \qquad (6.1.1.2)$$

where k is a positive parameter that characterizes the maximum sustainable population size and is referred to as the *carrying capacity* of the habitat. In the limit case $k \to \infty$, equation (6.1.1.2) becomes the Malthusian model (6.1.1.1).

The exact solution of equation (6.1.1.2) with the initial condition $u(t=0) = u_0$ is

$$u(t) = \frac{k}{1 + [(k/u_0) - 1]\exp(-bt)}, \quad t > 0. \qquad (6.1.1.3)$$

Below are the main properties of solution (6.1.1.3).

1°. If $u_0 > 0$, we have $u(t) > 0$ for all $t > 0$.

2°. The following limit relation holds: $\lim_{t \to \infty} = k$.

3°. The function $u(t)$ increases for $0 < u_0 < k$ and decreases for $u_0 > k$.

4°. The equilibrium $u = k$ is globally asymptotically stable.

Hutchinson's equation and its properties. The logistic law (6.1.1.2) describes well the growth dynamics of populations of single-celled organisms, but it does not apply to model the population dynamics of most mammals. This is because populations of such species are subject to intense cyclic fluctuations. To handle this situation, George Evelyn Hutchinson [227] suggested a more sophisticated model based on the nonlinear delay ODE

$$u'_t = bu(1 - w/k), \quad w = u(t - \tau). \tag{6.1.1.4}$$

It describes the dynamics of a population taking into account the period of maturation, when individuals are not capable of reproduction. Equation (6.1.1.4) includes the population density, $u = u(t) \geq 0$, reproductive rate, $b > 0$, saturation value, $k > 0$, and delay time, $\tau > 0$, characterizing the mean reproductive age of the species. The rate of population growth is directly proportional to the population size at the current time and the factor $(k - w)/k$; it is self-regulatory. In the limit case $\tau = 0$, equation (6.1.1.4) becomes the logistic differential equation (6.1.1.2). Introducing a positive τ into the delay equation brings about an oscillatory process. This is because the population growth does not stop immediately upon reaching the saturation level k, as it would do in the absence of delay, but the time τ after. Therefore, the population size exceeds the saturation level and, unable to maintain its maximum, starts to decrease. Having dropped to saturation k, it also does not stop but continues to decrease until it reaches its minimum size, after which growth begins again. Then the cycle repeats.

The change of variable $u = kv$ converts equation (6.1.1.4) to the simpler form

$$v'_t = bv[1 - v(t - \tau)]. \tag{6.1.1.5}$$

Below are a few qualitative and quantitative features of the solution to equation (6.1.1.5); see [241, 274, 283, 567] for details.

1°. The equilibrium $v = 0$ is unstable.

2°. The equilibrium $v = 1$ is asymptotically stable for $0 < b\tau \leq \pi/2$ and unstable for $b\tau > \pi/2$.

3°. For $0 < b\tau \leq \frac{37}{24}$ (the upper estimate can be improved [263]), all solutions to equation (6.1.1.5), except for the zero solution, tend to 1 as $t \to \infty$.

4°. For $b\tau > \pi/2$, there is a nontrivial periodic solution [242, 245, 274]. Let $v_*(t, \lambda)$ denote this solution, where $\lambda = b\tau$ and its period is $T_*(\lambda)$. If $\lambda = b\tau \gg 1$, the following asymptotic formulas hold:

$$T_*(\lambda) = \frac{e^\lambda + 1}{\lambda} + O\left(\frac{e^{-\lambda}}{\lambda}\right),$$

$$\max_{0 \leq t \leq T_*} v_*(t, \lambda) = e^{\lambda-1} + (2e)^{-1} + O(e^{-\lambda}),$$

$$\min_{0 \leq t \leq T_*} v_*(t, \lambda) = \exp\left[-e^\lambda + 2\lambda - 1 + \frac{1 + (1+\lambda)\ln\lambda}{\lambda} + O\left(\frac{\ln^2 \lambda}{\lambda^2}\right)\right].$$

Remark 6.1. *The stability and instability conditions for generalized Hutchinson equations are discussed in Subsection 1.3.4 (see examples 1.16 and 1.17).*

Consider the Cauchy problem for equation (6.1.1.5) with $b = 1$ and the initial condition

$$v(t) = 0.5, \quad -\tau \leq t \leq 0. \tag{6.1.1.6}$$

Figure 6.1 shows in solid lines numerical solutions of this problem for two different delay times, (a) $\tau = 0.5$ and (b) $\tau = 2$, which correspond to a stable and unstable equilibrium $v = 1$. The dashed line indicates the solution without delay, at $\tau = 0$. All solutions were obtained using the implicit second-order Runge–Kutta method with the Mathematica software package (see Subsections 5.1.4 and 5.1.7). It is apparent that the solution to the problem without delay rises monotonically and rapidly approaches the equilibrium state. The solution at $\tau = 0.5$ oscillates about the solution without delay and approaches it rapidly with t, which indicates a stable equilibrium. The solution at $\tau = 2$ also oscillates about the solution without delay, but its amplitude increases with t, which indicates an unstable mode.

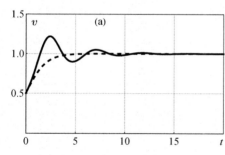 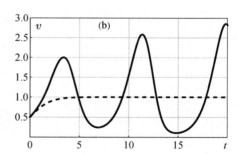

Figure 6.1. Numerical solutions to problem (6.1.1.5), (6.1.1.6) with $b = 1$ and two delay times: (a) $\tau = 0.5$ and (b) $\tau = 2$; the dashed line indicates the solution at $\tau = 0$.

The logistic model with limited food supply. It is easily seen from (6.1.1.2) that the average growth rate of the population, u'_t/u, is a linear function of the density. However, experiments on bacteria cultures did not confirm this, which resulted in the need to improve the logistic model (6.1.1.2). Consequently, a more sophisticated, food-limited model was suggested in [481], which is described by the ODE

$$u_t = bu\frac{k-u}{k+cu}, \tag{6.1.1.7}$$

where $u = u(t) \geq 0$ is the population density, $b > 0$ is the reproductive rate with unlimited food, $k > 0$ is the saturation value (when $u_t = 0$), and τ is the delay time, characterizing the mean reproductive age of the species. Although this model was obtained with no attention to the properties or possible biological interpretations of the new constant $c > 0$, the examination of the model [481] allows one to treat the ratio b/c as the rate of replacement of mass per unit mass in the existing population at saturation. This includes both the replacement of metabolic loss and dead organisms. Model (6.1.1.7) considers that a growing population will use food faster than a saturated population. This is due to the fact that during the growth phase of a population, food is consumed both for maintenance and growth. In contrast, when the population reaches saturation, food is mainly used for maintenance only. Just as in equation (6.1.1.2), the solution of equation (6.1.1.7) tends monotonically to k as $t \to \infty$. However, it was established experimentally (e.g., see [361]) that the population density usually tends to fluctuate about an equilibrium. In cases of convergence, it tends to a positive equilibrium, with such a convergence being rarely monotonic. To incorporate such oscillations in the food-limited population model (6.1.1.7), the study [183] suggested a model described by the following delay ODE:

$$u_t = bu \frac{k-w}{k+cw}, \quad w = u(t-\tau). \tag{6.1.1.8}$$

For properties of equation (6.1.1.8) and its solutions, see, for example, [51, 183, 184, 188, 486, 536].

6.1.2. Nicholson's Equation

Nicholson's blowflies equation is another quite common delay model [198]:

$$u'_t = pwe^{-\kappa w} - \delta u, \quad w = u(t-\tau). \tag{6.1.2.1}$$

It is used to model the behavior of a blowflies population and agrees well with the experiments described in [361]. Here $p > 0$ is the maximum possible per capita egg production rate (corrected for egg-to-adult survival), $1/\kappa > 0$ is the size at which the population reproduces at its maximum rate, $\delta > 0$ is per capita daily adult mortality rate, and $\tau > 0$ is the generation time, or the time taken from birth to maturity.

Equation (6.1.2.1) has two states of equilibrium: trivial $u_0 = 0$ and positive

$$u_* = \frac{1}{\kappa} \ln \frac{p}{\delta}, \tag{6.1.2.2}$$

which exists if $p > \delta$.

We will treat equation (6.1.2.1) in the region $t > 0$ with the initial condition

$$u = \varphi(t) \quad \text{at} \quad -\tau \leq t \leq 0. \tag{6.1.2.3}$$

Below we outline the key results of studying Nicholson's equation (6.1.2.1).

1°. Let $\varphi(t) \geq 0$ in (6.1.2.3). Then the solution to the associated Cauchy problem for equation (6.1.2.1) is nonnegative [485], that is, $u(t) \geq 0$ for $t > 0$.

2°. For given positive initial data, all solutions of equation (6.1.2.1) remain positive for all $t > 0$. Moreover, the following inequality holds [485]:

$$\limsup_{t \to \infty} u(t) \leq \frac{p}{e\delta\kappa}.$$

3°. Suppose that $p \leq \delta$. Then for any solution of equation (6.1.2.1) the relation $u(t) \to 0$ holds as $t \to \infty$ [485]. In other words, the trivial solution $u_0 = 0$ is a *global attractor*, meaning that it is *globally asymptotically stable*, regardless of τ.

4°. Suppose that $p > \delta$. Then there is no nontrivial solution $u(t)$ to equation (6.1.2.1) such that

$$\lim_{t \to \infty} u(t) = 0.$$

Moreover, all solutions to equation (6.1.2.1) are *uniformly stable* [485]; that is, there exists an $\eta > 0$ such that for any trajectory with positive initial values, the relation $\liminf_{t \to \infty} u(t) > \eta$ holds.

Remark 6.2. *The criteria of local asymptotic stability of the nontrivial equilibrium u_* defined by (6.1.2.2) can be obtained by directly computing the roots of the characteristic equation*

$$\lambda + \delta + \delta[\ln(p/\delta) - 1]e^{-\tau\lambda} = 0,$$

which corresponds to the solution of equation (6.1.2.1) linearized about the stationary solution.

5°. Suppose the bilateral inequality

$$1 < p/\delta < e^2$$

holds. Then the nontrivial stationary solution u_* is uniformly stable. If

$$1 < p/\delta < e,$$

the nontrivial stationary solution u_* is uniformly asymptotically stable [259].

6°. Suppose $p/\delta > e^2$. Then the positive equilibrium u_* of equation (6.1.2.1) is locally asymptotically stable if the following inequality holds [180]:

$$\tau < \frac{1}{\delta}\ln\left(\frac{c}{c-1}\right), \quad c = \ln\frac{p}{\delta} - 1.$$

Moreover, there exists a periodic solution (other than constant) if

$$\tau > \frac{\arccos(-1/c)}{\delta\sqrt{c^2 - 1}}, \quad c = \ln\frac{p}{\delta} - 1.$$

7°. The following proposition was proved in [549]. For $p/\delta > e^2$, the solution $u = u_*$ is locally asymptotically stable if $\tau \in (0, \tau_0)$ and unstable if $\tau > \tau_0$, where

$$\tau_0 = \frac{1}{\delta\sqrt{c^2 - 1}} \arcsin\frac{\sqrt{c^2 - 1}}{c}$$

and the constant c is defined in Item 6°.

8°. The condition
$$(e^{\delta\tau} - 1)\ln(p/\delta) < 1$$
guarantees that the positive equilibrium is globally asymptotically stable [485].

9°. The positive solution u_* of equation (6.1.2.1) is globally asymptotically stable [314] if either
$$1 < p/\delta \leq e$$
or
$$\frac{p}{\delta} > e \quad \text{and} \quad e^{-\delta\tau} > c\ln\frac{c^2+c}{c^2+1}, \quad c = \ln\frac{p}{\delta} - 1.$$

10°. The following two statements hold [201, 287]:

a) Suppose
$$p/\delta > e \quad \text{and} \quad \delta\tau e^{\delta\tau}[\ln(p/\delta) - 1] > e^{-1}.$$
Then all solutions of equation (6.1.2.1) oscillate about u_*.

b) Suppose
$$p/\delta > e^2 \quad \text{and} \quad \delta\tau e^{\delta\tau}[\ln(p/\delta) - 1] \leq e^{-1}.$$
Then there exists a solution of equation (6.1.2.1) non-oscillating about u_*.

Definition 1. The function $u(t)$ is said to be *non-oscillating* about a value K if the difference $u(t) - K$ is either positive or negative at sufficiently large t. Otherwise, the function $u(t)$ is said to be *oscillating* about K.

11°. Suppose that $p > \delta$ and $u(t)$ is a positive non-oscillating function about the solution u_* of equation (6.1.2.1). Then $\lim_{t\to\infty} u(t) = u_*$ [485].

Definition 2. A nonzero solution $u(t)$ to equation (6.1.2.1) is said to be *rapidly oscillating* about u_*, if there exist sequences $\{t_n\}$ and $\{t'_n\}$ such that $t_n \to \infty$ and $t'_n \to \infty$ and
$$t_n \neq t'_n, \quad u(t_n) = u(t'_n) = u_*, \quad |t_n - t'_n| \leq \tau, \quad n \geq 1.$$
Otherwise, the solution $u(t)$ is said to be *slowly oscillating* about u_*.

12°. The study [202] obtained some general results stated below.

If $1 < p/\delta < e$, then

a) equation (6.1.2.1) has positive solutions other than u_* that are not oscillating about u_*;

b) equation (6.1.2.1) has infinitely many positive solutions rapidly oscillating about u_*;

c) equation (6.1.2.1) does not have positive solutions slowly oscillating about u_*;

For $p/\delta = e$, all solutions of equation (6.1.2.1) other than u_* are non-oscillating about u_*.

For more details about the properties of Nicholson's equations and related more complicated nonlinear delay ODEs, see the review [53].

Consider the Cauchy problem for equation (6.1.2.1) subjected to the initial condition
$$u(t) = u_0 = 50, \quad -\tau \leq t \leq 0. \tag{6.1.2.4}$$

Following the recommendations of [198] based on an analysis of the experimental data from [361], we set the equation parameters as follows: $p = 10$, $\kappa = 0.1$, and $\tau = 15$. We solve the problem by the Gear method using the Mathematica package (see Subsection 5.1.7).

Figure 6.2 displays the graphs of solutions to problem (6.1.2.1), (6.1.2.4) at different values of the per capita daily mortality rate δ. The graphs are qualitatively different, which agrees with the above stability conditions for the equilibrium state u_*. Figure 6.2a illustrates asymptotic stability of u_*. Near the boundary of the asymptotic stability region, the solution enters a simple oscillatory mode with one local maximum and one local minimum per oscillation period (Figure 6.2b). As we move deeper into the instability region, we first observe a doubling of local maxima (Figure 6.2c) and then a chaotic mode (Figure 6.2d). Similar graphs at different values of the problem parameters were obtained and analyzed in [198].

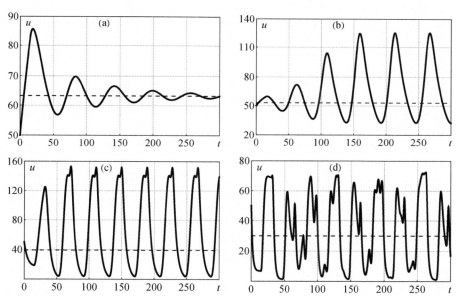

Figure 6.2. Solution $u(t)$ to problem (6.1.2.1), (6.1.2.4) at $\kappa = 0.1$, $p = 10$, $\tau = 15$ and different values of the parameter δ: (a) $\delta = 0.018$, (b) $\delta = 0.05$, (c) $\delta = 0.2$, (d) $\delta = 0.5$. The dashed lines indicate the equilibrium state $u_* = \frac{1}{\kappa} \ln \frac{p}{\delta}$.

6.1.3. Mackey–Glass Hematopoiesis Model

The article [330] describes the dynamics of a homogeneous population of mature circulating blood cells of density $u = u(t)$ using a delay ODE of the form

$$u_t = \beta_0 \frac{\theta^n w}{\theta^n + w^n} - \gamma u, \quad w = u(t - \tau), \qquad (6.1.3.1)$$

where β_0, θ, n and γ are some positive parameters.

The right-hand side of equation (6.1.3.1) with $n > 1$ and fixed u is a function of w with a single maximum. As τ increases, the initially stable equilibrium becomes unstable, and stable periodic solutions arise. As τ increases further, the system dynamics reveals a sequence of bifurcations. A chaotic mode is also observed. Model (6.1.3.1) and related models were investigated in [52, 305, 462, 469].

The study [329] analyzed clinical and laboratory data on periodic hematopoiesis[1] and found that the dynamics of periodic hematopoiesis arises in populations of pluripotent hematopoietic stem cells.[2]

Cells subdivide into those of the proliferative phase[3] (population density $v(t)$, cells/kg) and resting phase G_0 (population density $u(t)$, cells/kg). The cells of these phases differ as follows. In the proliferation stage, cells undergo mitosis[4] after a fixed time τ (days) from the start of the stage. In turn, the resting-phase cells can randomly leave the resting phase to either return to proliferation, at a rate of β (days^{-1}), or be completely removed from the process due to differentiation[5] into various hematopoietic cells (such as erythrocytes, lymphocytes, platelets, etc.), at a rate of δ (days^{-1}). Although proliferating cells can also be irrevocably excluded from any phase of the cell cycle at a rate of γ (days^{-1}), a 'normal' stem cell population is, by definition, characterized by the value $\gamma = 0$. The parameters γ, δ, and τ are constant in time and independent of the population size. Determining the numerical values of these parameters is difficult due to their number and insufficient data on the physiology and pathophysiology of stem cells (some attempts to estimate the parameters were made in [329]).

The rate of transition from the resting to proliferative phase depends on the number of cells in the resting phase, meaning that $\beta = \beta(u)$. When u is small, β reaches a maximum; when u increases, β decreases. The transition rate is expressed as follows:

$$\beta(u) = \frac{\beta_0 \theta^n}{\theta^n + u^n}, \tag{6.1.3.2}$$

where β_0 is the maximum rate of transition of cells from the resting phase G_0 to proliferation (days^{-1}), θ is the cell population density in the G_0 phase at which the transition rate is maximum (cells/kg), and n is a dimensionless number responsible for the sensitivity of the transition rate to the population size u in the resting phase. This choice of the function $\beta(u)$ is substantiated in [329].

The dynamics of the cell population in the resting phase G_0 is described by the delay ODE

$$u_t = -\delta u - \beta(u)u + 2\beta(w)we^{-\gamma\tau}, \quad w = u(t-\tau). \tag{6.1.3.3}$$

---

[1] Hematopoiesis is the process of formation, development and maturation of blood cells.

[2] Pluripotent stem cells can self-renew by dividing and developing to form the early embryo's three main layers of germ cells. Therefore, these are all cells of the adult body but not extra-embryonic tissues such as the placenta. Embryonic stem cells and induced pluripotent stem cells are pluripotent stem cells.

[3] Proliferation is the process of rapid increase of an organism's tissue through cell division.

[4] Mitosis is the process by which a single parent cell divides to make two new daughter cells. Each daughter cell receives a complete set of chromosomes from the parent cell.

[5] Differentiation is the process when young, immature (unspecialized) cells take on individual characteristics and reach their mature (specialized) form and function.

6.1. Models Described by Nonlinear Delay ODEs

The rate of change of the population density in the resting phase equals the sum of three terms. The first term is responsible for the irreversible loss of cells in the G_0 phase due to differentiation. The second term corrects the loss because of the cells' transition to proliferation. The third term takes into account the increment of the cell population due to the transition of proliferative cells to the G_0 phase from the preceding generation. The factor '2' indicates that cell proliferation occurs through mitosis. The exponential factor adjusts the probability of loss of proliferative cells.

The dynamics of the cell population in the proliferative stage is described by a similar delay ODE

$$v_t = -\gamma v + \beta(u)u - \beta(w)we^{-\gamma\tau}, \quad w = u(t-\tau). \tag{6.1.3.4}$$

The first term is responsible for the irreversible loss among the proliferative cells, the second one simulates the inflow of cells from the resting phase, while the third accounts for the outflow of cells from the proliferative to resting phase. Notably, equation (6.1.3.4) involves two unknown functions, u and v, while equation (6.1.3.3) involves only one, u.

Substituting (6.1.3.2) into (6.1.3.3) and (6.1.3.4) yields a system of two delay ODEs describing the dynamics of the production of pluripotent stem cells:

$$\begin{aligned} u_t &= -\delta u - \frac{\beta_0 \theta^n u}{\theta^n + u^n} + \frac{2\beta_0 \theta^n w}{\theta^n + w^n} e^{-\gamma\tau}, \\ v_t &= -\gamma v + \frac{\beta_0 \theta^n u}{\theta^n + u^n} - \frac{\beta_0 \theta^n w}{\theta^n + w^n} e^{-\gamma\tau}, \end{aligned} \tag{6.1.3.5}$$

where $w = u(t - \tau)$.

Consider system (6.1.3.5) with the initial conditions

$$u(t) = 6.25 \times 10^8, \quad v(t) = 0.69 \times 10^8, \quad -\tau \leq t \leq 0. \tag{6.1.3.6}$$

The initial conditions and the values of the parameters are in agreement with the estimates from [329] (see Figure 3 of the article). The solutions are obtained with the fourth-order Runge–Kutta method using Mathematica (see Subsections 5.1.4 and 5.1.7). Figure 6.3 displays qualitative changes of the cell number in the resting phase $u(t)$ (solid line) and proliferative phase $v(t)$ (dashed line) depending on the parameter γ responsible for irreversible cell removal from the population (e.g., due to death). Figure 6.3a depicts the decrease in the number of resting phase cells and a slight increase in the number of proliferative phase cells to some stable levels. Figure 6.3d corresponds to a more substantial decrease in the number of resting phase cells. In both cases, one can observe a decrease in the total number of stem cells $u(t) + v(t)$. Figures 6.3b and 6.3c correspond to periodic hematopoiesis.

The study [329] investigated the stability of steady-state solutions and provided graphs of solutions to system (6.1.3.5) with other initial data and values of the parameters. It was noted that the behavior of the system solutions is consistent with quantitative and qualitative properties of aplastic anemia and periodic hematopoiesis in humans. The coefficient of irreversible loss of stem cells was shown to influence the dynamics of their population size.

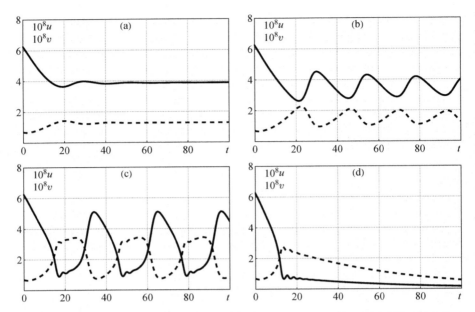

Figure 6.3. Solutions $u(t)$ (solid line) and $v(t)$ (dashed line) to problem (6.1.3.5), (6.1.3.6) at $\delta = 0.04$, $\beta_0 = 1.9$, $\theta = 1.74 \times 10^8$, $n = 3$, and $\tau = 2.6$ and different values of γ: (a) $\gamma = 0.18$, (b) $\gamma = 0.20$, (c) $\gamma = 0.23$, and (d) $\gamma = 0.28$.

6.1.4. Other Nonlinear Models with Delay

The simplest epidemiological model. The article [111] developed a simple model of the spread of infection from person to person using a vector (e.g., a malarial mosquito):

$$u'_t = \beta w(1 - u) - \lambda u, \quad w = u(t - \tau), \qquad (6.1.4.1)$$

where $u = u(t)$ is the relative number of infected individuals, $\beta > 0$ is the interaction factor, $\lambda > 0$ is the recovery rate, and $\tau > 0$ is the time it takes for the infectious agent to develop in a vector and make the vector infectious to susceptible individuals. The following assumptions were adopted to derive the delay ODE (6.1.4.1):

1°. The disease is not lethal and does not cause the formation of immunity; that is, the population consists of only infected, $u(t)$, and susceptible, $v = v(t)$, individuals.

2°. The population size is fixed, meaning that $u + v \equiv 1$.

3°. The infection rate is proportional to the number of contacts between susceptible humans and infectious vectors, i.e., to the product $vz = (1-u)z$, where $z = z(t)$ is the number of infectious vectors.

4°. The size of the vector population z is proportional to $w = u(t - \tau)$, i.e., to the number of infected people at time $t - \tau$.

Epidemic models for three groups of individuals (SIR models). The dynamics of the development of an epidemic in a population of variable size is described by the following system of delay ODEs for three groups of individuals: susceptible $u_1(t)$ (S), infected $u_2(t)$ (I), and non-susceptible (recovered with complete immunity or died) $u_3(t)$ (R):

$$u_1' = b - \mu u_1 - \beta u_1 w_2 + \gamma u_3,$$
$$u_2' = \beta u_1 w_2 - (\mu + \alpha + \lambda) u_2, \qquad (6.1.4.2)$$
$$u_3' = \lambda u_2 - (\mu + \gamma) u_3,$$

where $u_i = u_i(t)$ ($i = 1, 2, 3$), $w_2 = u_2(t - \tau)$, b is the population reproduction rate (number of individuals born per day), μ is the natural mortality rate, β is the transmission coefficient, γ is the reduction coefficient of immunity formed after disease (with $\gamma = 0$ for permanent immunity), α is the disease death rate, λ is the recovery rate, and $\tau > 0$ is the time it takes for the infectious agent to develop inside a vector and make the vector infectious to susceptible individuals.

Notably, model (6.1.4.2) is based on the simpler model without delay suggested in [14]. System (6.1.4.2) and related systems were studied, for example, in [49, 50, 328, 338, 508].

Remark 6.3. *Systems like (6.1.4.2) are also used to describe the spread of viruses, such as HIV or hepatitis B, inside a human body. In this case, $u_1(t)$, $u_2(t)$ and $u_3(t)$ are, respectively, the densities of uninfected cells, infected cells producing viruses, and active viruses. The delay τ is the time between the moment when a cell becomes infected and the moment when it starts producing viruses. These kinds of models are discussed, for example, in [119, 191, 213, 346, 357, 358].*

A simple climate model. The ocean and atmosphere usually interact through the following mechanism: a large-scale anomaly in the ocean surface temperature causes diabatic heating or cooling of the atmosphere, which changes atmospheric circulation and, consequently, the wind stress and heat fluxes at the ocean surface. In turn, changes in the wind stress change the thermal structure and circulation of the ocean, causing a series of positive feedbacks that amplify the initial anomaly in the ocean surface temperature. Therefore, oceanic and atmospheric circulations should generally be considered together, while the interaction of the two media is due to temperature changes on the ocean surface.

El Niño–Southern Oscillation (ENSO) is an irregular, periodic fluctuation in wind strength and sea surface temperatures across the eastern tropical Pacific Ocean, affecting the climate of much of the tropics and subtropics. The study [503] suggested a simple qualitative model to describe the Southern Oscillation based on strong positive feedback in the coupled ocean-atmosphere system and on nonlinear effects limiting the growth of unstable disturbances. A key element of the model is the use of delay to account for the effects of ocean waves propagating in a closed equatorial basin.

The climate model of ENSO [503] relies on an ODE with a cubic nonlinearity and a constant delay:

$$u_t' = u - u^3 - \alpha w, \quad w = u(t - \tau), \qquad (6.1.4.3)$$

where $u = u(t)$ is the growing disturbance amplitude, α is a coefficient characterizing the relative importance of the local and delayed wave effects, and τ is a constant delay time (wave transit time).

The nonlinear delay equation (6.1.4.3) has a few stationary states, each of which can become unstable and give rise to self-oscillations with period equal to double the delay time. The study [503] investigates the model stability and presents several numerical solutions. Oscillating solutions were found to arise at significant delay effects ($\alpha > \frac{1}{2}$) and sufficiently large delays ($\alpha \tau > 1$).

A model of regenerative machine tool vibration. The article [249] deals with a regenerative machine tool vibration model in the case of orthogonal cutting. One of the most essential effects causing poor surface quality during the cutting process is the vibration that occurs due to delay. Because of external disturbances, the cutting tool experiences a damped oscillation relative to the workpiece, thus making its surface uneven. Furthermore, after one revolution of the workpiece, the chip thickness changes. As a result, the cutting force depends not only on the tool edge current position relative to the workpiece but also on the delayed value of the displacement. The length of this delay is the time τ in which the workpiece makes one revolution. This process is called the regenerative effect. To study the delay-related properties of the system, the authors used a simple model with one degree of freedom. It was assumed that the tool moves in the vertical direction and possesses elasticity and viscosity, and all forces are directed vertically.

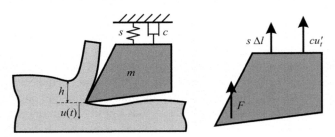

Figure 6.4. Free body diagram (horizontal forces ignored) of the one-degree-of-freedom mechanical model of the regenerative machine tool vibration. Notation: $\Delta l = l - l_0 + u(t)$ is the spring deformation, where l is the initial spring length and l_0 is the spring length in steady-state cutting.

Let $u(t)$ denote the vertical tool edge position, m the tool mass, s the tool coefficient of elasticity (more precisely, the stiffness of the spring modeling the tool elasticity), c the damping coefficient characterizing the viscous properties of the tool, h the current chip thickness, h_0 the chip thickness in steady-state cutting, and F the vertical component of the cutting force (see Figure 6.4). Then the motion of the tool can be described by the one-dimensional equation [249]:

$$u''_{tt} + 2\zeta\omega_n u'_t + \omega_n^2 u = -\frac{1}{m}\Delta F, \qquad (6.1.4.4)$$

where $\omega_n = \sqrt{s/m}$ is the natural angular frequency of the undamped free oscillations of the system, $\zeta = c/(2m\omega_n)$ is the so-called relative damping factor, and

$\Delta F = F(h) - F(h_0)$ is the increment of the cutting force. The cutting force in steady-state cutting equals $F(h_0) = -s(l - l_0)$, where l is the initial spring length and l_0 is the spring length in steady-state cutting. The quantity ΔF is determined by F as a function of the technological parameters, primarily as a function of the chip thickness h, which depend on the tool edge position u. It was established experimentally that
$$F(h) = Kdh^{3/4},$$
where d is the chip width and K is some coefficient. Expanding F in a power series about h_0 and retaining the four leading terms, we obtain
$$F(h) \approx Kd\left[h_0^{3/4} + \frac{3}{4}(h-h_0)h_0^{-1/4} - \frac{3}{32}(h-h_0)^2 h_0^{-5/4} + \frac{5}{128}(h-h_0)^3 h_0^{-9/4}\right].$$

Introducing the cutting force coefficient $k_1 = \frac{3}{4}Kdh_0^{-1/4}$, we write ΔF as a function of $\Delta h = h - h_0$ in the form
$$\Delta F(\Delta h) \approx k_1 \Delta h - \frac{1}{8}\frac{k_1}{h_0}(\Delta h)^2 + \frac{5}{96}\frac{k_1}{h_0^2}(\Delta h)^3.$$

The chip thickness increment Δh can be represented as the difference between the current tool edge position $u(t)$ and its delayed value $w = u(t - \tau)$, where the delay time $\tau = 2\pi/\Omega$ equals the time of one revolution of the workpiece, where Ω is the constant angular frequency of rotation of the workpiece. As a result, equation (6.1.4.4) becomes
$$u''_{tt} + 2\zeta\omega_n u'_t + \omega_n^2 u = f(u-w), \quad f(z) = -\frac{k_1}{m}\left(z - \frac{1}{8h_0}z^2 + \frac{5}{96h_0}z^3\right). \quad (6.1.4.5)$$

Using the dimensional quantities $\tilde{t} = \omega_n t$, $\tilde{u} = \frac{5}{12h_0}u$, $\tilde{\tau} = \omega_n \tau$, and $p = k_1/(m\omega_n^2)$ and omitting the tildes, we obtain
$$u''_{tt} + 2\zeta u'_t + u = f(u-w), \quad f(z) = -pz + \frac{3}{10}p(z^2 - z^3). \quad (6.1.4.6)$$

The study [249] showed the occurrence of bifurcations* when the parameters of the delay ODE (6.1.4.6) change.

Distribution of cells in an organism's tissue (equation with proportional argument). Consider a steady size distribution of cells, $u(x)$, of size x in an organism's tissue while assuming that the normalization condition
$$\int_0^\infty u(x)\,dx = 1$$
holds. A steady size distribution of cells can arise in a growing population if the rates of growth and division of cells are consistent, only depend on the cell size x,

---
*Bifurcations are qualitative changes in the properties of a system occurring as the system parameters change and characterized by the loss of stability of stationary solutions.

and are independent of time t. In this case, $u(x)$ is described by the following ODE with proportional argument [206]:

$$\frac{d}{dx}[g(x)u(x)] = -b(x)u(x) - (\alpha-1)u(x)\int_0^\infty b(x)u(x)dx + \alpha^2 b(\alpha x)w, \quad w = u(\alpha x),$$

where $g(x)$ is the rate of growth of cells in size (per second) and $b(x)$ is the rate at which cells of size x divide (cells per second) and form α new cells of size x/α. A cell typically divides into two, meaning that $\alpha = 2$. However, mathematically, other values of α can also be used with $\alpha > 1$ corresponding to cell division and increasing population, while $\alpha < 1$ corresponding to cell fusion and decreasing population.

The study [206] describes the properties of a solution to the equation in the simplest nontrivial case of $b(x) = b = \text{const}$ and $g(x) = c = \text{const}$:

$$u'_x = -au + a\alpha w, \quad a = \alpha b/c, \quad w = u(\alpha x).$$

The exact solution was obtained using the Laplace transform; it was shown that $u(x)$ tends to a normal distribution as $\alpha \to 1 + 0$.

6.2. Models of Economics and Finance Described by ODEs

6.2.1. The Simplest Model of Macrodynamics of Business Cycles

For the first time, a delay ODE in the field of economics was apparently used in [246] to describe the macrodynamics of business cycles in an isolated economic system (see also [168, 236]).

Assume that any investment process goes through three stages: (i) preparing and placing investment orders (purchase orders), i.e., orders for capital goods that ensure the reproduction or expansion of industrial equipment; the total volume of such orders per unit of time is denoted $v = v(t)$; (ii) producing capital goods; and (iii) delivering finished industrial equipment. The invested funds thus turn into the required industrial equipment after a certain time τ, which is responsible for the duration of the investment process, the average value of which is assumed to be 0.6 years. The study [246] derived the following differential equation to describe the associated process of changing the volume of investments over time:

$$v'_t = \frac{m}{\tau}(v - \bar{v}) - n(\bar{v} - c), \quad \bar{v} = v(t - \tau),$$

where c is the constant demand for restoration of the industrial equipment, while m and n are positive constants determined empirically (the values $m = 0.95$ and $n = 0.121$ were used in [246]). With the new notations

$$u(t) = v(t) - c, \quad a = m/\tau, \quad b = m/\tau + n,$$

we get the linear homogeneous delay ODE
$$u'_t = au - b\bar{u}, \quad \bar{u} = u(t-\tau),$$
which was treated in detail in Subsection 1.1.3.

6.2.2. Model of Interaction of Three Economical Parameters

Consider the following system of three ODEs without delay that describes the interdependence of the interest rate, $u = u(t)$, investment demand, $v = v(t)$, and price index, $w = w(t)$ [327]:
$$\begin{aligned} u'_t &= (v-a)u + w, \\ v'_t &= 1 - bv - u^2, \\ w'_t &= -u - cw, \end{aligned} \tag{6.2.2.1}$$

where $a \geq 0$ is the savings amount, $b \geq 0$ is the cost per investment, and $c \geq 0$ is the elasticity of demand of commercial markets. The change in the interest rate u is affected by the excess of investment demand over savings and the commodity price index. The change in investment demand v decreases as the cost of investment and the interest rate increase. The change in the price index w obeys the balance of supply and demand in the commercial market and is influenced by inflation.

The study [605] investigated a more complex mathematical model than (6.2.2.1), which considers the delay between price changes and interest rate changes:
$$\begin{aligned} u'_t &= (v-a)u + \bar{w}, \\ v'_t &= 1 - bv - u^2, \\ w'_t &= -u - cw, \end{aligned} \tag{6.2.2.2}$$

where $\bar{w} = w(t-\tau)$.

System (6.2.2.2) has an equilibrium $P_0 = (0, 1/b, 0)$. If $abc - c + b < 0$, then system (6.2.2.2) also has two more equilibria [96, 605]:
$$P_\pm = \left(\pm\sqrt{1 - ab - \frac{b}{c}},\; a + \frac{1}{c},\; \mp\frac{1}{c}\sqrt{1 - ab - \frac{b}{c}} \right),$$

where one should take either the upper or the lower signs.

The following two statements hold true [605].

Proposition 1. Suppose that $1 - ab - bc < 0$ and $abc - c - b > 0$. Then the equilibrium P_0 of system (6.2.2.2) is locally asymptotically stable for all $\tau \geq 0$.

Proposition 2. Suppose $1 - ab - bc < 0$, $abc - c + b > 0$, and $abc - c - b < 0$. Then there exists a τ_0 such that:
 (i) for $0 \leq \tau < \tau_0$, the equilibrium P_0 of system (6.2.2.2) is asymptotically stable;
 (ii) for $\tau = \tau_0$, a Hopf bifurcation occurs. This means that system (6.2.2.2) has a branch of periodic solutions bifurcating from P_0 near $\tau = \tau_0$.

The study [96] treated another model described by the system of delay ODEs

$$u'_t = (\bar{v} - a)u + w,$$
$$v'_t = 1 - bv - \bar{u}^2, \qquad (6.2.2.3)$$
$$w'_t = -\bar{u} - cw,$$

where $\bar{v} = v(t - \tau)$ and $\bar{u} = u(t - \tau)$. System (6.2.2.3) has the same equilibria as system (6.2.2.2).

See the articles [96, 605] for other results on the stability of equilibria and Hopf bifurcations for systems (6.2.2.2) and (6.2.2.3).

6.2.3. Delay Model Describing Tax Collection in a Closed Economy

Consider a fixed-price disequilibrium intermediate-run IS-LM model* augmented by a government budget constraint. The model is represented by the following system of equations [93] (see also [465–467]):

$$u'_t = \alpha[f_1(u,v) + \delta - f_2(u - z) - z],$$
$$v'_t = \beta[f_3(u,v) - w], \qquad (6.2.3.1)$$
$$w'_t = \gamma - z,$$

where α and β are positive constants, $u = u(t)$ is the income, $v = v(t)$ is the interest rate, $w = w(t)$ is the real money supply (prices are fixed at unity); $f_1(\dots)$ is the investment, δ is the constant government expenditure, $f_2(\dots)$ is the savings, $z = z(t)$ is the tax collection, and $f_3(\dots)$ is the liquidity preference function. The argument $u - z$ of the function f_2 is known as the disposable income (income after taxes). The first equation of system (6.2.3.1) represents the traditional disequilibrium adjustment in the product market, the second equation represents the disequilibrium dynamic adjustment in the money market, and the third equation defines the government's budget constraint.

In the simplest case, one should set

$$z = \mu u \qquad (6.2.3.2)$$

in system (6.2.3.1); μ is a common average tax rate.

The investment function f_1, savings function f_2, and liquidity preference function f_3 are assumed to possess the properties

$$\frac{\partial f_1}{\partial u} > 0, \quad \frac{\partial f_1}{\partial v} < 0, \quad 0 < f'_2 < 1, \quad \frac{\partial f_3}{\partial u} > 0, \quad \frac{\partial f_3}{\partial v} < 0.$$

---
*The IS-LM model (without delay), which stands for 'investment–savings' (IS) and 'liquidity preference–money supply' (LM), is a Keynesian macroeconomic model that shows how the market for economic goods (IS) interacts with the loanable funds market (LM) or money market. It is represented as a graph in which the IS and LM curves [231] intersect to show the short-run equilibrium between interest rates and output.

These functions can be defined, for example, as follows:

$$f_1(u,v) = A\frac{u^a}{v^b}, \quad f_2(u-z) = k(u-z), \quad f_3(u,v) = f_{31}(u) + f_{32}(v) = \gamma u + \frac{\lambda}{v - v_*},$$

where $A > 0$, $a > 0$, $b > 0$, $0 < k < 1$, $\gamma > 0$, and $\lambda > 0$; $v_* > 0$ is a fixed very small rate of interest generating the liquidity trap as v falls to the level $v_* > 0$ (i.e., $f_{32}(v) \to +\infty$ as $v \to v_*$).

The study [93] complicated model (6.2.3.1)–(6.2.3.2) by introducing a delay between the calculation and payment of taxes.* It was assumed that at each time, tax revenues $z = z(t)$ consist of two complementary components: one based on the current income and the other based on a past income, with the tax rate remaining the same:

$$z = \mu(1 - \varepsilon)u + \mu\varepsilon\bar{u}, \tag{6.2.3.3}$$

where $\bar{u} = u(t - \tau)$, τ is a fixed mean time lag of the income the current tax revenue is based on, ε is the income tax share of the delayed component, μ is a common average tax rate, $0 < \mu < 1$, and $\mu > f_2'$.

The work [93] shows that system (6.2.3.1)–(6.2.3.3) has a unique equilibrium point, which can be stable or unstable depending on the delay value. Moreover, under certain conditions, a sequence of intervals arises in which the regions of stability/instability alternate. Given a certain tax policy, it follows that the policy makers may face severe difficulties in stabilizing the economic system if they use the usual fiscal policy instruments.

6.3. Models and Delay PDEs in Population Theory

6.3.1. Preliminary Remarks

The physical meaning of diffusion. In population dynamics models, diffusion arises due to the tendency of a species to migrate to regions of lower population density [109]. In this case, for simplicity, one usually assumes that food is supplied continuously and uniformly in time and space. As a result, food becomes scarce in regions of high population density, and individuals will tend to migrate to areas of lower density to have a higher chance of survival. As noted in [190], most of the existing literature deals with the simplest situation where the movement of each individual is assumed to occur due to Fickian diffusion. It suggests that the population flow is proportional to the concentration gradient with a negative proportionality constant. The studies [72, 83, 353] treat the diffusion process from an ecological point of view.

Remark 6.4. *When introducing diffusion in a model with delay, many authors simply add the diffusion term in the delay ODE. However, it turned out that some difficulties may arise with this approach. The point is that although diffusion and temporal delay are related to*

---

*The delay was first paid attention to in [513]. It was shown that delay leads to a decrease in real (inflation-adjusted) government revenues.

space and time, respectively, they are not independent since individuals are not located at the same points in space at previous instants of time. Possible ways to overcome this problem by introducing a distributed (non-local) delay are discussed in [190].

Initial and boundary conditions. Suppose the equation of interest holds true in a domain $\mathbf{x} \in \Omega$ for $t > 0$. The initial condition has the form

$$u(\mathbf{x}, t) = \varphi(\mathbf{x}, t) \quad \text{at} \quad -\tau \leq t \leq 0. \tag{6.3.1.1}$$

Since the unknown function means population density, it is nonnegative. To ensure this, the initial conditions must also be nonnegative, that is, $\varphi(\mathbf{x}, t) \geq 0$. Suppose the environment is hostile so that all individuals that reach the boundary leave the population forever. In that case, one sets a homogeneous boundary condition of the first kind on the boundary $\partial\Omega$:

$$u(\mathbf{x}, t)|_{\partial\Omega} = 0. \tag{6.3.1.2}$$

If the population is isolated in Ω, implying that the individuals reaching the boundary 'rebound' from it and return to the population, then one sets a homogeneous boundary condition of the second kind:

$$\left.\frac{\partial u(\mathbf{x}, t)}{\partial n}\right|_{\partial\Omega} = 0, \tag{6.3.1.3}$$

where n is the outward normal to $\partial\Omega$. If individuals can cross the boundary, one uses a homogeneous boundary condition of the third kind:

$$\left[\frac{\partial u(\mathbf{x}, t)}{\partial n} + \sigma u(\mathbf{x}, t)\right]_{\partial\Omega} = 0, \tag{6.3.1.4}$$

where the coefficient σ characterizes the boundary crossing speed. If $\sigma > 0$, the flow of individuals goes outside the domain, and if $\sigma < 0$, it goes inside.

In what follows, for simplicity, we will often describe mathematical models using equations in one space variable x.

6.3.2. Diffusive Logistic Equation with Delay

The *diffusive logistic equation with delay* generalizes Hutchinson's equation (6.1.1.4) and has the form

$$u_t = au_{xx} + bu(1 - w/k), \quad w = u(x, t - \tau), \tag{6.3.2.1}$$

where $u = u(x, t) \geq 0$ is the population density, $b > 0$ is the growth rate coefficient, and k is the carrying capacity of the habitat. The delay time τ represents the average reproductive age of individuals, and $0 < a \ll 1$ is a parameter that characterizes the effect of diffusion acting equally on all individuals. The growth rate of the population is directly proportional to the population size at the current time, and the factor $(1 - w/k)$ determines the self-regulation mechanism.

Remark 6.5. *In the literature, equation (6.3.2.1) is known by different names, such as the diffusive logistic equation with delay, Fisher equation with delay, and diffusive Hutchinson equation. Sometimes, the diffusive equation (6.3.2.1) is simply referred to as Hutchinson's equation (e.g., see [167, 323]). However, the term 'Hutchinson's equation' is historically assigned to the delay ODE (6.1.1.4). In our opinion, the term 'diffusive logistic equation with delay' is the most suitable for equation (6.3.2.1). In addition, it is also known as the Fisher–KPP equation (Fisher–Kolmogorov–Petrovsky–Piskunov equation) with delay, since it was dealt with in the studies [162, 278] at $\tau = 0$.*

The change of variable $u = kv$ reduces equation (6.3.2.1) to the simpler form

$$v_t = av_{xx} + bv(1 - \bar{v}), \quad \bar{v} = v(x, t - \tau). \tag{6.3.2.2}$$

It was established in the study [323] that nonnegative solutions of the Dirichlet initial-boundary value problem for the delay PDE (6.3.2.1) on a finite interval $0 \leq x \leq L$ remain bounded at indefinitely large times. However, in related problems with several space variables, this only takes place when the delay time is not too large. It was shown in [167] that, in multi-dimensional problems in a finite domain with boundary conditions of the first or second kind and with a large delay time τ and small diffusion coefficient a, there are lots of solution trajectories such that the total mass of the population increases exponentially as $t \to \infty$.

Consider the initial-boundary value problem for equation (6.3.2.2) on an interval $0 \leq x \leq L$ subjected to the initial and boundary conditions

$$v = \varphi(x, t) \quad \text{at} \quad -\tau \leq t \leq 0, \quad v(0, t) = v(L, t) = 0. \tag{6.3.2.3}$$

Let $b_* = a\pi^2/L^2$. The following statements were proved in [502]:

1°. If $b < b_*$, the trivial (zero) solution to problem (6.3.2.2)–(6.3.2.3) is a global attractor of all nonnegative solutions to equation (6.3.2.2) for any $\tau \geq 0$.

2°. For any b such that $0 < b - b_* \ll 1$, equation (6.3.2.2) has a positive stationary solution v_b, and there exists a constant τ_0 such that the solution v_b is locally asymptotically stable for $0 \leq \tau \leq \tau_0$ and unstable for $\tau > \tau_0$. Moreover, there exists a sequence of numbers $\{\tau_n\}_{n=0}^{\infty}$ such that a Hopf bifurcation of the stationary solution arises at $\tau = \tau_n$.

Remark 6.6. *Equation (6.3.2.2) admits a traveling wavefront solution; see Subsection 3.1.3 for details.*

6.3.3. Delay Diffusion Equation Taking into Account Nutrient Limitation

The reaction-diffusion equation with constant delay under nutrient limitation is written as

$$u_t = au_{xx} + bu\frac{1 - w/k}{1 + cw/k}, \quad w = u(x, t - \tau), \tag{6.3.3.1}$$

where $b > 0$ is the population growth rate coefficient with unlimited food supply, $k > 0$ is the carrying capacity of the habitat, and $b/c > 0$ is the mass replacement rate per unit mass in the population at saturation. This equation generalizes the diffusive logistic equation (6.3.2.1) and coincides with it in the special case $c = 0$.

Equation (6.3.3.1) takes into account that a growing population consumes 'food' faster than the population at saturation. This is because, in the growth phase, food is consumed both for maintenance and growth. While at the saturation level, food is used for maintenance only. The delay τ characterizes the average reproductive age of the species and allows one to consider the experimentally confirmed fluctuations in the population density $u(x,t)$ about the equilibrium value $u = k$.

The change of variable $u = kv$ takes equation (6.3.3.1) to the simpler form

$$v_t = av_{xx} + bv\frac{1-\bar{v}}{1+c\bar{v}}, \quad \bar{v} = v(x, t-\tau). \tag{6.3.3.2}$$

The study [123] investigated the existence, uniqueness, and asymptotic stability on nonnegative equilibria of equation (6.3.3.1) under zero Dirichlet boundary conditions.

The article [502] investigated the existence and stability of positive stationary solutions and the existence of Hopf bifurcations of a positive stationary solution to equation (6.3.3.2) subjected to the initial and boundary conditions (6.3.2.3). The authors proved similar assertions to those for the diffusive logistic equation above.

6.3.4. Lotka–Volterra Type Diffusive Logistic Model with Several Delays

The Lotka–Volterra type reaction-diffusion model with several delays is described by the system of equations:

$$\begin{aligned} u_t &= a_1 u_{xx} + b_1 u(1 - c_1\bar{u}_1 + d_1\bar{v}_2), \\ v_t &= a_2 v_{xx} + b_2 v(1 + d_2\bar{u}_3 - c_2\bar{v}_4), \end{aligned} \tag{6.3.4.1}$$

where $u = u(x,t)$ and $v = v(x,t)$ are the unknown functions; $\bar{u}_i = u(x, t-\tau_i)$, $\bar{v}_j = v(x, t-\tau_j)$ ($i = 1, 3; j = 2, 4$); $\tau_i \geq 0$ and $\tau_j \geq 0$ are delay times.

System (6.3.4.1) generalizes the Fisher equation without delay and Hutchinson's equation by taking into account the interaction between two species (if $d_1 = d_2 = 0$, we get two independent diffusive logistic equations (6.3.2.1)). The unknown functions $u(x,t)$ and $v(x,t)$ and coefficients a_i, b_i, c_i ($i = 1, 2$) are all nonnegative and similar in physical meaning to the functions and coefficients in equation (6.3.2.1). As in the single equation, the delay times τ_1 and τ_4 characterize the average reproductive age of individuals while the delays τ_2 and τ_3 represent the time required for changes in the size of one population to bring about changes in the other population. All the delays are nonnegative and can be zero in some models. The terms with nonzero d_1 and d_2 distinguish the present model from a single equation, while the coefficients are responsible for the interaction between individuals of the two populations. In the case of cooperative interaction, when one species persists in the absence of the other and when the species mutually increase each other's growth rate, both coefficients d_1 and d_2 are positive. For competitive interaction, an increase in one population produces a decline in the other (for example, increasing the number of predators leads to decreasing the population of prey), and the coefficients d_1 and d_2 are both negative. For cooperative Lotka–Volterra delay models, see [224, 295]; for competitive models, see [151, 324, 377].

Remark 6.7. *Equilibria and traveling wavefront solutions of the system of Lotka–Volterra type delay PDEs (6.3.4.1) are discussed above in Subsection 3.1.3).*

The study [295] investigated a simpler system than (6.3.4.1):

$$\begin{aligned} u_t &= u_{xx} + bu(1 - \bar{u} + d_1 \bar{v}), & 0 < x < \pi, & \quad t > 0, \\ v_t &= v_{xx} + bv(1 + d_2 \bar{u} - \bar{v}), & 0 < x < \pi, & \quad t > 0, \end{aligned} \qquad (6.3.4.2)$$

where $\bar{u} = u(x, t - \tau)$ and $\bar{v} = v(x, t - \tau)$. It was equipped with the boundary and initial conditions

$$u(0, t) = u(\pi, t) = v(0, t) = v(\pi, t) = 0, \quad t \geq 0, \qquad (6.3.4.3)$$
$$u(x, t) = v(x, t) = 0.1\,(1 + t/\tau) \sin x, \quad -\tau \leq t \leq 0, \ 0 \leq x \leq \pi. \qquad (6.3.4.4)$$

System (6.3.4.2) is obtained from (6.3.4.1) if one sets $a_1 = a_2 = 1$, $b_1 = b_2 = b$, $c_1 = c_2 = 1$ and $\tau_1 = \cdots = \tau_4 = \tau$.

The following three statements were proved in [295]:

$1°$. For all $\tau > 0$, the trivial solution $u = v = 0$ of system (6.3.4.2) is stable for $b < 1$ and unstable for $b > 1$.

$2°$. If $d_1 d_2 < 1$ and $b = 1 + \varepsilon$ with $0 < \varepsilon \ll 1$, there exists a positive stationary solution of system (6.3.4.2) with the boundary conditions (6.3.4.3) such that $u \neq \text{const}$ and $v \neq \text{const}$.

$3°$. Under the condition of $2°$, there exists a number τ_b such that the stationary solution of system (6.3.4.2) is asymptotically stable for $0 \leq \tau < \tau_b$ and unstable for $\tau > \tau_b$.

The study [295] illustrates these assertions with graphs of solutions obtained using MATLAB with a combination of the method of steps and an implicit numerical method for integrating PDEs.

Figure 6.5 depicts the numerical solutions to problem (6.3.4.2)–(6.3.4.4) at the point $x = \pi/2$ obtained with Mathematica using a combination of the method of lines (with $N = 200$) and Gear method (see Subsections 5.1.7 and 5.2.2) with $d_1 = 0.4$, $d_2 = 0.7$, and different values of b and τ. Figure 6.5a refers to the stable trivial equilibrium, Figure 6.5b to a stable positive stationary solution, and Figure 6.5c to an unstable positive stationary solution. The thin dashed lines in Figure 6.5b correspond to a stationary solution with $u \approx 0.023$ and $v \approx 0.028$ at $x = \pi/2$.

6.3.5. Nicholson's Reaction-Diffusion Model with Delay

Nicholson's reaction-diffusion model with delay is described by the nonlinear equation

$$u_t = a \Delta u - \delta u + pw e^{-\kappa w}, \quad w = u(x, t - \tau), \qquad (6.3.5.1)$$

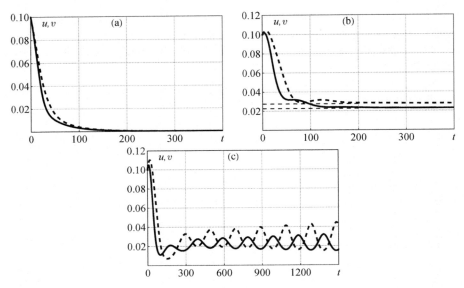

Figure 6.5. Numerical solutions $u = u(t)$ (solid line) and $v = v(t)$ (dashed line) to problem (6.3.4.2)–(6.3.4.3) at $x = \pi/2$ with $d_1 = 0.4$ and $d_2 = 0.7$. Cases: (a) $b = 0.98$, $\tau = 20$, (b) $b = 1.01$, $\tau = 20$, and (c) $b = 1.01$, $\tau = 30$. The dashed lines indicate stationary solutions at $x = \pi/2$.

where $p > 0$ is the maximum possible per capita egg production rate (corrected for egg-to-adult survival), $1/\kappa > 0$ is the size at which the population reproduces at its maximum rate, $\delta > 0$ is per capita daily adult mortality rate, time and density independent, and $\tau > 0$ is the generation time, or the time taken from birth to maturity.

Equation (6.3.5.1) generalizes the delay ODE (6.1.2.1). To analyze the population dynamics of individuals in a non-laboratory environment, one has to consider space inhomogeneity and employ space variables. In this context, a diffusion term needs to be included in the equation to describe the chaotic motion of individuals. When immature individuals are not subject to diffusion, while mature ones are, model (6.1.2.1) can naturally be generalized to the reaction-diffusion equation with delay (6.3.5.1).

Below we describe the stability conditions for solutions to Nicholson's reaction-diffusion equation with delay presented in the review part of the article [586]. Consider equation (6.3.5.1) subjected to the boundary condition of the first kind (6.3.1.2) and initial condition (6.3.1.1). Let λ_1 be the least eigenvalue of the auxiliary linear stationary problem

$$\Delta u + \lambda u = 0, \quad u|_{\partial\Omega} = 0.$$

The study [487] showed that if $p/\delta - 1 < a\lambda_1$, the trivial stationary solution $u = 0$ to the original nonstationary problem attracts all nonnegative solutions. If $p/\delta - 1 > a\lambda_1$, the solution $u = 0$ becomes unstable, and then there is only one positive equilibrium, $u^+(x)$, which attracts all positive solutions, provided that $e < p/\delta \leq e^2$.

Consider equation (6.3.5.1) subjected to the boundary condition of the second kind (6.3.1.3) and initial condition (6.3.1.1). An equilibrium of the corresponding delay ODE (6.1.2.1) determines an equilibrium of problem (6.3.5.1), (6.3.1.3), (6.3.1.1). The study [583] showed that if $0 < p/\delta \leq 1$, all positive solutions converge to $u=0$, and if $1 < p/\delta \leq e$, all nontrivial solutions converge to $u_* = \frac{1}{\kappa} \ln \frac{p}{\delta}$ regardless of $\tau \geq 0$. The study [586] proved that the solution u_* remains a global attractor for $e < p/\delta \leq e^2$ regardless of the value of τ. Also, it was shown in [583] that if $p/\delta > e^2$, the equilibrium u_* can be unstable, and a Hopf bifurcation can arise as τ increases.

The study [463] found sufficient conditions for oscillations of all positive solutions about the positive equilibrium and stated the following theorem.

Theorem. *For $p > e\delta$, any positive solution to equation (6.3.5.1) with the initial condition (6.3.1.1) oscillates about u_* if and only if*

1) the inequality

$$\delta\tau[\ln(p/\delta) - 1]e^{(\lambda_1 a + \delta)\tau} > 1/e$$

holds when the homogeneous boundary conditions of the first kind (6.3.1.2) is used;

2) the inequality

$$\delta\tau[\ln(p/\delta) - 1]e^{\delta\tau} > 1/e$$

holds when the homogeneous boundary condition of the second kind (6.3.1.3) or the third kind (6.3.1.4) is used.

6.3.6. Model That Takes into Account the Effect of Plant Defenses on a Herbivore Population

The study [504] investigated the effect of plant defenses on a herbivore population, taking into account spatial heterogeneity and delay effects. In many plants, particularly trees, the damage inflicted by herbivores causes changes in the chemical, physical and other properties of the leaves. These changes are known as the *induced defense*. The delay is responsible for the time it takes for the plant to prepare its induced defense. Herbivore-induced plant defense affects the stability and persistence of herbivore populations. For example, many herbivorous insect populations are characterized by outbreaks, when short periods of high population density and numerous leaf defects alternate with long periods of low population density.

The mathematical model and system of delay equations described below are based on the following four assumptions [504]:

$1°$. The changes in the induced plant defense at time t depend on the herbivore population density at time $t - \tau$.

$2°$. The level of induced defense depends on the herbivore population density and the level of the already existing defense.

$3°$. In the absence of induced changes in plants, the herbivore population obeys the logistic law with growth rate γ and habitat carrying capacity k.

$4°$. The seeds of some plants can travel in space due to various environmental factors, such as wind. Therefore, it is assumed that the induced defense and the herbivores travel randomly with diffusion coefficients a_1 and a_2, respectively.

Under these assumptions, the authors derived the following reaction-diffusion type system with delay [504]:

$$\frac{\partial u_1}{\partial t} = a_1 \frac{\partial^2 u_1}{\partial x^2} + (\alpha - \beta u_1)\frac{\bar{u}_2^n}{b^n + \bar{u}_2^n} - \mu u_1,$$
$$\frac{\partial u_2}{\partial t} = a_2 \frac{\partial^2 u_2}{\partial x^2} + \gamma u_2\left(1 - \frac{u_2}{k}\right) - m u_1 u_2,$$
(6.3.6.1)

where $u_1 = u_1(x,t)$ is the population density of plants with induced defense, $u_2 = u_2(x,t)$ is that of herbivores, $\bar{u}_2 = u_2(x, t-\tau)$, α is the maximum level of induced defense per plant, β is the per-unit reduction in the elicitation rate due to plant self-limitation, μ is the per-unit induction decay rate, m is the per-unit reduction in the growth rate of herbivores caused by the induction of defense, b is the half-maximum herbivore effectiveness of damage, and n is the herbivore damage effectiveness shape tuning parameter. The study [504] also described other constants appearing in system (6.3.6.1) as well as approximate numerical values of all constants with references to publications where the data was found.

It was found in [504] that large delays can lead to a population density reduction and an increased risk of extinction of herbivores, whereas moderate delays preserve the herbivore population density in a particular range. The authors obtained the minimum critical delay τ_* at which the herbivore population exhibits periodic outbreaks and showed that τ_* depends on the herbivore diffusion coefficient a_2 nonlinearly. The joint action of the delay and diffusion was found to increase the average herbivore population density during the outbreaks and, hence, increase the herbivores' viability.

6.4. Models and Delay PDEs Describing the Spread of Epidemics and Development of Diseases

6.4.1. Classical SIR Model of Epidemic Spread

The classical Kermack–McKendrick spatial homogeneous epidemic model is described by the system of three first-order ODEs [261]:

$$\begin{aligned} u_1'(t) &= -\beta u_1 u_2, \\ u_2'(t) &= \beta u_1 u_2 - \lambda u_2, \\ u_3'(t) &= \lambda u_2 \end{aligned}$$
(6.4.1.1)

with the initial conditions

$$u_1(0) = u_{10} > 0, \quad u_2(0) = u_{20} > 0, \quad u_3(0) = 0,$$

where $u_1(t)$, $u_2(t)$, and $u_3(t)$ are the population densities of susceptible, infectious, and removed (due to immunity or death) individuals, β is the contact rate (number of contacts of an infectious individual per unit time), and λ is the recovery rate of infected individuals (without counting their death). In system (6.4.1.1), the nonlinear term $\beta u_1 u_2$ is responsible for the incidence rate, while the removal is proportional to the share of infected individuals λu_2. Model (6.4.1.1) is known as the SIR model, because individuals (humans) move from susceptible to infectious and then to removed.

System (6.4.1.1) admits the first integral $u_1 + u_2 + u_3 = C$ (conservation of the total number of individuals), where C is the integration constant. On dividing all u_i by C, one can rewrite the conservation law in a dimensionless form.

The Kermack–McKendrick model (6.4.1.1) assumes that the population is well mixed so that the infection is transmitted instantaneously. However, due to the high mobility of people within one country or even worldwide, spatially homogeneous models do not adequately describe the spread of diseases. For the model to be more realistic, spatial effects must be included. If the environment is spatially continuous, random diffusion is often used to describe population mobility, leading to models based on reaction-diffusion equations (see [353]).

The introduction of a time delay in such models makes them more realistic. The delay in epidemiological models may occur for several different reasons. The most well-known ones include (i) the time it takes for the infectious agent to develop inside a vector and make the vector infectious to susceptible individuals and (ii) the latent period of disease, which is the time interval between when an individual is infected and when they become infectious, i.e., capable of transmitting pathogens to other susceptible individuals. Sometimes the latent period coincides with the incubation period, which is the time from the moment of infection to the first signs of the disease. However, in general, the two periods do not match. The study [111] introduced the effect of time delay into the epidemiological model under the assumption that the infection rate at time t is determined by the expression $\beta u_1(t) u_2(t - \tau)$ with τ considered as (i).

A Kermack–McKendrick type diffusive epidemic SIR model with delay can be represented by the system of equations

$$\frac{\partial u_1}{\partial t} = a_1 \frac{\partial^2 u_1}{\partial x^2} - \beta u_1 w_2,$$
$$\frac{\partial u_2}{\partial t} = a_2 \frac{\partial^2 u_2}{\partial x^2} + \beta u_1 w_2 - \lambda u_2, \qquad (6.4.1.2)$$
$$\frac{\partial u_3}{\partial t} = a_3 \frac{\partial^2 u_3}{\partial x^2} + \lambda u_2,$$

where $u_1 = u_1(x, t)$, $u_2 = u_2(x, t)$, $u_3 = u_3(x, t)$, and $w_2 = u_2(x, t - \tau)$; a_1, a_2, and a_3 are the diffusion coefficients or susceptible, infectious, and removed individuals.

A more complicated SIR model that takes into account population growth, natural mortality, and mortality due to a disease can be represented as

$$\frac{\partial u_1}{\partial t} = a_1 \frac{\partial^2 u_1}{\partial x^2} + b - \mu u_1 - \beta u_1 w_2,$$
$$\frac{\partial u_2}{\partial t} = a_2 \frac{\partial^2 u_2}{\partial x^2} + \beta u_1 w_2 - (\mu + \alpha + \lambda) u_2, \qquad (6.4.1.3)$$
$$\frac{\partial u_3}{\partial t} = a_3 \frac{\partial^2 u_3}{\partial x^2} + \lambda u_2 - \mu u_3,$$

where b is the birth rate, μ is the natural mortality rate, and α is the disease death rate. See [581] for a similar model.

The more general model known as the SIRS model [177], which generalizes model (6.1.4.2), assumes that susceptible people develop their immunity for only a limited time. This means that susceptible individuals become infected, then recover, becoming immune to the disease, and are removed from infectious, and then after a while, when immunity weakens, become susceptible again. To account for this, one adds the term γu_3 to the right-hand side of the first equation in (6.4.1.3) and subtracts γu_3 from the right-hand side of the third equation. As a result, we get

$$\frac{\partial u_1}{\partial t} = a_1 \frac{\partial^2 u_1}{\partial x^2} + b - \mu u_1 - \beta u_1 w_2 + \gamma u_3,$$

$$\frac{\partial u_2}{\partial t} = a_2 \frac{\partial^2 u_2}{\partial x^2} + \beta u_1 w_2 - (\mu + \alpha + \lambda) u_2, \qquad (6.4.1.4)$$

$$\frac{\partial u_3}{\partial t} = a_3 \frac{\partial^2 u_3}{\partial x^2} + \lambda u_2 - (\mu + \gamma) u_3,$$

where γ is the rate at which recovered individuals lose their immunity and return to the group of susceptible.

In the literature, there are models in which the bilinear incidence term $\beta u_1 w_2$ is replaced with a nonlinear term. This can be justified as explained below (in [87], it was shown for a model without delay).

The case where the bilinear term is a linearly increasing function of the number of infectious individuals may be valid for a small number of such individuals. However, this is hardly realistic for a large number of infectious individuals. In fact, the number of contacts of a susceptible person per unit time may not always grow linearly with an increase of w_2.

It seems much more realistic to introduce, instead of the bilinear term $\beta u_1 w_2$, a more complex nonlinear term of the form $g(w_2) u_1$, where the number of infectious individuals is determined by a nonlinear bounded function g, which eventually tends to a saturation level. Using such a function also allows one to consider 'psychological' effects: for a large number of infectious individuals, the infection force $g(w_2)$ can decrease with w_2, because the population may tend to reduce the number of contacts per unit time. As an example, the study [87] considered (for a model without delay) the function

$$g(w_2) = \frac{\beta w_2}{1 + \theta w_2}.$$

Thus, the incidence rate $\beta u_1 w_2$ is corrected by the factor $1/(1 + \theta w_2)$ and allows for the slowdown effect due to a change in the susceptibles' behavior. For diffusive equations with delay and a nonlinear infection rate, see, for example, [28, 97, 577, 581].

6.4.2. Two-Component Epidemic SI Model

Relying on the classical Kermack–McKendrick model, the study [54] developed a two-component epidemic model based on the following system of two nonlinear ODEs:

$$\begin{aligned} u_1'(t) &= b(u_1 + u_2)\left(1 - \frac{u_1 + u_2}{k}\right) - \beta \frac{u_1}{u_1 + u_2} u_2 - (\mu + m) u_1, \\ u_2'(t) &= \beta \frac{u_1}{u_1 + u_2} u_2 - (\mu + \alpha) u_2. \end{aligned} \quad (6.4.2.1)$$

Unlike (6.4.1.1), this model accounts for variable population size, disease mortality, and migration of individuals.

Equations (6.4.2.1) assume that the population consists of two groups: susceptible and infectious individuals with the respective population densities $u_1 = u_1(t)$ and $u_2 = u_2(t)$. The population reproduction obeys the logistic law with intrinsic growth rate b and carrying capacity k; β is the transmission coefficient, μ is the natural mortality rate, α is the disease death rate, and m is the migration rate of susceptible individuals. System (6.4.2.1) uses the nonlinear incidence term $\beta \frac{u_1}{u_1+u_2} u_2$. In this case, the infection rates are proportional to the ratio of susceptible-to-total population size $\frac{u_1}{u_1+u_2}$, where $u_1 + u_2$ is the total population, rather than the susceptible population size u_1. The term $(\mu + m)u_1$ characterizes the reduction in the susceptible population due to natural mortality and migration. The term $(\mu + \alpha)u_2$ is responsible for the decrease of the infectious population due to natural mortality and disease mortality. The study [228] investigated model (6.4.2.1) with $m = 0$ and called it a parasite–host model.

The studies [296, 306] dealt with a reaction-diffusion epidemic model of the type of (6.4.2.1) (see also [81] for a similar model):

$$\begin{aligned} \frac{\partial u_1}{\partial t} &= a_1 \frac{\partial^2 u_1}{\partial x^2} + \nu r_d (u_1 + w_2)(1 - u_1 - w_2) - \nu u_1 - r_0 \frac{u_1 w_2}{u_1 + w_2}, \\ \frac{\partial u_2}{\partial t} &= a_2 \frac{\partial^2 u_2}{\partial x^2} + r_0 \frac{u_1 w_2}{u_1 + w_2} - w_2, \end{aligned}$$

where $u_1 = u_1(x, t)$, $u_2 = u_2(x, t)$, and $w_2 = u_2(x, t - \tau)$; a_1 and a_2 are the diffusion coefficients, $\nu = \frac{\mu+m}{\mu+\alpha}$ is the ratio of the average life spans of susceptible and infectious individuals. The basic demographic reproductive number r_d and basic epidemiological reproductive number r_0 are determined by

$$r_d = \frac{b}{\mu + m}, \quad r_0 = \frac{\beta}{\mu + \alpha}.$$

The basic demographic reproductive number is the growth-to-death ratio in the absence of infection. The case $r_d > 1$ indicates population growth, while $r_d < 1$ indicates its extinction. The basic epidemiological reproductive number r_0 characterizes the contagiousness of an infectious disease; it is defined as the average number of susceptible individuals that one sick person can infect. The infection spreads widely when $r_0 > 1$ and does not spread when $r_0 < 1$. The case $r_0 = 1$ indicates a boundary situation when the process can go either way.

6.4.3. Epidemic Model of the New Coronavirus Infection

The study [610] developed a delay reaction-diffusion model close to the actual spread of the COVID-19 epidemic. It includes relapse, time delay, home quarantine, and a spatiotemporal heterogeneous environment that influences the spread of COVID-19. The model includes six groups of people: susceptible ($u_1 = u_1(\mathbf{x},t)$), exposed ($u_2 = u_2(\mathbf{x},t)$), quarantined at home ($u_3 = u_3(\mathbf{x},t)$), infected ($u_4 = u_4(\mathbf{x},t)$), quarantined in hospital ($u_5 = u_5(\mathbf{x},t)$), and temporarily recovered ($u_6 = u_6(\mathbf{x},t)$). The model with multiple spatial variables is described by the following mixed system of differential equations (PDEs and ODEs) with delay:

$$\frac{\partial u_1}{\partial t} = a_1 \Delta u_1 + b - \beta_1 \frac{u_1 u_2}{u_1 + u_2} - \beta_2 \frac{u_1 w_4}{u_1 + w_4} - \mu u_1,$$

$$\frac{\partial u_2}{\partial t} = a_2 \Delta u_2 + \beta_1 \frac{u_1 u_2}{u_1 + u_2} + \beta_2 \frac{u_1 w_4}{u_1 + w_4} + \rho_2 u_6 - (\mu + \omega_2 + p) u_2,$$

$$\frac{\partial u_3}{\partial t} = p u_2 - (\mu + \alpha_3 + \omega_3 + \sigma) u_3, \qquad (6.4.3.1)$$

$$\frac{\partial u_4}{\partial t} = a_4 \Delta u_4 + \omega_2 u_2 + \omega_3 u_3 + \rho_1 u_6 - (\mu + \alpha_4 + q) u_4,$$

$$\frac{\partial u_5}{\partial t} = q u_4 - (\mu + \alpha_5 + \nu) u_5,$$

$$\frac{\partial u_6}{\partial t} = a_6 \Delta u_6 + \sigma u_3 + \nu u_5 - (\mu + \rho_1 + \rho_2) u_6,$$

where $u_i = u_i(\mathbf{x},t)$ ($i = 1, \ldots, 6$), $w_2 = u_2(\mathbf{x}, t - \tau)$; $a_j > 0$ are diffusion coefficients ($j = 1, 2, 4, 6$), b is the population birth rate, $\beta_{1,2}$ are contact rates, $\omega_{2,3}$ are morbidity rates of the exposed and quarantined at home, p is the home quarantine rate, q is the hospitalization rate, ρ_1 is the relapse rate, ρ_2 is the recontact rate of the recovered, σ is the home quarantine leaving rate, ν is the hospital recovery rate, μ is the natural mortality rate, and $\alpha_{3,4,5}$ are the disease death rate among the quarantined at home, infected, and quarantined in hospital, respectively. The delay τ is responsible for the incubation period, which is the time from the infection to the first signs of the disease. During the incubation period, a person is unknown to have been infected or not, so no restrictions can be placed on them, which means that such a person may contact a susceptible one. It is noteworthy that the equations for u_3 and u_5 do not include diffusion since people in home quarantine or in the hospital are assumed to be immobile.

The work [610] gave approximate numerical values of the parameters of model (6.4.3.1) for the epidemics in China and the USA and investigated the stability of solutions.

Remark 6.8. *The diffusion coefficients of model (6.4.3.1) can depend on $\mathbf{x}$, so $a_j = a_j(\mathbf{x})$, where $j = \{1, 2, 4, 6\}$. In this case, the terms $a_j \Delta u_j$ should be replaced with $\nabla \cdot (a_j \nabla u_j)$, where ∇ is the gradient operator. The other model parameters can depend on both $\mathbf{x}$ and t; for example, $b = b(\mathbf{x}, t)$.*

6.4.4. Hepatitis B Model

Models of disease development are related to models of the spread of epidemics. Uninfected cells are treated as belonging to the susceptible component of the population, $u_1(x,t)$, infected cells are regarded as infectious individuals, $u_2(x,t)$, and free viral particles, $v(x,t)$, represent the third component of the population. In the case of hepatitis B, liver cells are considered immobile, and viral particles can move randomly. The study [540] used these assumptions to develop a model consisting of two ODEs and one reaction-diffusion equation:

$$\frac{\partial u_1}{\partial t} = b - \beta u_1 v - \mu_1 u_1,$$
$$\frac{\partial u_2}{\partial t} = \beta u_1 v - \mu_2 u_2, \qquad (6.4.4.1)$$
$$\frac{\partial v}{\partial t} = a \frac{\partial^2 v}{\partial x^2} + \gamma u_2 - \mu_3 v,$$

where b is the reproduction rate of uninfected cells, μ_1 is the death rate of uninfected cells, β is the coefficient responsible for the infection rate, γ is the virus reproduction rate, μ_2 is the death rate of infected cells, μ_3 is the rate of disappearance of free viruses, and a is the diffusion coefficient. The main difference between model (6.4.4.1) and the above epidemic models lies in the explicit consideration of an intermediate agent (virus) in the infection transmission process from an infected to an uninfected cell. The infection does not occur from contact between infected and susceptible individuals, as in epidemic models, but from contact between a susceptible cell and a virus. However, similar models also exist among epidemic models, for example, the model of the spread of malaria, when a malarial mosquito is involved in transmitting infection.

The study [541] developed an enhanced model that considers the intracellular delay between the moment a cell gets infected and when it starts producing new viral particles. The model is described by the following mixed system of delay differential equations (two ODEs and one PDE):

$$\frac{\partial u_1}{\partial t} = b - \beta u_1 v - \mu_1 u_1,$$
$$\frac{\partial u_2}{\partial t} = \beta \bar{u}_1 \bar{v} - \mu_2 u_2, \qquad (6.4.4.2)$$
$$\frac{\partial v}{\partial t} = a \frac{\partial^2 v}{\partial x^2} + \gamma u_2 - \mu_3 v,$$

where $\bar{u}_1 = u_1(x, t - \tau)$ and $\bar{v} = v(x, t - \tau)$. System (6.4.4.2) is treated on the interval $0 \leq x \leq 1$ and subjected to the standard initial conditions

$$u_i(x,t) = u_i^\circ(x) \geq 0 \quad (i = 1, 2), \quad v(x,t) = v^\circ(x) \geq 0 \quad \text{at} \quad -\tau \leq t \leq 0 \quad (6.4.4.3)$$

and homogeneous boundary conditions of the second kind

$$v_x(0,t) = v_x(1,t) = 0, \quad t > 0. \qquad (6.4.4.4)$$

The article [575] noted that using bilinear forms of the incidence term $\beta u_1 v$ is not always justified. The incidence rate is likely not a linear function over the entire range of u_1 and v. For example, a weaker-than-linear dependence on v may occur due to saturation at high virus concentrations. In this case, it makes sense to use an incidence term of the form $\frac{\beta u_1 v^p}{1+\theta v^q}$, where $p, q, \theta > 0$. The study dealt with the case $p = q = 1$. It obtained the value of the basic reproductive number $r_0 = \frac{b\gamma\beta}{\mu_1\mu_2\mu_3}$, which characterizes the average number of cells impaired by one infectious cell at the beginning of the infection process. For $r_0 < 1$, there is a unique equilibrium, which is asymptotically stable and corresponds to the absence of infection: $u_1 = b/\mu_1$, $u_2 = 0$, $v = 0$. For $r_0 > 1$, there are two equilibria: an unstable point of the absence of disease and a stable point with disease.

The study [210] suggested a model that generalizes the above two and the related models of [103, 211, 606] with nonlinear incidence rates. The model is represented by the following mixed system of differential equations (two ODEs and one PDE) with two delays

$$\frac{\partial u_1}{\partial t} = b - f(u_1, u_2, v)v - \mu_1 u_1,$$

$$\frac{\partial u_2}{\partial t} = f(\bar{u}_{11}, \bar{u}_{21}, \bar{v})\bar{v}e^{-c_1\tau_1} - \mu_2 u_2, \qquad (6.4.4.5)$$

$$\frac{\partial v}{\partial t} = a\frac{\partial^2 v}{\partial x^2} + \gamma \bar{u}_{22} e^{-c_2\tau_2} - \mu_3 v,$$

where $\bar{u}_{11} = u_1(x, t - \tau_1)$, $\bar{u}_{21} = u_2(x, t - \tau_1)$, $\bar{u}_{22} = u_2(x, t - \tau_2)$, and $\bar{v} = v(x, t - \tau_1)$. The delay τ_1 is the time between the moment a cell is infected and when it starts producing virions. The multiplier $e^{-c_1\tau_1}$ represents the probability of cell survival during the time τ_1, where c_1 is the death rate of infected but not yet virus-producing cells. The delay τ_2 is the time between when a virion forms and when it is capable of infecting; the immature virion's survival probability is determined by the factor $e^{-c_2\tau_2}$, and its average lifetime is $1/c_2$. The incidence function $f(x, y, z)$ is continuously differentiable in its arguments and satisfies the following three hypotheses:

1) $f(0, y, z) = 0$ for $y, z \geq 0$,
2) $f_x(x, y, z) > 0$ for $x > 0$, $y, z \geq 0$,
3) $f_y(x, y, z) \leq 0$, $f_z(x, y, z) \leq 0$ for $x, y, z \geq 0$.

System (6.4.4.5) is subjected to the initial conditions (6.4.4.3) and boundary conditions of the second kind (6.4.4.4).

The study [210] proved the existence, positiveness, and boundedness of solutions to the initial-boundary value problem (6.4.4.5), (6.4.4.3), (6.4.4.4). The basic reproductive number is expressed as

$$r_0 = \gamma(\mu_2\mu_3)^{-1} f(b/\mu_1, 0, 0) e^{-c_1\tau_1 - c_2\tau_2}.$$

For $r_0 \leq 1$, system (6.4.4.5) has a unique globally asymptotically stable equilibrium representing the absence of disease. For $r_0 > 1$, the equilibrium without disease is unstable; moreover, there is another point of equilibrium that corresponds to chronic infection and, under certain conditions, is globally asymptotically stable. These conditions are satisfied, in particular, for a linear incidence function, $f = \beta u_1$, and also for a more complicated Beddington–DeAngelis function, $f = \frac{\beta u_1}{1+\kappa_1 u_1+\kappa_2 v}$, and Crowley–Martin function $f = \frac{\beta u_1}{1+\kappa_1 u_1+\kappa_2 v+\kappa_1 \kappa_2 u_1 v}$, where $\kappa_1 \geq 0$ and $\kappa_2 \geq 0$.

6.4.5. Model of Interaction between Immunity and Tumor Cells

Following [31], we will consider a model of the interaction process between immune T cells and malignant tumor cells. T cells divide into T killers (cytotoxic T lymphocytes) and T helpers. T killers attack malignant cells, while T helper cells release various cytokines (small regulatory proteins), stimulating T killers. The growth of T helper and tumor cells obeys the logistic law. The destruction of tumor cells and T killers occurs in proportion to the product of the densities of their populations. A T helper turns into a T killer either by direct contact with a T killer or by contact with a cytokine released by a T helper. The transformation occurs with some delay τ, which appears in the terms describing the transformation of T helpers and the growth of the T killer population. The model assumes that a T killer can never turn back into a T helper and dies with some constant probability per unit time.

The suggested spatially homogeneous model of the interaction between immunity and tumor cells is described by the following system of three delay ODEs [31]:

$$\begin{aligned} u_1' &= b_1 u_1(1 - u_1/k_1) - c_1 u_1 u_2, \\ u_2' &= \beta u_1 w_3 - \mu u_2 - c_2 u_1 u_2, \\ u_3' &= b_2 u_3(1 - u_3/k_3) - \beta u_2 w_3, \end{aligned} \qquad (6.4.5.1)$$

where $u_1 = u_1(t)$, $u_2 = u_2(t)$, and $u_3 = u_3(t)$ are the population densities of tumor cells, T killer cells, and T helper cells, respectively; $w_3 = u_3(t-\tau)$; b_1 and b_2 are the natural growth rates of tumor and T helper cells; k_1 and k_3 are the maximum carrying capacities for the tumor and T helper cells; μ is the T-killer death rate; c_1 is the death rate of tumor cells when in contact with T killers, c_2 is the death rate of T killer cells when in contact with tumor cells, and β is the coefficient of transformation of T helpers into T killers.

The study [237] modified model (6.4.5.1) to consider only two cell groups: tumor and immune. The third equation of (6.4.5.1) was excluded from the analysis, and spatial heterogeneity was introduced by adding diffusion terms. As a result, the following reaction-diffusion system of delay PDEs was obtained:

$$\begin{aligned} \frac{\partial u_1}{\partial t} &= a_1 \frac{\partial^2 u_1}{\partial x^2} + b u_1(1 - u_1/k) - c_1 u_1 u_2, \\ \frac{\partial u_2}{\partial t} &= a_2 \frac{\partial^2 u_2}{\partial x^2} + \beta w_1 u_2 - \mu u_2 - c_2 u_1 u_2, \end{aligned} \qquad (6.4.5.2)$$

where $u_1 = u_1(x,t)$ and $u_2 = u_2(x,t)$ are the population densities of tumor and immune cells, respectively; $w_1 = u_1(x, t-\tau)$; a_1 and a_2 are diffusion coefficients,

b is the natural growth rate of tumor cells, k is the maximum carrying capacity for tumor cells, μ is the death rate of immune cells, c_1 is the death rate of tumor cells upon contact with immune cells, c_2 is the death rate of immune cells upon contact with tumor cells, and β is the coefficient of activation of immune cells. The activity of tumor cells leads to an appropriate immune response, which depends on the number of tumor cells, but occurs after some delay time τ. System (6.4.5.2) is supplemented with standard initial conditions and uniform boundary conditions of the second kind.

It is noteworthy that system (6.4.5.2) is a special case of the diffusive Lotka–Volterra type system with several delays (6.3.4.1).

The equilibrium $u_1^\circ = k$, $u_2^\circ = 0$ of this system is globally asymptotically stable for $\mu > k\beta$. If $\mu < k(\beta - c_2)$, system (6.4.5.2) has a unique positive equilibrium

$$u_1^\circ = \frac{\mu}{\beta - c_2}, \quad u_2^\circ = \frac{b}{c_1}\left(1 - \frac{\mu}{k(\beta - c_2)}\right). \tag{6.4.5.3}$$

Without delay, the equilibrium (6.4.5.3) is locally asymptotically stable. However, as shown in [237], the delay is essential in destabilizing this equilibrium state. The study established that there exists a critical value τ_* such that the equilibrium (6.4.5.3) is locally asymptotically stable for $\tau < \tau_*$ and unstable for $\tau > \tau_*$. Furthermore, the authors found that the delay also affects Hopf bifurcations' direction, stability, and frequency.

Remark 6.9. *The work [388] dealt with a system of delay PDEs similar to (6.4.5.2) where the term* $\beta w_1 u_2$ *in the second equation was replaced with* $\beta w_1 w_2$, *where* $w_i = u_i(x, t - \tau)$, $i = 1, 2$.

6.5. Other Models Described by Nonlinear Delay PDEs

6.5.1. Belousov–Zhabotinsky Oscillating Reaction Model

The Belousov–Zhabotinsky reaction is a class of oscillating chemical reactions that serve as a classic example of non-equilibrium thermodynamics. Some reaction parameters (such as color, the concentration of components, temperature, and others) change periodically, forming a complex spatiotemporal structure of the reaction medium. For example, temporal fluctuations in the color of a homogeneous solution caused by fluctuations in the concentrations of intermediates were first described in [45]. The study investigated the process of catalytic oxidation of citric acid with potassium bromate in the presence of cerium ions. The works [597, 598] described various organic acids and metal ions that can be used in such reactions. The most complex mechanism of such reactions was thoroughly studied in [157, 158].

Following [351], we will briefly outline the reaction. It can be roughly divided into two processes: process I and process II. In the reaction, when the bromide ion Br^- is above some critical concentration, process I occurs. In it, the bromate ion BrO_3^- is reduced to bromine Br_2, with bromous acid $HBrO_2$ as an intermediary, and the malonic acid $CH_2(COOH)_2$ is brominated. During this process there is little oxidation of the cerium ion Ce(III) (or in the case of iron, of the ferrous state). Process I thus uses up the bromide. When the concentration of bromide falls sufficiently low, process II takes over. In this, the bromous acid and the bromate ion produce a radical bromate species $BrO_2^{\bullet}$, which oxidizes the cerium ion Ce(III) to the Ce(IV) form (or in the iron case the ferrous to the ferric state), with bromous acid generated autocatalytically. When all of the Ce(III) has been oxidized to Ce(IV) and the bromide ion concentration is low, the Ce(IV) then reacts with the bromomalonic acid to produce the cerium ion Ce(III) and bromide again. When the bromide passes the critical concentration, process I takes over again and the cycle is repeated.

The study [592] found, using ferroin as the catalyst and malonic acid, that if the reacting mixture is placed in a thin flat layer, about 2 mm thick, then circular space-time waves arise in it. A reaction-diffusion model for studying such waves was proposed in [159], with its dimensionless and simplified form suggested in [351]. A more complicated model that takes into account the effects of delay in the formation of bromous acid and generalizes the dimensionless model [351] is described by the following reaction-diffusion system of equations [301, 327, 520, 572, 600]:

$$u_t = u_{xx} + u(1 - u - b\bar{v}), \quad \bar{v} = v(x, t - \tau),$$
$$v_t = v_{xx} - cuv, \qquad (6.5.1.1)$$

where $u = u(x,t)$ and $v = v(x,t)$ are the dimensionless concentrations of bromous acid and bromide ions, respectively ($0 \le u \le 1$, $0 \le v \le 1$); $b > 0$ and $c > 0$ are some dimensionless parameters whose ranges are specified, for example, in [351]: b ranges from 5 to 50, and c ranges from 2.5 to 12.5.

6.5.2. Mackey–Glass Model of Hematopoiesis

The dynamics of a homogeneous population of mature circulating blood cells can be described using the following reaction-diffusion equation generalizing the delay ODE (6.1.3.1):

$$u_t = au_{xx} - \gamma u + \beta_0 \frac{\theta^n w}{\theta^n + w^n}, \quad w = u(t - \tau). \qquad (6.5.2.1)$$

Introducing a diffusion term allows one to consider the movement of cells from a region of high concentration to a region of low concentration.

The study [545] obtained the following results for equation (6.5.2.1) with $a = 1$ and homogeneous boundary conditions of the second kind:

1°. If $0 < \beta_0/\gamma \le 1$, then $u \to 0$ as $t \to \infty$ uniformly in x.

2°. If $1 < \beta_0/\gamma \le \frac{n}{n-1}$, then $u_* = \theta[(\beta_0 - \gamma)/\gamma]^{1/n}$ is a unique positive equilibrium, and any solution tends to u_* as $t \to \infty$ uniformly in x.

3°. If $\frac{n}{n-1} < \beta_0/\gamma$ and $\beta_0^{-1}\gamma\tau[\beta_0(n-1) - n\gamma] > e^{-\gamma\tau - 1}$, then any solution oscillates about the equilibrium u_*.

4°. If $\beta_0/\gamma > 1$, then any solution to equation (6.5.2.1) non-oscillating about the equilibrium $u = u_*$ tends to u_* as $t \to \infty$ uniformly in x (see Definition 1 in Item 10° of Subsection 6.1.2).

The study [448] proved the following theorem.

Theorem. *There exists a $\lambda^* > 0$ such that for any $\lambda \geq \lambda^*$, equation (6.5.2.1) has a positive monotonous front $u = U(z)$, $z = x + \lambda t$, traveling from $u = 0$ to $u = u_*$, provided that either pair of the following conditions hold:*
1) $1 < \beta_0/\gamma \leq \infty$ and $0 < n \leq 1$;
2) $1 < \beta_0/\gamma \leq n/(n-1)$ and $n > 1$.

The following system of two delay reaction-diffusion equations describes the dynamics of pluripotent stem cells production and generalizes the system of delay ODEs (6.1.3.5):

$$u_t = au_{xx} - \delta u - \frac{\beta_0 \theta^n u}{\theta^n + u^n} + \frac{2\beta_0 \theta^n w}{\theta^n + w^n} e^{-\gamma \tau}, \qquad (6.5.2.2)$$

$$v_t = av_{xx} - \gamma v + \frac{\beta_0 \theta^n u}{\theta^n + u^n} - \frac{\beta_0 \theta^n w}{\theta^n + w^n} e^{-\gamma \tau},$$

where $u = u(x, t)$ is the population density of cells in the resting phase G_0, $w = u(x, t-\tau)$, and $v = v(x, t)$ is the population density of cells in the proliferative phase.

Equations of the form (6.5.2.1) and (6.5.2.2) were studied, for example, in [302, 303, 376, 546].

6.5.3. Model of Heat Treatment of Metal Strips

The process of heat treatment of metal strips can be described using the equation [527, 544, 569, 612]:

$$u_t = au_{xx} + g(w_1)u_x + c[f(w_2) - u], \quad t > 0, \quad 0 < x < 1,$$
$$w_1 = u(x, t - \tau_1), \quad w_2 = u(x, t - \tau_2),$$

subjected to the homogeneous boundary conditions of the first kind

$$u(0, t) = u(1, t) = 0 \quad \text{for} \quad t \geq 0,$$

and the initial condition

$$u(x, t) = \varphi(x, t) \quad \text{at} \quad -\tau_{\max} \leq t \leq 0 \quad (0 < x < 1).$$

Here, $u(x, t)$ is the temperature distribution in the metal strip; $\tau_1 > 0$ and $\tau_2 > 0$ are delay times, $\tau_{\max} = \max\{\tau_1, \tau_2\}$, $g(w_1)$ is the strip speed, and $f(w_2)$ is the distributed source function.

The process proceeds as follows [612]. First, a metal strip enters the furnace and undergoes heat treatment. Then, the heater controller provides the desired spatial temperature distribution, and the speed controller controls the speed of the strip passing through the furnace. Finally, temperature sensors along the strip transmit information to a computer that generates appropriate signals for the heater and speed controllers. Thus, a certain time passes between the moment of reading the temperature values and the arrival of signals at the controllers, which is taken into account using the delay times.

6.5.4. Food Chain Model

The study [273] proposed a delay reaction-diffusion model to describe a simple food chain consisting of $n+1$ living species, n types of dissolved organic and inorganic substances, and detritus. The living species include zooplankton, z, phytoplankton, u_n, and microorganisms, u_i, $i = 1, \ldots, n-1$. The substances (nutrients) are denoted v_j, $j = 1, \ldots, n$, and detritus, the remains of all living organisms involved, is denoted d.

It is assumed to that microorganism u_i ($i = 1, \ldots, n-1$) feeds on substance v_i, phytoplankton u_n feeds on substance v_n, and zooplankton z feeds on phytoplankton u_n. Substance v_1 consists of dissolved organic substances resulting from the partial decay of dead organisms d, as well as the vital activity of phytoplankton u_n and zooplankton z. Substance v_j, $j = 2, \ldots, n$, is a metabolic product of microorganism u_{j-1}. The modeling of $n+1$ levels of living organisms, detritus, and n substances is carried out in terms of their nitrogen content. We assume that the flow of substance from level to level changes according to the Lotka–Volterra hypothesis as the product of interacting components, which leads to the nonlinearity of the system of differential equations. The study [273] proposed the following system of delay reaction-diffusion equations to describe the behavior of this simple food chain:

$$\frac{\partial u_i}{\partial t} = a_{ui} \frac{\partial^2 u_i}{\partial x^2} + u_i(x, t - \tau_{ui}) U_i(v_i(x, t - \tau_{ui}))$$
$$- u_i(x, t - \tau_{ei}) E_i(v_i(x, t - \tau_{ei})) - u_i(x, t) M_i(v_i(x, t)), \quad i = 1, \ldots, n-1;$$

$$\frac{\partial u_n}{\partial t} = a_{un} \frac{\partial^2 u_n}{\partial x^2} + u_n(x, t - \tau_{un}) U_n(v_n(x, t - \tau_{un}))$$
$$- u_n(x, t - \tau_{en}) E_n(v_n(x, t - \tau_{en})) - u_n(x, t) M_n(v_n(x, t))$$
$$- z(x, t - \tau_{uz}) U_z(u_n(x, t - \tau_{uz}));$$

$$\frac{\partial v_1}{\partial t} = a_{v1} \frac{\partial^2 v_1}{\partial x^2} + z(x, t - \tau_{ez}) E_z(u_n(x, t - \tau_{ez}))$$
$$+ u_n(x, t - \tau_{en}) E_n(v_n(x, t - \tau_{en})) + K d(x, t - \tau_{k1})$$
$$- u_1(x, t - \tau_{u1}) U_1(v_1(x, t - \tau_{u1}));$$

$$\frac{\partial v_j}{\partial t} = a_{vj} \frac{\partial^2 v_j}{\partial x^2} + u_{j-1}(x, t - \tau_{ej-1}) E_{j-1}(v_{j-1}(x, t - \tau_{ej-1}))$$
$$- u_j(x, t - \tau_{uj}) U_j(v_j(x, t - \tau_{uj})), \quad j = 2, \ldots, n;$$

$$\frac{\partial z}{\partial t} = a_1 \frac{\partial^2 z}{\partial x^2} + z(x, t - \tau_{uz}) U_z(u_n(x, t - \tau_{uz}))$$
$$- z(x, t - \tau_{ez}) E_z(u_n(x, t - \tau_{ez})) - z(x, t) M_z(u_n(x, t));$$

$$\frac{\partial d}{\partial t} = a_2 \frac{\partial^2 d}{\partial x^2} + \sum_{i=1}^{n} u_i M_i(v_i) + z M_z(u_n) - K d(x, t - \tau_{k1});$$

$$0 < x < 1, \quad t > 0.$$

The system is subjected to the homogeneous boundary conditions of the second kind

$$\left.\frac{\partial u_i}{\partial x}\right|_{x=0,1} = \left.\frac{\partial v_j}{\partial x}\right|_{x=0,1} = \left.\frac{\partial z}{\partial x}\right|_{x=0,1} = \left.\frac{\partial d}{\partial x}\right|_{x=0,1} = 0$$

and the initial conditions at $-\tau_{\max} \leq t \leq 0$:

$$u_i = \varphi_i(x) \geq 0, \quad i = 1, \ldots, n; \quad v_j = \varphi_{n+j}(x) \geq 0, \quad j = 1, \ldots, n;$$
$$z = \varphi_{2n+1}(x) \geq 0; \quad d = \varphi_{2n+2}(x) \geq 0 \quad (0 \leq x \leq 1),$$

where $u_i = u_i(x,t)$, $v_j = v_j(x,t)$, $z = z(x,t)$, and $d = d(x,t)$ are the concentrations* of recyclable matter in microorganisms, available nutrients, zooplankton, and detritus, respectively; τ_{ui} and τ_{uz} are delay times in the consumption of substances by the ith organism ($i = 1, \ldots, n$) and zooplankton, τ_{ei} and τ_{ez} are time delays in excreting substances by the ith organism ($i = 1, \ldots, n$) and zooplankton, τ_{k1} is a delay time in the decay of detritus, $\tau_{\max}$ is the maximum delay coefficient of all components in the food chain, and K is some constant. The functions U_i and U_z determine the rates of consumption of substances by the ith organism ($i = 1, \ldots, n$) and zooplankton, functions E_i and E_z are the rates of excretion of substances by the ith organism ($i = 1, \ldots, n$) and zooplankton, and functions M_i and M_z are the death rates of the ith organism ($i = 1, \ldots, n$) and zooplankton. The work [273] investigates the stability of solutions to this problem and gives examples of numerical modeling.

6.5.5. Models of Artificial Neural Networks

Delay reaction-diffusion equations and systems of such equations are widespread in the mathematical theory of artificial neural networks. The results are applied to signal and image processing, pattern recognition problems, work of associative machine memory, and determining the speed of moving objects. Delays arise in artificial neural networks due to the finite switching speed of amplifiers and the finite speed of signal propagation between neurons.

Many artificial neural network models rely on ordinary differential equations with delay (e.g., see [19, 85, 86, 319, 571, 607] and references therein). However, in some instances, the introduction of a diffusion term is required, making it possible to consider the motion of electrons in an asymmetric magnetic field. The studies [297, 318, 320, 539, 542, 584] deal with a class of artificial neural network models that combines the Hopfield neural network with cellular neural networks. This class of models is described by the system of reaction-diffusion equations

$$\frac{\partial u_i}{\partial t} = \sum_{k=1}^{3} \frac{\partial}{\partial x_k}\left(a_{ik}\frac{\partial u_i}{\partial x_k}\right) - b_i u_i + \sum_{j=1}^{n} c_{ij} f_j(u_j) + \sum_{j=1}^{n} d_{ij} g_j(\bar{u}_{ji}) + I_i(t),$$
$$\bar{u}_{ji} = u_j(\mathbf{x}, t - \tau_{ij}), \quad \mathbf{x} \in \Omega, \quad t > 0,$$
(6.5.5.1)

---

*For clarity, we use the same symbols to denote the substances and their concentrations.

where $u_i = u_i(\mathbf{x}, t)$ is the state of the ith neuron in the network ($i = 1, \ldots, n$), $a_{ik} = a_{ik}(\mathbf{x}, t, \mathbf{u}) \geq 0$ are smooth functions modeling the transmission diffusion operators along the ith neuron, $b_i > 0$ is the speed at which the potential of the ith neuron isolated from the network and external influence reaches the resting state, c_{ij} and d_{ij} are constants characterizing communication between neurons, $f_j(u_j)$ and $g_j(\bar{u}_{ji})$ are activation functions of the jth neuron, $I_i(t)$ is the external input on the ith neuron, $\tau_{ij} = \tau_{ij}(t)$ are delay times, and Ω is a closed bounded domain in $\mathbb{R}^3$ with boundary $\partial \Omega$. The initial conditions are as follows:

$$u_i(\mathbf{x}, t) = \varphi_i(\mathbf{x}, t) \quad \text{at} \quad -\max_{i,j} \tau_{ij} \leq t \leq 0, \quad \mathbf{x} \in \Omega.$$

The boundary conditions can be of the first kind

$$u_i(\mathbf{x}, t) = 0 \quad \text{for} \quad \mathbf{x} \in \partial \Omega, \quad t \geq 0,$$

or the second kind

$$\frac{\partial u_i}{\partial n} = 0 \quad \text{for} \quad \mathbf{x} \in \partial \Omega, \quad t \geq 0.$$

Remark 6.10. *The article [580] deals with a related system of the form (6.5.5.1) involving unknown functions with proportional delays:* $\bar{u}_{ji} = u_j(x, p_i t)$ $(0 < p_i < 1)$.

The study [493] considers a class of bidirectional associative memory neural networks described by the delay reaction-diffusion system

$$\frac{\partial u_i}{\partial t} = \sum_{k=1}^{3} \frac{\partial}{\partial x_k}\left(a_{ik} \frac{\partial u_i}{\partial x_k}\right) - b_i u_i + \sum_{j=1}^{m} c_{ji} f_j(\bar{v}_{ji}) + I_i(t), \quad i = 1, \ldots, n;$$

$$\frac{\partial v_j}{\partial t} = \sum_{k=1}^{3} \frac{\partial}{\partial x_k}\left(a^*_{jk} \frac{\partial v_j}{\partial x_k}\right) - b^*_j v_j + \sum_{i=1}^{n} d_{ij} g_i(\bar{u}_{ij}) + J_j(t), \quad j = 1, \ldots, m;$$

$$\bar{u}_{ij} = u_i(\mathbf{x}, t - \tau^*_{ij}), \quad \bar{v}_{ji} = v_j(\mathbf{x}, t - \tau_{ji}), \quad \mathbf{x} \in \Omega, \quad t > 0.$$

The authors find global exponential stability conditions for the system as well as conditions for the existence of periodic solutions.

Remark 6.11. *A differential-difference diffusion (thermal conduction) model with a finite relaxation time, which leads to a PDE with a delay in the diffusion term, was discussed in Section 2.3.*

Remark 6.12. *The study [537] investigated a system of hyperbolic equations with proportional delay.*

References

[1] Abazari R., Ganji M. Extended two-dimensional DTM and its application on nonlinear PDEs with proportional delay. *Int. J. Comput. Math.*, 2011, Vol. 88, pp. 1749–1762.

[2] Abbaoui K., and Cherruault Y. Convergence of Adomian's method applied to differential equations. *Computers Math. Applic.*, 1994, Vol. 28, No. 5, pp. 103–109.

[3] Ablowitz M.J., Clarkson P.A. *Solitons, Nonlinear Evolution Equations and Inverse Scattering*. Cambridge: Cambridge Univ. Press, 1991.

[4] Adimy M., Crauste F., Ruan S. A mathematical study of the hematopoiesis process with applications to chronic myelogenous leukemia. *SIAM J. Appl. Math.*, 2005, Vol. 65, No. 4, pp. 1328–1352.

[5] Adomian G. A review of the decomposition method in applied mathematics. *J. Math. Anal. Appl.*, 1988, Vol. 135, pp. 501–544.

[6] Adomian G. *Nonlinear Stochastic Operator Equations*. Orlando: Academic Press, 1986.

[7] Adomian G. *Solving Frontier Problems of Physics: The Decomposition Method*. Boston: Kluwer, 1994.

[8] Aiello W.G., Freedman H.I., Wu J. Analysis of a model representing stage-structured population growth with state-dependent time delay. *SIAM J. Appl. Math.*, 1992, Vol. 52, No. 3, pp. 855–869.

[9] Aksenov A.V., Polyanin A.D. Methods for constructing complex solutions of nonlinear PDEs using simpler solutions. *Mathematics*, 2021, Vol. 9, No. 4, 345.

[10] Alshina E.A., Kalitkin N.N., Koryakin P.V. Diagnostics of singularities of exact solutions in computations with error control, *Zh. Vychisl. Mat. Mat. Fiz.*, 2005, Vol. 45, No. 10, pp. 1837–1847 (in Russian).

[11] Ambartsumyan V.A. On the fluctuation of the brightness of the Milky Way. *Dokl. Akad. Nauk SSSR*, 1944, Vol. 44, pp. 223–226.

[12] Ames W.F., Lohner J.R., Adams E. Group properties of $u_{tt} = [f(u)u_x]_x$. *Int. J. Non-Linear Mech.*, 1981, Vol. 16, No. 5–6, pp. 439–447.

[13] Aminikhah H., Dehghan P. GDTM-Pade technique for solving the nonlinear delay differential equations. *Comput. Research Progress Appl. Science & Eng.*, 2015, Vol. 1, No. 4, pp. 112–125.

[14] Anderson R.M., May R.M. Population biology of infectious diseases: Part I. *Nature*, 1979, Vol. 280, pp. 361–367.

[15] Andreev V.K., Kaptsov O.V., Pukhnachov V.V., Rodionov A.A. *Applications of Group-Theoretical Methods in Hydrodynamics*. Kluwer, Dordrecht, 1998.

[16] Andrews G.A. *Number Theory*. Philadelphia: Saunders, 1971.

[17] Antaki P.J. Analysis of hyperbolic heat conduction in a semi-infinite slab with surface convection. *Int J. Heat Mass Transfer*, 1997, Vol. 40, No. 13, pp. 3247–3250.

[18] Arecchi F.T., Giacomelli G., Lapucci A., Meucci R. Dynamics of a CO_2 laser with delayed feedback: The short-delay regime. *Phys. Rev. A*, 1991, Vol. 43, No. 9, pp. 4997–5004.

[19] Arik S. Global asymptotic stability of a larger class of neural networks with constant time delay. *Phys. Lett. A*, 2003, Vol. 311, pp. 504–511.

[20] Arrigo D., Broadbridge P., Hill J.M. Nonclassical symmetry solutions and the methods of Bluman–Cole and Clarkson–Kruskal. *J. Math. Phys.*, 1993, Vol. 34, pp. 4692–4703.

[21] Asl F.M., Ulsoy A.G. Analysis of a system of linear delay differential equations. *ASME J. Dynamic Systems, Measurement and Control*, 2003, Vol. 125, No. 2, pp. 215–223.

[22] Azbelev N.V., Simonov P.M. Stability of equations with a delayed argument. *Izvestiya Vuzov. Matematika*, 1997, No. 6, pp. 3–16 (in Russian).

[23] Azbelev N.V., Simonov P.M. Stability of equations with a delayed argument. II. *Izvestiya Vuzov. Matematika*, 2000, No. 4, pp. 3–13 (in Russian).

[24] Azbelev N.V., Simonov P.M. *Stability of Differential Equations with Aftereffect*. London: Taylor & Francis, 2002.

[25] Azizbekov E., Khusainov D.Ya. Solution of a heat equation with delay. *Visnik Kiivs'kogo nats. univer. im. Tarasa Shevchenka, Kibernetika*, 2012, No. 12, pp. 4–12 (in Russian).

[26] Bahgat M.S.M. Approximate analytical solution of the linear and nonlinear multipantograph delay differential equations. *Physica Scripta*, 2020, Vol. 95, No. 5, 055219.

[27] Bahsi M.M., Cevik M. Numerical solution of pantograph-type delay differential equations using perturbation-iteration algorithms. *J. Applied Math.*, 2015, Vol. 2015, 139821.

[28] Bai Z., Wu S.-L. Traveling waves in a delayed SIR epidemic model with nonlinear incidence. *Appl. Meth. Comput.*, 2015, Vol. 263, pp. 221–232.

[29] Baker C.T.H., Paul C.A.H, Willé D.R. Issues in the numerical solution of evolutionary delay differential equations. *Adv. Comput. Math.*, 1995, Vol. 3, pp. 171–196.

[30] Baker G.A. (Jr.), Graves–Morris P. *Padé Approximants, 2nd ed.* Cambridge Univ. Press, Cambridge, 1996.

[31] Banerjee S., Sarkar R. R. Delay-induced model for tumor–immune interaction and control of malignant tumor growth. *Biosystems*, Vol. 91, No. 1, pp. 268–288.

[32] Banks H.T., Kappel F. Spline approximations for functional differential equations. *J. Differ. Equations*, 1979, Vol. 34, pp. 496–522.

[33] Barry D.A., Parlange J.-Y., Li L., Prommer H., Cunningham C.J., Stagnitti F. Analytical approximations for real values of the Lambert W-function. *Math. Comput. Simul.*, 2000, Vol. 53, No. 1–2, pp. 95–103.

[34] Basse B., Wake G.C., Wall D.J.N., van Brunt B. On a cell-growth model for plankton. *Math. Med. Biol.*, 2004, Vol. 21, No. 1, pp. 49–61.

[35] Bateman H. and Erdélyi A., *Tables of Integral Transforms. Vols. 1 and 2*, McGraw-Hill Book Co., New York, 1954.

[36] Bekela A.S., Belachew M.T., Wole G.A. A numerical method using Laplace-like transform and variational theory for solving time-fractional nonlinear partial differential equations with proportional delay. *Adv. Difference Equations*, 2020, Vol. 2020, 586.

[37] Bellman R., Cooke K.L. *Differential-Difference Equations*. Academic Press, New York, 1963.

[38] Bellen A. One-step collocation for delay differential equations. *J. Comput. Appl. Math.*, 1984, Vol. 10, pp. 275–283.

[39] Bellen A., Guglielmi N., Torelli L. Asymptotic stability properties of θ-methods for the pantograph equation. *Appl. Numer. Math.*, 1997, Vol. 24, pp. 279–293.

[40] Bellen A., Maset S., Zennaro M., Guglielmi N. Recent trends in the numerical solution of retarded functional differential equations. *Acta Numerica*, 2009, Vol. 18, p. 1–110.

[41] Bellen A., Zennaro M. Numerical solution of delay differential equations by uniform corrections to an implicit Runge–Kutta method. *Numer. Math.*, 1985, Vol. 47, pp. 301–316.

[42] Bellen A., Zennaro M. *Numerical Methods for Delay Differential Equations*. Oxford: Oxford University Press, 2013.

[43] Bellman R. On the computational solution of differential-difference equations. *J. Math. Anal. Appl.*, 1961, Vol. 2, pp. 108–110.

[44] Bellman R., Cooke K.L. On the computational solution of a class of functional differential equations. *J. Math. Anal. Appl.*, 1965, Vol. 12, pp. 495–500.

[45] Belousov B.P. Periodically acting reaction and its mechanism. In: *Collection of Abstracts in Radiation Medicine for 1958*. Medgiz, Moscow, 1959 (in Russian).

[46] Benhammouda B., Vazquez-Leal H., Hernandez-Martinez L. Procedure for exact solutions of nonlinear pantograph delay differential equations. *British J. Math. & Comp. Science*, 2014, Vol. 4, No. 19, pp. 2738–2751.

[47] Brazhnikov A.M., Karpychev V.A., Lykova A.V. On an engineering method for calculating heat conduction processes. *Inzh.-fiz. zhurn.*, 1975, Vol. 28, No. 4, pp. 677–680 (in Russian).

[48] Bratsun D.A., Zakharov A.P. On the numerical calculation of spatially distributed dynamical systems with time delay. *Vestnik Permskogo Universiteta: Matematika. Mekhanika. Informatika*, 2012, No. 4, pp. 32–42 (in Russian).

[49] Beretta E., Takeuchi Y. Global stability of an SIR epidemic model with time delays. *J. Math. Biol.*, 1995, Vol. 33, pp. 250–260.

[50] Beretta E., Takeuchi Y. Convergence results in SIR epidemic model with varying population sizes. *Nonl. Anal.*, 1997, Vol. 28, pp. 1909–1921.

[51] Berezansky L., Braverman E. On oscillation of a food-limited population model with time delay. *Abstr. Appl. Anal.*, 2003, Vol. 1, pp. 55–66.

[52] Berezansky L., Braverman E. Mackey–Glass equation with variable coefficients. *Comput. Math. Appl.*, 2006, Vol. 51, pp. 1–16.

[53] Berezansky L., Braverman E., Idels L. Nicholson's blowflies differential equations revisited: Main results and open problems. *Appl. Math. Modelling*, 2010, Vol. 34, pp. 1405–1417.

[54] Berezovsky F., Karev G., Song B., Castillo-Chavez C. A simple epidemic model with surprising dynamics. *Math. Biosci. Eng.*, 2005, Vol. 2, No. 1, pp. 133–152.

[55] Bertrand J., Bertrand P., Ovarlez J. The Mellin transform. In: *The Transforms and Applications Handbook, 2nd ed.* (ed. Poularikas A.D.), Boca Raton: CRC Press, 2000.

[56] Bhalekar S., Daftardar-Gejji V. Convergence of the new iterative method. *Int. J. Differ. Equations*, 2011, Vol. 2011, 989065.

[57] Bhalekar S., Patade J. Analytical solutions of nonlinear equations with proportional delays. *Appl. Comput. Math.*, 2016, Vol. 15, No. 3, pp. 331–445.

[58] Bhrawy A.H., Abdelkawy M.A., Mallawi F. An accurate Chebyshev pseudospectral scheme for multi-dimensional parabolic problems with time delays. *Bound. Value Probl.*, 2015, 103.

[59] Blakely J.N., Corron N.J. Experimental observation of delay-induced radio frequency chaos in a transmission line oscillator. *Chaos*, 2004, Vol. 14, pp. 1035–1041.

[60] Bluman G.W., Cole J.D. The general similarity solution of the heat equation. *J. Math. Mech.*, 1969, Vol. 18, pp. 1025–1042.

[61] Bluman G.W., Cole J.D. *Similarity Methods for Differential Equations*. New York: Springer, 1974.

[62] Bluman G.W., Temuerchaolu, Sahadevan R. Local and nonlocal symmetries for nonlinear telegraph equation. *J. Math. Phys.*, 2005, Vol. 46, 023505.

[63] Bocharov G.A., Marchuk G.I., Romanyukha A.A. Numerical solution by LMMs of stiff delay differential systems modelling an immune response. *Numer. Math.*, 1996, Vol. 73, pp. 131–148.

[64] Bocharov G.A., Rihan F.A. Numerical modelling in biosciences using delay differential equations. *J. Comput. Appl. Math.*, 2000, Vol. 125, No. 1–2, pp. 183–199.

[65] Bodnar M. On the differences and similarities of the first order delay and ordinary differential equations. *J. Math. Anal. Appl*, 2004, Vol. 300, pp. 172–188.

[66] Bodnar M. Differential equations with time delay. MIM Colloquium, 2016, https://www.mimuw.edu.pl/sites/default/files/seminaria/kolokwium-12-2016.pdf.

[67] Boese F.G. Stability with respect to the delay: On a paper of K. L. Cooke and P. van den Driessche. *J. Math. Anal. Appl.*, 1998, Vol. 228, No. 2, pp. 293–321.

[68] Borges E.P. A possible deformed algebra and calculus inspired in nonextensive thermostatistics. *Physica A*, 2004, Vol. 340, pp. 95–101.

[69] Borwein J.M., Corless R.M. Emerging tools for experimental mathematics. *Am. Math. Mon.*, 1999, Vol. 106, No. 10, pp. 889–909.

[70] Bratsun D., Zakharov A. Adaptive numerical simulations of reaction-diffusion systems with history and time-delayed feedback. In: *ISCS 2013: Interdisciplinary Symposium on Complex Systems: Emergence, Complexity and Computation*, Vol. 8, pp. 70–81. Berlin: Springer, 2014.

[71] Breda D., Maset S., Vermiglio R. *Stability of Linear Delay Differential Equations. A Numerical Approach with MATLAB*. New York: Springer, 2015.

[72] Britton N.F. *Reaction-Diffusion Equations and Their Applications to Biology*. New York: Academic Press, 1986.

[73] Brown P.N., Hindmarsh A.C., Petzold L.R. Using Krylov methods in the solution of large-scale differential-algebraic systems. *SIAM J. Sci. Comput.*, 1994, Vol. 15, pp. 1467–1488.

[74] Brown P.N., Hindmarsh A.C., Petzold L.R. Consistent initial condition calculation for differential-algebraic systems. *SIAM J. Sci. Comput.*, 1998, Vol. 19, pp. 1495–1512.

[75] Brunner H., Huang Q., Xie H. Discontinuous Galerkin methods for delay differential equations of pantograph type. *SIAM J. Numer. Anal.*, 2010, Vol. 48, No. 5, pp. 1944–1967.

[76] Brunner H., Maset S. Time transformations for delay differential equations. *Discrete Contin. Dyn. Syst.*, 2009, Vol. 25, No. 3, pp. 751–775.

[77] Bueler E, Butcher E. A. Stability of periodic linear delay-differential equations and the Chebyshev approximation of fundamental solutions. *Technical report, No. 02–03.* Department of Math. Sciences, University of Alaska Fairbanks, 2002.

[78] Butzer P.L., Jansche S. A direct approach to the Mellin transform. *J. Fourier Anal. Appl.*,1997, Vol. 3, No. 4, pp. 325–376.

[79] Cahlon B., Schmidt D. Stability criteria for certain second order delay differential equations. *Dyn. Continuous Discrete Impulsive Systems*, 2003, Vol. 10, pp. 593–621.

[80] Cahlon B., Schmidt D. Stability criteria for certain second-order delay differential equations with mixed coefficients. *J. Comput. Appl. Math.*, 2004, Vol. 170, No. 1, pp. 79–102.

[81] Cai Y. et al. Spatiotemporal dynamics in a reaction-diffusion epidemic model with a time-delay in transmission. *Int. J. Bifurc. Chaos Appl. Sci. Eng.*, 2015, Vol. 25, No. 8, 1550099.

[82] Calogero F., Degasperis A. *Spectral Transform and Solitons: Tolls to Solve and Investigate Nonlinear Evolution Equations.* Amsterdam: North-Holland Publ., 1982.

[83] Cantrell R.S., Cosner C. *Spatial Ecology via Reaction-Diffusion Equations.* Chichester: John Wiley & Sons, 2003.

[84] Canuto C., Hussaini M.Y., Quarteroni A., Zang T.A. *Spectral Methods: Evolution to Complex Geometries and Applications to Fluid Dynamics.* Berlin: Springer, 2007.

[85] Cao J. New results concerning exponential stability and periodic solutions of delayed cellular neural networks. *Phys. Lett. A*, 2003, Vol. 307, pp. 136–147.

[86] Cao J., Liang J., Lam J. Exponential stability of high-order bidirectional associative memory neural networks with time delays. *Physica D*, 2004, Vol. 199, No. 3–4, pp. 425–436.

[87] Capasso V., Serio G. A generalization of the Kermack–McKendrick deterministic epidemic model. *Math. Biosci.*, 1978, Vol. 42, pp. 43–61.

[88] Carslow H.S., Jaeger J.C. *Conduction of Heat in Solids, 2nd ed.* Oxford University Press, Oxford, 1959.

[89] Carslaw H.S., Jaeger J.C. *Conduction of Heat in Solids.* Oxford: Clarendon Press, 1984.

[90] Casal A.C., Díaz J.I., Vegas J.M. Blow-up in some ordinary and partial differential equations with time-delay. *Dynam. Syst. Appl.*, 2009, Vol. 18, pp. 29–46.

[91] Cattaneo C. Sulla conduzione de calore. *Atti Semin. Mat. Fis. Univ. Modena*, 1948, Vol. 3, pp. 3–21.

[92] Cattaneo C. Sur une forme de l'équation de la chaleur éliminant le paradoxe d'une propagation instantanée *Comptes Rendus*, 1958, Vol. 247, pp. 431–433.

[93] Cesare L.D., Sportelli M. A dynamic IS-LM model with delayed taxation revenues. *Chaos, Solitons & Fractals*, 2005, Vol. 25, pp. 233–244.

[94] Chattopadhyay J., Sarkar R.R., el Abdllaoui A. A delay differential equation model on harmful algal blooms in the presence of toxic substances. *IMA J. Math. Med. Biol.*, 2002, Vol. 19, No. 2, pp. 137–161.

[95] Chen S., Shi J., Wei J. A note on Hopf bifurcations in a delayed diffusive Lotka–Volterra predator–prey system. *Comput. & Math. Appl.*, 2011, Vol. 62, pp. 2224–2245.

[96] Chen X., Liu H., Xu Ch. The new result on delayed finance system. *Nonlinear Dyn.*, 2014, Vol. 78, pp. 1989–1998.

[97] Cheng Y., Lu D., Zhou J., Wei J. Existence of traveling wave solutions with critical speed in a delayed diffusive epidemic model. *Adv. Differ. Equations*, 2019, 494.

[98] Cherniha R.M. Conditional symmetries for systems of PDEs: New definitions and their application for reaction-diffusion systems. *J. Phys. A: Math. Theor.*, 2010, Vol. 43, 405207.

[99] Cherniha R., Davydovych V. *Nonlinear Reaction-Diffusion Systems: Conditional Symmetry, Exact Solutions and Their Applications in Biology*. Cham: Springer, 2017.

[100] Cherniha R., Serov M., Pliukhin O. *Nonlinear Reaction-Diffusion-Convection Equations: Lie and Conditional Symmetry, Exact Solutions and Their Applications*. Boca Raton: Chapman & Hall/CRC Press, 2018.

[101] Cherniha R., Davydovych V. New conditional symmetries and exact solutions of the diffusive two-component Lotka–Volterra system. *Mathematics*, 2021, Vol. 9, No. 16, 1984.

[102] Cherruault Y. Convergence of Adomian's method. *Kybernetes*, 1989, Vol. 18, No. 2, pp. 31–38.

[103] Chí N.C., Vales E.Á, Almeida G.G. Analysis of a HBV model with diffusion and time delay. *J. Appl. Math.*, 2012, 578561.

[104] Chuma J., van den Driessche P. A general second-order transcendental equation. *App. Math. Notes*, 1980, Vol. 5, pp. 85–96.

[105] Clarkson P.A. Nonclassical symmetry reductions of the Boussinesq equation. *Chaos, Solitons & Fractals*, 1995, Vol. 5, pp. 2261–2301.

[106] Clarkson P.A., Kruskal M.D. New similarity reductions of the Boussinesq equation. *J. Math. Phys.*, 1989, Vol. 30, No. 10, pp. 2201–2213.

[107] Clarkson P.A., Ludlow D.K., Priestley T.J. The classical, direct and nonclassical methods for symmetry reductions of nonlinear partial differential equations. *Methods Appl. Anal.*, 1997, Vol. 4, No. 2, pp. 173–195.

[108] Clarkson P.A., McLeod J.B., Olver P.J., Ramani R. Integrability of Klein–Gordon equations. *SIAM J. Math. Anal.*, 1986, Vol. 17, pp. 798–802.

[109] Cohen D.S., Rosenblat S. Multi-species interactions with hereditary effects and spatial diffusion. *J. Math. Biol.*, 1979, Vol. 7, pp. 231–241.

[110] Cook R.J., Milonni P.W. Quantum theory of an atom near partially reflecting walls. *Phys. Rev. A*, 1987, Vol. 35, No. 12, pp. 5081–5087.

[111] Cooke K.L. Stability analysis for a vector disease model. *Rocky Mountain J. Math.*, 1979, Vol. 9, pp. 31–42.

[112] Cooke K.L., Grossman Z. Discrete delay, distributed delay and stability switches. *J. Math. Anal. Appl.*, 1982, Vol. 86, pp. 592–627.

[113] Cooke K.L., van den Driessche P. On zeroes of some transcendental equations. *Funkcialaj Ekvacioj*, 1986, Vol. 29, pp. 77–90.

[114] Corless R.M., Gonnet G.H., Hare D.E.G., Jeffrey D.J., Knuth D.E. On the Lambert W function. *Adv. Comput. Math.*, 1996, Vol. 5, pp. 329–359.

[115] Cryer C.W. Numerical methods for functional differential equations. In: *Delay and Functional Differential Equations and Their Applications* (ed. Schmitt K.), New York: Academic Press, 1972, pp. 17–101.

[116] Cryer C.W., Tavernini L. The numerical solution of Volterra functional differential equations by Euler's method. *SIAM J. Numer. Anal.*, Vol. 9, No. 1, pp. 105–129.

[117] Cui B.T., Yu Y.H., Lin S.Z. Oscillations of solutions of delay hyperbolic differential equations. *Acta Math. Appl. Sinica*, 1996, Vol. 19. pp. 80–88.

[118] Cui S., Xu Z. Interval oscillation theorems for second order nonlinear partial delay differential equations. *Differ. Equat. Appl.*, 2009, Vol. 1. No. 3, pp. 379–391.

[119] Culshaw R.V., Ruan A. A delay-differential equation model of HIV infection of $CD4^+$ T-cells. *Math. Biosci.*, 2000, Vol. 165, pp. 27–39.

[120] Daftardar-Gejji V., Jafari H. An iterative method for solving nonlinear functional equations. *J. Math. Anal. Appl.*, 2006, Vol. 316, pp. 753–763.

[121] Damsen R.A., Al-Odat M.Q., Al-Azab T.A., Shannak B.A., Aa-Hussien F.M. Numerical investigations and validation of hyperbolic heat conduction model applied to fast precooling of a slab food product. *J. Indian Inst. Sci.*, 2006, Vol. 86, pp. 695–703.

[122] Dankwerts P.V. *Gas–Liquid Reactions*. New York: McGraw-Hill, 1970.

[123] Davidson F.A., Gourley S.A. The effects of temporal delays in a model for a food-limited, diffusing population. *J. Math. Anal. Appl.*, 2001, Vol. 261, pp. 633–648.

[124] De Nevers K., Schmitt K. An application of the shooting method to boundary value problems for second order delay equations. *J. Math. Anal. Appl.*, 1971, Vol. 36, pp. 588–597.

[125] Dehghan M., Shakeri F. The use of the decomposition procedure of Adomian for solving a delay differential equation arising in electrodynamics. *Phys. Scripta*, 2008, Vol. 78, No. 6, 065004.

[126] Demirel Y. *Nonequilibrium Thermodynamics: Transport and Rate Processes in Physical, Chemical and Biological Systems, 2nd ed.* Amsterdam: Elsevier, 2007.

[127] Derfel G., Grabner P.J., Tichy R.F. On the asymptotic behaviour of the zeros of the solutions of a functional-differential equation with rescaling. In: *Indefinite Inner Product Spaces, Schur Analysis, and Differential Equations* (eds. Alpay D., Kirstein B.), pp. 281–295. Cham: Birkhäuser, 2018.

[128] Derfel G., van Brunt B., Wake G.C. A cell growth model revisited. *Functional Differential Equations*, 2012, Vol. 19, No. 1–2, pp. 71–81.

[129] Diblík J., Fečkan M., Pospíšil M. Representation of a solution of the Cauchy problem for an oscillating system with two delays and permutable matrices. *Ukr. Math. J.*, 2013, Vol. 65, pp. 64–76.

[130] Ditkin V.A., Prudnikov A.P. *Integral Transforms and Operational Calculus*. Fizmatlit, Moscow, 1974 (in Russian).

[131] Dmitriev A.S., Kislov V.Ya. *Stochastic Oscillations in Radiophysics and Electronics*. Nauka, Moscow, 1989 (in Russian).

[132] Doetsch G., *Einführung in Theorie und Anwendung der Laplace-Transformation*, Birkhäuser Verlag, Basel–Stuttgart, 1958.

[133] Doha H., Bhrawy A.H., Baleanu D., Hafez R.M. A new Jacobi rational — Gauss collocation method for numerical solution of generalized pantograph equations. *Appl. Numer. Math.*, 2014, Vol. 77, pp. 43–54.

[134] Dorodnitsyn V.A. On invariant solutions of a nonlinear heat equation with a source. *Zh. vychisl. matem. i matem. fiz.*, 1982, Vol. 22, No. 6, pp. 1393–1400.

[135] Dorodnitsyn V.A., Kozlov R., Meleshko S.V., Winternitz P. Linear or linearizable first-order delay ordinary differential equations and their Lie point symmetries. *J. Phys. A: Math. Theor.*, 2018, Vol. 51, No. 20, 205203.

[136] Dorodnitsyn V.A., Kozlov R., Meleshko S.V., Winternitz P. Lie group classification of first-order delay ordinary differential equations. *J. Phys. A: Math. Theor.*, 2018, Vol. 51, No. 20, 205202.

[137] Doyle P.W., Vassiliou P.J. Separation of variables for the 1-dimensional non-linear diffusion equation. *Int. J. Non-Linear Mech.*, 1998, Vol. 33, No. 2, pp. 315–326.

[138] Driver R.D. *Ordinary and Delay Differential Equations.* New York: Springer, 1977.

[139] Dubinov A.E., Dubinova I.D., Saikov S.K. *Lambert W-Function and Its Application in Problems of Mathematical Physics.* FGUP RFYaTs-VNIIEF, Sarov, 2006 (in Russian).

[140] Ebert D., Lipsitch M., Mangin K.L. The effect of parasites on host population density and extinction: Experimental epidemiology with Daphnia and six microparasites. *American Naturalist*, 2000, Vol. 156, No. 5, pp. 459–477.

[141] Efendiev M., van Brunt B., Wake G.C., Zaidi A.A. A functional partial differential equation arising in a cell growth model with dispersion. Math. Meth. Appl. Sci., 2018, Vol. 41, No. 4, pp. 1541–1553.

[142] Eigen M., Schuster P. *The Hypercycle: A Principle of Natural Self-Organization.* Berlin: Springer, 1979.

[143] El-Safty A., Salim M.S., El-Khatib M.A. Convergence of the spline functions for delay dynamic system. *Int. J. Comput. Math.*, 2003, Vol. 80, No. 4, pp. 509–518.

[144] Elsgol'ts L.E., Norkin S.B. *Introduction to the Theory and Application of Differential Equations with Deviating Arguments.* New York: Academic Press, 1973.

[145] EqWorld. *Integral Transforms.* https://eqworld.ipmnet.ru/en/auxiliary/aux-inttrans.htm (accessed: January 19, 2023).

[146] Erneux T. *Applied Delay Differential Equations.* New York: Springer, 2009.

[147] Estévez P.G., Qu C.Z. Separation of variables in nonlinear wave equations with variable wave speed. *Theor. Math. Phys.*, 2002, Vol. 133, pp. 1490–1497.

[148] Estévez P.G., Qu C., Zhang S. Separation of variables of a generalized porous medium equation with nonlinear source. *J. Math. Anal. Appl.*, 2002, Vol. 275, pp. 44–59.

[149] Evans D.J., Raslan K.R. The Adomian decomposition method for solving delay differential equation. *Int. J. Comput. Math.*, 2005, Vol. 82, No. 1, pp. 49–54.

[150] Ezzinbi K., Jazar M. Blow-up results for some nonlinear delay differential equations. *Positivity*, 2006, Vol. 10, pp. 329–341.

[151] Faria T. Stability and bifurcation for a delayed predator–prey model and the effect of diffusion. *J. Math. Anal. Appl.*, 2001, Vol. 254, pp. 433–463.

[152] Faria T., Trofimchuk S. Nonmonotone travelling waves in a single species reaction-diffusion equation with delay. *J. Differ. Equations*, 2006, Vol. 228, pp. 357–376.

[153] Fedorov A.V. Ocean-atmosphere coupling. In: *Oxford Companion to Global Change* (eds. Goudie A. and Cuff D.), pp. 369–374. Oxford: Oxford University Press, 2008.

[154] Feldstein M.A., Neves K.W. High-order methods for state-dependent delay differential equations with nonsmooth solutions. *SIAM J. Numer. Anal.*, 1984, Vol. 21, pp. 844–863.

[155] Ferguson T.S. Lose a dollar or double your fortune. In: *Proceedings of the 6th Berkeley Symposium on Mathematical Statistics and Probability*, Vol. III (eds. L.M. Le Cam, J. Neyman, E.L. Scott), pp. 657–666. Berkeley: Univ. California Press, 1972.

[156] Ferreira J.A., Da Silva P.M. Energy estimates for delay diffusion-reaction equations. *J. Comput. Math.*, 2008, Vol. 26, No. 4, pp. 536–553.

[157] Field R.J., Koros E., Noyes R.M. Oscillations in chemical systems. II. Thorough analysis of temporal oscillation in the bromate-cerium-malonic acid system. *J. Am. Chem. Soc.*, 1972, Vol. 94, No. 25, pp. 8649–8664

[158] Field R.J., Noyes R.M. Oscillations in chemical systems. IV. Limit cycle behavior in a model of a real chemical reaction. *J. Chem. Phys.*, 1974, Vol. 60, No. 5, pp. 1877–1884.

[159] Field R.J., Noyes R.M. Oscillations in chemical systems. V. Quantitative explanation of band migration in the Belousov–Zhabotinskii reaction. *J. Amer. Chem. Soc.*, 1974, Vol. 96, No. 7, pp. 2001–2006.

[160] Fife P.C. *Mathematical Aspects of Reaction and Diffusion Systems*. Berlin: Springer, 1979.

[161] Finlayson B.A. *The Method of Weighted Residuals and Variational Principles*. New York: Academic Press, 1972.

[162] Fisher R.A. The wave of advance of advantageous genes. *Ann Eugenics*, 1937, Vol. 191, pp. 295–298.

[163] Fletcher C.A. *Computational Galerkin Methods*. Springer, New York, 1984.

[164] Fornberg B. *A Practical Guide to Pseudospectral Methods*. Cambridge: Cambridge University Press, 1996.

[165] Fort J., Méndez V. Wavefronts in time-delayed reaction-diffusion systems. Theory and comparison to experiment. *Rep. Prog. Phys.*, 2002, Vol. 65, pp. 895–954.

[166] Fox L., Mayers D., Ockendon J.R., Tayler A.B. On a functional differential equation. *IMA J. Appl. Math.*, 1971, Vol. 8, pp. 271–307.

[167] Friesecke G. Exponentially growing solutions for a delay-diffusion equation with negative feedback. *J. Differ. Equations*, 1992, Vol. 98, pp. 1–18.

[168] Frisch R, Holme H. The characteristic solutions of a mixed difference and differential equation occurring in Economic Dynamics. *Econometrica*, 1935, vol. 3, pp. 225–239.

[169] Galaktionov V.A., Posashkov S.A. On new exact solutions of parabolic equations with quadratic nonlinearities. *Zh. Vych. Matem. & Mat. Fiziki*, 1989, Vol. 29, No. 4, pp. 497–506 (in Russian).

[170] Galaktionov V.A., Posashkov S.A. Exact solutions and invariant subspace for nonlinear gradient-diffusion equations. *Zh. Vych. Matem. & Mat. Fiziki*, 1994, Vol. 34, No. 3, pp. 374–383 (in Russian).

[171] Galaktionov V.A. Quasilinear heat equations with first-order sign-invariants and new explicit solutions. *Nonlinear Anal. Theor. Meth. Appl.*, 1994, Vol. 23, pp. 1595–621.

[172] Galaktionov V.A. Invariant subspaces and new explicit solutions to evolution equations with quadratic nonlinearities. *Proc. Roy. Soc. Edinburgh, Sect. A*, 1995, Vol. 125, No. 2, pp. 225–246.

[173] Galaktionov V.A., Posashkov S.A., Svirshchevskii S.R. On invariant sets and explicit solutions of nonlinear evolution equations with quadratic nonlinearities. *Dif. & Integral Equations*, 1995, Vol. 8, No. 8, pp. 1997–2024.

[174] Galaktionov V.A., Posashkov S.A., Svirshchevskii S.R. Generalized separation of variables for differential equations with polynomial nonlinearities. *Differ. Equations*, 1995, Vol. 31, No. 2, pp. 233–240.

[175] Galaktionov V.A., Svirshchevskii S.R. *Exact Solutions and Invariant Subspaces of Nonlinear Partial Differential Equations in Mechanics and Physics*. Boca Raton: Chapman & Hall/CRC Press, 2007.

[176] Galovic S., Kotoski D. Photothermal wave propagation in media with thermal memory. *J. Applied Physics*, 2003, Vol. 93, No. 5, pp. 3063–3070.

[177] Gan Q., Xu R, Yang P. Travelling waves of a delayed SIRS epidemic model with spatial diffusion. *Nonlinear Anal.: Real World Appl.*, 2011, Vol. 12, pp. 52–68.

[178] Gantmakher F.R. *Theory of Matrices, 5th ed.* Fizmatlit, Moscow 2010 (in Russian).

[179] Gaver D.P. An absorption probability problem. *J. Math. Anal. Appl.*, 1964, Vol. 9, pp. 384–393.

[180] Giang D., Lenbur Y., Seidman T. Delay effect in models of population growth. *J. Math. Anal. Appl.*, 2005, Vol. 305, pp. 631–643.

[181] Gomez A., Sergei S. Monotone traveling wavefronts of the KPP-Fisher delayed equation. *J. Differ. Equations*, 2011, Vol. 250, 1767–1787.

[182] Gopalsamy K. *Stability and Oscillations in Delay Differential Equations of Population Dynamics*. New York: Springer, 1992.

[183] Gopalsamy K., Kulenovic M.R.S., Ladas G. Time lags in a 'food-limited' population model. *Appl. Anal.*, 1988, Vol. 31, pp. 225–237.

[184] Gopalsamy K., Kulenovic M.R.S., Ladas G. Environmental periodicity and time delays in a 'food-limited' population model. *J. Math. Anal. Appl.*, 1990, Vol. 147, pp. 545–555.

[185] Goriely A., Hyde C. Necessary and sufficient conditions for finite time singularities in ordinary differential equations. *J. Differ. Equations*, 2000, Vol. 161, pp. 422–448.

[186] Goryachenko V.D. *Qualitative Methods in Dynamics Nuclear Reactors*. Energoatomizdat, Moscow, 1983 (in Russian).

[187] Gourley S.A. Wave front solutions of a diffusive delay model for population of Daphnia magna. *Comput. Math. Appl.*, 2001, Vol. 42, pp. 1421–1430.

[188] Gourley S.A., Chaplain M.A.J. Travelling fronts in a food-limited population model with time delay. *Proc. Roy. Soc. Edin. A*, 2002, Vol. 132, pp. 75–89.

[189] Gourley S.A., Kuang Y. Wavefronts and global stability in time-delayed population model with stage structure. *Proc. Roy. Soc. London A*, 2003, Vol. 459, pp. 1563–1579.

[190] Gourley S.A, So J. W.-H., Wu J.H. Nonlocality of reaction-diffusion equations induced by delay: Biological modeling and nonlinear dynamics. *J. Math. Sci.*, 2004, Vol. 124, No. 4, pp. 5119–5153.

[191] Gourley S.A., Kuang Y., Nagy J.D. Dynamics of a delay differential equation model of hepatitis B virus infection. *J. Biol. Dynam.*, 2008, Vol. 2, No. 2, pp. 140–153.

[192] Grimm L.J., Schmitt K. Boundary value problems for differential equations with deviating arguments. *Aequationes Math.*, 1970, Vol. 4, pp. 176–190.

[193] Grindrod P., Pinotsis D.A. On the spectra of certain integro-differential-delay problems with applications in neurodynamics. *Physica D: Nonlinear Phenomena*, 2011, Vol. 240, No. 1, pp. 13–20.

[194] Grover D., Sharma D., Singh P. Accelerated HPSTM: An efficient semi-analytical technique for the solution of nonlinear PDE's. *Nonlinear Engineering*, 2020, Vol. 9, pp. 329–337.

[195] Grundland A.M., Infeld E. A family of non-linear Klein–Gordon equations and their solutions. *J. Math. Phys.*, 1992, Vol. 33, pp. 2498–2503.

[196] Guglielmi N. Stability of one-leg θ-methods for the variable coefficient pantograph equation on the quasi-geometric mesh. *IMA J. Numer. Anal.*, 2003, Vol. 23, pp. 421–438.

[197] Guo L., Grimsmo A., Kockum A.F., Pletyukhov M., Johansson G. Giant acoustic atom: A single quantum system with a deterministic time delay. *Phys. Rev. A*, 2017, Vol. 95, 053821.

[198] Gurney W.S.C., Blythe S.P., Nisbet R.M. Nicholson's blowflies revisited. *Nature*, 1980, Vol. 287, pp. 17–21.

[199] Gülsu M., Sezer M. A Taylor polynomial approach for solving differential-difference equations. *J. Comput. Appl. Math.*, 2006, Vol. 186, No. 2, pp. 349–364.

[200] Gülsu M., Sezer M. A Taylor collocation method for solving high-order linear pantograph equations with linear functional argument. *Numer. Methods Partial Differ. Equations*, 2011, Vol. 27, pp. 1628–1638.

[201] Györi I., Ladas G. *Oscillation Theory of Delay Differential Equations with Applications*. New York: Clarendon Press, 1991.

[202] Györi I., Trofimchuk S. On the existence of rapidly oscillatory solutions in the Nicholson blowflies equation. *Nonlinear Anal.*, 2002, Vol. 48, No. 7, pp. 1033–1042.

[203] Hairer E., Nørsett S.P., Wanner G. *Solving Ordinary Differential Equations I: Nonstiff Problems*. 2nd Ed. Springer-Verlag, Berlin, 1993.

[204] Hairer E., Wanner G. *Solving Ordinary Differential Equations II: Stiff and Differential-Algebraic Problems*. 2nd Ed. Springer-Verlag, Berlin, 1996.

[205] Hale J.K., Verduyn Lunel S.M. *Introduction to Functional Differential Equations*. New York: Springer, 1993.

[206] Hall A.J., Wake G.C. A functional differential equation arising in the modelling of cell growth. *J. Aust. Math. Soc. Ser. B*, 1989, Vol. 30, pp. 424–435.

[207] Hall A.J., Wake G.C, Gandar P.W. Steady size distributions for cells in one dimensional plant tissues. *J. Math. Biol.*, 1991, Vol. 30, pp. 101–123.

[208] Hara T., Rinko M., Morii T. Asymptotic stability condition for linear differential-difference equations with delays. *Dynamic Systems Appl.*, 1997, Vol. 6, pp. 493–506.

[209] Hastings A., Gross L. (eds.). *Encyclopedia of Theoretical Ecology*. Berkeley: University of California Press, 2012.

[210] Hattaf K., Yousfi N. A generalized HBV model with diffusion and two delays. *Comput. Math. Appl.*, 2015, Vol. 69, No 1, pp. 31–40.

[211] Hattaf K., Yousfi N. Global dynamics of a delay reaction-diffusion model for viral infection with specific functional response. *Comput. Appl. Math.*, 2015, Vol. 34, No. 3, pp. 807–818.

[212] He Q., Kang L., Evans D.J. Convergence and stability of the finite difference scheme for nonlinear parabolic systems with time delay. *Numer. Algorithms*, 1997, Vol. 16, pp. 129–153.

[213] Herz A.V., Bonhoeffer S., Anderson R.M., May R.M., Nowak M.A. Viral dynamics in vivo: Limitations on estimates of intracellular delay and virus decay. *Proc. Nat. Acad. Sci. USA*, 1996, Vol. 93, pp. 7247–7251.

[214] Higham D.J. Highly continuous Runge–Kutta interpolants. *ACM Trans. Math. Soft.*, 1991, Vol. 17, pp. 368–386.

[215] Hindmarsh A., Taylor A. *User Documentation for IDA: A Differential-Algebraic Equation Solver for Sequential and Parallel Computers.* Lawrence Livermore National Laboratory report, UCRL-MA-136910, 1999.

[216] Hou C.-C., Simos T.E., Famelis I.T. Neural network solution of pantograph type differential equations. Math. Meth. Appl. Sci., 2020, Vol. 43, No. 6, pp. 3369–3374.

[217] Van Der Houwen P.J., Sommeijer B.P. On the stability of predictor-corrector methods for parabolic equations with delay. *IMA J. Numer. Anal.*, 1986, Vol. 6, pp. 1–23.

[218] Hu J., Qu C. Functionally separable solutions to nonlinear wave equations by group foliation method. *J. Math. Anal. Appl.*, 2007, Vol. 330, pp. 298–311.

[219] Huang C., Vandewalle S. Unconditionally stable difference methods for delay partial differential equations. *Numer. Math.*, 2012, Vol. 122, No. 3, pp. 579–601.

[220] Huang D.J., Ivanova N.M. Group analysis and exact solutions of a class of variable coefficient nonlinear telegraph equations. *J. Math. Phys.*, 2007, Vol. 48, No. 7, 073507.

[221] Huang D.J., Zhou S. Group properties of generalized quasi-linear wave equations. *J. Math. Anal. Appl.*, 2010, Vol. 366, pp. 460–472.

[222] Huang D.J., Zhou S. Group-theoretical analysis of variable coefficient nonlinear telegraph equations. *Acta Appl. Math.*, 2012, Vol. 117, No. 1, pp. 135–183.

[223] Huang D.J., Zhu Y., Yang Q. Reduction operators and exact solutions of variable coefficient nonlinear wave equations with power nonlinearities. *Symmetry*, 2017, Vol. 9, No. 1, 3.

[224] Huang J., Zou X. Traveling wavefronts in diffusive and cooperative Lotka–Volterra system with delays. *J. Math. Anal. Appl.*, 2002, Vol. 271, pp. 455–466.

[225] Huang W.Z. On asymptotic stability for linear delay equations. *Differ. & Integral Equations*, 1991, Vol. 4, No. 6, pp. 1303–1310.

[226] Huo H.-F., Li W.-T. Positive periodic solutions of a class of delay differential system with feedback control. *Appl. Math. Comput.*, 2004, Vol. 148, pp. 35–46.

[227] Hutchinson G.E. Circular causal systems in ecology. *Ann. N. Y. Acad. of Sci.*, 1948, Vol. 50, No. 4, pp. pp. 221–246.

[228] Hwang T.-W., Kuang Y. Deterministic extinction effect of parasites on host populations. *J. Math. Biol.*, 2003, Vol. 46, pp. 17–30.

[229] Ibragimov N.H. (ed.). *CRC Handbook of Lie Group Analysis of Differential Equations, Vol. 1, Symmetries, Exact Solutions and Conservation Laws*. Boca Raton: CRC Press, 1994.

[230] In't Hout K.J. A new interpolation procedure for adapting Runge–Kutta methods to delay differential equations. *BIT*, 1992, Vol. 32, pp. 634–649.

[231] Investopedia. *IS-LM model definition*. https://www.investopedia.com/terms/i/islm-model.asp (accessed: 19 January 2023).

[232] Iserles A. On the generalized pantograph functional differential equation. *Eur. J. Appl. Math.*, 1993, Vol. 4, No. 1, pp. 1–38.

[233] Isik O.R., Turkoglu T. A rational approximate solution for generalized pantograph-delay differential equations. *Math. Meth. Appl. Sci.*, 2016, Vol. 39, No. 8, pp. 2011–2024.

[234] Ismagilov R.S., Rautian N.A., Vlasov V.V. Examples of very unstable linear partial functional differential equations. *arXiv:1402.4107v1 [math.AP]*, 2014.

[235] Jackiewicz Z., Zubik-Kowal B. Spectral collocation and waveform relaxation methods for nonlinear delay partial differential equations. *Appl. Numer. Math.*, 2006, Vol. 56, pp. 433–443.

[236] James R.W., Belz M.H. The significance of characteristic solutions of mixed difference and differential equations. *Econometrica*, 1938, Vol. 6, pp. 326–343.

[237] Jia Yu. Bifurcation and pattern formation of a tumor–immune model with time-delay and diffusion. *Math. Comput. Simul.*, 2020, Vol. 178, pp. 92–108.

[238] Jia H., Xu W., Zhao X., Li Z. Separation of variables and exact solutions to nonlinear diffusion equations with x-dependent convection and absorption. *J. Math. Anal. Appl.*, 2008, Vol. 339, pp. 982–995.

[239] Jimbo M., Kruskal M.D., Miwa T. Painlevé test for the self-dual Yang–Mills equation. *Phys. Lett., Ser. A*, 1982, Vol. 92, No. 2, pp. 59–60.

[240] Johansson F. Computing the Lambert W function in arbitrary-precision complex interval arithmetic. *Numer. Algorithms*, 2020, Vol. 83, No. 1, pp. 221–242.

[241] Jones G.S. Asymptotic behavior and periodic solutions of a nonlinear differential-difference equation. *Proc. Nat. Acad. Sci. USA*, 1961, Vol. 47, pp. 879–882.

[242] Jones G. The existence of periodic solutions of $f'(x) = -\alpha f(x-1)[1+f(x)]$. *J. Math. Anal. Appl.*, 1962, Vol. 5, pp. 435–450.

[243] Jordan P.M., Dai W., Mickens R.E. A note on the delayed heat equation: Instability with respect to initial data. *Mech. Research Comm.*, 2008, Vol. 35, pp. 414–420.

[244] Jou D., Casas-Vázquez J., Lebon G. *Extended Irreversible Thermodynamics, 4th ed.* Springer, 2010.

[245] Kakutati S., Marcus L. On the non-linear difference-differential equation $y'(t) = (A - By(t - \tau))y(t)$. *Annals Math. Studies*, 1958, Vol. 41, pp. 1–18.

[246] Kalecki M. A macrodynamic theory of business cycles. *Econometrica*, 1935, Vol. 3, pp. 327–344.

[247] Kalitkin N.N. *Numerical Methods*. Nauka, Moscow, 1978 (in Russian).

[248] Kalitkin N.N., Koryakin P.V. *Numerical Methods. 2. Methods of Mathematical Physics*. Akademiya, Moscow, 2013 (in Russian).

[249] Kalmár-Nagy T., Stépán G., Moon F.C. Subcritical HOPF bifurcation in the delay equation model for machine tool vibrations. *Nonlinear Dyn.*, 2001, Vol. 26, pp. 121–142.

[250] Kalospiros N.S., Edwards B.J., Beris A.N. Internal variables for relaxation phenomena in heat and mass transfer. *Int. J. Heat Mass Transfer*, 1993, Vol. 36, pp. 1191–1200.

[251] Kantorovich L.V., Krylov V.I. *Approximate Methods of Higher Analysis*, Fizmatgiz, Moscow, 1962 (in Russian).

[252] Kamenskii G.A. Boundary value problems for differential equations with deviating arguments. *Nauk. Dokl. Vyssh. Shkoly Fiz. Mat. Nauki*, 1958, Vol. 2, pp. 60–66 (in Russian).

[253] Kaminski W. Hyperbolic heat conduction equation for materials with a nonhomogeneous inner structure. *J. Heat Transfer*, 1990, Vol. 112, No. 3, pp. 555–560.

[254] Kamke E. *Differentialgleichungen: Lösungsmethoden und Lösungen, II, Partielle Differentialgleichungen Erster Ordnung für eine gesuchte Funktion*. Akad. Verlagsgesellschaft Geest & Portig, Leipzig, 1965.

[255] Kamke E. *Differentialgleichungen: Lösungsmethoden und Lösungen, I, Gewöhnliche Differentialgleichungen*. B.G. Teubner, Leipzig, 1977.

[256] Kaptsov O.V., Verevkin I.V. Differential constraints and exact solutions of nonlinear diffusion equations. *J. Phys. A: Math. Gen.*, 2003, Vol. 36, No. 5, pp. 1401–1414.

[257] Kar A., Chan C.L., Mazumder J. Comparative studies on nonlinear hyperbolic and parabolic heat conduction for various boundary conditions: Analytic and numerical solutions. *Int. J. Heat Transfer*, 1992, Vol. 114, pp. 14–20.

[258] Karakoç F., Bereketoğlu H. Solutions of delay differential equations by using differential transform method. *Int. J. Comput. Math.*, 2009, Vol. 86, No. 5, pp. 914–923

[259] Karakostas G., Philos Ch., Sficas Y. Stable steady state of some population model. *J. Dynam. Differ. Eq.*, 1992, Vol. 4, No. 2, pp. 161–190.

[260] Kato T., McLeod J.B. Functional-differential equation $y' = ay(\lambda t) + by(t)$. *Bull. Amer. Math. Soc.*, 1971, Vol. 77, No. 6, pp. 891–937.

[261] Kermack W.O., McKendrick A.G. A contribution to the mathematical theory of epidemics. *Proc. Roy. Soc. A*, 1927, Vol. 155, pp. 700–721.

[262] Kashchenko I.S., Kashchenko S.A. Dynamics of the logistic equation with delay and a large coefficient of spatially distributed control. *Zhurn. vych. mat. i mat. fiz.*, 2014, Vol. 54, No. 5, pp. 766–778 (in Russian).

[263] Kashchenko S.A. Asymptotic behavior of solutions to the generalized Hutchinson equation. *Model. i analiz inform. sistem*, 2012, Vol. 19, No. 3, pp. 32–62 (in Russian).

[264] Kashchenko S.A. Cyclical risks and systems with delay. In: *Risk Management. Risk, Sustainability, Synergy*. Nauka, Moscow, 2000 (in Russian).

[265] Kashchenko S.A., Maiorov V.V. *Wave Memory Models, 2nd ed.* URSS, Moscow, 2013 (in Russian).

[266] Khusainov D.Y., Ivanov A.F., Shuklin G.V. On a representation of solutions of linear delay systems. *Differ. Equations*, 2005, Vol. 41, pp. 1054–1058.

[267] Khusainov D.Y., Shuklin G.V. Linear autonomous time-delay system with permutation matrices solving. *Stud. Univ. Žilina, Math. Ser.*, 2003, Vol. 17, No. 1, pp. 101–108.

[268] Khusainov D.Y., Shuklin G.V. Relative controllability in systems with pure delay. *Int. Appl. Mechanics*, 2005, Vol. 41, No. 2, pp. 210–221.

[269] Khusainov D.Y., Diblík J., Růžičková M., Lukáčová J. Representation of a solution of the Cauchy problem for an oscillating system with pure delay. *Nonlinear Oscillations*, 2008, Vol. 11, pp. 276–285.

[270] Khusainov D.Y., Ivanov A.F., Kovarzh I.V. Solution of one heat equation with delay. *Nonlinear Oscillations*, 2009, Vol. 12, No. 2, pp. 260–282.

[271] Khusainov D.Y., Pokojovy M., Azizbayov E.I. On classical solvability for a linear 1D heat equation with constant delay. *Konstanzer Schriften in Mathematik*, 2013, No. 316, ISSN 1430-3558 (see also arXiv:1401.5662v1 [math.AP], 2014, https://arxiv.org/pdf/1401.5662.pdf).

[272] Khusainov D., Pokojovy M., Reinhard R. Strong and mild extrapolated L^2-solutions to the heat equation with constant delay. *SIAM J. Math. Anal.*, 2015, Vol. 47, No. 1, pp. 427–454.

[273] Kmet T. Modelling and simulation of food network. *Proceedings 22nd European Conference on Modelling and Simulation, Nicosia, Cyprus*, 2008, pp. 157–164.

[274] Kolesov A.Yu., Rosov N.Kh. The theory of relaxation oscillations for Hutchinson's equation. *Mathematics*, 2011, Vol. 202, No. 6, pp. 829–858.

[275] Kolmanovskii V., Myshkis A. *Applied Theory of Functional Differential Equations.* Dordrecht: Kluwer, 1992.

[276] Kolmanovskii V., Myshkis A. *Introduction to the Theory and Applications of Functional Differential Equations.* New York: Springer, 1999.

[277] Kolmanovskii V.B., Nosov V.R. *Stability of Functional Differential Equations.* London: Academic Press, 1986.

[278] Kolmogorov A.N., Petrovskii I.G., Piskunov N.S. A study of the diffusion equation with increase in the amount of substance, and its application to a biological problem. *Bull. Moscow Univ. Math. Mech.*, 1937, Vol. 1, pp. 1–26 (in Russian).

[279] Koto T. Stability of Runge–Kutta methods for the generalized pantograph equation. *Numer. Math.*, 1999, Vol. 84, pp. 233–247.

[280] Korn G.A., Korn T.M. *Mathematical Handbook for Scientists and Engineers, 2nd ed.* New York: Dover Publ., 2000.

[281] Krainov A.Yu., Min'kov L.L. *Numerical Solution Methods for Problems of Heat and Mass Transfer.* STT, Tomsk, 2016 (in Russian).

[282] Krasnosel'skii M.A., Vainikko G.M., Zabreiko P.P., Ruticki Ya.B., Stet'senko V.Ya. *Approximate Solution of Operator Equations.* Nauka, Moscow, 1969 (in Russian).

[283] Kuang Y. *Delay Differential Equations with Applications in Population Dynamics.* San Diego: Academic Press, 2012.

[284] Kuang J., Cong Y. *Stability of Numerical Methods for Delay Differential Equations.* Beijing: Science Press, 2005.

[285] Kudinov V.A., Kudinov I.V. On a method for obtaining an exact analytical solution of the hyperbolic heat equation based on the use of orthogonal methods. *Vestn. Sam. gos. tekhn. un-ta. Ser. fiz.-mat. nauki*, 2010, Vol. 21, No. 5, pp. 159–169 (in Russian).

[286] Kudryashov N.A. *Methods of Nonlinear Mathematical Physics.* Izd. Dom Intellekt, Dolgoprudnyi, 2010 (in Russian).

[287] Kulenovic M.R.S., Ladas G. Linearized oscillations in population dynamics. *Bull. Math. Biol.*, 1987, Vol. 49, No. 5, pp. 615–627.

[288] Kwong M.K., Ou C. Existence and nonexistence of monotone traveling waves for the delayed Fisher equation. *J. Differ. Equations*, 2010, Vol. 249, pp. 728–745.

[289] Lagerstrom P.A. *Matched Asymptotic Expansions. Ideas and Techniques.* New York: Springer, 1988.

[290] Lang R., Kobayashi K., External optical feedback effects on semiconductor injection laser properties. *IEEE J. Quantum Electron.*, 1980, Vol. 16, No. 3, pp. 347–355.

[291] Langley J.K. A certain functional-differential equation. *J. Math. Anal. Appl.*, 2000, Vol. 244, No. 2, pp. 564–567.

[292] Lavrentev M.A., Shabat B.V. *Method of the Theory of Functions of a Complex Variable.* Nauka, Moscow, 1973 (in Russian).

[293] Lekomtsev A., Pimenov V. Convergence of the scheme with weights for the numerical solution of a heat conduction equation with delay for the case of variable coefficient of heat conductivity. *Appl. Math. Comput.*, 2015, Vol. 256, pp. 83–93.

[294] Li D., Liu M.Z. Runge–Kutta methods for the multi-pantograph delay equation. *Appl. Math. Comput.*, 2005, Vol. 163, pp. 383–395.

[295] Li W.-T., Yan X.-P., Zhang C.-H. Stability and Hopf bifurcation for a delayed cooperation diffusion system with Dirichlet boundary conditions. *Chaos, Solitons & Fractals*, 2008, Vol. 38, No. 1, pp. 227–237.

[296] Li J., Sun G.-Q., Jin Z. Pattern formation of an epidemic model with time delay. *Physica A: Statistical Mechanics and its Applications*, 2014, Vol. 403, pp. 100–109.

[297] Liang J., Cao J. Global exponential stability of reaction-diffusion recurrent neural networks with time-varying delays. *Phys. Lett. A.*, 2003, Vol. 314, pp. 434–442.

[298] Liao S.J. A kind of approximate solution technique which does not depend upon small parameters—II. An application in fluid mechanics. *Int. J. Non-Linear Mech.*, 1997, Vol. 32, No. 5, pp. 815–822.

[299] Liao S.J. *Beyond Perturbation: Introduction to the Homotopy Analysis Method.* Boca Raton: CRC Press, 2004.

[300] Liao S.J. *Homotopy Analysis Method in Nonlinear Differential Equations.* Berlin: Springer, 2012.

[301] Lin G., Li W.-T. Travelling wavefronts of Belousov–Zhabotinskii system with diffusion and delay. *Appl. Math. Lett.*, 2009, Vol. 22, pp. 341–346.

[302] Ling Z., Lin Z. Traveling wavefront in a Hematopoiesis model with time delay. *Applied Mathematics Letters*, 2010, Vol. 23, pp. 426–431.

[303] Ling Z., Zhu L. Traveling wavefronts of a diffusive hematopoiesis model with time delay. *Appl. Math.*, 2014, Vol. 5, pp. 2172–2718.

[304] Luo L., Wang Y. Oscillation for nonlinear hyperbolic equations with influence of impulse and delay. *Int. J. Nonlinear Sci.*, 2012, Vol. 14. No. 1, pp. 60–64.

[305] Liu B. New results on the positive almost periodic solutions for a model of hematopoiesis. *Nonlinear Anal. Real World Appl.*, 2014, Vol. 17, pp. 252–264.

[306] Liu P.-P. Periodic solutions in an epidemic model with diffusion and delay. *Appl. Math. Comput.*, 2015, Vol. 265, pp. 275–291.

[307] Liu C.-S., Liu Y. Comparison of a general series expansion method and the homotopy analysis method. *Modern Phys. Letters B*, 2010, Vol. 24, No. 15, pp. 1699–1706.

[308] Liu C.-S. Basic theory of a kind of linear functional differential equations with multiplication delay, 2018, arXiv:1605.06734v4 [math.CA].

[309] Liu Y. On the θ-method for delay differential equations with infinite lag. *J. Comput. Appl. Math.*, 1996, Vol. 71, pp. 177–190.

[310] Liu M.Z., Li D. Properties of analytic solution and numerical solution of multi-pantograph equation. *Appl. Math. Comput.*, 2004, Vol. 155, No. 3, pp. 853–871.

[311] Liu H., Sun G. Implicit Runge–Kutta methods based on Lobatto quadrature formula *Int. J. Comput. Math.*, 2005, Vol. 82, No. 1, pp. 77–88.

[312] Liu M.Z., Yang Z.W., Xu Y. The stability of modified Runge–Kutta methods for the pantograph equation. *Math. Comput.*, 2006, Vol. 75, No. 25, pp. 1201–1215.

[313] Liu B., Zhou X., Du Q. Differential transform method for some delay differential equations. *Applied Mathematics*, 2015, Vol. 6, pp. 585–593.

[314] Liz E., Tkachenko V., Trofimchuk S. A global stability criterion for scalar functional differential equation. *SIAM J. Math. Anal.,* 2003, Vol. 35, No. 3, pp. 596–622.

[315] Liz E., Tkachenko V., Trofimchuk S. A global stability criterion for a family of delayed population models. *Quart. Appl. Math.*, 2005, Vol. 63, pp. 56–70.

[316] Long F.-S., Meleshko S.V. On the complete group classification of the one-dimensional nonlinear Klein–Gordon equation with a delay. *Math. Methods Appl. Sciences*, 2016, Vol. 39, No. 12, pp. 3255–3270.

[317] Longtin A., Milton J.G. Modelling autonomous oscillations in the human pupil light reflex using non-linear delay-differential equations. *Bull. Math. Biol.*, 1989, Vol. 51, No. 5, pp. 605–624.

[318] Lou X.-Y., Cui B.-T. Asymptotic synchronization of a class of neural networks with reaction-diffusion terms and time-varying delays. *Comput. Math. Appl.*, 2006, Vol. 52, pp. 897–904.

[319] Lu H.T., Chung F.L., He Z.Y. Some sufficient conditions for global exponential stability of delayed Hopfield neural networks. *Neural Networks*, 2004, Vol. 17, pp. 537–544.

[320] Lu J.G. Global exponential stability and periodicity of reaction-diffusion delayed recurrent neural networks with Dirichlet boundary conditions. *Chaos, Solitons and Fractals*, 2008, Vol. 35, pp. 116–125.

[321] Lu X. Monotone method and convergence acceleration for finite-difference solutions of parabolic problems with time delays. *Numer. Methods Partial Differ. Equations*, 1995, Vol. 11, pp. 591–602.

[322] Lu X. Combined iterative methods for numerical solutions of parabolic problems with time delays. *Appl. Math. Comput.,* 1998, Vol. 89, pp. 213–224.

[323] Luckhaus S. Global boundedness for a delay differential equation. *Trans. Am. Math. Soc.*, 1986, Vol. 294, No. 2, pp. 767–774.

[324] Lv G., Wang M. Traveling wave front in diffusive and competitive Lotka–Volterra system with delays. *Nonlinear Anal.: Real World Appl.*, 2010, Vol. 11, pp. 1323–1329.

[325] Lykov A.V. *Thermal Conduction Theory*. Vysshaya Shkola, Moscow 1967 (in Russian).

[326] Ma J., Chen Y. Study for the bifurcation topological structure and the global complicated character of a kind of nonlinear finance system (I). *Appl. Math. Mech.*, 2001, Vol. 22, No. 11, pp. 1240–1251.

[327] Ma S. Traveling wavefronts for delayed reaction-diffusion systems via a fixed point theorem. *J. Differ. Equations*, 2001, Vol. 171, 294–314.

[328] Ma W., Song M., Takeuchi Y. Global stability of an SIR epidemic model with time delay. *Appl. MAth. Lett.*, 2004, Vol. 17, pp. 1141–1145.

[329] Mackey M.C. Unified hypothesis for the origin of aplastic anemia and periodic hematopoiesis. *Blood*, 1978, Vol. 51, pp. 941–956.

[330] Mackey M.C., Glass L. Oscillation and chaos in physiological control system. *Science*, 1977, Vol. 197, pp. 287–289.

[331] Mahaffy J.M., Bélair J., Mackey M.C. Hematopoietic model with moving boundary condition and state dependent delay: Applications in erythropoiesis. *J. Theor. Biol.*, 1998, Vol. 190, No. 2, pp. 135–146.

[332] Mahler K. On a special functional equation, *J. London Math. Soc.*, 1940, Vol. 1, No. 2, pp. 115–123.

[333] Malakhovski E., Mirkin L. On stability of second-order quasi-polynomials with a single delay. *Automatica*, 2006, Vol. 42, No. 6, pp. 1041–1047.

[334] Maple Programming Help. *Numeric Delay Differential Equation Examples.* http://www.maplesoft.com/support/help/Maple/view.aspx?path=examples/NumericDDEs (accessed: January 19, 2023).

[335] Martín J.A., Rodríguez F., Company R. Analytic solution of mixed problems for the generalized diffusion equation with delay. *Math. Comput. Modelling*, 2004, Vol. 40, pp. 361–369.

[336] MATLAB Documentation. *Delay Differential Equations.* http://www.mathworks.com/help/matlab/delay-differential-equations.html (accessed: 19 January 2023).

[337] Maxwell J.C. On the dynamical theory of gases. *Phil. Trans. Royal Soc.*, 1867, Vol. 157, pp. 49–88.

[338] McCluskey C.C. Complete global stability for an SIR epidemic model with delay—distributed or discrete. *Nonlinear Anal. Real World Appl.*, 2010, Vol. 11, pp. 55–59.

[339] Mei M., So J., Li M., Shen S. Asymptotic stability of travelling waves for Nicholson's blowflies equation with diffusion. *Proc. Roy. Soc. Edinburgh Sect. A*, 2004, Vol. 134, pp. 579–594.

[340] Mead J., Zubik-Kowal B. An iterated pseudospectral method for delay partial differential equations. *Appl. Numer. Math.*, 2005, Vol. 55, pp. 227–250.

[341] Meleshko S.V., Moyo S. On the complete group classification of the reaction-diffusion equation with a delay. *J. Math. Anal. Appl.*, 2008, Vol. 338, pp. 448–466.

[342] Miklin S.G., Smolitskii Kh.L. *Approximate Solution Methods for Differential and Integral Equations.* Nauka, Moscow, 1965 (in Russian).

[343] Miller W. (Jr.). Mechanism for variable separation in partial differential equations and their relationship to group theory. In: *Symmetries and Nonlinear Phenomena* (eds. D. Levi, P. Winternitz). London: World Scientific, 1989.

[344] Miller W. (Jr.), Rubel L.A. Functional separation of variables for Laplace equations in two dimensions. *J. Phys. A.*, 1993, Vol. 26, pp. 1901–1913.

[345] Mitra K., Kumar S., Vedavarz A., Moallemi M.K. Experimental evidence of hyperbolic heat conduction in processed meat. *J. Heat Transfer*, 1995, Vol. 117, No. 3, pp. 568–573.

[346] Mittler J.E., Sulzer B., Neumann A.U., Perelson A.S. Influence of delayed viral production on viral dynamics in HIV-1 infected patients. *Math. Biosci.*, 1998, Vol. 152, pp. 143–163.

[347] Myshkis A.D. General theory of differential equations with a delayed argument. *UMN*, 1949, Vol. 4, No. 5(33), pp. 99–141 (in Russian).

[348] Myshkis A.D. *Linear Differential Equations with a Delayed Argument*. Nauka, Moscow, 1972 (in Russian).

[349] Morris G.R., Feldstein A., Bowen E.W. The Phragmén-Lindelöf principle and a class of functional differential equations. In: *Ordinary Differential Equations* (ed. L. Weiss), pp. 513–540. San Diego: Academic Press, 1972.

[350] Murphy G.M. *Ordinary Differential Equations and Their Solutions*. New York: D. Van Nostrand, 1960.

[351] Murray J.D. On traveling wave solutions in a model for Belousov–Zhabotinskii reaction. *J. Theor. Biol.*, 1976, Vol. 56, pp. 329–353.

[352] Murray J.D. Spatial structures in predator-prey communities—a nonlinear time delay diffusional model. *Math. Biosci.*, 1976, Vol. 30, pp. 73–85.

[353] Murray J.D. *Mathematical Biology, 3rd ed.* New York: Springer, 2002.

[354] Nayfeh A.H. *Introduction to Perturbation Techniques*. New York: Wiley–Interscience, 1981.

[355] Nayfeh A.H. *Perturbation Methods*. New York: Wiley–Interscience, 2000.

[356] Naudt J. The q-exponential family in statistical physics. *J. Phys.: Conf. Ser.*, 2010, Vol. 201, 012003.

[357] Nelson P.W., James D.M., Perelson A.S. A model of HIV-1 pathogenesis that includes an intracellular delay. *Math. Biosci.*, 2000, Vol. 163, pp. 201–215.

[358] Nelson P.W., Perelson A.S. Mathematical analysis of delay differential equation models of HIV-1 infection. *Math. Biosci.*, 2002, Vol. 179, No. 1, pp. 73–94.

[359] Neves K.W., Feldstein M.A. Characterisation of jump discontinuities for state-dependent delay differential equations. *J. Math. Anal. Appl.*, 1976, Vol. 56, pp. 689–707.

[360] Neves K.W., Thompson S. Software for the numerical-solution of systems of functional-differential equations with state-dependent delay. *Appl. Numer. Math.*, 1992, Vol. 9, pp. 385–401.

[361] Nicholson A.J. An outline of the dynamics of animal populations. *Australian J. Zoology*, 1954, Vol. 2, No. 1, pp. 9–65.

[362] Novikov S.P., Manakov S.V., Pitaevskii L.B., Zakharov V.E. *Theory of Solitons. The Inverse Scattering Method*. New York: Plenum Press, 1984.

[363] Nucci M.C., Clarkson P.A. The nonclassical method is more general than the direct method for symmetry reductions. An example of the Fitzhugh–Nagumo equation. *Phys. Lett. A*, 1992, Vol. 164, pp. 49–56.

[364] Onanov G.G., Skubachevskii A.L. Differential equations with deviating arguments in stationary problems of solid mechanics. *Prilkadnaya Mekhanika*, 1979, Vol. 15, No. 5, pp. 30–47 (in Russian).

[365] Oberhettinger F., Badii L. *Tables of Laplace Transforms*. New York: Springer, 1973.

[366] Oberle H.J., Pesch H.J. Numerical treatment of delay differential equations by Hermite interpolation. *Num. Math.*, 1981, Vol. 37, pp. 235–255.

[367] Ockendon J.R., Tayler A.B. The dynamics of a current collection system for an electric locomotive. *Proc. R. Soc. Lond. A*, 1971, Vol. 332, pp. 447–468.

[368] Olver P.J. *Application of Lie Groups to Differential Equations, 2nd ed.* Springer, New York, 2000.

[369] Olver F.W.J., Lozier D.W., Boisvert R.F., Clark C.W. (eds.). *NIST Handbook of Mathematical Functions*. Cambridge: Cambridge Univ. Press, 2010.

[370] Ordóñez-Miranda J., Alvarado-Gil J.J. Thermal wave oscillations and thermal relaxation time determination in a hyperbolic heat transport model. *Int. J. Thermal Sciences*, 2009, Vol. 48, pp. 2053–2062.

[371] Oron A., Rosenau P. Some symmetries of the nonlinear heat and wave equations. *Phys. Lett. A*, 1986, Vol. 118, pp. 172–176.

[372] Ovsiannikov L.V. *Group Properties of Differential Equations*. Izd-vo SO AN USSR, Novosibirsk, 1962 (in Russian); English translation by G. Bluman, 1967.

[373] Owren B., Zennaro M. Derivation of efficient, continuous, explicit Runge–Kutta methods. *SIAM J. Sci. Stat. Comp.*, 1992, Vol. 13, pp. 1488–1501.

[374] Ozisik M.N., Tzou D.Y. On the wave theory in heat conduction, *ASME J. Heat Transfer*, 1994, Vol. 116, No. 3, pp. 526–535.

[375] Patade J., Bhalekar S. Analytical solution of pantograph equation with incommensurate delay. *Phys. Sci. Rev.*, 2017, Vol. 2, No. 9, 20165103.

[376] Pan X., Shu H., Wang L., Wang X.-S. Dirichlet problem for a delayed diffusive hematopoiesis model. *Nonlinear Anal.: Real World Appl.*, 2019, Vol. 48, pp. 493–516.

[377] Pao C. Global asymptotic stability of Lotka–Volterra competition systems with diffusion and time delays. *Nonlinear Anal.: Real World Appl.*, 2004, Vol. 5, No. 1, pp. 91–104.

[378] Pao C.V. Numerical methods for systems of nonlinear parabolic equations with time delays. *J. Math. Anal. Appl.*, 1999, Vol. 240, No. 1, pp. 249–279.

[379] Pao C.V. Finite difference reaction-diffusion systems with coupled boundary conditions and time delays. *J. Math. Anal. Appl.*, 2002, Vol. 272, pp. 407–434.

[380] Paul C.A.H. Developing a delay differential equation solver. *Appl. Numer. Math.*, 1992, Vol. 9, pp. 403–414.

[381] Paul C.A.H. Designing efficient software for solving delay differential equations. *J. Comput. Appl. Math.*, 2000, Vol. 125, No. 1–2, pp. 287–295.

[382] Peiraviminaei A., Ghoreishi F. Numerical solutions based on Chebyshev collocation method for singularly perturbed delay parabolic PDEs. *Math. Meth. Appl. Sci.*, 2014, Vol. 37, pp. 2112–2119.

[383] Pike R., Sabatier P. (eds.). *Scattering: Scattering and Inverse Scattering in Pure and Applied Science, Vols. 1–2*. San Diego: Academic Press, 2002.

[384] Pimenov V.G. *Difference Methods for Solving Partial Differential Equations with Heredity*. Izd-vo Ural'skogo Universiteta, Ekaterinburg, 2014 (in Russian).

[385] Pimenov V.G., Numerical methods for solving the heat equation with delay. *Vestn. Udmurtsk. Un-ta. Matem. Mekh. Kompyut. Nauki*, 2008, No. 2, pp. 113–116 (in Russian).

[386] Pimenov V.G. *Numerical Methods for Solving Equations with Heredity.* Yurait, Moscow, 2021 (in Russian).

[387] Pimenov V.G., Tashirova E.E. Numerical methods for solving a hereditary equation of hyperbolic type. *Proc. Steklov Inst. Math.*, 2013, Vol. 281, pp. s126–s136.

[388] Piotrowska M.J., Foryś U. A simple model of carcinogenic mutations with time delay and diffusion. *Math. Biosci. Eng.*, 2013, Vol. 10, No. 3, pp. 861–872.

[389] Pinney E. *Ordinary Difference-Differential Equations.* University of California Press, Berkeley, 1958.

[390] Polyanin A.D., Vyazmin A.V. Differential-difference models and equations of heat conduction and diffusion with a finite relaxation time. *Teor. osnovy khim. tekhnologii*, 2013, Vol. 47, No. 3, pp. 27–278 (in Russian).

[391] Polyanin A.D., Zaitsev V.F. *Nonlinear Equations or Mathematical Physics, Vol. 1.* Yurait, Moscow, 2017 (in Russian).

[392] Polyanin A.D., Zaitsev V.F. *Nonlinear Equations or Mathematical Physics, Vol. 2.* Yurait, Moscow, 2017 (in Russian).

[393] Polyanin A.D. Exact solutions to the Navier–Stokes equations with generalized separation of variables. *Doklady Physics*, 2001, Vol. 46, No. 10, pp. 726–731.

[394] Polyanin A.D. Construction of exact solutions in implicit form for PDEs: New functional separable solutions of non-linear reaction-diffusion equations with variable coefficients. *Int. J. Non-Linear Mech.*, 2019, Vol. 111, pp. 95–105.

[395] Polyanin A.D. Comparison of the effectiveness of different methods for constructing exact solutions to nonlinear PDEs. Generalizations and new solutions. *Mathematics*, 2019, Vol. 7, No. 5, 386.

[396] Polyanin A.D. Functional separable solutions of nonlinear convection-diffusion equations with variable coefficients. *Commun. Nonlinear Sci. Numer. Simul.*, 2019, Vol. 73, pp. 379–390.

[397] Polyanin A.D. Functional separable solutions of nonlinear reaction-diffusion equations with variable coefficients. *Applied Math. Comput.*, 2019, Vol. 347, pp. 282–292.

[398] Polyanin A.D. Generalized traveling-wave solutions of nonlinear reaction-diffusion equations with delay and variable coefficients. *Appl. Math. Lett.*, 2019, Vol. 90, pp. 49–53.

[399] Polyanin A.D. Construction of functional separable solutions in implicit form for non-linear Klein–Gordon type equations with variable coefficients. *Int. J. Non-Linear Mech.*, 2019, Vol. 114, pp. 29–40.

[400] Polyanin A.D. Functional separation of variables in nonlinear PDEs: General approach, new solutions of diffusion-type equations. *Mathematics*, 2020, Vol. 8, No. 1, 90.

[401] Polyanin A.D., Kutepov A.M., Vyazmin A.V., Kazenin D.A. *Hydrodynamics, Mass and Heat Transfer in Chemical Engineering.* London: Taylor & Francis, 2002.

[402] Polyanin A.D., Manzhirov A.V. *Handbook of Mathematics for Engineers and Scientists.* Boca Raton–London: Chapman & Hall/CRC Press, 2007.

[403] Polyanin A.D., Manzhirov A.V. *Handbook of Integral Equations, 2nd ed.* Boca Raton: CRC Press, 2008.

[404] Polyanin A.D., Nazaikinskii V.E. *Handbook of Linear Partial Differential Equations for Engineers and Scientists, 2nd ed.* Boca Raton: Chapman & Hall/CRC Press, 2016.

[405] Polyanin A.D., Shingareva I.K. Nonlinear problems with blow-up solutions: Numerical integration based on differential and nonlocal transformations, and differential constraints. *Appl. Math. Comput.*, 2018, Vol. 336, pp. 107–137.

[406] Polyanin A.D., Shingareva I.K. Non-linear problems with non-monotonic blow-up solutions: Non-local transformations, test problems, exact solutions, and numerical integration. *Int. J. Non-Linear Mech.*, 2018, Vol. 99, pp. 258–272.

[407] Polyanin A.D., Shingareva I.K. Application of non-local transformations for numerical integration of singularly perturbed boundary-value problems with a small parameter. *Int. J. Non-Linear Mech.*, 2018, Vol. 103, pp. 37–54.

[408] Polyanin A.D., Sorokin V.G. Reaction-diffusion equations with delay: Mathematical models and qualitative features. *Vestnik NIYaU MIFI*, 2017, Vol. 6, No. 1, pp. 41–55 (in Russian).

[409] Polyanin A.D., Sorokin V.G. Reaction-diffusion equations with delay: Numerical methods and test problems. *Vestnik NIYaU MIFI*, 2017, Vol. 6, No. 2, pp. 126–142 (in Russian).

[410] Polyanin A.D., Sorokin V.G. On the stability and instability of solutions to reaction-diffusion and more complex equations with delay. *Vestnik NIYaU MIFI*, 2018, Vol. 7, No. 5, pp. 389–404 (in Russian).

[411] Polyanin A.D., Sorokin V.G. Exact solutions of pantograph-type nonlinear partial differential equations with variable delay. *Vestnik NIYaU MIFI*, 2020, Vol. 9, No. 4, pp. 315–328 (in Russian).

[412] Polyanin A.D., Sorokin V.G. Nonlinear delay reaction-diffusion equations: Traveling-wave solutions in elementary functions. *Appl. Math. Lett.*, 2015, Vol. 46, pp. 38–43.

[413] Polyanin A.D., Sorokin V.G. New exact solutions of nonlinear wave type PDEs with delay. *Appl. Math. Lett.*, 2020, Vol. 108, 106512.

[414] Polyanin A.D., Sorokin V.G. A method for constructing exact solutions of nonlinear delay PDEs. *J. Math. Anal. Appl.*, 2021, Vol. 494, No. 2, 124619.

[415] Polyanin A.D., Sorokin V.G. Construction of exact solutions to nonlinear PDEs with delay using solutions of simpler PDEs without delay. *Commun. Nonlinear Sci. Numer. Simul.*, 2021, Vol. 95, 105634.

[416] Polyanin A.D., Sorokin V.G. Nonlinear pantograph-type diffusion PDEs: Exact solutions and the principle of analogy. *Mathematics*, 2021, Vol. 9, No. 5, 511.

[417] Polyanin A.D., Sorokin V.G. Reductions and exact solutions of Lotka–Volterra and more complex reaction-diffusion systems with delays. *Appl. Math. Lett.*, 2022, Vol. 125, 107731.

[418] Polyanin A.D., Sorokin V.G. Reductions and exact solutions of nonlinear wave-type PDEs with proportional and more complex delays. *Mathematics*, 2023, Vol. 11, No. 3, 516.

[419] Polyanin A.D., Sorokin V.G., Vyazmin A.V. Exact solutions and qualitative features of nonlinear hyperbolic reaction-diffusion equations with delay. *Theor. Found. Chem. Eng.*, 2015, Vol. 49, No. 5, pp. 622–635.

[420] Polyanin A.D., Sorokin V.G., Vyazmin A.V. Reaction-diffusion models with delay: Some properties, equations, problems, and solutions. *Theor. Found. Chem. Eng.*, 2018, Vol. 52, No. 3, pp. 334–348.

[421] Polyanin A.D., Zaitsev V.F. *Handbook of Exact Solutions for Ordinary Differential Equations*, 2nd ed. Boca Raton–New York: CRC Press, 2003.

[422] Polyanin A.D., Zaitsev V.F. *Handbook of Nonlinear Partial Differential Equations*, 2nd ed. Boca Raton: CRC Press, 2012.

[423] Polyanin A.D., Zaitsev V.F. *Handbook of Ordinary Differential Equations: Exact Solutions, Methods, and Problems*. Boca Raton–London: CRC Press, 2018.

[424] Polyanin A.D., Zaitsev V.F., Moussiaux A. *Handbook of First Order Partial Differential Equations*. Taylor & Francis, London, 2002.

[425] Polyanin A.D., Zaitsev V.F., Zhurov A.I. *Solution Methods for Nonlinear Equations of Mathematical Physics and Mechanics*. Fizmatlit, Moscow, 2005 (in Russian).

[426] Polyanin A.D., Zhurov A.I. Exact separable solutions of delay reaction-diffusion equations and other nonlinear partial functional-differential equations. *Commun. Nonlinear Sci. Numer. Simul.*, 2014, Vol. 19, No. 3, pp. 409–416.

[427] Polyanin A.D., Zhurov A.I. Exact solutions of linear and nonlinear differential-difference heat and diffusion equations with finite relaxation time. *Int. J. Non-Linear Mech.*, 2013, Vol. 54, pp. 115–126.

[428] Polyanin A.D., Zhurov A.I. Functional constraints method for constructing exact solutions to delay reaction-diffusion equations and more complex nonlinear equations. *Commun. Nonlinear Sci. Numer. Simul.*, 2014, Vol. 19, No. 3, pp. 417–430.

[429] Polyanin A.D., Zhurov A.I. Generalized and functional separable solutions to nonlinear delay Klein–Gordon equations. *Commun. Nonlinear Sci. Numer. Simul.*, 2014, Vol. 19, No. 8, pp. 2676–2689.

[430] Polyanin A.D., Zhurov A.I. *Methods of Separation of Variables and Exact Solutions to Nonlinear Equations of Mathematical Physics*. IPMech RAN, Moscow, 2020 (in Russian).

[431] Polyanin A.D., Zhurov A.I. Multi-parameter reaction-diffusion systems with quadratic nonlinearity and delays: New exact solutions in elementary functions. *Mathematics*, 2022, Vol. 10, No. 9, 529.

[432] Polyanin A.D., Zhurov A.I. New generalized and functional separable solutions to nonlinear delay reaction-diffusion equations. *Int. J. Non-Linear Mech.*, 2014, Vol. 59, pp. 16–22.

[433] Polyanin A.D., Zhurov A.I. Nonlinear delay reaction-diffusion equations with varying transfer coefficients: Exact methods and new solutions. *Appl. Math. Lett.*, 2014, Vol. 37, pp. 43–48.

[434] Polyanin A.D., Zhurov A.I. Non-linear instability and exact solutions to some delay reaction-diffusion systems. *Int. J. Non-Linear Mech.*, 2014, Vol. 62, pp. 33–40.

[435] Polyanin A.D., Zhurov A.I. The functional constraints method: Application to nonlinear delay reaction-diffusion equations with varying transfer coefficients. *Int. J. Non-Linear Mech.*, 2014, Vol. 67, pp. 267–277.

[436] Polyanin A.D., Zhurov A.I. The generating equations method: Constructing exact solutions to delay reaction-diffusion systems and other non-linear coupled delay PDEs. *Int. J. Non-Linear Mech.*, 2015, Vol. 71, pp. 104–115.

[437] Polyanin A.D., Zhurov A.I. Separation of variables in PDEs using nonlinear transformations: Applications to reaction-diffusion type equations. *Applied Math. Letters*, 2020, Vol. 100, 106055.

[438] Polyanin A.D., Zhurov A.I. Separation of Variables and Exact Solutions to Nonlinear PDEs. Boca Raton: CRC Press, 2022.

[439] Pospíšil M. Representation and stability of solutions of systems of functional differential equations with multiple delays. *Electron. J. Qual. Theory Differ. Equations*, 2012, No. 54, pp. 1–30.

[440] Prudnikov A.P., Brychkov Y.A., Marichev O.I. *Integrals and Series, Vol. 4, Direct Laplace Transform*. New York: Gordon & Breach, 1992.

[441] Prudnikov A.P., Brychkov Y.A., Marichev O.I. *Integrals and Series, Vol. 5, Inverse Laplace Transform*. New York: Gordon & Breach, 1992.

[442] Pucci E., Saccomandi G. Evolution equations, invariant surface conditions and functional separation of variables. *Physica D*, 2000, Vol. 139, pp. 28–47.

[443] Pucci E., Salvatori M.C. Group properties of a class of semilinear hyperbolic equations. *Int. J. Non-Linear Mech.*, 1986, Vol. 21, pp. 147–155.

[444] Pue-on P., Meleshko S.V. Group classification of second-order delay ordinary differential equations. *Commun. Nonlinear Sci. Numer. Simul.*, 2010, Vol. 15, pp. 1444–1453.

[445] Qu C.Z., Zhang S.L., Liu R.C. Separation of variables and exact solutions to quasilinear diffusion equations with the nonlinear source. *Physica D*, 2000, Vol. 144, pp. 97–123.

[446] Quarteroni A., Valli A. *Numerical Approximation of Partial Differential Equations*. Berlin: Springer, 2008.

[447] Rach R. A convenient computational form for the Adomian polynomials. *J. Math. Anal. Appl.*, 1984, Vol. 102, pp. 415–419.

[448] Ramirez-Carrasco C., Molina-Garay J. Existence and approximation of traveling wavefronts for the diffusive Mackey–Glass equation. *Aust. J. Math. Anal. Appl.*, 2021, Vol. 18, No. 1, pp. 1–12.

[449] Raslan K.R., Abu Sheer Z.F. Comparison study between differential transform method and Adomian decomposition method for some delay differential equations. *Int. J. Phys. Sci.*, 2013, Vol. 8, No. 17, pp. 744–749.

[450] Rebenda J., Šmarda Z. A differential transformation approach for solving functional differential equations with multiple delays. *Commun. Nonlinear Sci. Numer. Simul.*, 2017, Vol. 48, pp. 246–257.

[451] Reutskiy S.Y. A new collocation method for approximate solution of the pantograph functional differential equations with proportional delay. *Appl. Math. Comput.*, 2015, Vol. 266, pp. 642–655.

[452] Reyes E., Rodríguez F., Martín J.A. Analytic-numerical solutions of diffusion mathematical models with delays. *Comput. Math. Appl.*, 2008, Vol. 56, pp. 743–753.

[453] Reyes E., Castro M.Á., Sirvent A., Rodríguez F. Exact solutions and continuous numerical approximations of coupled systems of diffusion equations with delay. *Symmetry*, 2020, Vol. 12, 1560.

[454] Robinson R.W. Counting labeled acyclic digraphs. In: *New Directions in the Theory of Graphs* (ed. F. Harari), pp. 239–273. New York: Academic Press, 1973.

[455] Rodríguez F., Roales M., Marín J.A. Exact solutions and numerical approximations of mixed problems for the wave equation with delay. *Appl. Math. Comput.*, 2012, Vol. 219, No. 6, pp. 3178–3186.

[456] Roetzel W., Putra N. Das S.K. Experiment and analysis for non-Fourier conduction in materials with non-homogeneous inner structure. *Int. J. Thermal Sciences*, 2003, Vol. 42, No. 6, pp. 541–552.

[457] Romanovskii Yu.M., Stepanova N.V., Chernavskii D.S. *Mathematical Biophysics*. Nauka, Moscow, 1984 (in Russian).

[458] Rossovskii L.E. Elliptic functional differential equations with contractions of arguments. *Doklady Math.*, 2006, Vol. 74, pp. 809–811.

[459] Rossovskii L.E. Elliptic functional differential equations with contractions and extensions of independent variables of the unknown function. *J. Math. Sciences*, 2017, Vol. 223, No. 4, pp. 351–493.

[460] Ruan S. Delay differential equations in single species dynamics. In: *Delay Differential Equations and Applications. NATO Science Series (II. Mathematics, Physics and Chemistry)* (eds. Arino O., Hbid M., Dads E.A.), Vol. 205, pp. 477–517. Dordrecht: Springer, 2006.

[461] Sakar M.G., Uludag F., Erdogan F. Numerical solution of time-fractional nonlinear PDEs with proportional delays by homotopy perturbation method. *Appl. Math. Model.*, 2016, Vol. 40, pp. 6639–6649.

[462] Saker S.H. Oscillation and global attractivity of hematopoiesis model with delay time. *Appl. Math. Comput.*, 2003, Vol. 136, pp. 27–36.

[463] Saker S.H. Oscillation of continuous and discrete diffusive delay Nicholson's blowflies models. *Appl. Math. Comput.*, 2005, Vol. 167, pp. 179–197.

[464] Samarskii A.A., Gulin A.V. *Numerical Methods*. Nauka, Moscow, 1989 (in Russian).

[465] Sasakura K. On the dynamic behavior of Schinasi's business cycle model. *J. Macroecon.*, 1994, Vol. 16, pp. 423–444.

[466] Schinasi G.J. A nonlinear dynamic model of short run fluctuations. *Rev. Econ. Stud.*, 1981, Vol. 48, pp. 649–656.

[467] Schinasi G.J. Fluctuations in a dynamic, intermediate-run IS-LM model: Applications of the Poicar–Bendixon theorem. *J. Econ. Theory*, 1982, Vol. 28, pp. 369–375.

[468] Schmitt K. On solution of differential equations with deviating arguments. *SIAM J. Appl. Math.*, 1969, Vol. 17, pp. 1171–1176.

[469] Schiesser W.E. *Time Delay ODE/PDE Models: Applications in Biomedical Science and Engineering*. Boca Raton: CRC Press, 2019.

[470] Sedaghat S., Ordokhani Y., Dehghan M. Numerical solution of the delay differential equations of pantograph type via Chebyshev polynomials. *Commun. Nonlinear Sci. Numer. Simul.*, 2012, Vol. 17, pp. 4815–4830.

[471] Sezer M., Akyüz-Daşcioğlu A. A Taylor method for numerical solution of generalized pantograph equations with linear functional argument. *J. Comput. Appl. Math.*, 2007, Vol. 200, pp. 217–225.

[472] Sezer M., Kaynak M. Chebyshev polynomial solutions of linear differential equation. *Int. J. Math. Educ. Sci. Technol.*, 1996, Vol. 27, No. 4, pp. 607–618.

[473] Sezer M., Yalçinbaş S., Gülsu M. A Taylor polynomial approach for solving generalized pantograph equations with nonhomogeneous term. *Int. J. Comput. Math.*, 2008, Vol. 85, No. 7, pp. 1055–1063.

[474] Sezer M., Yalçinbaş S., Sahin N. Approximate solution of multi-pantograph equation with variable coefficients. *J. Comput. Appl. Math.*, 2008, Vol. 214, pp. 406–416.

[475] Shakeri F., Dehghan M. Application of the decomposition method of Adomian for solving the pantograph equation of order m. *Z. Naturforsch.*, 2010, Vol. 65a, pp. 453–460.

[476] Shampine L.F. Interpolation for Runge–Kutta methods. *SIAM J. Numer. Anal.*, 1985, Vol. 22, pp. 1014–1026.

[477] Shampine L.F., Gahinet P. Delay-differential-algebraic equations in control theory. *Appl. Numer. Math.*, 2006, Vol. 56, No. 3–4, pp. 574–588.

[478] Shampine L.F., Thompson S. Solving DDEs in Matlab. *Appl. Numer. Math.*, 2001, Vol. 37, No. 4, pp. 441–458.

[479] Shampine L.F., Thompson S. Numerical solution of delay differential equations. In: *Delay Differential Equations*. Boston: Springer, 2009.

[480] Sidorov A.F., Shapeev V.P., Yanenko N.N. *Method of Differential Constraints and Its Applications in Gas Dynamics*. Nauka, Novosibirsk, 1984 (in Russian).

[481] Smith F.E. Population dynamics in Daphnia magna. *Ecology*, 1963, Vol. 44, pp. 651–663.

[482] Smith H.L. *An Introduction to Delay Differential Equations with Applications to the Life Sciences*. New York: Springer, 2010.

[483] Smith H.L., Zhao X.Q. Global asymptotic stability of travelling waves in delayed reaction-diffusion equations. *SIAM J. Math. Anal.*, 2000, Vol. 31. pp. 514–534.

[484] Skubachevskii A.L. Boundary-value problems for elliptic functional-differential equations and their applications. *Russian Math. Surveys*, 2016, Vol. 71, No. 5, pp. 801–906.

[485] So J.W.-H., Yu J.S. Global attractivity and uniform persistence in Nicholson's blowflies. *Differ. Equations Dynam. Syst.*, 1994, Vol. 2, No. 1, pp. 11–18.

[486] So J.W.-H., Yu J.S. On the uniform stability for a 'food-limited' population model with time delay. *Proc. Roy. Soc. Edin. A*, 1995, Vol. 125, pp. 991–1002.

[487] So J.W.-H., Yang Y. Dirichlet problem for the diffusive Nicholson's blowflies equation. *J. Differ. Equations*, 1998, Vol. 150, pp. 317–348.

[488] So J.W.-H., Zou X. Traveling waves for the diffusive Nicholson's blowflies equation. *Appl. Math. Comput.*, 2001, Vol. 122, No. 3, pp. 385–392.

[489] Sobolev S.L., Transport processes and traveling waves in systems with local nonequilibrium, *Soviet Physics Uspekhi*, 1991, Vol. 34, No. 3, pp. 217–229.

[490] Sobolev S.L., Influence of local nonequilibrium on the rapid solidification of binary alloys, *Technical Physics*, 1998, Vol. 43, No. 3, pp. 307–313.

[491] Sokal A. Roots of a formal power series, with applications to graph enumeration and q-series, 2011; http://www.maths.qmul.ac.uk/~pjc/csgnotes/sokal/.

[492] Solodushkin S.I., Yumanova I.F., Staelen R.D. First-order partial differential equations with time delay and retardation of a state variable. *J. Comput. Appl. Math.*, 2015, Vol. 289, pp. 322–330.

[493] Song O.K., Cao J.D. Global exponential stability and existence of periodic solutions in BAM networks with delays and reaction diffusion terms. *Chaos, Solitons & Fractals*, 2005, Vol. 23, No. 2, pp. 421–430.

[494] Sophocleous C., Kingston J.G. Cyclic symmetries of one-dimensional non-linear wave equations. *Int. J. Non-Linear Mech.*, 1999, Vol. 34, pp. 531–543.

[495] Sorokin V.G. Exact solutions of some nonlinear ordinary differential-difference equations. *Vestnik NiYaU MIFI*, 2015, Vol. 4, No. 6, pp. 493–500 (in Russian).

[496] Sorokin V.G. Exact solutions of some nonlinear partial differential equations with delay and systems of such equations. *Vestnik NiYaU MIFI*, 2016, Vol. 5, No. 3, pp. 199–219 (in Russian).

[497] Sorokin V.G. Exact solutions of nonlinear telegraph equations with delay. *Vestnik NIYaU MIFI*, 2019, Vol. 8, No. 5, pp. 453–464 (in Russian).

[498] Sorokin V.G., Polyanin A.D. Nonlinear partial differential equations with delay: Linear stability/instability of solutions, numerical integration. *J. Phys. Conf. Ser.*, 2019, Vol. 1205, 012053.

[499] Sorokin V.G., Vyazmin A.V. Nonlinear reaction–diffusion equations with delay: Partial survey, exact solutions, test problems, and numerical integration. *Mathematics*, 2022, Vol. 10, No. 11, 1886.

[500] Stépán G. *Retarded Dynamical Systems: Stability and Characteristic Functions*. New York: Longman Scientific & Technical, 1989.

[501] Stuart A.M., Floater M.S. On the computation of blow-up. *Eur. J. Appl. Math.*, 1990, Vol. 1, No. 1, pp. 47–71.

[502] Su Y., Wei J., Shi J. Hopf bifurcation in a reaction-diffusion population model with delay effect. *J. Differ. Equations*, 2009, Vol. 247, pp. 1156–1184.

[503] Suarez M.J., Schopf P.S. A delayed action oscillator for ENSO. *J. Atmos. Sci.*, 1988, Vol. 45, pp. 3283–3287.

[504] Sun G.Q., Wang S.L., Ren Q., Jin Z., Wu Y.-P. Effects of time delay and space on herbivore dynamics: Linking inducible defenses of plants to herbivore outbreak. *Sci. Rep.*, 2015, Vol. 5, 11246.

[505] Sun Z.Z. On the compact difference scheme for heat equation with Neumann boundary conditions. *Numer. Methods Partial Differ. Equations*, 2009, Vol. 25, pp. 1320–1341.

[506] Sun Z., Zhang Z. A linearized compact difference scheme for a class of nonlinear delay partial differential equations. *Appl. Math. Model.*, 2013, Vol. 37, pp. 742–752.

[507] Taganov I.N. *Modeling of Mass and Energy Transfer Processes*. Khimiya, Leningrad, 1979 (in Russian).

[508] Takeuchi Y., Ma W., Beretta E. Global asymptotic properties of a delay SIR epidemic model with finite incubation times. *Nonlinear Anal.*, 2000, Vol. 42, pp. 931–947.

[509] Tang C., Zhang C. A fully discrete θ-method for solving semi-linear reaction-diffusion equations with time-variable delay. *Math. Comput. Simul.*, 2021, Vol. 179, pp. 48–56.

[510] Tanthanuch J. Symmetry analysis of the nonhomogeneous inviscid Burgers equation with delay. *Commun. Nonlinear Sci. Numer. Simulat.*, 2012, Vol. 17, 4978–4987.

[511] Tanthanuch J., Meleshko S.V. Application of group analysis to delay differential equations. In: *Nonlinear Acoustics at the Beginning of the 21st Century* (eds. Rudenko O.V., Sapozhnikov O.A.), pp. 607–610. Moscow: MSU, 2002.

[512] Tanthanuch J., Meleshko S.V. On definition of an admitted Lie group for functional differential equations. *Commun. Nonlinear Sci. Numer. Simul.*, 2004, Vol. 9, pp. 117–125.

[513] Tanzi V. Inflation, lags in collection, and the real value of tax revenue. *IMF—Staff Papers*, 1977, Vol. 24, pp. 154–167.

[514] Tikhonov A.N., Samarskii A.A. *Equations of Mathematical Physics*. Dover Publ., New York, 1990.

[515] Titov S.S. A method of finite-dimensional rings for solving nonlinear equations of mathematical physics. In: *Aerodynamics* (ed. T.P. Ivanova), Saratov Univ., Saratov, 1988, pp. 104–110 (in Russian).

[516] Titov S.S. On solutions of nonlinear partial differential equations of the form of a simple variables. *Chisl. Met. Mech. Sploshnoi Sredy*, 1977, Vol. 8, pp. 586–599 (in Russian).

[517] Titov S.S., Ustinov V.A. Investigation of polynomial solutions to the equations of motion of gases through porous media with integer isentropic exponent. In: *Approximate Methods of Solution of Boundary Value Problems in Solid Mechanics*, AN USSR, Ural. Otd-nie Inst. Matematiki i Mekhaniki, pp. 64–70, 1985 (in Russian).

[518] Tohidi E., Bhrawy A.H., Erfani K. A collocation method based on Bernoulli operational matrix for numerical solution of generalized pantograph equation. *Appl. Math. Model.*, 2013, Vol. 37, pp. 4283–4294.

[519] Trofimchuk E., Tkachenko V., Trofimchuk S. Slowly oscillating wave solutions of a single species reaction-diffusion equation with delay. *J. Differ. Equations*, 2008, Vol. 245, pp. 2307–2332.

[520] Trofimchuk E., Pinto M., Trofimchuk S. Traveling waves for a model of the Belousov–Zhabotinsky reaction. *J. Differ. Equations*, 2013, Vol. 254, pp. 3690–3714.

[521] Tzou D.Y. *Macro- to Microscale Heat Transfer: The Lagging Behavior*. Washington: Taylor & Francis, 1997.

[522] Vagina M.Yu., Kipnis M.M. Stability of the zero solution of a differential equation with delays. *Matem. zametki*, 2003, Vol. 74, No. 5, pp. 786–789 (in Russian).

[523] Valeev K.G. Linear differential equations with a delay linearly dependent on the argument. *Sibirskii matem. zhurnal*, 1964, Vol. 5, No. 2, pp. 290–309 (in Russian).

[524] Valluri S.R., Jeffrey D.J., Corless R.M. Some applications of the Lambert W function to physics. *Can. J. Phys.*, 2000, Vol. 78, No. 9, pp. 823–831.

[525] Van Brunt B., Wake G.C. A Mellin transform solution to a second-order pantograph equation with linear dispersion arising in a cell growth model. *Eur. J. Appl. Math.*, 2011, Vol. 22, No. 2, pp. 151–168.

[526] Van Brunt B., Zaidi A.A., Lynch T. Cell division and the pantograph equation. *ESAIM Proc. Surveys*, 2018, Vol. 62, pp. 158–167.

[527] Vandewalle S., Gander M.J. Optimized overlapping Schwarz methods for parabolic PDEs with time-delay. In: *Domain Decomposition Methods in Science and Engineering*. Springer, Berlin, 2005, pp. 291–298.

[528] Veberic D. Lambert W function for applications in physics. *Computer Physics Communications*, 2012, Vol. 183, No. 12, pp. 2622–2628.

[529] Vedavarz A., Kumar S., Moallemi M.K. Significance of non-Fourier heat waves in conduction. *ASME J. Heat Transfer*, 1994, Vol. 116, No. 1, pp. 221–224.

[530] Verblunsky S. On a class of differential-difference equations. *Proc. London Math. Soc.*, 1956, Vol. s3–6, No. 3, pp. 355–365.

[531] Vernotte P. Les paradoxes de la théorie continue de l'équation de la chaleur. *Comptes Rendus*, 1958, Vol. 246, pp. 3154–3155.

[532] Vernotte P. Some possible complications in the phenomena of thermal conduction. *Comptes Rendus*, 1961, Vol. 252, pp. 2190–2191.

[533] Villasana M., Radunskaya A. A delay differential equation model for tumor growth. *J. Math. Biol.*, 2003, Vol. 47, pp. 270–294.

[534] Vladimiriv V.S. *Equations of Mathematical Physics*. Fizmatlit, Moscow, 1971 (in Russian).

[535] Vladimirov A., Turaev D. Model for passive mode locking in semiconductor lasers. *Phys. Rev. A*, 2005, Vol. 72, 033808.

[536] Wan A., Wei J. Hopf bifurcation analysis of a food-limited population model with delay. *Nonlinear Anal. Real World Appl.*, 2010, Vol. 11, pp. 1087–1095.

[537] Wan P., Sun D., Chen D., Zhao M., Zheng L. Exponential synchronization of inertial reaction-diffusion coupled neural networks with proportional delay via periodically intermittent control. *Neurocomputing*, 2019, Vol. 356, pp. 195–205.

[538] Wang J., Meng F., Liu S. Interval oscillation criteria for second order partial differential equations with delays. *J. Comput. Appl. Math.*, 2008, Vol. 212. No. 2, pp. 397–405.

[539] Wang K., Teng Z., Jiang H. Global exponential synchronization in delayed reaction-diffusion cellular neural networks with the Dirichlet boundary conditions. *Math. Comput. Model.*, 2010, Vol. 52, pp. 12–24.

[540] Wang K., Wang W. Propagation of HBV with spatial dependence. *Math. Biosci.*, 2007, Vol. 210, pp. 78–95.

[541] Wang K., Wang W., Song S. Dynamics of an HBV model with diffusion and delay. *J. Theor. Biol.*, 2008, Vol. 253, pp. 36–44.

[542] Wang L., Gao Y. Global exponential robust stability of reaction-diffusion interval neural networks with time-varying delays. *Physics Letters A*, 2006, Vol. 350, pp. 342–348.

[543] Wang L., Zhang C. Zeros of the deformed exponential function. *Adv. Math.*, 2018, Vol. 332, pp. 311–348.

[544] Wang P.K.C. Asymptotic stability of a time-delayed diffusion system. *J. Appl. Mech.*, 1963, Vol. 30, pp. 500–504.

[545] Wang X., Li Z. Dynamics for a type of general reaction-diffusion model. *Nonlinear Analysis*, 2007, Vol. 67, pp. 2699–2711.

[546] Wang X., Zhang H., Li Z. X. Oscillation for a class of diffusive hematopoiesis model with several arguments. *Acta Math. Sin. Eng. Ser.*, 2012, Vol. 28, No. 11, pp. 2345–2354.

[547] Wang Z.-Q., Wang L.-L. A Legendre–Gauss collocation method for nonlinear delay differential equations. *Discrete Contin. Dyn. Syst. Ser. B*, 2010, Vol. 13, No. 3, pp. 685–708.

[548] Wazwaz A.M., Raja M.A.Z., Syam M.I. Reliable treatment for solving boundary value problems of pantograph delay differential equation. *Rom. Rep. Phys.*, 2017, Vol. 69, 102.

[549] Wei J., Li M. Hopf bifurcation analysis in a delayed Nicholson blowflies equation. *Nonlinear Anal.*, 2005, Vol. 60, No. 7, pp. 1351–1367.

[550] Weiss J., Tabor M., Carnevalle G. The Painlevé property for partial differential equations. *J. Math. Phys.*, 1983, Vol. 24, No. 3, pp. 522–526.

[551] Weiss J. The Painlevé property for partial differential equations. II: Bäcklund transformation, Lax pairs, and the Schwarzian derivative. *J. Math. Phys.*, 1983, Vol. 24, No. 6, pp. 1405–1413.

[552] Weiss S. On the controllability of delay-differential systems. *SIAM J. Control*, 1967, Vol. 5, No. 4, pp. 575–587.

[553] Weisstein E.W. *CRC Concise Encyclopedia of Mathematics, 2nd ed.*, Boca Raton: CRC Press, 2003.

[554] Welfert B.D. Generation of pseudospectral differentiation matrices I. *SIAM J. Numer. Anal.*, 1997, Vol. 34, pp. 1640–1657.

[555] Willé D.R., Baker C.T.H. The tracking of derivative discontinuities in systems of delay-differential equations. *Appl. Numer. Math.*, 1992, Vol. 9, No. 3–5, pp. 209–222.

[556] Willé D.R., Baker C.T.H. DELSOL—a numerical code for the solution of systems of delay differential equations. *Appl. Numer. Math.*, 1992, Vol. 9, pp. 223–234.

[557] Winitzki S. Uniform approximations for transcendental functions. In: *Computational Science and Its Applications—ICCSA 2003* (eds. Kumar V., Gavrilova M.L., Kenneth Tan C.J., L'Ecuyer P.); Lecture Notes in Computer Science, Vol. 2667, pp. 780–789. Berlin–Heidelberg: Springer, 2003.

[558] Wolfram Language Documentation. *The Numerical Method of Lines*. http://reference.wolfram.com/language/tutorial/NDSolveMethodOfLines.html (accessed: 19 January 2023).

[559] Wolfram Language Documentation. *NDSolve*. http://reference.wolfram.com/language/ref/NDSolve.html (accessed: 19 January 2023).

[560] Wolfram Language Documentation. *Delay Differential Equations*. http://reference.wolfram.com/mathematica/tutorial/NDSolveDelayDifferentialEquations.html (accessed: 19 January 2023).

[561] Wolfram Language Documentation. *"ExplicitRungeKutta" Method for NDSolve*. http://reference.wolfram.com/language/tutorial/NDSolveExplicitRungeKutta.html (accessed: 19 January 2023).

[562] Wolfram Language Documentation. *"ImplicitRungeKutta" Method for NDSolve*. http://reference.wolfram.com/language/tutorial/NDSolveImplicitRungeKutta.html (accessed: 19 January 2023).

[563] Wolfram Language Documentation. *IDA Method for NDSolve*. http://reference.wolfram.com/language/tutorial/NDSolveIDAMethod.html (accessed: 19 January 2023).

[564] Wolfram Language Documentation. *Norms in NDSolve*. http://reference.wolfram.com/language/tutorial/NDSolveVectorNorm.html (accessed: 19 January 2023).

[565] Wolfram Language Documentation. *Numerical Solution of Differential Equations*. http://reference.wolfram.com/language/tutorial/NumericalSolutionOfDifferentialEquations.html (accessed: 19 January 2023).

[566] Wolfram Language Documentation. *FindRoot*. http://reference.wolfram.com/language/ref/FindRoot.html (accessed: 19 January 2023).

[567] Wright E.M. A non-linear difference-differential equation. *J. Reine Angew. Math.*, 1955, Vol. 194, pp. 66–87.

[568] Wright E.M. Solution of the equation $ze^z = a$. *Proc. R. Soc. Edinb.*, 1959, Vol. 65, pp. 193–203.

[569] Wu J. *Theory and Applications of Partial Functional Differential Equations.* New York: Springer-Verlag, 1996.

[570] Wu J.H. *Introduction to Neural Dynamics and Signal Transmission Delay,* Berlin: de Gruyter, 2002.

[571] Wu J., Campbell S. A., Bélair J. Time-delayed neural networks: Stability and oscillations. In: *Encyclopedia of Computational Neuroscience,* pp. 2966–2972. New York: Springer, 2015.

[572] Wu J., Zou X. Traveling wave fronts of reaction-diffusion systems with delay. *J. Dynamics & Differ. Equations,* 2001, Vol. 13, No. 3, pp. 651–687.

[573] Wu F., Wang Q., Cheng X., Chen X. Linear θ-method and compact θ-method for generalised reaction-diffusion equation with delay. *Int. J. Differ. Equations,* 2018, Vol. 2018, 6402576.

[574] Xiaoxin L., XiaoJun W. Stability for differential-difference equations. *J. Math. Anal. Appl.,* 1991, Vol. 174, pp. 84–102.

[575] Xu R., Ma Z.E. An HBV model with diffusion and time delay. *J. Theor. Biol.* 2009, Vol. 257, pp. 499–509.

[576] Xu Y., Liu M. H-stability of Runge–Kutta methods with general variable stepsize for pantograph equation. *Appl. Math. Comput.,* 2004, Vol. 148, pp. 881–892.

[577] Xu Z. Traveling waves in a Kermack–McKendrick epidemic model with diffusion and latent period. *Nonlinear Anal.,* 2014, Vol. 111, pp. 66–81.

[578] Yalçinbaş S., Aynigül M., Sezer M. A collocation method using Hermite polynomials for approximate solution of pantograph equations. *J. Franklin Inst.,* 2011, Vol. 348, pp. 1128–1139.

[579] Yanenko N.N. Compatibility theory and methods for integrating systems of nonlinear partial differential equations. *Trudy IV Vsesoyuz mat. s'ezda,* Vol. 2, pp. 613–621, Nauka, Leningrad, 1964 (in Russian).

[580] Yang C. Modified Chebyshev collocation method for pantograph-type differential equations. *Appl. Numer. Math.,* 2018, Vol. 134, pp. 132–144.

[581] Yang J., Liang S., Zhang Yi. Travelling waves of a delayed SIR epidemic model with nonlinear incidence rate and spatial diffusion. *PLoS ONE,* 2011, Vol. 6, No. 6, e21128.

[582] Yang X., Song Q., Cao J., Lu J. Synchronization of coupled Markovian reaction-diffusion neural networks with proportional delays via quantized control. *IEEE Trans. Neural Netw. Learn. Syst.,* 2019, Vol. 30, No. 3, pp. 951–958.

[583] Yang Y., So J.W.-H. Dynamics for the diffusive Nicholson blowflies equation. In: *Dynamical Systems and Differential Equations* (eds. Chen W., Hu S.), Vol. 2, pp. 333–352. Springfield, MO: Southwest Missouri State University, 1998.

[584] Yang Z., Xu D. Global dynamics for non-autonomous reaction-diffusion neural networks with time-varying delays. *Theor. Comput. Sci.,* 2008, Vol. 403, pp. 3–10.

[585] Yi S., Nelson P.W., Ulsoy A.G. *Time-Delay Systems: Analysis and Control Using the Lambert W Function.* Singapore: World Scientific, 2010.

[586] Yi T., Zou X. Global attractivity of the diffusive Nicholson blowflies equation with Neumann boundary condition: A non-monotone case. *F. Differ. Equations,* 2008, Vol. 245, pp. 3376–3388.

[587] Yuanhong Yu. Stability criteria for linear second order delay differential systems. *Acta Math. Appl. Sinica,* 1988, Vol. 4, pp. 109–112.

[588] Yusufoğlu E. An efficient algorithm for solving generalized pantograph equations with linear functional argument. *Appl. Math. Comput.*, 2010, Vol. 217, pp. 3591–3595.

[589] Yüzbaşi S., Şahin N., Sezer M. A Bessel collocation method for numerical solution of generalized pantograph equations. *Numer. Methods Partial Differ. Equations*, 2011, Vol. 28, pp. 1105–1123.

[590] Yüzbaşi S., Sezer M. An exponential approximation for solutions of generalized pantograph-delay differential equations. *Appl. Math. Model.*, 2013, Vol. 37, pp. 9160–9173.

[591] Zaidi A.A., Van Brunt B., Wake G.C. Solutions to an advanced functional partial differential equation of the pantograph type. *Proc. R. Soc. A.*, 2015, Vol. 471, 20140947.

[592] Zaikin A.N., Zhabotinsky A.M. Concentration wave propagation in two-dimensional liquid-phase self-oscillating system. *Nature*, 1970, Vol. 225, pp. 535–537.

[593] Zaitsev V.F., Polyanin A.D. Exact solutions and transformations of nonlinear heat and wave equations. *Doklady Math.*, 2001, Vol. 64, No. 3, pp. 416–420.

[594] Zakharov A.P., Bratsun D.A. An adaptive algorithm for storing fields in the calculation of the dynamics of a continuous medium with hereditary or delayed feedback. *Vychislitel'naya mekhanika sploshnykh sred*, 2013, Vol. 6, No. 2, pp. 198–206 (in Russian).

[595] Zennaro M. Natural continuous extensions of Runge–Kutta methods. *Math. Comp.*, 1986, Vol. 46, pp. 119–133.

[596] Zennaro M. P-stability properties of Runge–Kutta methods for delay differential equations. *Numer. Math.*, 1986, Vol. 49, pp. 305–318.

[597] Zhabotinskii A.M. Periodic oxidative reactions in the liquid phase. *Dokl. AN SSSR*, 1964, Vol. 157, No. 2, pp. 392–395 (in Russian).

[598] Zhabotinskii A.M. Periodic processes of oxidation of malonic acid in a solution (a study of the Belousov reaction kinetics). *Biofizika*, 1964, Vol. 9, pp. 306–310 (in Russian).

[599] Zhang F., Zhang Y. State estimation of neural networks with both time-varying delays and norm-bounded parameter uncertainties via a delay decomposition approach. *Commun. Nonlinear Sci. Numer. Simul.*, 2013, Vol. 18, No. 12, pp. 3517–3529.

[600] Zhang G.-B. Asymptotics and uniqueness of traveling wavefronts for a delayed model of the Belousov–Zhabotinsky reaction, *Applicable Analysis*, 2020, Vol. 99, No. 10, pp. 1639–1660.

[601] Zhang Q., Li D., Zhang C., Xu D. Multistep finite difference schemes for the variable coefficient delay parabolic equations. *J. Differ. Equations Appl.*, 2016, Vol. 22, No. 6, pp. 745–765.

[602] Zhang Q., Zhang C. A new linearized compact multisplitting scheme for the nonlinear convection-reaction-diffusion equations with delay. *Commun. Nonlinear Sci. Numer. Simul.*, 2013, Vol. 18, No. 12, pp. 3278–3288.

[603] Zhang S.L., Lou S.Y., Qu C.Z. New variable separation approach: Application to nonlinear diffusion equations. *J. Phys. A: Math. Gen.*, 2003, Vol. 36, pp. 12223–12242.

[604] Zhang W., Fan M. Periodicity in a generalized ecological competition system governed by impulsive differential equations with delays. *Math. Comput. Model.*, 2004, Vol. 39, No. 4–5, pp. 479–493.

[605] Zhang X., Zhu H. Hopf bifurcation and chaos of a delayed finance system. *Complexity*, 2019, Vol. 2019, 6715036.

[606] Zhang Y., Xu Z. Dynamics of a diffusive HBV model with delayed Beddington–DeAngelis response. *Nonlinear Anal: Real World Appl.*, 2014, Vol. 15, pp. 118–139.

[607] Zhao H. Exponential stability and periodic oscillatory of bidirectional associative memory neural network involving delays. *Neurocomputing*, 2006, Vol. 69, pp. 424–448.

[608] Zhdanov R.Z. Separation of variables in the non-linear wave equation. *J. Phys. A*, 1994, Vol. 27, pp. L291–L297.

[609] Zhivotovskii L.L. Absolute stability of solutions to differential equations with several delays. *Trudy seminara po teorii dif. uravn. s otkl. argumentom*, 1969, Vol. 7, pp. 82–91 (in Russian).

[610] Zhu C.-C., Zhu J. Dynamic analysis of a delayed COVID-19 epidemic with home quarantine in temporal-spatial heterogeneous via global exponential attractor method. *Chaos, Solitons & Fractals*, 2021, Vol. 143, 110546.

[611] Zhurov A.I., Polyanin A.D. Symmetry reductions and new functional separable solutions of nonlinear Klein–Gordon and telegraph type equations. *J. Nonlinear Math. Physics*, 2020, Vol. 27, No. 2, pp. 227–242.

[612] Zubik-Kowal B. Delay partial differential equations. *Scholarpedia*, 2008, Vol. 3, No. 4, 2851.

[613] Zwillinger D. *Handbook of Differential Equations, 3rd ed.* New York: Academic Press, 1997.

Index

A

abbreviations
  KPP (Kolmogorov–Petrovsky–Piskunov) equation, 345
  ODE (ordinary differential equation), 16
  ODEs (ordinary differential equations), 16
  PDE (partial differential equation), 16
  PDEs (partial differential equations), 16
  SI (susceptible, infectious) epidemic model, 353
  SIR (susceptible, infectious, removed) epidemic model, 350, 351
  SIRS (susceptible, infectious, removed, susceptible) epidemic model, 352
additive separable solution, 90, 130, 139–154, 170, 180, 189, 199, 214, 221, 254, 265
Adomian decomposition method, xii, 76–78
advanced differential equations, 4, 29, 131, 251
analytical methods
  approximate, see approximate analytical methods
  exact solutions to nonlinear delay PDEs, 127–272
  integral transforms, xii, 55–64, 91
  Laplace transform, 56
  Laplace transform, solution example, 59
  matched asymptotic expansions, see method of matched asymptotic expansions
  Mellin transform, 60
  Mellin transform, solution example, 61–63
  solution of Cauchy problems, see Cauchy problems
  solution of initial value problems, see Cauchy problems
  solution of initial-boundary value problems, see initial-boundary value problems
  standard, 129
    difficulties in using, 129, 166

antiperiodic function, 164, 169, 196
approximate analytical methods, 80
  Adomian decomposition, 76–78
  asymptotic, 71, 127
  Bubnov–Galerkin, 82
  collocation, 82
  delay ODEs, 55
  expansion of nonlinear operator, 75
  Galerkin, general scheme, 81
  Galerkin-type projection, xii, 80
  homotopy analysis, 78
  iterative, 74
  least squares, 82
  matched asymptotic expansions, 71–74
  minimization of root mean square error, 84
  moments, 82
  Padé approximants, 69
  perturbation-iteration algorithms, 79
  power series solution, 64–68
  principles for selecting test problems, 309
  projection, 80
  regular expansion in small parameter, 69–71
  representation as linear combinations of basis functions, 80
  steps, see method of steps
  successive approximations, 74
approximate solutions, see approximate analytical methods
approximating function, 81, 83
asymptotic boundary conditions, 135, 138
asymptotic stability of solutions, 47, 333
  local, criteria, 331
  sufficient conditions, 110
attractor, global, 331, 345, 349

B

basis functions, 80–82, 285
Belousov–Zhabotinsky delay reaction-diffusion model, 137
Belousov–Zhabotinsky oscillating reaction model, 359

biharmonic operator, 151
blow-up problems, 23, 24
　suppression of singularities, 24, 25
boundary condition, *see* boundary conditions
boundary conditions, xii, 67, 71, 73, 92, 97, 101, 104
　boundary value problems, 67
　Dirichlet, 97, 102
　first kind, 83, 91, 97, 99, 344, 349
　homogeneous, 61, 80, 92, 95, 98, 344, 349
　initial-boundary value problems, 91, 96, 102
　mixed, 97, 99
　most common, 97, 99
　Neumann, 97
　nonhomogeneous, 80, 92, 96
　periodic, 116
　Robin, 97
　second kind, 97, 99, 344, 349
　special, 124
　third kind, 97, 99, 289, 344
boundary value problems, *see also* initial-boundary value problems, 71, 104, 108, 120, 123, 135, 287, 288
　first, 288
　linear, 80, 99, 108, 120, 288, 289
　mixed, 67
　nonlinear, 80, 288
　nonlinear boundary conditions, 288
　second, 288
　third, 288
　two-point, 71, 80
Bubnov–Galerkin method, 82, 154

C

carrying capacity (of habitat), 327, 344, 349, 353
Cattaneo–Vernotte differential law (model), 113, 119
Cauchy problems, 2, 91, 329
　auxiliary, 288, 289
　for delay ODEs, 2, 13
　　numerical integration, 273, 278, 279, 283
　　test problems, 293–295
　for first-order ODEs with constant delay, 13
　for higher-order delay ODEs, 28
　for linear delay ODEs
　　exact solutions, 13–16
　for linear systems of delay ODEs, 38–41

for Nicholson equation, 330
for nonlinear delay ODEs, 16, 17, 69, 76
for nonlinear ODEs with proportional delay
　exact solutions, 20–22
for ODEs with proportional delay, 18
for ODEs with several delays, 3, 22, 23
　existence and uniqueness of solutions, 23
for second-order delay ODEs
　exact solutions, 29–34
characteristic equation, 4, 35, 42–49, 51–53, 109, 331
　roots, 39, 43–45, 52, 53, 110
Chebyshev nodes, 82
classical diffusion equation, 113
classical heat equation, 113
classical SIR model of epidemic spread, 350
classical solution, 35, 86
collocation method, xii, 80, 82–84
　spectral, 285
comparison of solutions
　numerical vs approximate analytical, 84
　numerical vs exact
　　nonlinear delay Klein–Gordon type wave equations, 322
　　nonlinear delay reaction-diffusion equations, 316
　　test problems, 293, 309, 316
composite asymptotic solution, 73
conservation law
　energy, 113
　total number of individuals, 351
construction of exact solutions, 155
　delay ODEs, 1–41, 55–68
　delay PDEs, 85–125, 127–199, 201–269
　description and examples, 201
　linear transformations, 162
　methods for nonlinear delay PDEs, 201
　nonlinear delay PDEs
　　examples, 130, 139–141, 156–159
　　using solutions to simpler non-delay PDEs, 201
　using particular solutions, 86
construction of functional separable solutions
　linear transformations, 135
　nonlinear delay equations, examples, 133
construction of generalized separable solutions
　linear transformations, 135
　nonlinear delay equations, examples, 130

construction of test problems, xii, 127, 293
   delay PDEs, 293
      examples, 293–296, 311–315
      direct method, 315
Cooke–van den Driessche method, 49

D

D-partition, 46–49
D-partition method, 46
data storage optimization in RAM, 305
delay differential equations, 1, 85, 127, 201
   analytical methods, 127–272
   applications, 327–363
   numerical methods, 273–308
   ordinary, *see* delay ODEs
   partial, *see* delay PDEs
delay diffusive logistic equation, 135, 136, 310, 344–346
   limited food conditions, 136
   with limited food, 136
delay equation
   Fisher, 135, 310, 316, 345
   Klein–Gordon type, *see* delay Klein–Gordon type wave equations
delay equations
   applications, 327
   diffusive logistic, *see* delay diffusive logistic equation
   ordinary differential, *see* delay ODEs
   partial differential, *see* delay PDEs
   reaction-diffusion, *see* delay reaction-diffusion equations
delay Fisher equation, 135, 310, 316, 345
delay Klein–Gordon type wave equations, 129, 148, 194
   comparison of numerical and exact solutions, 322
   forms of exact solutions, 150–152
   linear transformations, 163
   nonlinear, 206
      exact solutions, 163–165, 194–199, 263–269
   test problems, 313
delay logistic equation, *see* Hutchinson equation
   diffusive, *see* delay diffusive logistic equation
delay models, 18, 24, 62, 128, 327–363
   artificial neural networks, 362
   Belousov–Zhabotinsky oscillating reaction, 358

   Belousov–Zhabotinsky reaction-diffusion, 137
   competitive, 347
   cooperative, 347
   differential-difference thermal conduction, 115, 118
   diffusive logistic, 344
   Eigen–Schuster, 247
   epidemic, for three groups of individuals (SIR), 336
   epidemic, new coronavirus infection, 354
   food chain, 361
   food-limited logistic, 330
   heat treatment of metal strips, 360
   hepatitis B, 355
   Hutchinson, 328
   immunity–tumor interaction, 357
   induced defense, 349
   interaction of three economical parameters, 341
   Kermack–McKendrick diffusive, 351
   Lotka–Volterra type diffusive logistic, 346
   Mackey–Glass hematopoiesis, 333, 359
   macrodynamics of business cycles, 340
   Nicholson blowflies, 137, 330
   Nicholson reaction-diffusion, 347
   regenerative machine tool vibration, 338
   SI, 353
   simple climate, 337
   simple epidemiological, 336
   SIR, 336, 351, 352
   tax collection in closed economy, 342
   thermal conduction with finite relaxation time, 115
   two-component epidemic (SI), 353
delay ODEs, xi, xii, 1–84
   advanced, 3
   approximate analytical solution methods, 55–64
      Adomian decomposition, 75
      Bubnov–Galerkin method, 82
      collocation method, 82
      combination of basis functions, 80
      expansion of nonlinear operator, 75
      Galerkin-type methods, 80–82
      homotopy analysis, 78
      least squares method, 82
      matched asymptotic expansions, 71–74
      method of moments, 82
      minimization of root mean square error, 84

delay ODEs (continued)
  Padé approximant solutions, 69
  perturbation-iteration algorithms, 79
  power series solutions, 64–69
  regular expansion in small parameter, 69–71
  successive approximations, 74, 75
  Cauchy problem, 2
    exact solutions, 24–26, 29–34
    existence and uniqueness of solutions, 23
    qualitative features, 2
    solution by method of steps, 13, 16, 23
  comparison of numerical and exact solutions, 293–296
  first-order, 1–3
    with several constant delays, 3
  first-order, linear, 4
    characteristic equation, 4
    Lambert W function, 4–12
    stability of solutions, 43–45, 50
  first-order, nonlinear, 12
    reduction to linear delay ODEs, 12, 13
  linear systems, 38, 40
    Cauchy problem, 39
    exact solutions, 39, 41
  neutral, 3
  nonlinear
    linear stability analysis, 52–55
  nonlinear models
    delay food-limited model, 330
    Hutchinson equation, 328
  nth-order, 28
  nth-order, linear, 34–38
  numerical integration, 273
    first-order Euler method, 278
    Heun method, 281
    implicit Runge–Kutta method, 290
    midpoint method, 280
    modified method of steps, 278
    modified shooting method, 289
    qualitative features, 275
    Runge–Kutta methods, 281–283, 285
    shooting method, 287
    spectral collocation methods, 285
    stiff systems, 290
  proportional delay, 17
    exact solutions, 20, 26, 27
  second-order, 28
  second-order, linear, 29–34
    stability of solutions, 45, 46, 51
  several proportional delays, 21
  several variable delays, 22
  stability of solutions, 42, 50–52
    Cooke–van den Driessche method, 49, 50
    D-partition method, 46–49
  test problems, 293
delay ordinary differential equations, see delay ODEs
delay partial differential equations, see delay PDEs
delay PDEs, xii, 85
  applications, 327
    Belousov–Zhabotinsky oscillating reactions, 358
    development of diseases, 350
    food chain model, 361
    heat treatment of metal strips, 360
    hepatitis B model, 355
    immunity–tumor iteration, 357
    Mackey–Glass model of hematopoiesis, 359
    models of artificial neural networks, 362
    population theory, 343–349
    spread of epidemics, 350
  construction of test problems, 309, 310, 315
  global instability of solutions, 270, 272
  Hadamard ill-posedness of delay problems, 271, 272
  linear, 85–126
    boundary conditions, 96, 97
    differential-difference heat equations, 115
    dispersion equation, 90
    formal solution, 86
    hyperbolic heat equation, 113
    initial-boundary value problems, see initial-boundary value problems
    principle of linear superposition, 86
    properties, 85
    properties of solutions, 86–89
    separable solutions, 89, 90
    stability and instability conditions, 109, 112
    Stokes problem, 116–118
    sufficient conditions for asymptotic stability, 110
  nonlinear, 127–272
    additive separable solutions, 139–154
    analytical solution methods, 127–272
    Belousov–Zhabotinsky, 137

delay PDEs (*continued*)
   constructing exact solutions, 202–211
   diffusive logistic, *see* delay diffusive logistic equation
   exact solutions, 127–272
   functional separable solutions, 159–162
   generalized separable solutions, 154–159
   Klein–Gordon, *see* delay Klein–Gordon type wave equations
   linear transformations to construct solutions, 162–165
   Lotka–Volterra, 138
   most common types of exact solutions, 130
   multiplicative separable solutions, 139–154
   Nicholson blowflies, 137, 330
   reaction-diffusion, *see* delay reaction-diffusion equations
   separable solutions, 139–153
   separable solutions, examples of equations, 139–154
   solution via solutions to non-delay PDEs, 201–211
   solution with method of functional constraints, 166–199
   solution with method of invariant subspaces, 154–159
   states of equilibrium, 130
   traveling wave front solutions, 134–139
   traveling wave solutions, 131–134
  nonlinear systems, 212–224
   solution method, *see* generating equations method
  nonlinear systems homogeneous in unknowns, 247
   exact solutions, 248–250
  nonlinear with proportional delay, 257
   exact solutions, 257–263, 267–269
  numerical integration, 296–308
   finite difference methods, 302
   finite difference scheme, 302
   method of lines, 297–300
   method of lines, using Mathematica, 300, 301
   time-domain decomposition, 296, 297
  proportional arguments, 250, 253
   principle of analogy of solutions, 250–252
  quasilinear, 167, 253, 296
   exact solutions, 167–176

  quasilinear with proportional delay, 253
   exact solutions, 253–257, 263–266
  reaction-diffusion system with several delays, *see* Lotka–Volterra system
  selection of test problems, 309
  test problems, *see* test problems
delay problems, 2
  boundary value
   see boundary value problems, 67
  Cauchy
   see Cauchy problems, 2
  initial value
   see Cauchy problems, 2
  initial-boundary value
   see initial-boundary value problems, xi
delay reaction-diffusion equation(s), 141
  hyperbolic, 129, 199
  linear
   proportional argument, 119
  Lotka–Volterra type systems, 225–250
   exact solutions, 226, 239, 240, 246
   reductions, 226, 239
  Nicholson blowflies delay model, 137
  nonlinear, 134, 137, 140, 151, 166
   comparison of numerical and exact solutions, 316
   Mackey–Glass model of hematopoiesis, 359
   numerical integration, 296–308
   proportional arguments, 251–269
   separable solutions, 141–148, 156–162, 167–194
   test problems, 310
  nonlinear systems, 213–215, 218–222
   artificial neural network models, 362
   food chain model, 361
   generalizations, 222–224
  proportional delay, 134
  quasilinear, 167, 253
  quasilinear systems, 215–218
  traveling wave solutions, 131
  with nutrient limitation, 345
delayed cosine, *see* delayed cosine function
delayed cosine function, xv, 30
  matrix, 41
delayed exponential, *see* delayed exponential function
delayed exponential function, xv, 15, 16, 95, 100, 105
  matrix, 40
delayed sine, *see* delayed sine function

delayed sine function, xv, 30
  matrix, 41
difference scheme, *see* finite-difference scheme
differential constraint, 167, 227, 236
differential equation
  advanced, 4, 29, 131, 251
  functional, *see* functional differential equation
  neutral, 3, 23, 29
  ordinary, *see* delay ODEs
  pantograph, *see* pantograph equation
  partial, *see* delay PDEs
differential-difference heat equation, 113
  exact solutions, 115, 116
  finite relaxation time, 115
  initial-boundary value problem, 118
  Stokes problem, 116, 118
diffusion equation, *see also* delay reaction-diffusion equations
  classical, 113
  hyperbolic, 113, 114
    derivation, 113
    relaxation time, 114
  linear, 202, 208
  nonlinear, 251, 252, 270
    constant delay, 270, 345
    proportional delay, 257
  nonlinear with proportional delay, 257
  quasilinear
    proportional delay, 253
  quasilinear with proportional delay, 253
diffusion, physical meaning, 343
direct method for constructing test problems, 315
direct method of symmetry reductions, 129
Dirichlet boundary conditions, *see* boundary conditions of first kind
dispersion equation, 90, 153, 320

E

eigenvalue problem, 93
  eigenfunctions, 82, 93–95, 99–101, 104–109, 121–123, 125
  eigenvalues, 41, 82, 94, 99, 104–109, 123
  linear homogeneous, 98, 104, 108, 122
elliptic operator, 151
energy conservation law, 113
equation, *see* delay ODEs, delay PDEs
  advanced, *see* advanced differential equations

characteristic, *see* characteristic equation
delay, *see* delay differential equations
delay reaction-diffusion, *see* delay reaction-diffusion equations
differential, *see* differential equation
diffusion, *see* diffusion equation
dispersion, *see* dispersion equation
heat, *see also* diffusion equation
  classical, 113
  delay, 90
  differential-difference, *see* differential-difference heat equation
  hyperbolic, *see* hyperbolic heat equation
Helmholtz, *see* Helmholtz equation
Hutchinson, *see* Hutchinson equation
logistic, *see* delay logistic equation
Malthusian, 327
neutral, *see* neutral differential equations
Nicholson, 137, 295, 330, 332
ordinary differential, *see* delay ODEs
pantograph, *see* pantograph equation
partial differential, *see* delay PDEs
quasilinear, *see* delay PDEs, quasilinear
reaction-diffusion, *see* delay reaction-diffusion equations
equilibrium, 46, 114, 130, 138, 331, 333, 341, 343, 349, 359
  asymptotically stable, 328, 332, 356, 358
  stable, 46, 329, 334, 347
  unstable, 46, 328, 329
error function, 124
estimates of thermal and diffusion relaxation times, 114
Euler method, xii
  continuous, 280
  explicit, 275, 280, 285
  first-order, 279, 285
  implicit, 280, 285, 292
exact solution
  see exact solutions, xi
exact solution methods, *see* analytical methods
  delay ODEs, 55
  delay PDEs, 127
exact solutions
  additive separable, *see* additive separable solution
  closed form, 130
  construction, *see* construction of exact solutions
  definition, 129

exact solutions (*continued*)
  delay Klein–Gordon type wave equations, *see also* delay Klein–Gordon type wave equations, 151, 152, 163–165, 194–199, 263–269
  delay ODEs, *see also* delay ODEs
    Cauchy problems, 24–26, 29–34
    linear systems, 39, 41
    proportional delay, 20, 26, 27
  delay PDEs, *see also* delay PDEs
    nonlinear systems, 248–250
  delay reaction-diffusion equations, 131, 141–148, 156–162, 167–194
    nonlinear systems, 212, 225
  functional separable, *see* functional separable solution
  generalized separable, *see* generalized separable solution
  linear delay ODEs
    Cauchy problems, 13–16
  Lotka–Volterra system
    in elementary functions, 226, 233–239, 241–246
  multiplicative separable, *see* multiplicative separable solution
  nonlinear delay PDEs, 127–272
    most common types, 130
    multiplicative separable solutions, 139–154
    proportional delay, 257–263, 267–269
  nonlinear ODEs with proportional delay
    Cauchy problems, 20–22
  nonlinear wave-type equations
    proportional delay, 263
  one-dimensional differential-difference heat equation, 115
  quasilinear delay PDEs, 167–176
    proportional delay, 253–257, 263–266
  second-order delay ODEs
    Cauchy problems, 29–34
  Stokes problems, 115–118
  traveling wave, *see* traveling wave solution
existence and uniqueness of solutions, 23
explicit finite-difference scheme, 305

F

factors leading to need to consider delay, xi, 127
finite difference methods, 302
  approximation order, 304
  boundary nodes, 302
  conditional approximation, 304
  conditionally stable, 304
  convergent with order p, 304
  finite difference scheme, 302
    explicit, 305
    higher-order, 307
    implicit, 306
    special, 308
    weighted, 307
    well-defined, 305
  finite-difference differentiation operator, 303
  grid nodes, 302
  initial layer, 302
  inner nodes, 302
  residual, 303
  spatial layer, 302
  spatio-temporal grid, 302
  stable scheme, 304
  stencil, 303
  temporal layer, 302
  theory, basic concepts and definitions, 302
  unconditional approximation, 304
  unconditionally stable, 304
  uniform grid, 302
finite-difference differentiation operator, 274, 303
finite-difference scheme, 303
  explicit, 305
  higher-order, 307
  implicit, 306
  two special schemes for linear problem, 308
  weighted, 307
first initial-boundary value problem, 91
  differential-difference heat equation, 118
  hyperbolic equation with proportional delay, 124
  parabolic equation with proportional delay, 120
  solutions, 112
  statement of problem, 91, 120, 124
first-order delay ODEs, *see also* delay ODEs, 274, 297
  advanced equations, 4
  characteristic equation, 4
  constant delay, 1
    Cauchy problem, 1, 13–17
    delayed exponential function, 15
    solution by method of steps, 14, 15

first-order delay ODEs (*continued*)
  exact solutions, 12–17, 20–22
  existence and uniqueness of solutions, 23, 24
  exponential solutions, 4
  Lambert W function, *see* Lambert W function
  linear systems, 38
    Cauchy problem, 39
    exact solutions, 39, 40
  neutral equations, 3
  proportional delay, 18, 20–22
    Cauchy problem, 18, 26–28
  qualitative features, 2
  several constant delays, 3
  several delays, 22
  stability of solutions, 42–45, 50, 52–55
  suppression of singularities, 24–26
  variable delay, 17
    Cauchy problem, 18
    stretched exponential function, 19, 20
first-order equations, *see* first-order delay ODEs
Fisher equation, *see* delay Fisher equation
fixed point, *see* equilibrium
food chain model, 361
formal solution, 35, 86, 94
function
  antiperiodic, 164, 169, 196
  approximating, 81, 83
  delayed cosine, xv, 30
    matrix, 41
  delayed exponential, xv, 15, 16, 95, 100, 105
    matrix, 40
  delayed sine, xv, 30
    matrix, 41
  error, 124
  grid, 273–276, 280, 281, 303, 308
    continuous approximation, 274, 303
    discrete, 302
    residual, 275
  Lambert W, *see* Lambert W function
  outer, 159, 160
  stretched cosine, xv, 32
  stretched exponential, xv, 19, 20, 31–33
  stretched sine, xv, 32
functional constraints, *see* method of functional constraints
  degenerate, 167
  first kind, 167, 212

second kind, 167, 212
functional differential equation, xiii, 3, 18, 22
functional separable solution, 130, 159, 179, 181–193, 207
  employing linear transformations, 162, 163
  examples of constructing, 160–162
  inner functions, 159
  nonlinear delay PDEs, 163–165, 259, 261, 262
  outer function, 159
  quasilinear diffusion equations, 253
  systems of nonlinear delay PDEs, 220
  transformation of unknown function, 160
functions
  basis, 80–82, 285
  inner, 159

G

Galerkin method, *see* Galerkin-type projection methods
Galerkin-type projection methods, xii, 80
  approximating function, 81
  basis functions, 81
  Bubnov–Galerkin method, 82, 154
  Chebyshev nodes, 82
  collocation method, xii, 82–84
    spectral, 285
  general scheme of application, 81
  least squares method, 82
  method of moments, 82
  residual, 81
Gear method, 290, 293–295, 301, 316–326, 333, 347
  description, 292
  fourth-order, 292
  implicit, 290
  multi-step, 290, 292
  second-order, 292
  third-order, 292
generalized Hutchinson equation, 53, 329
generalized separable solution, 130, 154, 159, 162–164, 166, 171–174, 184, 189, 191
  method of functional constraints, 166, 167
  method of invariant subspaces, 155–159
  nonlinear delay Klein–Gordon type wave equation, 198
  nonlinear Lotka–Volterra system, 225, 227, 229, 232, 236
  systems of delay PDEs, 212, 213, 215–218
  transformation of unknown function, 160

generating equations method, *see* generating equations method
  application examples, 213–215
  general description, 212
  generalizations, 222
    systems with higher-order equations, 224
    systems with many equations, 224
    systems with many space variables, 224
    systems with two constant delays, 223
    systems with variable delays, 223
  principle of constructing delay systems, 212
  solution of nonlinear systems, 218–222
  solution of quasilinear systems, 215–218
global attractor, 331, 345, 349
gradient operator, 113, 354
grid function, 273–276, 280, 281, 303, 308
  continuous approximation, 274, 303
  discrete, 302
  residual, 275

H

Hadamard ill-posedness, 271
  delay initial value problem, 271
  delay initial-boundary value problem, 91, 96, 109, 271, 272
Hayes theorem, 12, 44, 54, 110
heat equation
  classical, 113, 117
  constant delay, 90
    linear problem, 168–172
  differential-difference, *see* differential-difference heat equation
  hyperbolic, 113, 115, 117, 119
    derivation, 113, 114
  linear, 225, 242, 245–247, 250, 271
  parabolic, 117
Helmholtz equation, 102, 107, 188
hepatitis B model, 355
Heun method, 68, 84, 281
homotopy analysis method, xii, 78, 79
Hutchinson equation, 53, 54, 295, 328, 345
  diffusive, 345
  generalization, 344, 346
  generalized, 53, 329
  properties, 298
hyperbolic diffusion equation, 113
hyperbolic heat equation, 113, 117, 119
  derivation, 113, 114

I

implicit finite-difference scheme, *see also* finite-difference scheme
implicit Runge–Kutta method, *see also* Runge–Kutta methods
induced defense, 349, 350
initial conditions, 15, 69, 92, 296, 330
  Cauchy problem for delay ODE, 2
  Cauchy problem for linear delay ODE, 3
  Cauchy problem for nth-order delay ODE, 29
  delay reaction-diffusion equations, 296
  general, 16, 17, 92, 97
  homogeneous, 16, 21
  linear delay parabolic PDEs in n variables, 102
  linear initial-boundary value problems, 92, 97, 102, 120, 122, 296
  linear ODE with proportional delay, 18
  linear parabolic PDEs with proportional delays, 120, 122
  linear system of delay ODEs, 39
  nonhomogeneous, 92, 98
  nonlinear ODE with proportional delay, 20, 21, 70, 75
  ODEs with several constant delays, 22
  ODEs with several proportional delays, 21
  ODEs with several variable delays, 22
  population theory models, 344
  series expansion in eigenfunctions, 99
  special, 14
  test problems, 312–315
  various models, 360, 362
initial data, *see* initial conditions
initial layer, *see also* finite difference methods, 304
initial value problems, *see* Cauchy problems
initial-boundary value problems, xi, 91
  boundary conditions, *see also* boundary conditions, 97
    first kind, 97, 99
    homogeneous, 92
    mixed, 97, 99
    nonhomogeneous, 92, 97
    second kind, 97, 99
    third kind, 97, 99
  delay diffusive logistic equation, 345
  delay, general properties and qualitative features, 91
  differential-difference heat equations, 118, 119

408 INDEX

initial-boundary value problems (*continued*)
  eigenvalue problem, 93
  first
    linear, statement of problem, 91, 120
    linear, with constant delay, 91
    linear, with proportional delay, 120
  first (Dirichlet), 97
  five main types, 97, 99, 100
  Hadamard ill-posedness, 271
  hepatitis B model, 356
  linear
    delay hyperbolic equations, 107–109
    delay parabolic equations, 91–96
    hyperbolic equations with proportional delay, 124–126
    instability conditions for solutions, 111
    necessary and sufficient stability conditions, 109
    nonhomogeneous, instability, 110
    parabolic equations with proportional delay, 120, 121
    stability conditions for solutions, 107–113
    sufficient conditions for asymptotic stability, 110
    with constant delay, 91–112
    with proportional delay, 119–126
    with two proportional arguments, 124
  Lotka–Volterra type system, 322
  numerical integration, 296
  second (Neumann), 97
  series solutions, 94, 99, 105, 108, 121
  solution as sum of three functions, 92
  solution by separation of variables, 96
  solutions, 101, 118, 123
  Sturm–Liouville problem, 93
  third (Robin), 97
inner functions, *see also* functional separable solutions
inner nodes, *see also* finite difference methods
inner region, *see* method of matched asymptotic expansions
instability conditions, 46, 54, 96, 109, 111, 112, 329
  global, 271
instability of solutions, xi
  delay ODEs, 42, 45, 51, 53
  delay PDEs, 96, 109, 111, 112, 270
  global, 270–272
  linear, 320, 326

integral transform, xii, 55–64, 91
  Laplace, 56
    solution example, 59
  Mellin, 60
    solution example, 61–63
  solution of linear problems, 55
inverse Laplace transform, 56
  definition, 56
  finitely many singular points, 58
  infinitely many singular points, 58
  rational functions, 57
iterative methods, 74

L

Lambert W function, 4
  complex-valued branches, 6, 8, 9
    asymptotic formula, 6
  in complex plane, 6
    contour lines, 10
  negative branch, 5
    asymptotic formula, 6
  on real axis, 5
  positive branch, 5
    approximate formulas, 6
    asymptotic formula, 6
    Taylor series expansion, 5
  principal branch, 5
  properties, 6
  real-valued branches, 5
Laplace equation, 102, 188
Laplace operator, xvi, 102, 107, 113, 188, 225, 302
Laplace transform, 56, 109, 340
  basic properties, 57
  definition, 56
  inverse, 56
  table, 58
least squares method, 82
linear boundary value problems, *see also* boundary value problems, 99, 108, 120, 288, 289
linear delay ODEs, *see also* delay ODEs, *see also* Cauchy problem, 4, 15, 59
  nth-order, 34, 46
  properties of solutions, 35
  second-order
    stability of solutions, 45
    stability of solutions, 42, 43
      general remarks, 42
      Hayes theorem, 44

linear delay ODEs (*continued*)
   small delays, 43
   stability and instability conditions, 45, 46
linear delay PDEs, *see also* delay PDEs, 85
   additive separable solutions, 90
   differential-difference heat equation, 115
      initial-boundary value problem, 118, 119
      Stokes problem, 115–118
   homogeneous, 85
   hyperbolic heat equation, 113, 114
   hyperbolic with proportional delay
      initial-boundary value problems, 124–126
   initial-boundary value problems, *see also* initial-boundary value problems
   multi-dimensional parabolic
      initial-boundary value problem, 102–107
   multiplicative separable solutions, 89, 90
   one-dimensional hyperbolic
      initial-boundary value problem, 107–109
   one-dimensional parabolic
      initial-boundary value problem, 91–101
   parabolic with proportional delay
      initial-boundary value problems, 120–124
   properties, 85
   properties of delay problems, 91
   proportional delay
      initial-boundary value problems, *see also* initial-boundary value problems
   solving using particular solutions, 86–89
linear initial-boundary value problems, *see* initial-boundary value problems
linear systems of delay ODEs, 38, 60
   first-order, 38
      Cauchy problem, 39
      exact solution, 40
   second-order, 40
      Cauchy problem, 40
      exact solution, 41
linear transformations, 162, 230
   constructing functional separable solutions, 162
   constructing generalized separable solutions, 162
   exact solutions to nonlinear delay PDEs, 163–165
   for nonlinear delay Klein–Gordon type wave equations, 163
logistic differential equation, *see* delay diffusive logistic equation

logistic equation, *see* delay diffusive logistic equation
Lotka–Volterra diffusive model with several delays, 138, 346
Lotka–Volterra system, 225
   reduction to Helmholtz equation, 226–229
   reduction to nonstationary system, 236–238
      solutions in elementary functions, 238, 239
   reduction to nonstationary system and heat equation, 246
   reduction to stationary system, 229, 230
      solutions in elementary functions, 233–237
   reduction to stationary system and Shrödinger equation, 239, 240
      solutions in elementary functions, 241–246
   simplest solutions, 226

M

Mackey–Glass hematopoiesis model, 333, 359
Malthusian coefficient of linear growth, 327
Malthusian equation, 327
Malthusian exponential model, 327
Malthusian parameter of population growth, 327
Mathematica (software), 127
   comparison of numerical and exact solutions, 293, 316, 322
   integration of stiff systems of delay ODEs, 290
   numerical solutions, 329, 333, 335, 347
   solution procedure for delay problems, 300, 301
Mellin transform, 60
   basic properties, 61
   definition, 60
   inverse, 60
   relation to Laplace transform, 61
   table, 62
method, *see also* methods
   Adomian decomposition, xii, 76–78
   analytical, *see* analytical methods
   approximate analytical, *see* approximate analytical methods
   backward differentiation formula, *see* Gear method

method (*continued*)
  Bubnov–Galerkin, 82, 154
  collocation, *see* collocation method
  Cooke–van den Driessche, 49
  D-partition, 46–49
  direct, 315
    constructing test problems, 315
    symmetry reductions, 129
  Euler, *see* Euler method
  functional constraints, *see* method of functional constraints
  Gear, *see* Gear method
  generating equations, *see* generating equations method
  Heun, *see* Heun method
  homotopy analysis, 78
  invariant subspaces, 155–159
  matched asymptotic expansions, 71–74
  minimization of root mean square error, 84
  moments, 82
  numerical, *see* numerical method, *see* numerical method
  Padé approximant solutions, 69
  power series solutions, 64–69
  regular expansion in small parameter, 69–71
  shooting, xii, 68, 84, 287–289
    modified, 289
  steps, *see* method of steps
  successive approximations, 74, 75
method of functional constraints, xii, 146, 159, 166–199, 212
  general description, 166, 167
method of invariant subspaces, 155
  generalized separable solutions, 155–159
method of lines, xii, 290, 299, 301, 302, 316–326, 347
  numerical solution of delay problems with Mathematica, 300, 301
  reduction of delay PDE to system of delay ODEs, 297
method of matched asymptotic expansions, 71
  application to boundary value problem, 71–72
  composite asymptotic solution, 73
  inner region, 72
  matching condition, 72
  outer region, 72
  solution example, 73
method of minimization of root mean square error, 84
method of moments, 82
method of regular expansion in small parameter, 69
method of steps, xii, 13–17, 23, 140, 239, 277
  for equations of neutral type, 23
  for first-order ODEs with constant delay, 13
  for linear problems with constant delay, 14
  for nonlinear problems with constant delay, 16
  for ODEs with several constant delays, 17
  for ODEs with variable delay, 23
  generalization, 297
  inapplicable for solution, 23, 277
  modified, 278
  proof of existence and uniqueness of solutions, 23
method of successive approximations, 74
method of time-domain decomposition, 296
methods
  construction of exact solutions, *see* construction of exact solutions
  exact solution, *see* exact solution methods
  finite difference, *see* finite difference methods
  Galerkin, *see* Galerkin-type projection methods
  Galerkin-type projection, *see* Galerkin-type projection methods
  perturbation-iteration algorithms, 79
  Runge–Kutta, *see* Runge–Kutta methods
  solution, *see* solution methods
mixed-type solutions, 216, 218
model, *see* delay models
model of immunity–tumor interaction, 357
model of heat treatment of metal strips, 360
model of interaction of three economical parameters, 341
model of regenerative machine tool vibration, 338
modified method of steps, 278, 279
modified shooting method, 289
multiplicative separable solution, 89, 90, 115, 130, 139–154, 162, 176, 178, 188, 195, 199, 213, 218–221, 248, 255–258, 266

N

necessary and sufficient stability conditions, 109

INDEX

Neumann boundary conditions, *see* boundary conditions of second kind
neutral differential equations, 3, 23, 29
Nicholson blowflies delay model, 137
Nicholson delay reaction-diffusion model, 347
Nicholson equation, 137, 295, 330, 332
nonlinear delay models, *see* delay models
nonlinear delay ODEs, *see also* delay ODEs, 12
   general form
      stability conditions, 54
   linearization, 12, 13
   solution by method of steps, 16, 17
   stability analysis, 52
   stability conditions, 53, 54
nonlinear delay PDEs, *see also* delay PDEs, 127
   constructing exact solutions, 203–207, 209–211
   functional separable solutions, 159
      construction with linear transformations, 162, 164, 165
      examples of construction, 160–162
      inner functions, 159
      outer function, 159
      transfromation of unknown, 160
   generalized separable solutions, 154
      construction with linear transformations, 162–165
      method of invariant subspaces, 155–159
      preset coordinate functions, 154
   generating equations method, 212–215
      generalizations, 222, 224
   Hadamard ill-posedness of delay problems, 271, 272
   Klein–Gordon type wave equations, *see* delay Klein–Gordon type wave equations
   Lotka–Volterra type systems, *see* Lotka–Volterra system
   most common types of exact solutions, 130
   nonlinear systems
      exact solutions, 218–222
   parameter replacement with functions, 201, 202, 207–209
   principle of analogy of solutions, *see* principle of analogy of solutions
   quasilinear systems
      exact solutions, 216–218
   reaction-diffusion equations, *see* delay reaction-diffusion equations

separable solutions, 139
   equations with constant delay, 139–145, 147–152
   equations with many delays, 146, 147
   equations with variable delay, 146, 149
   generalizations, 149, 151–153
solution instability, 270
solution methods, 201
   generating equations method, 212–215
   parameter replacement with functions, 201, 207
solution using method of functional constraints, 166
   description, 166
   exact solutions, 167–199
systems homogeneous in unknowns, 247
   exact solutions, 248–250
traveling wave front solutions, 134–139
traveling wave solutions, 131–139
nonlinear problems, 16
   boundary value, *see* boundary value problems
   Cauchy, *see* Cauchy problems
   initial-boundary value, *see* initial-boundary value problems
nonlinear systems of delay ODEs, 222, 225, 238, 247, 298
   proportional delay, 253
nonlinear systems of delay PDEs, 212–214
   any number of equations, 224
   any number of space variables, 224
   exact solutions, 218–222
   homogeneous in unknowns, 247
      exact solutions, 247–250
   Lotka–Volterra, *see* Lotka–Volterra system
   nth-order equations, 224
   two constant delays, 223
   variable delay, 223
numerical integration, *see also* numerical method
   Cauchy problems
      delay ODEs, 273, 278, 279, 283
   delay ODEs, 273
      first-order Euler method, 278
      Heun method, 281
      implicit Runge–Kutta method, 290
      midpoint method, 280
      modified method of steps, 278
      modified shooting method, 289
      qualitative features, 275
      Runge–Kutta methods, 281–283, 285

numerical integration (*continued*)
    shooting method, 287
    spectral collocation methods, 285
    stiff systems, 290
  delay PDEs, 296–308
    finite difference methods, 302
    finite difference scheme, 302
    method of lines, 297–300
    method of lines, using Mathematica, 300, 301
    time-domain decomposition, 296, 297
  delay reaction-diffusion equations, 296–308
  initial-boundary value problems, 296
numerical method, 154
  backward differentiation formula, *see* Gear method
  BDF, *see* Gear method
  collocation, xii, 80, 82
    spectral, 285
  delay differential equations, 273–308
  delay ODEs, 273
  Euler, xii
    continuous, 280
    explicit, 275, 280, 285
    first-order, 279, 285
    implicit, 280, 285, 292
  Gear, 290, 293–295, 301, 316–326, 333, 347
    description, 292
    fourth-order, 292
    implicit, 290
    multi-step, 290, 292
    second-order, 292
    third-order, 292
  Heun, 68, 84, 281
  interpolant of, 280
  midpoint, 280, 281
  qualitative features, 275
  Runge–Kutta, *see also* Runge–Kutta method, 281–283, 285, 290
    continuous, 283
    explicit, 283
    family, 282
    fourth-order, 283, 285, 293–296, 301, 322–326, 335
    general scheme, 282
    implicit, 283, 290, 298, 329
    second-order, 283, 293–296, 301, 316–326, 329
    stability, 285
  shooting, xii, 68, 84, 287–289
    modified, 289
  steps, *see* method of steps
numerical methods, *see* numerical method
numerical solution, 25, 27, 68, 84, 277, 279, 288
  methods, *see* numerical method
numerical solution methods, *see* numerical method

O

ODE, *see* delay ODEs
ODEs, *see* delay ODEs
operator
  biharmonic, 151
  continuous, 303
  delay linear differential, 34, 78, 85, 86
  elliptic, 151
  finite-difference, 274, 300, 308
  gradient, 113, 354
  Laplace, *see* Laplace operator
  linear differential, 89, 97, 102, 149, 152, 158
  nonlinear, 74, 75, 155, 157, 158
  transmission diffusion operator, 363
oscillating reaction model, Belousov–Zhabotinsky, 358
outer function, 159, 160
outer region, *see* method of matched asymptotic expansions

P

Padé approximant, 69
pantograph equation, *see also* pantograph-type equation, 18, 23
pantograph-type equation, xi, 77
PDE, *see* delay PDEs
PDEs, *see* delay PDEs
physical meaning of diffusion, 343
principle of analogy of solutions, 250–253
  nonlinear PDEs with proportional delay
    exact solutions, 257–263
  quasilinear PDEs with proportional delay
    exact solutions, 250–257
  wave-type equations with proportional delay
    exact solutions, 263–269
principle of linear superposition, 34, 36, 86
principles for selecting test problems, 309

problem
  eigenvalue, *see* eigenvalue problem
  seealso problems, 93
  singular perturbation, 71
  Stokes, 116, 117
  Sturm–Liouville, 93
problems
  boundary value, *see* boundary value problems
  Cauchy, *see* Cauchy problems
  for delay differential-difference heat equations, 118
  for delay reaction-diffusion equations, 245
  for linear hyperbolic equations with constant delay, 107
  for linear hyperbolic equations with proportional delay, 124
  for linear parabolic equations with constant delay, 91
  for linear parabolic equations with proportional delay, 120
  initial-boundary value, *see* initial-boundary value problems
properties of delay problems, 91
properties of Lambert W function, 4, 6
properties of Laplace transform, 57
properties of linear operators, 152
properties of Mellin transform, 61
properties of numerical methods, 274, 303
properties of solutions to linear delay equations, 35, 43, 85, 91, 340
properties of solutions to nonlinear delay equations, 129, 330, 332
properties of zeros of stretched exponential function, 19

Q

qualitative features of delay differential equations, xi, xii
qualitative features of delay ODEs, 1, 2
qualitative features of delay PDEs, 130, 309, 311
qualitative features of delay problems, 91
qualitative features of numerical integration of delay ODEs, 275
qualitative features of solutions to delay ODEs, 19
qualitative features of traveling wave front solutions, 135
quasi-geometric grid, 284

quasi-polynomial, 35, 37, 46, 49, 50, 60
quasilinear delay equations, 167, 263
quasilinear systems of delay ODEs, 137
quasilinear systems of delay PDEs, 215
  exact solutions, 216–218

R

rational functions, inverse Laplace transform, 57
reaction-diffusion equations, *see* delay reaction-diffusion equations
reaction-diffusion logistic equation, 135, 136
reaction-diffusion model, 128
  delay
    Belousov–Zhabotinsky, 137
    Lotka–Volterra type, 137, 346
    Nicholson, 347
reaction-diffusion system, 139
  delay PDEs, 357, 359, 363
  Lotka–Volterra type
    several delays, 139, 225, 228
  two delays, 247, 249
reductions of Lotka–Volterra system, 225, 226
  different diffusion coefficients, 226
  equal diffusion coefficients, 239
  single delay, to nonstationary system of delay ODEs, 236
  three delays, to Helmholtz equation, 226
  three delays, to stationary system, 229
region of instability, 44, 49, 54, 320
region of stability, 44, 49, 54
relaxation time, 114–116, 119, 363
residual grid function, 275
residue, 55
rest point, 130
Robin boundary conditions, *see* boundary conditions of third kind
Runge–Kutta methods, xii, 281, 285
  continuous, 283
  explicit, M-staged, 283
  fourth-order, 281, 293, 294, 301, 323–326
  general scheme, 282
  implicit, 290, 298, 329
    for stiff systems, 290
  second-order, 293, 294, 301, 316–326
  stability, 285

S

self-similar problem, 124

self-similar problem for linear PDE with two proportional arguments, 124
separable solution
  additive, *see* additive separable solution
  multiplicative, *see* multiplicative separable solution
shooting method, xii, 68, 84, 287–289
  modified, 289
SI model, 353
simple climate model, 337
simplest epidemiological model, 336
simplest model of macrodynamics of business cycles, 340
singular perturbation problems, 71
SIR model, 336, 351, 352
solution, xi
  additive separable, *see* additive separable solution, 130, 139–154, 170, 180, 189, 199, 214, 221, 254, 265
  Cauchy problem
    by method of steps, 14, 23
    first-order delay ODE, 14, 16, 17, 25
    first-order ODE with proportional delay, 18, 26
    first-order ODE with several proportional delays, 21
    first-order ODE with several variable delays, 22, 23
    nth-order delay ODE, 34
    second-order delay ODE, 28–30
    second-order ODE with proportional delay, 31–33
    second-order ODE with two proportional delays, 33
    stability of solutions, 42
    systems of delay ODEs, 38, 41
  composite asymptotic, 73
  delay initial-boundary value problems
    instability conditions, 112, 113
    stability conditions, 109, 113
  delay parabolic equations
    asymptotic stability conditions, 110
    instability conditions, 96, 111
    stability conditions, 96
  exact, *see* exact solutions
  existence and uniqueness, 23
  formal, 35, 86, 94
  functional separable, *see* functional separable solution
  generalized separable, *see* generalized separable solution
  global instability, 270–272
  Hutchinson type equations
    instability condition, 54, 329
    stability conditions, 53, 54, 329
  instability, *see* instability of solutions
  instability conditions, *see* instability conditions
  linear delay ODEs
    asymptotic stability, 51
  multiplicative separable, 89, *see* multiplicative separable solution, 115, 130, 139–154, 162, 176, 178, 188, 195, 199, 213, 218–221, 248, 255–258, 266
  nonlinear delay ODEs
    asymptotic stability, 53
    instability, 53, 295
    stability conditions, 55
  nonlinear delay PDEs
    global instability, 271, 272
    instability, 270
  numerical, *see* numerical solution
  traveling wave, *see* traveling wave solution
  traveling wave front, 134–139
solution instability, *see* instability of solutions
solution methods, xii
  analytical, *see* analytical methods
  approximate analytical, *see* approximate analytical methods
solution smoothing, 2
solutions, *see also* solution
  approximate
    linear combinations of basis functions, 80
    power series, 64
  delay initial-boundary value problems
    special form, 109
    sum of solutions to simpler problems, 92, 96
  linear problems
    delayed exponential function, 15
  systems of nonlinear delay PDEs
    mixed-type, 216, 218
spectral collocation methods, 285, 286
stability analysis, 42
  linear delay ODEs, 43, 47
  nonlinear delay ODEs, 52–55
stability conditions, 46, 53–55, 96, 109, 329, 333, 348
  global, 363

stability of solutions, *see also* instability of solutions, 1, 335, 339, 354, 362
  asymptotic, 47, 333
    local, criteria, 331
    sufficient conditions, 110
  delay ODEs, 1, 42, 43
    with respect to initial data, 275
    with respect to right-hand side, 275
  linear, 320
  linear delay ODEs
    Cooke–van den Driessche method, 49
    D-partition method, 46
    general remarks, 42
    region of stability, 47, 49
  linear initial-boundary value problems, 109, 112
    necessary and sufficient conditions, 109
  linear ODEs with several constant delays, 50
  linear ODEs with single constant delay, 43–46
  nonlinear delay ODEs, 52
    stability and instability theorems, 53
stability conditions, *see* stability conditions
state of equilibrium, *see* equilibrium
stationary point, *see* equilibrium
stencil, 303
step size, 273, 293, 298
  constant, 273, 276, 280, 283
  spatial, 302, 305, 307, 322
  temporal, 298, 301, 302, 307
  variable, 273, 274, 281, 302, 322
stiff problems, *see* stiff systems
stiff systems, 290, 298
  delay ODEs
    implicit Runge–Kutta schemes, 290
    solution using Mathematica, 290
Stokes problem, 116, 117
stretched cosine function, xv, 32
stretched exponential function, xv, 19, 20, 31–33
stretched sine function, xv, 32
Sturm–Liouville problem, 93
sufficient conditions for asymptotic stability of solutions, 110

suppression of singularities, 24
  blow-up problems, 24, 25
  first-order delay ODEs, 24–26

T

temporal layer, 300, 302, 305, 308, 316
test problems
  construction, 293
  construction with direct method, 315
  delay Klein–Gordon type wave equations, 322
  delay PDEs, 316
    comparison of numerical and exact solutions, 316, 322–326
    construction, 293
  delay reaction-diffusion equations, 316
    examples, 311–315
  exact solutions for delay PDEs
    examples, 293–296, 311–315
  main principles for selection, 309
transform
  integral, xii, 55
  inverse Laplace, 56–58
  inverse Mellin, 60, 61
  Laplace, *see* Laplace transform
  Mellin, *see* Mellin transform
traveling wave front solution, 134–139
traveling wave solution, xii, 90, 130, 131, 134
two-component epidemic SI model, 353

U

uniqueness of solutions, 1, 23

W

wave equation, 119, 315
  delay Klein–Gordon type, *see* delay Klein–Gordon type wave equations
wave equations
  nonlinear, Klein–Gordon type, delay, 107
wave-type delay PDEs, 299
weighted finite-difference scheme, 307